Catalysis in Confined Frameworks

Catalysis in Confined Frameworks

Synthesis, Characterization, and Applications

Edited by *Hermenegildo Garcia and Amarajothi Dhakshinamoorthy*

WILEY-VCH

Editors

Prof. Hermenegildo Garcia
Universidad Politécnica de Valencia
Av. De los Naranjos s/n
Valencia
SP, 46022

Dr. Amarajothi Dhakshinamoorthy
Madurai Kamaraj University
School of Chemistry
Madurai, Tamil Nadu
IN, 625 021

Cover Images: © Stanislaw Pytel/Getty Images; © athree23/Pixabay

Library of Congress Card No.: applied for

British Library Cataloguing-in-Publication Data
A catalogue record for this book is available from the British Library.

Bibliographic information published by the Deutsche Nationalbibliothek
The Deutsche Nationalbibliothek lists this publication in the Deutsche Nationalbibliografie; detailed bibliographic data are available on the Internet at <http://dnb.d-nb.de>.

Print ISBN: 978-3-527-35089-6
ePDF ISBN: 978-3-527-83925-4
ePub ISBN: 978-3-527-83926-1
oBook ISBN: 978-3-527-83927-8

Typesetting Straive, Chennai, India
Printing and Binding: CPI Group (UK) Ltd, Croydon, CR0 4YY

C9783527350896_161023

Contents

Preface

Heterogeneous catalysis by solids requires adsorption of substrates and reagents on the active sites of the surface of the material. Since catalysis is a surface phenomenon, one general way to increase the activity of solids is to increase their surface area. Among the various strategies to prepare solids with large surface area, one that has been widely exploited is the use of micro-/mesoporous materials. The presence of pores allows the ingress of substrates and reagents from the external medium to the interior of the solid, where the reaction takes place. Very often, the external surface area of the particles is comparatively smaller with respect to the internal surface. Crystalline microporous solids like zeolites are since the 1950s very successful catalysts employed in oil refining and petrochemistry in which hydrocarbons are transformed into high-quality fuels or bulk chemicals for chemical industry.

The presence of porosity not only increases the solid surface area but also can result in additional effects due to the confinement of the active sites inside a limited space that do not happen for analogous sites on homogeneous catalysts or even on the external surface of solids. The present book will illustrate the various aspects related to porous solids. Among the various types of porous solids, metal–organic frameworks (MOFs) have been considered as one of the most versatile, since they can be prepared for any di-, tri-, and tetra-positive metal cations and with a wide range of organic linkers having carboxylate, nitrogenated, or other chemical compositions. Coordinatively exchangeable positions around the metal ions or functional groups of the linker not compromised with the MOF structure can act as Lewis or Brönsted acids or bases, promoting acid-catalyzed reactions and oxidation reactions. In addition, the empty voids in the lattice can host a guest, such as a metal complex or nanoparticle that can become a catalytic site. In Chapter 1, Marti-Gastaldo and coauthors illustrate the various possibilities to use MOFs as solid catalysts. Confinement can result in a variety of effects that may result in selectivity. Engineering of the pore chemical environment and the presence of functional groups should make it possible to gain control on the chemical reaction as summarized in Chapter 2 by Srivastava and coauthors. Characterization of porous materials containing organic units, such as MOFs and covalent organic frameworks (COFs), are discussed by Oveisi, Daliran, and Peng in Chapter 3.

One of the strategies to modulate the interior of MOFs and adjust their properties is by using more than one organic ligand, and this topic is covered in Chapter 4 by Masoomi and Hashemi. Depending on the step on which the geometrical confinement around the active site occurs, mass transfer and diffusion of substrates and reagents to these internal sites have to be possible. This requires that the dimensions of the reacting molecules have to be smaller than pore dimensions. Among the various reaction types, those that are very challenging are the ones that require the geometrical control of the reaction site to obtain stereoisomers. Confinement is a particularly appealing strategy to develop stereoselective catalysts, since the active sites experience spatial limitations to the approaching direction of reagents and on the mechanistic steps requiring size enlargement. For this reason, MOFs have been widely explored to promote diastereoselective catalysis (Chapter 5 focusing on acid-catalyzed diastereoselective reactions by Llabrés, Navalón, and Garcia) and even as enantioselective catalysts (Chapter 6 by Morsali and Berijani).

While spatial constraints play a key role in stereoselective catalysis, they can also be used to limit the growth of incorporated guests, particularly metal nanoparticles. Since the catalytic activity of metal nanoparticles depends on their size and the reaction conditions favor their growth, there is a need to arrest nanoparticle dimensions. This can be achieved by using the internal void space of MOFs and complexation with functional groups to stabilize metal nanoparticles of a few nanometers in diameter. This topic is specifically covered in Chapter 7 by Li and coauthors including the synergistic effects of the cooperation between the metal nanoparticle and the MOF. Besides metal nanoparticles, molecular complexes incarcerated inside porous solids can also be the catalytic active sites. Durand, Serp, and Gouygou summarize the various issues related to the catalytic activity of complexes incorporated inside porous solids in Chapter 8.

Confinement of active sites within porous materials requires the diffusion of reagents inside the particles and diffusion of the products formed outside the pores. This diffusion control can result in special selectivity when the pore of the sizes does not allow the ingress of reagents (substrate shape selectivity), diffusion outside of the product (product shape selectivity), or even formation of certain bulky transition states (transition state selectivity). This topic is summarized in Chapter 9 by Dhakshinamoorthy and Garcia, and the area is further completed for important oxidation reactions in Chapter 10 by Kholdeeva.

While most of the book chapters deal with MOFs that certainly are versatile catalysts, other chapters focus on other materials that were used as catalysts before MOFs as it is the case of porous aluminophosphate zeolites with tailorable pore dimensions and structures (Chapter 11 by Srivastava and coauthors) or carbon nanotubes in which the curvature of the walls and the special internal orbital cloud determines specific catalytic properties (Chapter 14 by Masoomi and Hashemi).

Two final chapters deal with photochemical reactions in nanoporous titania, particularly those having highly ordered mesopores with hexagonal packing (Chapter 12 by Kumar Vatti and Selvam) and the specific reaction of glycerol dehydration by supported heteropolyacids (Chapter 13 by Selvam and coauthors).

Overall, the book provides a balanced view of the current state of the art in the area of heterogeneous catalysis on porous solids that is presently dominated by MOFs that are being applied for a large variety of reactions such as acid catalysts, including the effect of pore size, internal geometry, and confinement, but in which other types of materials such as zeolites, silicoaluminophosphates, and carbons have also interesting valuable properties that may result in high efficiency.

31st July 2023
Valencia, Spain

Hermenegildo Garcia
Amarajothi Dhakshinamoorthy

1

Engineering of Metal Active Sites in MOFs

Carmen Fernández-Conde, María Romero-Ángel, Ana Rubio-Gaspar, and Carlos Martí-Gastaldo

PhD Researchers on Functional Inorganic Materials Team (FuniMat), Instituto de Ciencia Molecular (ICMol) - Universidad de Valencia, C/ Catedràtic José Beltrán, 2, Vivero 1, 46980, Paterna, Spain

From their appearance, metal–organic frameworks (MOFs), have been an interesting field of research, partly due to the vast possibilities these materials offer [1]. The fact that they can be designed chemically to serve specific applications has already been proved in diverse areas such as gas separation [2], encapsulation [3] carbon capture [4], or catalysis [5–8].

Regarding catalysis, MOFs present many characteristics that make them potential candidates to carry out relevant catalytic processes as well as understanding the mechanisms through which these reactions take place. First of all, MOFs are heterogeneous catalysts, which can have an impact on separation, recyclability, and the possibility to operate in continuous mode, key aspects from an industrial standpoint. Additionally, heterogeneous catalysts often exhibit lower deactivation rates, compared to their homogeneous counterparts, mainly due to the spatial separation of the active sites avoiding its aggregation. However, there are some issues that limit the potential of traditional heterogeneous catalysts. In some cases, diffusion might prevent reagents from reaching the catalytic center, which is often poorly defined and understood. This lack of understanding also impacts the possibilities of modification of the catalyst, resulting in little room for improvement in these catalysts.

MOFs can serve as a platform for overcoming these drawbacks due to their inherent properties. Among the advantages of MOFs, it stands the isolation of the catalytic center and its uniform distribution along the framework, the sizeable porosity that enhances diffusion or the crystalline structure of these materials, which opens the door to the use of advance of characterization techniques. For all of this, research in MOFs as catalysts has become an attractive area. In this chapter, we will focus on metal-based catalysis in MOFs, which have already shown promising results in diverse catalytic processes. We intend to provide the reader with a general perspective of the different strategies used for engineering catalytic active sites in these porous, molecular frameworks. This chapter is divided into three different sections, attending to which component of the framework (metal node, organic

Catalysis in Confined Frameworks: Synthesis, Characterization, and Applications, First Edition.
Edited by Hermenegildo Garcia and Amarajothi Dhakshinamoorthy.

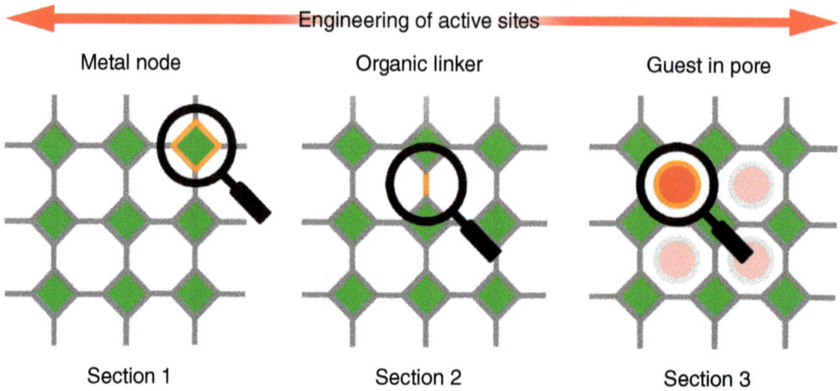

Figure 1.1 Schematic view of the three sections of this book chapter focused on engineering of active sites based on metal node, organic linker, or encapsulated guest. Source: Figure produced by the authors of the chapter.

linker, or guest) is responsible for the catalytic activity. Finally, different chemical incorporation strategies, characterization techniques, and examples of chemical reactivity of these reactive sites will be provided. This way, the structure of this chapter is shown in Figure 1.1.

1.1 Metal Node Engineering

This section will examine the different approaches used to obtain an active catalytic site in the nodes of MOFs, with the aim to be used in metal-based reactivity. Thus, this section of the chapter is divided into three main blocks, as represented in Figure 1.2. The first one comprises MOFs with intrinsically active metal nodes,

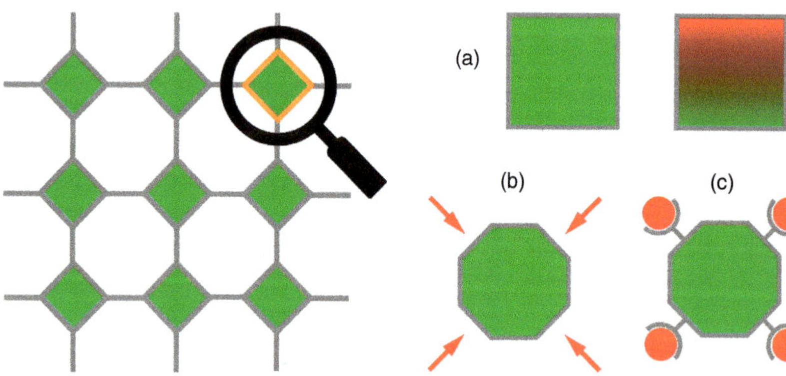

Figure 1.2 Schematic representation outlining the first section of the chapter: metal-based reactivity based on the engineering of the metallic cluster of MOFs through different strategies: intrinsic active metal sites with one or more than one metal (a), implantation of reactivity through the creation of defects (b), and incorporation of a metallic unit into the node cluster (c). Source: Figure produced by the authors of the chapter.

which can be further divided depending on the composition of the node. This way, this section will go through MOFs containing only one metal and MOFs with more than one (both mixed-metal and heterometallic MOFs).

The focus of the second part of the chapter deals with the introduction of defects as a way to increase the node performance in catalytic processes. Finally, the third part discusses different approaches that have been used to attach metallic units to the framework nodes. For each sub-section, we describe the different strategies used to tailor activity, the characterization techniques required for controlling and rationalizing activity, and different examples to illustrate their application in heterogeneous catalysis.

1.1.1 Frameworks with Intrinsically Active Metal Nodes

1.1.1.1 Metal–Organic Frameworks with Only One Metal

One of the main reasons why MOFs have attracted that much attention in recent years is their great chemical versatility. In fact, as this chapter explains, there are many ways to introduce specific functionalities to our frameworks, ultimately leading to better-performing materials. Here, we will turn our attention to the different kinds of homometallic clusters that can be found within MOFs, as well as their potential applicability in catalysis. In this part of the chapter, the use of MOFs containing only one metal, the so-called homometallic MOFs, and its application in the catalysis field will be briefly outlined.

In Figure 1.3, some of the most representative clusters are depicted, showing some of the extensive possibilities for constructing the framework. Being metallic nodes as extensive as they are, it is quite challenging to try to explain all their associated reactivity. In fact, MOFs, with certain kind of metals, are being used for redox and

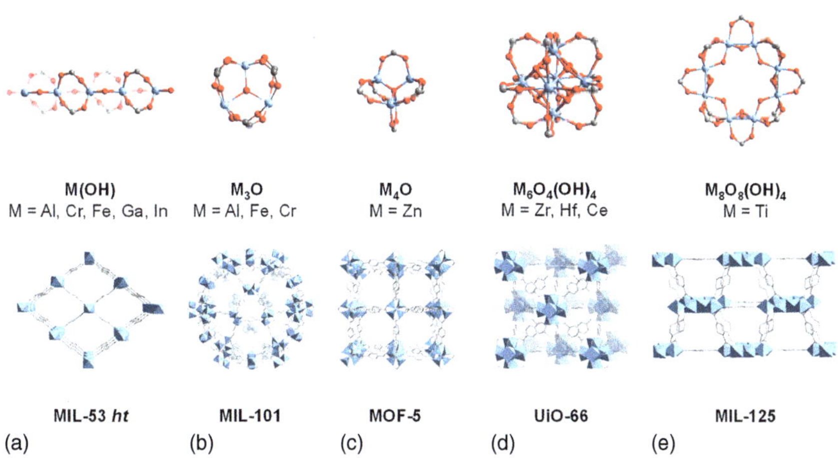

Figure 1.3 Representation of some of the most representative metal nodes that can be found (a) as well as the framework they form upon coordination with a binding organic ligand (b). Source: Reproduced with permission from Yang et al. [9]/American Chemical Society.

photocatalysis. Because of that, we will only consider the reactivity associated with Lewis and Bronsted-based catalysis.

In many cases, the reactivity of the metal cluster mainly comes from uncoordinated metal sites, the so-called open metal site (OMS). The catalytic power of these unsaturated centers is closely related to the Lewis acidity of the metal. This way, highly charged cations at the nodes, associated with high Lewis acidity, are good candidates for carrying out reactions based on Lewis acidity. In order to increase its reactivity, different activation processes can be applied to the pristine material. Furthermore, coordinated molecules can have Brønsted acidic character, as it has been proposed for adsorbed water in MOF-808-SO_4 [10].

Enhancing acidity-based reactivity in homometallic MOFs: creating open metal sites. Some post-synthetic treatments can be used to obtain a higher-performing catalyst and there exist different strategies that will be outlined here. The first one consists of a thermal reduction treatment to incorporate new OMS into MOFs as demonstrated by Serre et al. [11]. This strategy corresponds to Figure 1.4a which modifies the MIL-100(Fe) MOF. On the cluster of this framework, two of the terminal molecules coordinated to two Fe centers are water, while the third Fe can bear different molecules, mainly F- or OH-, depending on the synthetic conditions used to obtain it. The water molecules are easily removed, leaving two uncoordinated Fe centers at temperatures higher than 100 °C under vacuum or a gas stream. In their work, researchers demonstrated the possibility of thermally reducing the framework, appearing FeII OMS when heating the framework above 150 °C with a helium stream followed by 12 hours vacuum. Moreover, this work proved the relationship between unsaturated iron sites and the strength of

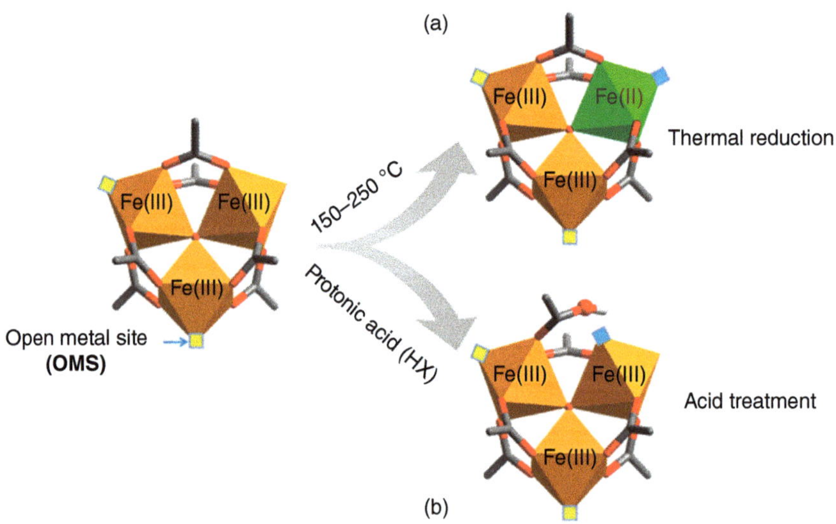

Figure 1.4 Schematic representation of the different proposed activation pathways by thermal reduction (a) and by the acid treatment (b). Source: Reproduced with permission from Wei et al. [12]/American Chemical Society.

interaction with gases, which has an impact on the performance of the framework for preferential gas sorption.

Another approach was followed by De Vos et al. [13] as a way to tune the catalytic properties of MIL-100(Fe). In this work, they treat the pristine framework with protonic acids, particularly CF3COOH and HClO4. The 1,3,5-benzenetricarboxylic acid, which serves as a ligand, is displaced from the cluster, resulting in the appearance of an OMS and an additional Brønsted acidic site, as represented in Figure 1.4b. In this case, a 2-fold increase in both Lewis and Brønsted acid sites after the acidic treatment of the framework is observed.

In other cases, the goal of the post-synthetic treatment might not be the creation of unsaturated sites within the node but maximizing the acidic character of a metal. This strategy was followed in the case of Lin et al. [14], who developed a synthetic procedure for designing a strongly Lewis acidic MOF. Particularly, they carried out a two-step transformation to the parent MOF-808 to obtain unsaturated Zr sites connected to triflate units, excellent withdrawing groups that maximize the Zr center's electron affinity, thus increasing its acidity.

Characterizing the active site in metal–organic frameworks. This part of the section will be centered on how to characterize the OMS introduced as well as the Brønsted acidity shown in the previous examples. Toward this goal, the use of probe molecules that interact differently with the different acidities present in the material has proven to be especially useful. This way, Fourier-transformed infrared spectroscopy (FTIR) spectra of adsorbed CO can give information on the formation of OMS and the presence of O-H stretching associated with Brønsted acidity [11, 13]. In the work of Lin et al. [14], in Figure 1.5, they could perform Density Functional Theory (DFT) studies to evaluate the most stable grafting mode of the triflate units. Additionally, in this case, Lewis acidity of the resulting material was quantified by N-methylacridone Fluorescence and compared with the pristine material, being an additional proof of the increased Lewis acidity.

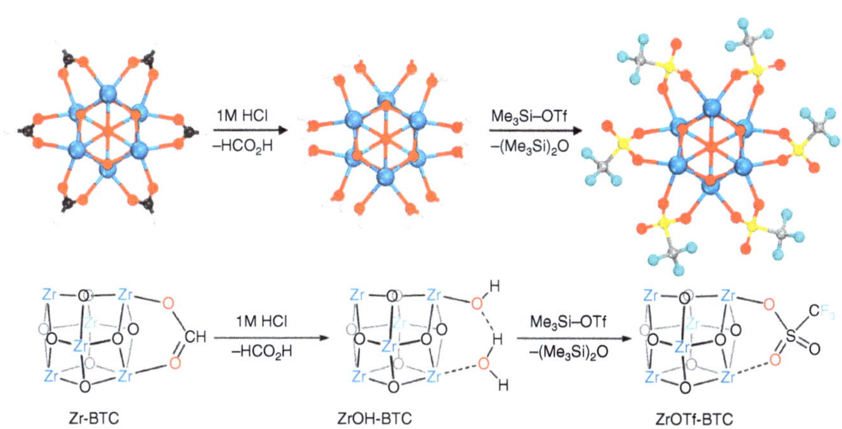

Figure 1.5 Synthetic procedure of the strongly acidic ZrOTf.BTC MOF through a two-step strategy. Source: Reproduced with permission from Ji et al. [14]/American Chemical Society.

Chemical reactivity in homometallic metal–organic frameworks. Acid catalysis has applications in many fields and is one of the most usual types. Amongst the reactions that can be used, the isomerization of a-pinene oxide [15], the cyclation of citronellal [16], the Friedel–Crafts acylation [17], or epoxide ring opening [18] reactions are common in the literature. Apart from its intrinsic value to synthetize certain products, the catalytic performance can be used to discriminate the different kinds of acidities found in a material, as exemplified by Corma et al. [19]

1.1.1.2 Metal–Organic Frameworks with more than One Metal in its Cluster

The thought of combining two or more metals in only one framework arises from the idea of having a material that combines their different characteristics of them. Bearing this in mind, ideally, these new materials can be useful to perform some applications in a more efficient way compared to homometallic MOFs. This combination of various metals can influence different applications such as gas sorption, catalysis, and sensing. Thus, for example, the incorporation of different metals (Mg/Ga, Mg/Fe, or Mg/V, among others) in the CPM-200s materials, showed an important effect on the CO_2 uptake capacity [20]. Also, the *doping* of the materials has also been studied [21, 22], demonstrating a clear effect on the electronic structure of the material, which is closely related to its performance in sensing and photovoltaic applications.

When it comes to catalysis, the use of a MOF with more than one metal results advantageous as both characteristics of these can be incorporated in one structure, which will be discussed later. However, for these MOFs having more than one metal, a distinction should be made according to the way these two or more metals are distributed in the structure. In fact, they can be statistically distributed along the framework, in which case, we would be talking *about mixed metal–organic frameworks* or, on the contrary, both metals are taking specific locations and are homogeneously distributed, then the term *heterometallic metal–organic framework* will be used. The achievement of this well-ordered disposition of metals is often restricted to the formation of clusters composed by more than one metal in specific proportions. The organization of the metals within the material structure can have a great impact on the catalytic performance of them. However, through the common characterisation techniques, it can be difficult to determine the metal distribution found in the material.

Mixed-metal or heterometallic MOF. In order to rationalize our catalytic findings and evaluate the effect of the distribution of metals in the material, one must be certain about it, often requiring specific characterization techniques. While some often-used analyses give information on the total amount of each metal, as it can be the case with inductively coupled plasma (ICP) or Energy-Dispersive X-Ray Spectroscopy (EDX) measurements, the need of exploring the exact positions of the metals usually requires more advanced techniques. In this way, the presence of domains or aggregates of one metal, which would prevent a homogeneous distribution, can be discarded.

Particularly, the heterometallic nature of clusters can be studied through Pair Distribution Function (PDF) analysis. There are many works devoted to an in-depth

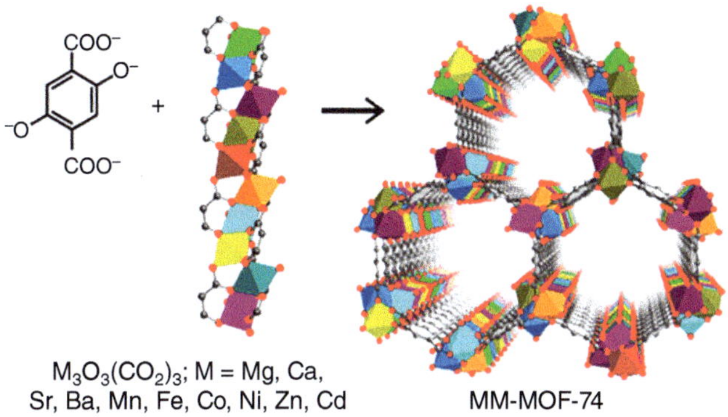

Figure 1.6 Representation of the family of mixed-metal MOF-74 showing all the metals that could be inserted in the structure. Source: Reproduced with permission from Wang et al. [27]/American Chemical Society.

analysis of this technique as well as its applicability to study different aspects of MOFs [23–25]. In summary, PDF analysis provides information about the distribution of distances between pairs of atoms from which information of the short-range order, or local structure of materials, can be extracted. As an example, PDF analysis was used in a work by Martí-Gastaldo et al. [26] to demonstrate the heterometallic nature of the cluster of a new family of MOFs.

Mixed-metal–organic frameworks. Most of the examples found in the literature for MOFs with more than one metal belong to the mixed-metal family. In this kind of materials, as previously explained, the metals are randomly distributed along the framework. One example of this is the work by Yaghi et al. [27] where they introduce different combinations of metals including Mg, Fe, Ni, or Co in the structure of MOF-74, obtaining the mixed-metal MOF-74, MM-MOF-74, schematically shown in Figure 1.6.

There are some other works involving mixed-metal MOFs [20, 28–30] and the influence of having the two metals has been evaluated in some aspects, such as flexibility [31, 32]. However, the influence of having two metals from a catalytic standpoint is still understudied, probably due to the difficulty of assessing the influence of each metallic center on the catalytic reaction.

Heterometallic Metal–Organic Frameworks. Heterometallic MOFs have been studied from different standpoints. Computationally, there have been studies shedding some light on the energetic stability and structural feasibility of different families of bimetallic heterometallic MOFs [30]. However, when it comes to the applications and advantages that heterometallic MOFs can offer, many works have failed to assess the difference that a particular distribution of metal can make. Instead, results are often presented without a direct comparison with homometallic materials, leaving aside the potential heterometallic materials can bring. However, some works demonstrate that this differentiation can be the source of radically different catalytic behaviors and should, by no means, be overlooked. Such is the case with the heterometallic MUV-101(Fe) [26], which is capable of degrading nerve agent

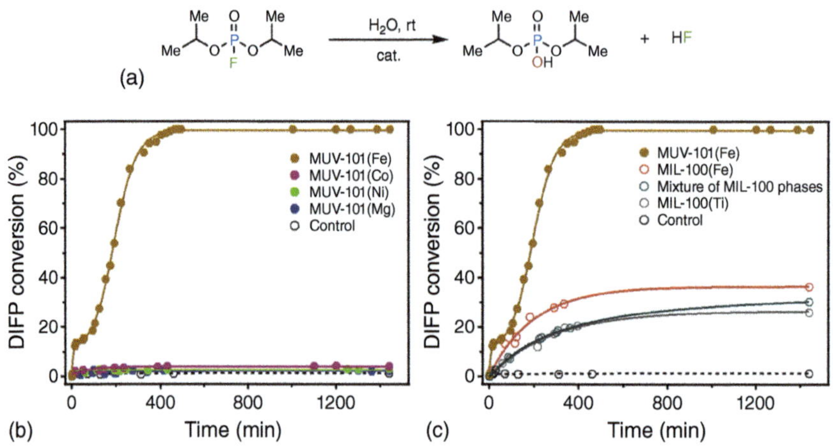

Figure 1.7 Different kinetic profiles for the degradation of a nerve simulant agent with heterometallic and homometallic MOF, as well as with the physical mixture of both. (a) Scheme of the hydrolytic degradation of DIFP, the chosen reaction to test the material reactivity in nerve agent degradation. (b) Hydrolysis profile of the different heterometallic titanium MOFs tested (c) Hydrolysis kinetics using heterometallic MUV-101(Fe), homometallic MIL(Fe) and MIL-100(Ti) and their physical mixture. Source: Reproduced with permission from Castells-Gil et al. [26]/Elsevier.

simulants, a function inaccessible for its isoreticular homometallic counterparts, MIL-100(Fe) [33], MIL-100(Ti) [34] and the physical mixture of both (Figure 1.7).

Another example, in which the exact disposition of the different metals contained in the MOF and its relationship with activity is studied, was presented by Deng et al. [35]. In their case, they could find that precoding the coordination number of heterometals in a MOF affected the electrochemical properties of MOFs, increasing its performance to carry out oxygen evolution reactions (OERs), as shown in Figure 1.8.

1.1.2 Introducing Defectivity as a Powerful Tool to Tune Metal-node Catalytic Properties in MOFs

The structure of MOFs, as many crystalline structures, often deviates from its ideal structure creating the so-called defects. These defects lead to changes in their mechanical and physical properties, porosity [36, 37], and density of OMSs. Ultimately, these defects can be somehow beneficial to the final application of the framework, tailoring its properties, for which a control and understanding of the defect engineering process is needed. Defect engineering of MOFs has already found applicability in gas storage [38, 39], or catalysis [13, 40], which will be the focus of this section.

Regarding defectivity, UiO family has been greatly studied due to its exceptional stability. In fact, the maintenance of this stability upon introduction of defects is probably the main reason why this family is often used as a model to exploit the defect engineering in MOFs. There are two basic classifications for defects in MOFs:

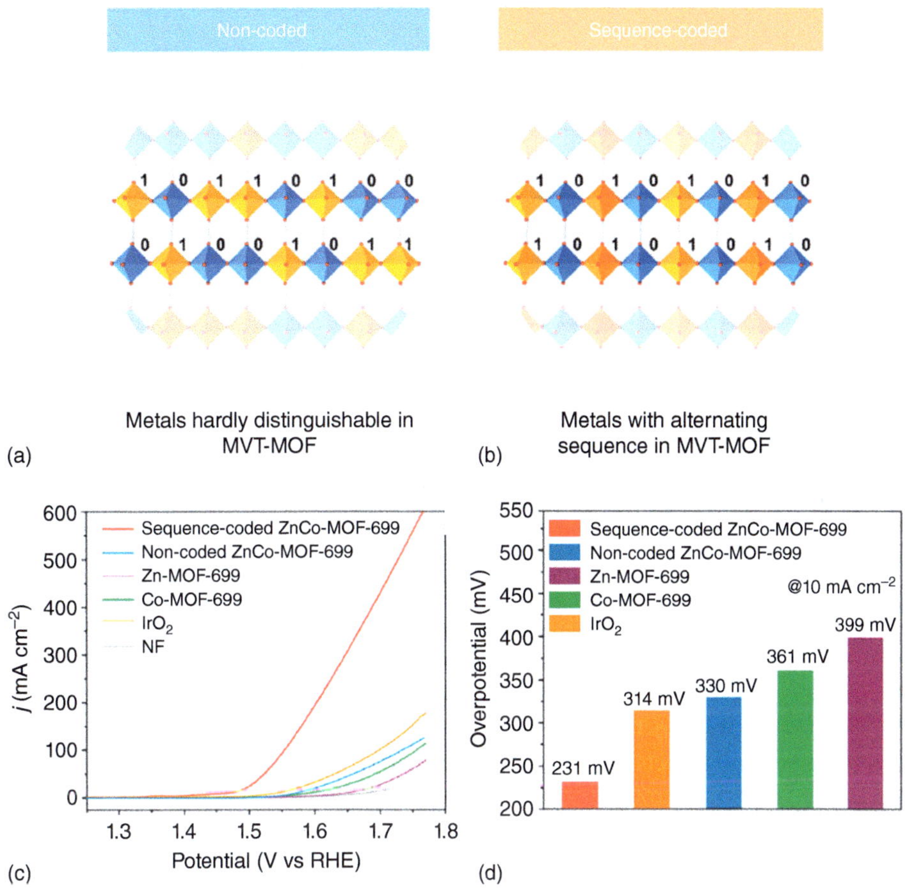

Figure 1.8 Representation of non-coded (a) and coded (b) MOFs and its electrochemical consequences: LSV (Linear sweep voltammetry) curves of different catalysts and electrodes in the OER test under the same conditions (c) and comparison of overpotentials (d). Source: Reproduced with permission from Jia et al. [35]/Elsevier.

the missing linker defects (MLD), or missing cluster defects (MCD), depending on the structural part involved in creating the defect, as illustrated in Figure 1.9 [41].

MLD originates from the loss of an organic linker from the framework while MCD is caused by the loss of the complete cluster. To compensate these vacant positions, different molecules can be incorporated the so-called caping ligands which include modulators, water molecules, or anions such as fluoride or chloride [42].

Chemical Incorporation. Briefly, the incorporation of defects to a MOF can be accomplished through two main strategies: *de novo* synthesis or by a post-synthetic treatment. The first one is the most widely used and relies on the addition of modulators to the reaction used to synthetize MOFs. Modulators are monocarboxylate species that connect to the metallic nodes as an organic linker would, but they only offer one-side connectivity, thus preventing two clusters from being connected. Typical modulator ligands include formic acid, acetic acid, trifluoroacetic acid, and

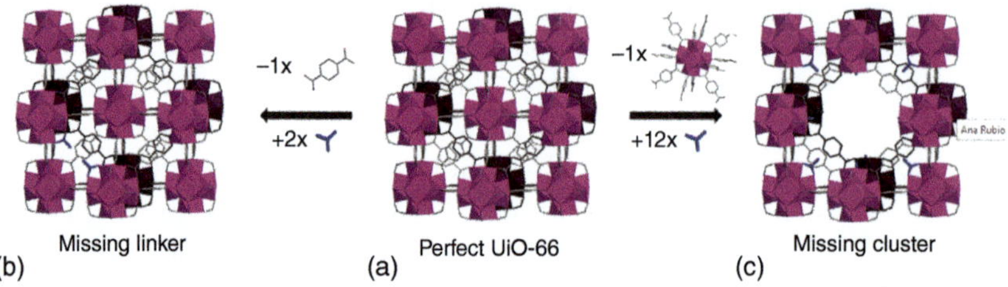

Figure 1.9 Schematic view of the generation of defects in UiO-66: from a perfect structure (a) to a missing-linker (b) and missing-cluster (c) defectivity. Source: Reproduced with permission from Taddei [41]/Elsevier.

difluoroacetic acid. It is beyond the scope of this chapter, but one should bear in mind that there are some factors influencing the creation of defects with modulators such as concentration, connectivity or acidity of them [42, 43].

When it comes to post-synthetic treatments to incorporate defectivity in MOFs, many methods rely on the acid/basic treatment of the pristine frameworks. Through this treatment, for example, defects were created in MIL-100(Fe) by the reprotonation of some 1,3,5-benzenetricarboxylic acid (BTC) linkers, removing them from the structure [13]. Additionally, monocarboxylate ligands can be incorporated into the structure in a Solvent-Assisted Ligand Incorporation fashion [44], where the framework is immersed in a solution containing the desired ligand.

Characterization Techniques. Many different techniques can be used to study the presence of defects in a MOF. One of the most used is Thermogravimetric Analysis (TGA), which provides an estimation of the number of defects present by comparing the theoretical weight loss of an ideal material with the real weight loss of the defective one. This can be correlated with the number of missing ligands per cluster, as reported by Lamberti et al. [45], shown in Figure 1.10.

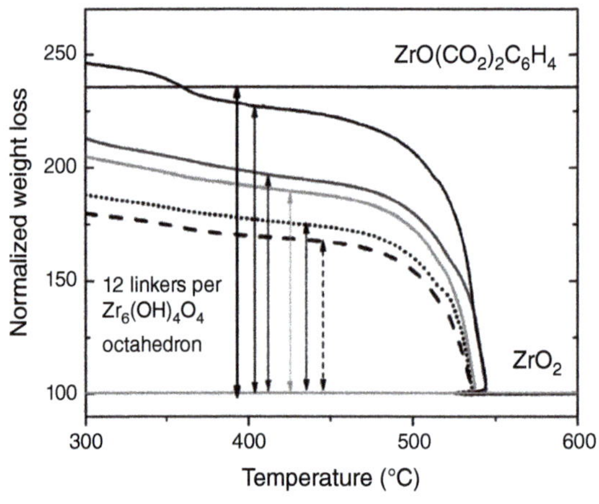

Figure 1.10 High-temperature part of the TGA curves for defective UiO-66 samples from which an estimation of the defects was made. Source: Reproduced with permission from Valenzano et al. [45]/American Chemical Society.

On other occasions, characterization techniques of the acid/base properties of a certain material can inform us about their defectivity. Thus, using probe molecules such as CO [13], pyridine [46], or CD_3CN [47], has been used to confirm the presence of defects as well as studying their newly introduced chemical properties.

X-ray diffraction patterns generally provide little information about defects. On the contrary, high-resolution neutron scattering has been employed to obtain evidence of the presence of caping ligands, proving missing linker defects [38].

Other techniques such as single-crystal X-ray diffraction, Extended X-Ray Absorption Fine Structure (EXAFS), or N_2 adsorption can also provide valuable information to study defective materials. However, with all the characterization techniques shown in this section, no information from the distribution of defects along the framework can be extracted. Recently, much effort has been dedicated to this issue, resulting in the development of a low-dose high-resolution electron microscopy (HRTEM) technique with electron crystallography [48]. The resultant images are shown in Figure 1.11 and show the types, distributions, and correlations as well as the exact three-dimensional structure of the defects in a UiO-66 framework.

Chemical Reactivity. The incorporation of defects can have a great impact on the catalytic properties of the MOF. Probably the most intuitive one could be related to enhancing diffusion inside the framework meaning that reactants can reach catalytic sites more easily. However, the chemistry of defective MOFs is different, and this can also be the cause of increased catalytic behavior. For example, MLD can become a new catalytic center through different pathways. It may be the case where the caping ligand is a solvent or formate/acetate molecule that may be removed upon activation, thus giving place to a Lewis acid site or, otherwise, the caping ligand may be a catalytic active specie, for example with Bronsted acidic character. Another option is to use these vacant sites to incorporate new ligands, with additional functionalities, or clusters to serve as catalytic centers, being this very similar to the topic that will be discussed in Section 1.2.2. These concepts are schematically illustrated in Figure 1.12 [42].

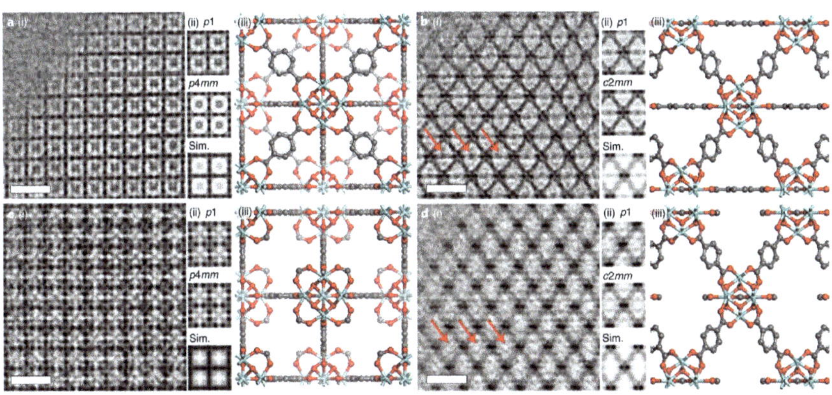

Figure 1.11 HRTEM analysis and structural models of perfect and missing-linker regions in defective UiO-66 along the [001] (a), [110] (b), [001] (c), and [110] (d) zone axes. Source: Reproduced with permission from Liu et al. [48], ©2019/Springer Nature.

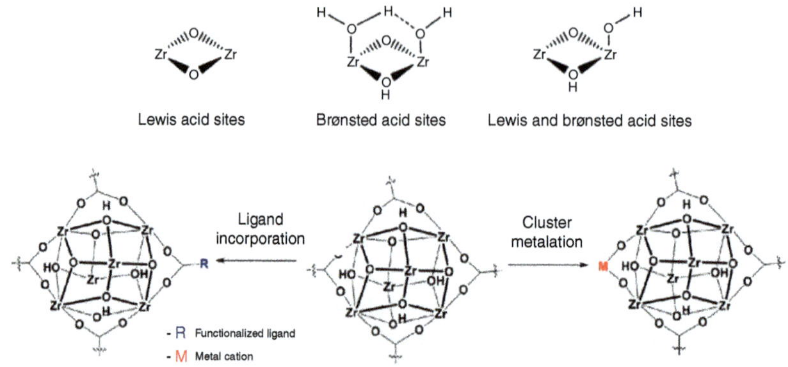

Figure 1.12 Illustration of the different types of reactivity associated with the engineering of defects and the possibilities to incorporate new catalytically active sites. Source: Adapted and reprinted with permission from Feng et al. [42]. Copyright 2021 American Chemical Society.

The enhanced reactivity as Lewis acids of defective MOFs has been exploited, for instance, in the citronellal cyclization reaction by the group of De Vos et al. [47]. In this work, the modulator approach with trifluoroacetic acid was followed in a UiO-66(Zr) MOFs, leading to a structure with a large number of OMSs upon activation.

The incorporation of defects can also provide the framework with more Brønsted acidic sites. This was exemplified by Farha et al. [40] in a study in which they correlate defects leading to inherent metal-bound hydroxides/water sites that can act as Brønsted acids with catalytic activity for the styrene oxide ring-opening reaction. In other cases, these metal-bound species can be displaced to incorporate reagents which would be impossible in the absence of defects. This strategy was used to carry out the hydrolysis of methylparaoxon [49], a phosphate-based nerve simulant upon displacement of water molecules connected to the nodes, as shown in Figure 1.13.

1.1.3 Incorporating Metals to Already-Synthetized Metal–Organic Frameworks: Isolating the Catalytic Site

In catalysis, the active site is often found on the atomic scale. For this reason, in order to maximize the material–reactivity relation, the more isolated our atoms can be, the higher the activity is expected compared to materials where atoms are aggregated. In fact, not only does aggregation prevent atoms from being exposed to the reaction media, thus inaccessible to reagents, but it can also deactivate them [50], as previously seen in the introduction of the chapter. Due to these, the use of single-metal-sites [51, 52] in catalysis has become really attractive, as it would contribute to overcoming these issues and MOFs are ideal candidates to incarnate them.

However, the introduction of some metals that might be interesting from a catalytic standpoint in MOFs by direct synthesis can sometimes prove challenging

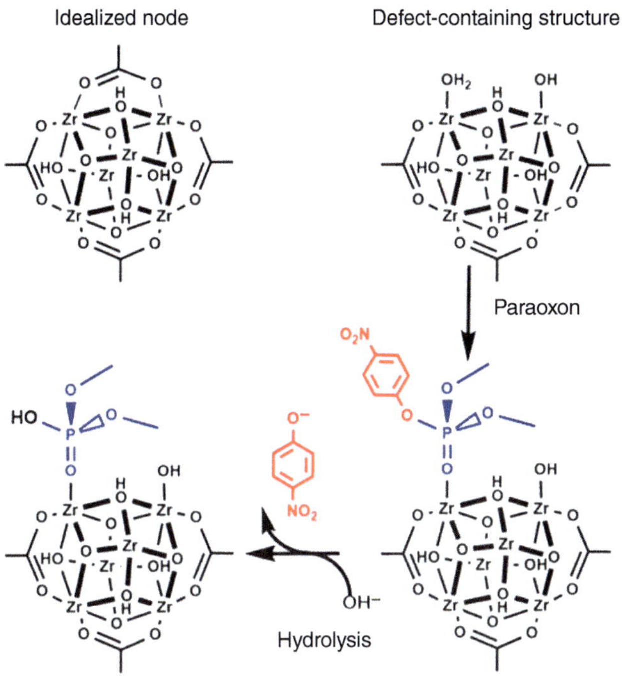

Figure 1.13 Representation of the reactivity enabled by missing-linker defectivity in UiO-66: coordination of the substrate into the node and subsequent attack of hydroxy anions. Source: Adapted and reprinted 49 from Yang et al. [9]. Copyright 2015 American Chemical Society.

This is due to the low solubility of its precursors, their high tendency to form oxides or several phases, or the low crystallinity of the resultant material. As a way to overcome this, new metals can be incorporated into an already-formed framework (*a posteriori*) through post-synthetic treatments [53].

Traditionally, the grafting of different metal species to a material has been done in diverse materials such as alumina, zirconia, or silica [54], often relying on the presence of –OH pending groups. Despite its great importance, however, the exact knowledge of the nature of the catalytic site still remains diffuse for many of the solids cited above. Furthermore, at other times, the leaching of these active species out of the support prevents its usefulness.

Recently, MOFs have been proposed as platforms to carry out this metal insertion and serving some of the already-commented advantages of MOFs (chemical versatility, sizable pores, high porosity). In this way, single-metal sites could be incorporated to MOFs.

Moreover, the incorporation of metals into MOFs offers additional advantages: these materials' crystalline structure can open the door to many techniques that would ultimately shed some light on the relationship between structure and activity as well as the mechanism through which the reaction takes place. Therefore, this finally paves the way toward a rational catalyst design. In this section, we will be turning our attention to how new metals can be incorporated to MOFs clusters and its advantages regarding catalysis.

Chemical incorporation: synthesizing MOFs with more than one metal. The incorporation of metallic units to the nodes of the framework post-synthetically can be divided into two big groups depending on whether a metal in the cluster is exchanged or, otherwise, an additional metal is incorporated to the metal node.

1.1.4 Metal Exchange

Chemical incorporation through metal exchange can be a method to integrate metals in the node that could not have been introduced by means of direct synthesis. This methodology is based on soaking of crystals in a solution where the metal is dissolved. This methodology was used in a work by Dinca et al. [55], in which they obtain MOF-5 with a great variety of metals, including reduced metal cations which are rare in MOF chemistry, mainly due to incompatibility of their oxidation states with the material synthetic conditions. The metals that were successfully incorporated in these MOFs are shown in Figure 1.14.

1.1.5 Attaching Metallic Units to the MOF

Over the years, two main strategies have been used to incorporate these metallic units into the framework: *Atomic Layer deposition (ALD)in MOFs (AIM)* or *Solvothermal deposition in MOFs (SIM)*, depending on the media that metalation takes place. Atomic layer deposition is a subclass of the chemical vapor deposition technique, in which "the precursor molecules deposit only at chemically reactive surface sites, being these reactions self-limiting" [56]. ALD is used for the "atomic-scale deposition of films on the surface of a chosen substrate" [57]. These ultra-thin films can be applied to different fields from which semiconductors stand out [58], where an exact control of homogeneity is crucial (Figure 1.15).

When applying ALD to MOFs, in AIM processes, some prerequisites have to be met: mesoporosity, thermal, and hydrolytic stability as well as functional groups to graft the metal in the correct disposition. The first example of AIM was carried out by Farha and co-workers [56] using NU-1000 as platform. This material had to be pretreated with HCl/DMF to remove the benzoic acid units initially coordinated to the pore and substitute them with -OH units, capable of carrying out metalation. Experimentally, the MOF powder was placed inside an ALD reactor with the metal

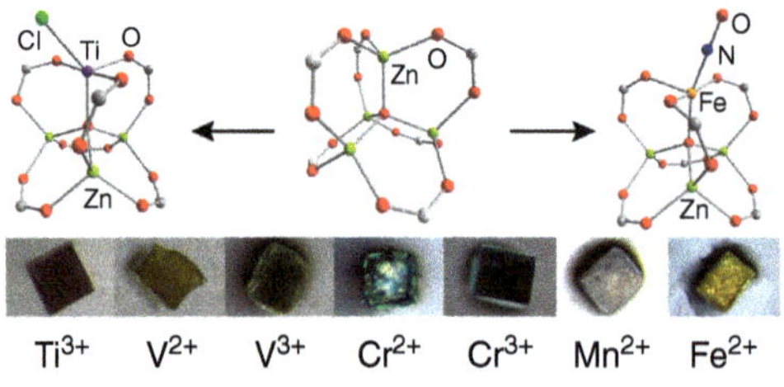

Figure 1.14 Illustration of all the possible metallic cations that can be introduced in the cluster of MOF-5 by using the metal exchange strategy. Source: Reproduced with permission from Brozek and Dincă [55], ©2013/American Chemical Society.

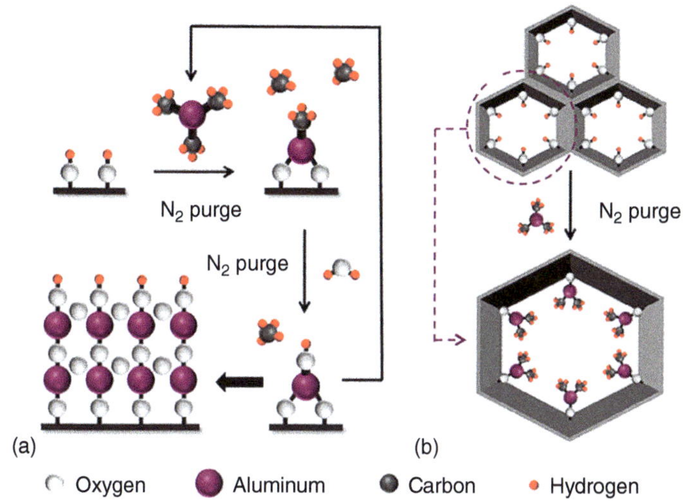

Figure 1.15 Illustration of the film ALD deposition on a surface (a) and the metalation by ALD in a MOF (AIM) (b). Source: Reproduced with permission from Mondloch et al. [56]/American Chemical Society.

salt submitted to temperatures around 120–150°C. After some exposure time, there is a N_2 purge to remove unreacted metal precursors.

In the case of solvothermal deposition in MOFs, the metalation takes place through the condensed phase. The experimental process is quite similar to AIM, being the metal precursor added in solution to the activated MOF, all in a non-coordinating solvent. Afterward, some washing is done to remove ungrafted metals [59]. It is important to highlight here that, in many cases, dry conditions are needed to carry out a successful grafting through SIM [60] (Figure 1.16).

Characterization Techniques of AIM and SIM. The insertion of different metals into the nodes of a MOF can be characterized in different ways. An interesting approach to check that the grafting has taken place is to study the FTIR -OH

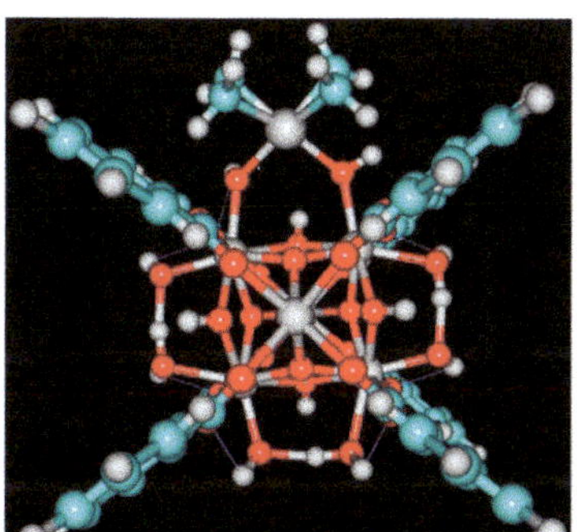

Figure 1.16 Representation of the grafting of Ir to the Zr_6 clusters of UiO-66. Source: Reproduced with permission from Yang et al. [59]/American Chemical Society.

or -OH_2 vibration bands which should diminish or disappear upon metal grafting. This was observed by Gates et al. [49] in their work on grafting of $Ir(CO)_2$ complexes in UiO-66 and NU-100 MOFs [59]. Complementarily, the number of metals inserted can be studied through inductively coupled plasma-mass spectrometry (ICP-MS) and give an estimated number of grafted metals per node.

X-ray absorption spectroscopy (XAS) and PDF studies can be really useful to gain some insight into the local structure of the supported metals. As an example, X-Ray Absorption Near Edge Structure (XANES) and EXAFS were used to support the grafting of Mo into NU-1000 [60] to be monomeric or at most few Mo atom clusters. In other cases, such as in the work from where Figure 1.17 is extracted [61], PDF is used to study the metal grafting mode to the cluster.

DFTs are also very important studies in assessing the accessible coordination modes of the metal [60, 61] and the energy it would take to interchange between them if it applies. Moreover, DFT studies might suggest the mechanism through which the catalytic reaction occurs through the new grafted material, as is the case of a Zn/Cu grafted NU-1000 MOF used for the selective hydrogenation of propyne [62].

Chemical Reactivity. This methodology allows the incorporation of chemical reactivity to the material that can be as diverse as the metals grafted. This way, many different reactions have been proposed for exploiting these chemically engineering metal sites in MOFs. Especially interesting are the cases where non-abundant metals are integrated in MOFs because their catalytic activity can be maximized with only a small amount of metal needed.

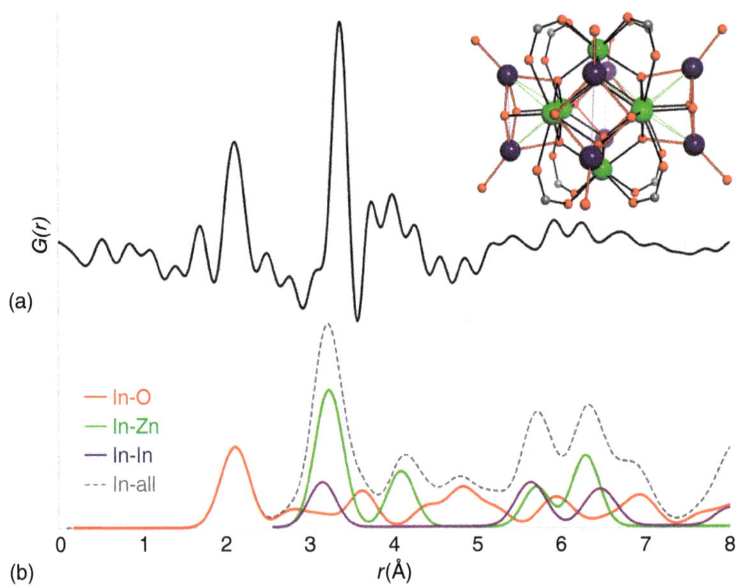

Figure 1.17 Differential PDF corresponding to the new atom–atom distances formed upon the incorporation of the In unit in the MOF through AIM methodologies (a), partial pair PDFs calculated based on the model for In-loaded NU-1000 structure predicted by DFT (b). Source: Reproduced with permission from Kim et al. [61]/American Chemical Society.

One of these examples could be the incorporation of a Molybdenum (VI) oxide catalyst in the clusters of a MOF via solvothermal deposition (SIM) [60]. With this material, the epoxidation of cyclohexene could be carried out selectively, being the epoxide and the ring-opened diol the main products. It is important to highlight that these oxomolybdenum species are active in homogeneous catalysis for the synthesis of epoxides but suffer from deactivation easily. In the past, these oxides have already been grafted to silica and alumina but, in these cases, there was evidence of considerable leaching. However, with the introduction of these Mo species into MOFs, this leaching seems to be suppressed, as the Hot-Filtration Tests carried out suggest, which is shown in Figure 1.18b.

Another example of catalyst based, but this time on AIM, was presented by Delferro et al. [52]. They grafted different metallic species to the node of NU-1000 and observed distinct reactivities depending on the metal grafted. While Zn afforded the isomerization of the product, Cu yielded a selective hydrogenation, as shown in Figure 1.19 [62].

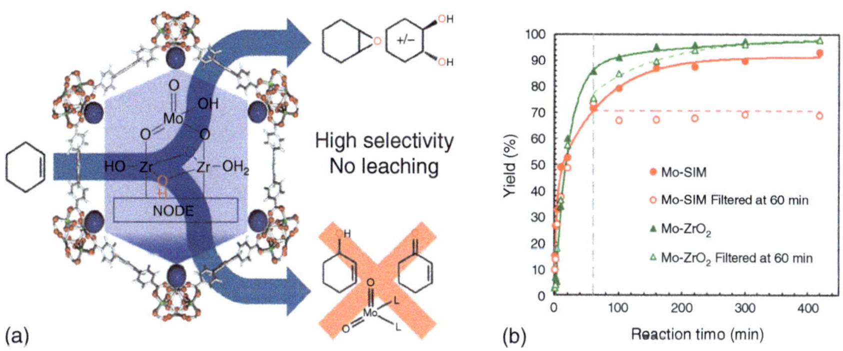

Figure 1.18 Illustration of the grafting of Mo species into the node and its consequences from a catalytic standpoint (a) and Hot-Filtration Tests used to discard the leaching of Mo into the reaction media (b). Source: Reproduced with permission from Noh et al. [60]/American Chemical Society.

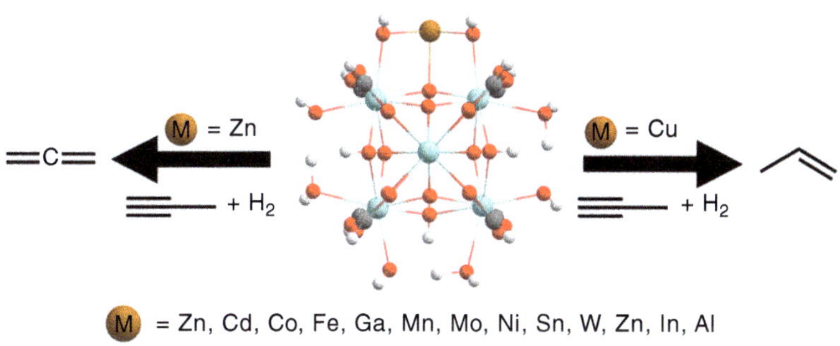

Figure 1.19 Illustration of the different reactivities enabled by the grafting of different metals to the metal–organic framework cluster. Source: Reproduced with permission from Hackler et al. [62]/American Chemical Society.

1.1.6 Grafting of Organometallic Complexes into the MOF Nodes

Since the appearance of organometallic chemistry, their advantages have been clear as these compounds allow a well-defined metallic catalytic site, which makes them highly efficient for catalytic processes as well as establishing structure-activity relations. Therefore, they have been proposed to catalyze industrially relevant processes such as the synthesis of polyethylenes with the organometallic Ziegler–Natta complexes. However, this type of compound usually operates in a homogeneous catalytic way fashion, which presents disadvantages in its practical use. On the view of having both, advantages of organometallic compounds (well-defined single site) and heterogeneous catalyst (easy separation and recyclability), researchers have worked toward the heterogenization of these by grafting them to solid supports thus getting the best of both worlds, being Jean Marie Basset one of the pioneers in the area.

Lots of effort has been put into trying to incorporate these organometallic compounds to the frameworks and can be carried out through different strategies, such as:

- Grafting the organometallic complex to the node.
- Inserting the organometallic compound through the ligand (Section 1.2 of this chapter).

At this point, it is worth mentioning that some grafted compounds can have different kind of ligands in their coordination sphere. In some cases, these ligands are one of those commonly used in organometallic chemistry, while other times these ligands are purely inorganic compounds such as hydroxides. In this direction, it is sometimes difficult to draw a line between the grafting of metals (that were covered in Section 1.2.2) and organometallic species. An example where different kind of ligands are coordinated to a metal grafted to a MOF-node is present in the work by Morsali et al. [63] as shown in Figure 1.20.

Chemical Incorporation of Organometallic Compounds to MOFs. One powerful approach to incorporate organometallic complexes to MOFs is through Surface Organometallic Chemistry (SOMC) [64, 65], which considers that the surface of the support can act as a ligand. In this way, grafting the metal complex to the surface of the MOF which ideally has a known structure and a specific chemical reactivity associated.

Traditionally, this SOMC approach has already been implemented in silica, alumina, and magnesia where –OH groups usually serve as anchoring points [66]. However, due to the advantages that MOFs can offer, great effort has been put to try to incorporate this SOMC approach to them.

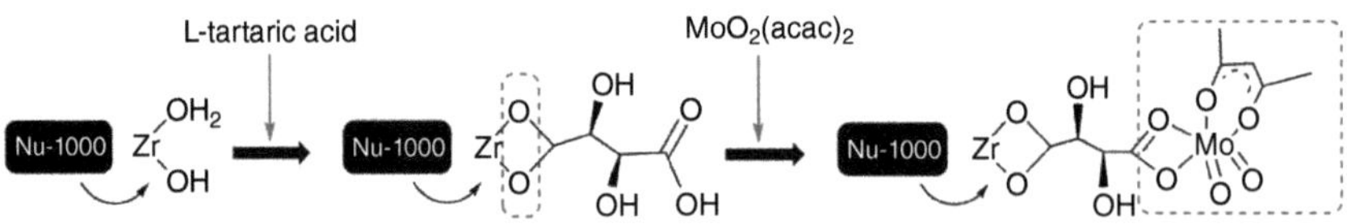

Figure 1.20 Insertion of L-tartaric acid and subsequently grafting with Mo. Source: Adapted and reprinted with permission from Berijani et al. [63]. Copyright 2011 Royal Society of Chemistry.

Different metals can be incorporated to the clusters to obtain the desired reactivity. Usually, the chemical incorporation of them consists of the addition of the organometallic complex (itself or dissolved in an organic solvent) to the previously activated framework. Afterward, it is common that the material is washed with organic solvents and evacuated to eliminate any unreacted species.

There are many examples of different organometallic compounds being grafted to MOFs' nodes. Lin et al. [67] reported the insertion of an isolated magnesium alkyl catalyst connected through the metalation of $Zr_3(\mu_3\text{-}OH)$ sites in the Secondary Building Unit (SBU) of the MOF. Another beautiful example is presented by Eddaoudi, Basset et al. [68] in a work in which, through a SOMC approach, a tungsten compound is grafted in a Zr-NU-1000 compound (Figure 1.21).

Characterization Techniques of Organometallic compounds in MOFs. As previously seen in previous sections, some commonly used techniques can apply to the characterization of these grafted organometallic compounds. Thus, while ICP-MS and EDX can quantify the metals incorporated, SEM-EDX mapping can assess its distribution. NMR techniques can be especially useful in this case, as they can be used to study if the ligand is still present and coordinated to the metal [68]. Moreover, FT-IR can give valuable information regarding the decrease of the Zr-OH band and the appearance of a M-ligand band upon grafting.

These modified frameworks can also be studied by means of single crystal X-ray diffraction, from which a complete vision of how the newly grafted organic compound is positioned with respect to the MOF can be extracted [50].

Computational studies can also be used to explore the more favorable grafting mechanism, taking into account that, as in many organometallic processes, many grafting scenarios could be possible.

Chemical Reactivity of Organometallic Compounds in MOFs. As seen in the last section, AIM and SIM reactivity depends mainly on the metal inserted giving rise to varied catalytic applications. It is also the case for the reactivity of organometallic compounds grafted to metallic clusters: the reactivity is so varied that we will only give a few examples of works published in the area. In fact, as we commented before,

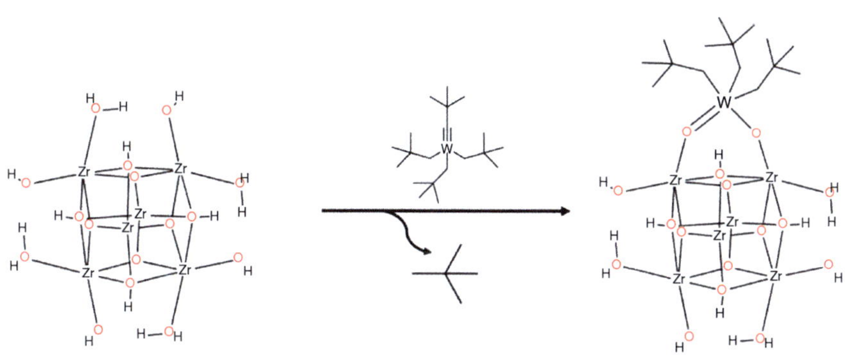

Figure 1.21 Grafting of a tungsten specie into the Zr nodes of NU-1000. Source: Adapted and reprinted with permission from Thiam et al. [68]. Copyright 2020 American Chemical Society.

the whole idea behind this post-synthetic treatment is to incorporate the same chemical reactivity found in organometallic chemistry to MOFs.

In this way, the strategy of SOMC has been applied to MOFs resulting in systems with higher catalytic turnover when compared to other catalysts based on the same metal. This is the case of the W complex grafted presented in the previously mentioned work by Basset and Eddaoudi et al. [68], showing better performances than other W-based catalysts. Particularly, this catalyst could be activated in the presence of an olefin, rendering an active carbene specie, active in the metathesis of olefins with good selectivity toward the primary products.

This example is of especial interest as it shows the potential advantages of this strategy. Apart from obtaining a competitive heterogeneous catalyst, the grafting mechanism and the actual structure of the precatalyst grafted W complex could be studied computationally, as shown in Figure 1.22. A deeper understanding of how this catalyst works could be obtained through this, and ultimately, improve the efficiency of industrial processes based on WO_3/SiO_3, where there is currently a lack of understanding.

Another example is the research carried out by Lin et al. [67]. In this case, the supported specie was a magnesium-alkyl which served as a catalyst in hydroboration and hydroamination reactions. This catalyst was demonstrated to be effective in hydroboration reactions with such low loadings as 0.05 mol% Mg and suitable to diverse substrates. Moreover, its recyclability and crystallinity maintenance were examined with excellent results. Although the position of the Mg grafted could not

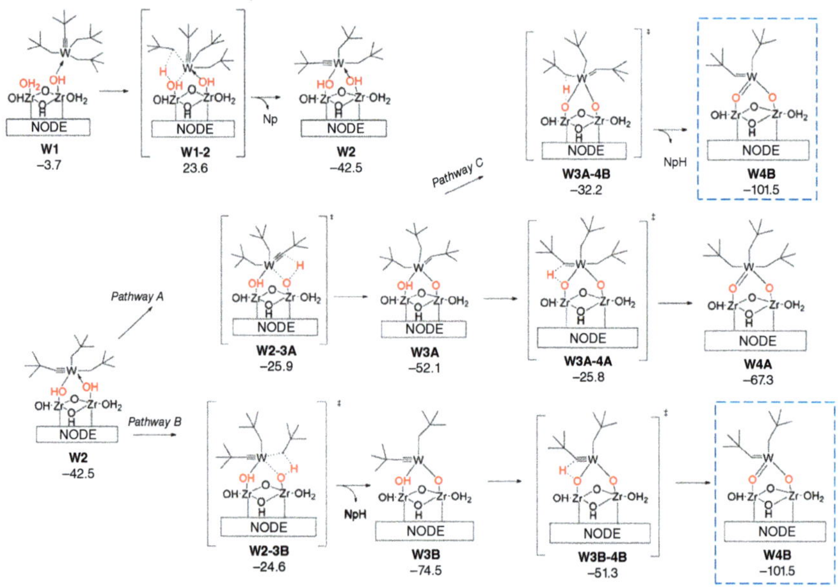

Figure 1.22 Grafting of the W-species to the node and the energy differences in kcal/mol calculated by DFT. Source: Reprinted with permission from Thiam et al. [68]. Copyright 2020 American Chemical Society.

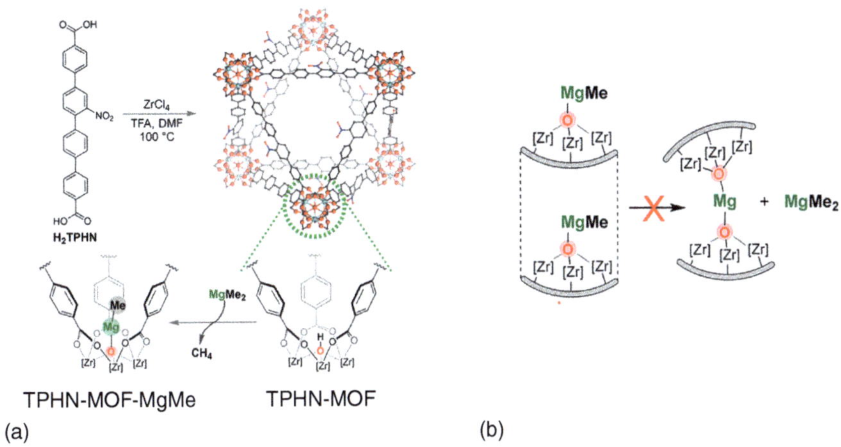

Figure 1.23 Synthesis of the TPHN-MOF and subsequent grafting in the metallic nodes of the structure of an alky-magnesium unit, THPN-MOF-MgMe (a) and benefits regarding the isolation of the active site within the framework ultimately preventing Schlenk-type ligand redistribution reactions (b). Source: Adapted and reprinted with permission from Manna et al. [67]. Copyright 2016 American Chemical Society.

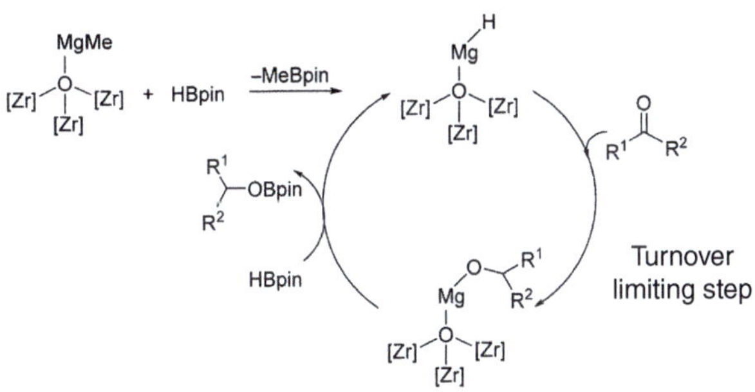

Figure 1.24 Proposed catalytic cycled of the organometallic Mg-grafted inserted into TPHN-MOF. Source: Adapted and reprinted with permission Manna et al. [67].

be determined by single-crystal diffraction due to disorder, the researchers proposed the structure shown in Figure 1.23.

All of this enabled the proposal of a tentative mechanism for the TPHN-MOF-Mg catalyzing hydroboration reactions, shown in Figure 1.24. In particular, they attributed the catalytic activity of the catalyst to the Mg atom, capable of activating the aldehyde unit.

1.2 Ligand Engineering

Through the last few years, researchers have made a lot of efforts to enhance the catalytic properties of MOFs [69–73]. One way to do that is by introducing a catalytic metal site, or several, into the structure [74]. This or these catalytic metal sites can

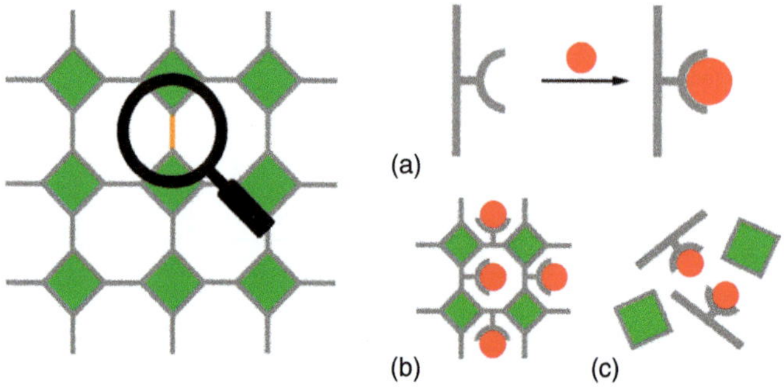

Figure 1.25 Schematic representation of the different linker metalation procedures: premetalation of the ligand (a), post-synthetic metalation (b), and introduction of the metal by direct synthesis (c). Source: Figure produced by the authors of the chapter.

be present in different parts of the framework, such as in the metal cluster [75], in the organic ligands [76], attached to the pores [77], or randomly distributed along the assembly [78], as mentioned in the introduction.

The grafting of diverse catalysts into the organic parts of the MOFs opens a new avenue for catalysis. As it is well known and studied, homogeneous and heterogeneous catalysts incorporated or embedded into the frameworks can give rise to more efficient catalytic processes, and particularly, MOF catalysts with active sites at organic linkers have attracted the attention in the last few years.

As it will be more detailed next, this part of the chapter will consist of the introduction of metal sites in the organic ligands of the MOF. Here, it will be described the different strategies to introduce the active metal sites into the organic parts of the framework, the variety of organic ligands that can be modified for that task, and the role of themselves in the catalytic process (Figure 1.25).

1.2.1 Ligands as Active Metal Sites

1.2.1.1 Creating Metal Sites in the Organic Linkers. Types of Ligands

Conceiving ligands as active sites for catalytic reactions is a well-studied area in the field of MOFs [79, 80]. MOFs are excellent platforms used in heterogeneous catalysis due to their tunability of the pore sizes, their high activity, and selectivity [81, 82]. Their organic part, the ligands, can also be modified by introducing different substituents like, for example, metal complex-based catalytic species which will improve the structural stabilization of the catalytic sites, create synergistic environments, or help in the interaction with substrates. Therefore, catalytic efficiency and selectivity are enhanced [83]. Introducing active metal moieties in the ligands is an easier technique than others used for the incorporation of metal nodes in MOFs, especially if it is done by post-synthetic modifications (PSM). But this can be achieved by both, direct synthesis and PSM. The insertion of single-metal sites in the ligand can be done before, after, or during MOF synthesis via coordination with their heteroatoms owning to lone pairs of electrons [84]. Direct methods are challenging due to the possible deactivation of the induced OMSs while introducing the metals, but on the contrary, as mentioned before, PSM are a more respectful, easier, and safer approach to generate highly catalytic active sites in the ligands after the MOF synthesis. This approach also includes the grafting of inorganic metal species or complexes onto the organic ligands and the exchange of metallolinkers.

Moreover, in MOFs with one metal in the structure acting as a catalytic single metal site, it is possible to include additional metals in some specific linkers denoted as metalloligands. These ligands are mainly constituted by N-donor atoms such as porphyrins [85], bipyridines [86], terpyridines [87], imines [88], salen [89], or diketiminates [90]. In this way, many homogeneous metal complexes that have catalytic active sites established can be attached to the metalloligands and provide various heterogeneous catalysts sites in the MOF that can work at the same time or not. This offers a bridge between heterogeneous and homogeneous catalysis that integrates their respective advantages into the MOF (Figure 1.26).

The first part of this chapter will be focused on the description of some examples of these kind of ligands that are the most commonly used for catalysis and how they are used for it. It is also described how they can interact with other single-metal sites that can be present in the MOF.

Porphyrin ligands. "Porphyrins are [18π]-electron heteroaromatic compounds in which the aromatic character of the underlying tetrapyrrole moiety, and the reactivity of the functional groups present in their side chains, governs their rich chemistry." [92]

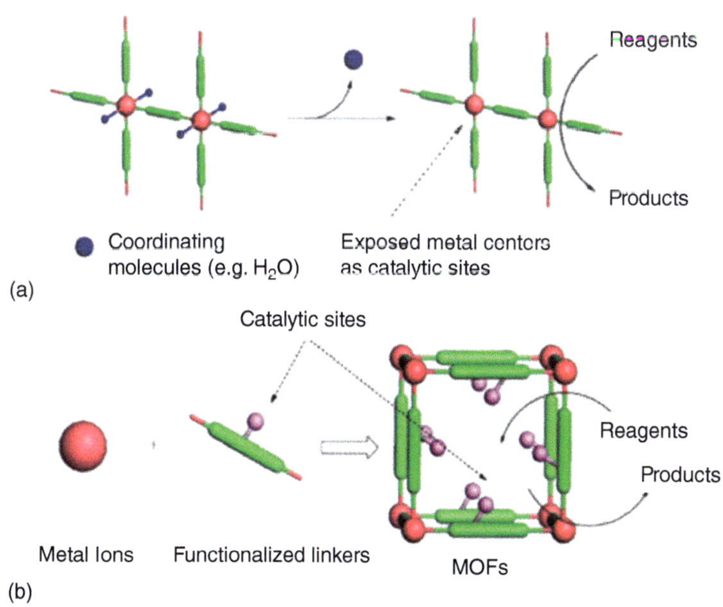

Figure 1.26 Illustration of the incorporation of active metal sites in MOFs: coordinatively unsaturated metallic nodes serving as catalytic sites (a) and incorporation of active catalytic sites into the ligands of MOFs (b). Source: Reproduced with permission from Ma et al. [91]/Royal Society of Chemistry.

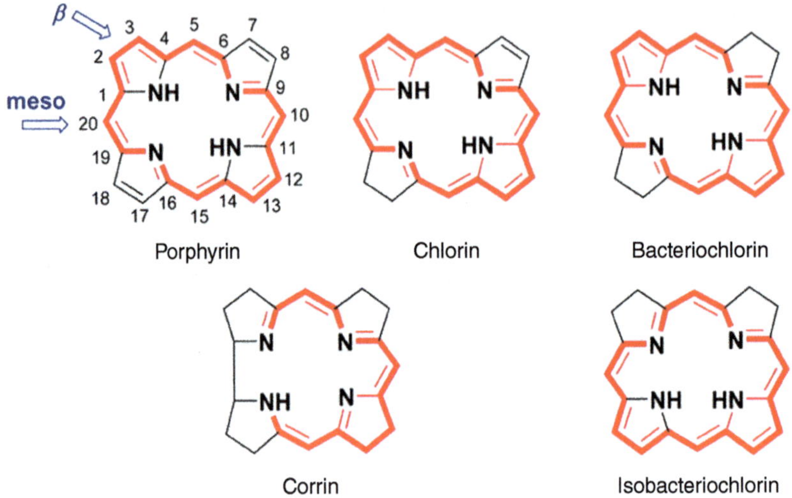

Figure 1.27 The different classification of macrocycles and their coordinative positions. Source: Reproduced with permission from Senge et al. [92]/Royal Society of Chemistry.

As Senge et al. commented in their review, porphyrins are organic compounds composed of a four-interconnected pyrrole macrocycles. Normally, these macrocycles can be divided into three groups: porphyrins, chlorins, and bacteriochlorins. These organic compounds have different positions in which they can coordinate with different metal ions. These positions are the free nitrogens present in the central ring of the macrocycle, the β positions of the pyrrole, and the meso positions of the porphyrin (Figure 1.27).

These kind of metalloligands are widely used due to their intrinsic catalytic properties, as they are able to oxidize different organic molecules in mild conditions [93]. Also, the rings of the porphyrins are easy to modify by adding pyridines and carboxylic acids converting them into interesting building blocks for MOFs. Up to date, a large number of porphyrinic-MOFs have been synthesized with different modifications in the ligand such as the metal centers (Co, Cu, Ni, Zn, Fe, Mn) [94] or the different substituents along the axial positions of the metalloligand (F^-, OH^-, O–CH_3 among others) [92]. However, the principal problem of these complexes is that only a few of them show good stability for catalytic reactions [95], an issue that has been trying to be solved for a while [96–100].

Chemical Reactivity. The incorporation of porphyrins in MOFs can provide some advantages in catalysis, as they have intrinsic catalytic properties as mentioned before, and more particularly, if modifications are added to the ligand such as the incorporation of metals. Some examples of works published for catalysis with porphyrins include epoxidation reaction of olefins as exemplified in the work of Fujii et al. where they performed a Marcus plot analysis for the epoxidation reaction of the oxoiron (IV) porphyrin π-cation radical complex with alkene, to demonstrate the mechanism of the electron transfer, which occurs from the Highest Occupied Molecular Orbital (HOMO) of the alkene to the porphyrin π-radical orbital [101].

The hydroxylation of linear and cyclic alkanes is another typical reaction where a metalloporphyrin can be used as catalyst. As is the case in the work of Suslik et al. where they use a MOF called PIZA-3 with a metalloporphyrin as ligand to run the reaction of epoxidation of alkenes [102].

In another work, Hod et al. use a material that contains functionalized Fe-porphyrins as catalytically competent, redox-conductive linkers to perform electrocatalytic CO_2 reduction [103].

Bipyridyl ligands. Bipyridil ligands are metalloligands that contain bipyridine groups. Bipyridine is an organic compound consisting of two pyridyl rings. It is a bidentate chelating ligand that can coordinate with different transition metals giving rise to diverse complexes. There exist six different isomers, but two are the most used: 2,2′-bipyridine and 4,4′-bipyridine (Figure 1.28).

The bipyridyl ligands offer different and more versatile single-metal sites when anchored to them, than those anchored to porphyrin heterocycles [105]. Also, for this kind of metalloligands, single-metal sites can be directly synthesized or incorporated by PSM. The last case presents a problem as the metal salts of the precursors can form aggregates on the surface of the MOFs thus blocking the pores channels. Also, the formation of single-metal sites can be randomly distributed along the framework [106]. On the other hand, for the direct synthesis, these drawbacks are overcome as the metal species incorporated allow relative isolation and uniform distribution of the active sites along the MOF [107]. So, for these kind of ligands, the introduction of single-metal sites works better by direct synthesis approaches rather than by PSM, contrary to what is expected from what was said above [108].

Chemical Reactivity. The bipyridil ligands are also well used in catalytic processes, as they can host a large variety of metals as single-metal sites, as it is shown by Chen et al. where they use two different complexes of bipyridil ligand with Ru and Pt and they incorporated these two complexes into a robust MOF of Zr(VI) by mix-and-max approach. The incorporation of the complexes $[Ru(dcbpy)(bpy)_2]^{2+}$ and $Pt(dcbpy)Cl_2$ allows the facile arrangement of the photosensitizer and the reduction catalyst very close to each other which allows the promotion of the electron transfer between them, this improves the hydrogen evolution reactivity upon visible light irradiation [109].

Imine ligands. Imine ligands refer to ligands that contain a C=N bond in their structure. The unique electronic properties of this group make it a very interesting candidate for catalysis. The lone pair of electrons of the nitrogen atom can coordinate

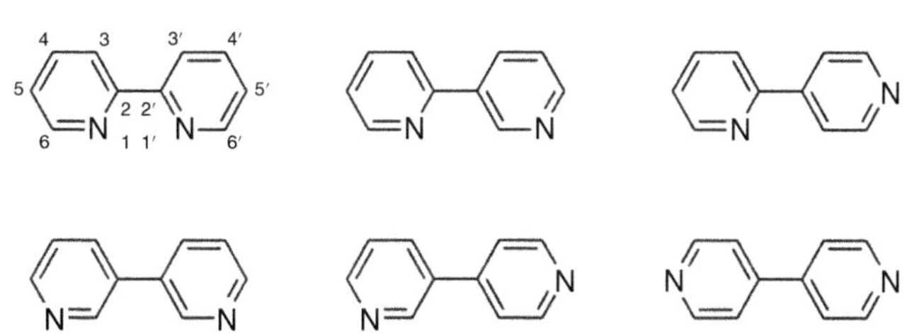

Figure 1.28 The different isomers of bipyridine. Source: Reproduced from Bipyridine [104]/Wikipedia/Public domain CC BY 3.0.

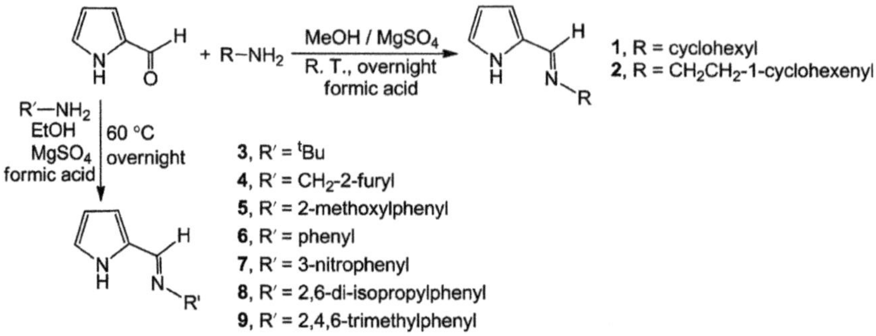

Figure 1.29 Synthesis of imine ligands. Source: Reproduced with permission from Chen et al. [110]/Elsevier.

to a large variety of metals. Imine ligands can be classified into three groups such as monodentate, bidentate, and polydentate imine ligands (Figure 1.29).

This type of metalloligands is commonly used because of its high stability in most organic solvents. It is obtained through a Schiff-based chemistry approach where a primary amine reacts with an aldehyde or a ketone under specific conditions, resulting in a replacement of the carbonyl group by an imine group. These Schiff reactions are an efficient and environmentally safe approach to generate homochirality. They are interesting because their nitrogen atoms can coordinate to different transition metal atoms, which makes them very convenient to perform as catalysts. Indeed, these kinds of complexes can be used as effective catalysts for organic reactions, mostly for asymmetric catalytic reactions. They also offer a large number of possibilities for post-synthetic transformations by metalation or conversion of the imine groups [111].

Chemical Reactivity. The use of the imine ligands in catalysis is quite frequent, since they provide a high stability. One example of the use of these kind of ligands in catalysis is the allylic alkylation reaction, as shown in the work of Ellman et al. where they use a *P,N*-sulfinyl imine ligand with Pd to form the complex. This material was utilized to carry out the catalytic reaction mentioned before with a high enantioselectivity (94%) [112].

Salen ligands. Chiral salen ligands are obtained from the condensation of diamines and aryl aldehydes. It was first described by Hugo Schiff in 1864 [113]. They are normally called Schiff base ligands and can coordinate metals through imino nitrogen and through different groups linked to the aldehyde. They are able to stabilize metals with different oxidation states which make them good candidates to perform as catalysts. This kind of ligand has four coordination sites and two axial ones. They are very similar to porphyrin ligands but can be prepared more easily, by simply condensation of amines and aldehydes in different solvents, and preferably with some dehydrating agent to favor the reaction (Figure 1.30).

Researchers normally use this kind of metalloligands with modifications on pyridyl or carboxylic groups, turning them into excellent building blocks for the formation of MOFs, and it has inspired the preparation of organic molecular switches-based MOFs. Playing with the chirality of these ligands, homochiral

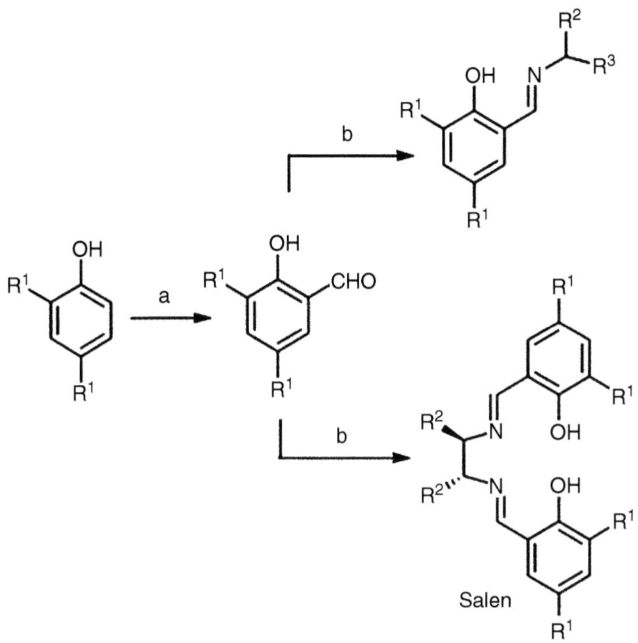

Figure 1.30 Preparation of Salen ligands. Source: Reprinted with permission from Cozzi [113]. Copyright 2004 Royal Society of Chemistry.

porous MOFs can be obtained by starting from enantiopure diamines, which give rise to chiral metallosalen ligands. These type of connectors have been considered one of the most important asymmetric catalysts to date [114–118].

Chemical Reactivity. A large number of people also use Salen ligands in catalysis, as they are very easy to modify with metals with different oxidation states. One example of a catalytic reaction performed through them is the Katsuki–Jacobsen epoxidation reaction. For that matter, Jacobsen et al. use a salen metalloligand to demonstrate the direct correlation of the asymmetric epoxidation reaction with the electronic properties of the ligand substituents [119].

Another interesting catalysis is the intramolecular acyl-transfer reaction. In this case, Nguyen et al. use a MOF with (salen) Mn complexes as ligand to carry out the reaction demonstrating that this metalloligand can enhance the catalyst stability and selectivity and allows its separation and reuse [120].

Metal-Free Organic Ligands. Although most mofs-based catalysts are catalytically active due to the metal or metals acting as active sites, there are also examples of metal-free ligand catalysis. The organic part of the MOF can be functionalized with different heterogeneous catalysts, as there are diverse functional groups that can act as active sites such as sulfoxy, pyridil, amino, and sulfonic groups, among others. Besides metal nodes, the functional groups of the ligands can also act as catalytic sites introducing acid and base catalytic activity [121]. The current catalytic processes use metals and metal oxides, however, there are different metal-free heterogeneous catalysts more efficient as they are less energy demanding and can compete with known conventional catalysts and even perform better than them.

Chemical Reactivity. As commented before, some organic ligands, and particularly, the majority of mentioned before, can act as catalysts by themselves, without the help of any metal. One example of this is the work published by Sanders et al. in which they use a cyclic porphyrin trimer as catalyst for catalytic acyl transfer. They create synthetic systems where the binding sites are positioned in a way where the substrate molecules are close in proximity, and they can catalyze reactions simply by their binding properties (Figure 1.31) [123].

1.2.1.2 Cooperation Between Single-Metal Sites and Metalloligands

The cavities on assemblies, like MOFs pores, –organic clusters, or some organic ligands, can stabilize reactive species for photocatalysis [124]. Also, they can enhance catalytic reactivities and selectivity by isolating guest molecules. So, bearing this in mind, the cooperativity among the different assemblies of the framework previously described and metalloligands can be used as homogeneous catalysts inside MOFs [125]. Combining their properties, they enhance the catalytic performance by having various heterogeneous catalysts in the same structure acting at the same time [126–128]. Immobilizing metal complexes into the organic ligands is one of the most convenient methodologies to prepare a MOF with different metal sites. Moreover, it can also offer high stability against the deactivation of the metal site.

Chemical Reactivity. Tandem reactions are frequently used in catalytic reactions due to their tuneable open metal centers (see Section 1.1), functional organic linkers, and active guest species in their pores (see Section 1.3). There are many examples as they are shown in the review of Cao et al. where they divide the tandem reactions into three categories: "(i) open metal centers and functional organic linkers, (ii) active guest sites in the pores and active sites in the MOF structure, and (iii) bimetallic nanoparticles (NPs) on MOF supports" [124].

1.2.1.3 Ligand Accelerated Catalysis (LAC)

Chiral ligands have gained a lot of attention due to the effects that they have on metal catalysis, which have been discovered and studied during the search of new metal-catalyzed asymmetric reactions. All of this has also led to the discovery of ligand-accelerated catalysis (LAC) [129] which consists of the addition of a specific ligand to an existing catalyzed process, improving the reaction time.

Chemical Reactivity. This phenomenon works for both homogeneous and heterogeneous catalysis. LAC is used in most of all catalytic reactions with transition metals. Some examples are guided by ligand exchange processes, which makes it possible that one single ligand could activate different catalysts inside a mixture of metals.

This phenomenon was discovered in 1988 by Sharpless and is found in the catalytic asymmetric reactions in Figure 1.32 [130]. But sometimes, there is a crowding around the catalytic sites, and this causes a deceleration rate, which is not desirable.

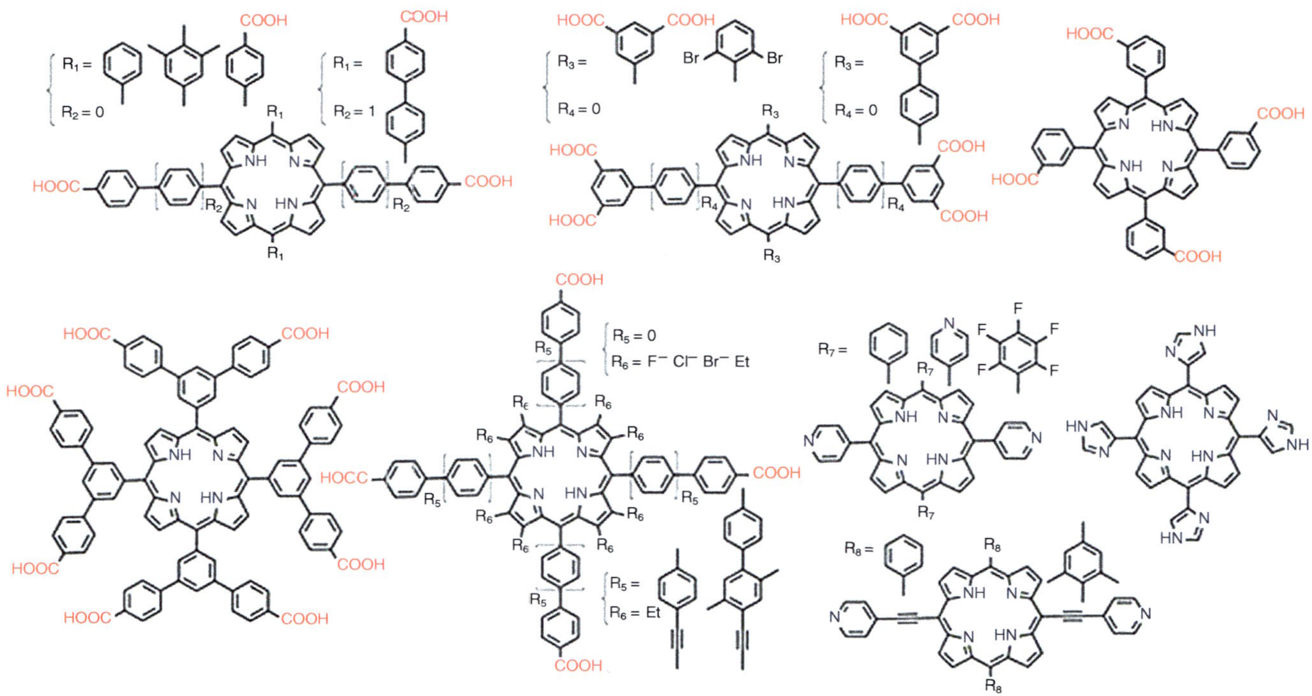

Figure 1.31 Summary of all metalloligands normally used in catalysis. Source: Reproduced with permission from Wei et al. [122]/American Chemical Society.

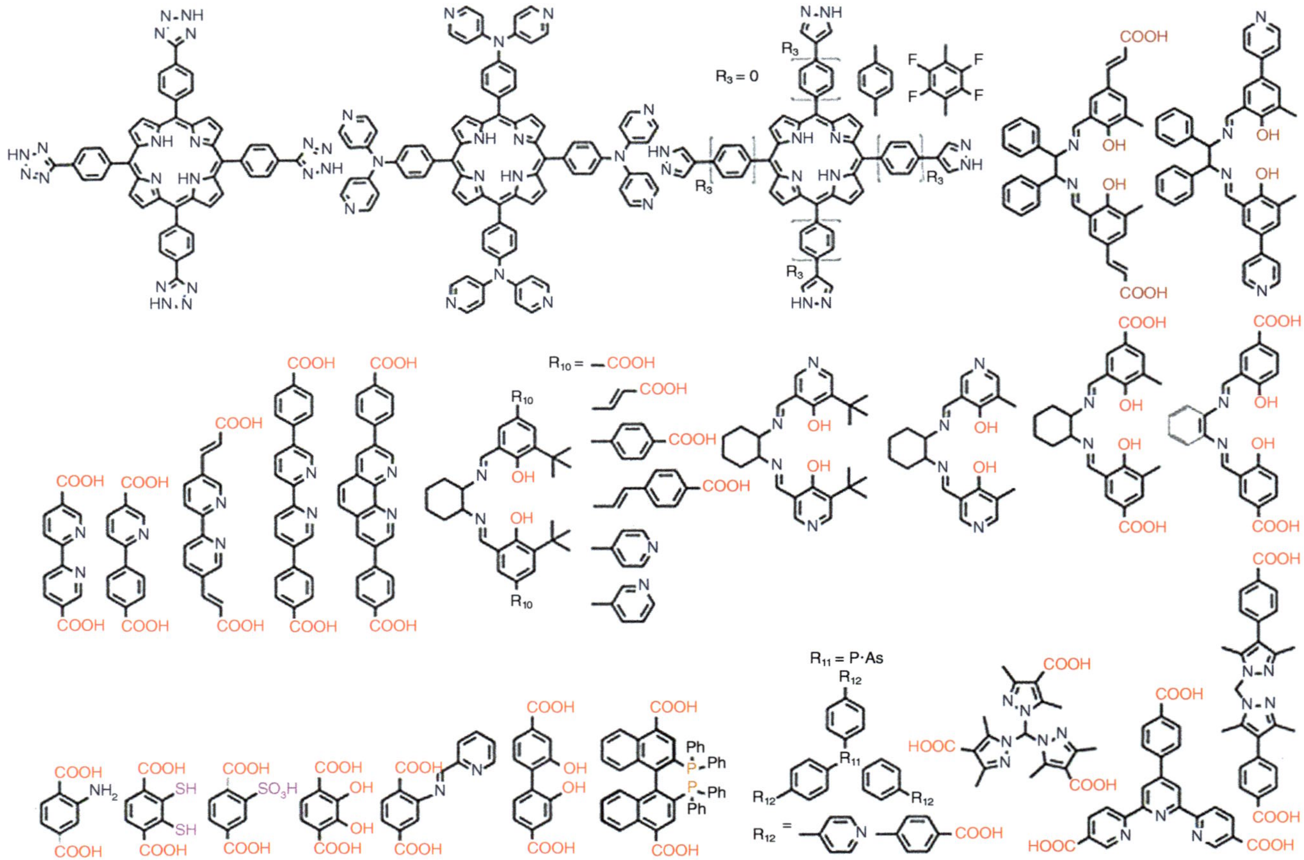

Figure 1.31 (*Continued*)

Entry	Olefin	Ligand, % ee,[b] confgn of diol[c] sign of $[\alpha]_D^{25}$	Time (h)
1	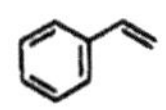	**1**, 62, *R*-(–) **1**, 60, *R*-(–) **2**, 53.6, *S*-(+)	3[d] 7[e] 7[e]
2	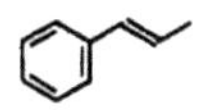	**1**, 65, *R*,*R*-(–) **2**, 55.4, *S*,*S*-(+)	5[d] 12[e]
3	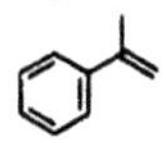	**1**, 33, *R*-(–)	1.5[d]
4	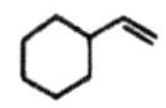	**1**, 46, *R*-(+)	1[d]
5	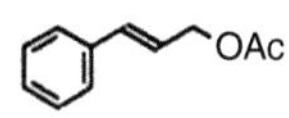	**1**, 76, *R*,*R*-(+)	7[d]
6	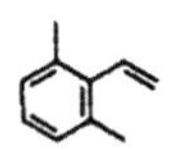	**1**, 65, (–)	3[d]
7	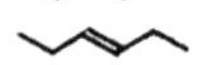	**1**, 20, *R*,*R*-(+)	17[d]
8	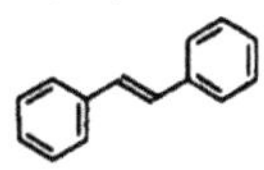	**1**, 88, *R*,*R*-(+) **1**, 85, *R*,*R*-(+) **2**, 78.5, *S*,*S*-(–)	7[d] 15[e] 17[e]

Figure 1.32 Some examples of ligands used in LAC. [a]All reactions were performed essentially as described for the molar scale process with stilbene. Specific notes and exceptions: (1) 1–5 mmol of olefin; (2) in small (ca. 7 ml) screw-cap vials (avoid rubber septa); (3) a temperature of 0 °C was maintained by storing the vials in an ice-bath for the duration of the reaction; (4) either 1 or 2 M in olefin as indicated (i.e. *d* or *e*). In all cases the isolated yield of the diol was 80–95%. [b]The enantiomeric excesses were determined by HPLC separation of the mono MTPA ester (entry 3), bis MTPA esters (entries 1, 2, 4–7), or the bis acetates (entry 8) in all cases by using a chiral Pirkle column (type 1A, preparative version) and eluting with *i*-PrOH/hexane. [c]The absolute configurations of the diols were established. For case 6, the correlation is not yet accomplished. Rotations were measured in EtOH except entry 5 which was taken in $CHCl_3$. [d][olefin] = 1 M and [Os] = 4×10^{-3} M. [e][olefin] = 2 M and [Os] = 4×10^{-3} M. Source: Reproduced with permission from Jacobsen et al. [130]/American Chemical Society.

1.2.2 Introduction of Metals by Direct Synthesis

The second part of this chapter is about the most common techniques used for the introduction of the metals into de organic ligands of the framework. *In situ* metalation, premetalation, and post-synthetic metalations are commonly used strategies to introduce metal active sites in the organic parts of MOFs.

This part of the chapter in particular will focus on the incorporation of the metal in the ligand by direct synthesis or by premetalation of the ligand and its subsequent incorporation into the framework that sometimes works better than the metalation by PSM, as it is more respectful with the conditions of the formation of the MOF.

The spatial arrangement of linkers can be elucidated by several techniques, but the choice of them will depend on the properties of the material. In these two last sections, we will also introduce some characterization techniques depending on the needs of each type of metalated ligand. It must be reminded that sometimes to obtain quantitative information about the ligands in the structure it is necessary to add an internal standard.

1.2.2.1 In-situ Metalation

In-situ metalation of the ligands in MOFs is a less used practice since it is not possible to know in advanced where the metal (that is desired to be in the ligand) will coordinate. There are a few reported works showing how to control the exact amount of metal added to the reaction to obtain a good incorporation into the ligand. Porphyrinic MOFs are the most used materials for this practice as they can have different coordination sites for the introduction of metals.

Chemical Reactivity. The in-situ metalation is not the most common technique used for catalytic processes but there are some examples where it is utilized as, for example, Zhang et al. reported a controllable synthesis of UNLPF-10, a porphyrinic MOF functionalized via *in-situ* metalation. They modified the metal/ligand ratio in order to control the extent of metalation within the MOF by varying this variable. The MOF obtained, exhibited channel-like pores along the crystallographic axes, as well as an enhanced performed in the photocatalytic experiments of oxygenation of sulfide. The results showed an increase in the rate of photo-oxygenation as the metalation ratio increased [131].

Characterization Techniques. In order to characterize the ligands that have been metalated *in situ*, in which you do not know what structure can be formed due to the addition of various metals at a time, Powder X-Ray Diffraction (PXRD) and Scanning Electron Microscopy, Transmission Electron Microscopy (SEM/TEM) are some of the characterization techniques than can help us solve this problem. This can be seen in the work previously mentioned by Zhang et al. where they use these techniques to characterize their material. With PXRD, they can determine the crystalline structure of the frameworks. This technique is also useful to visualize bond lengths, angles, and the coordination space at an atomic level. Furthermore, Single-Crystal X-Ray Diffraction (SCXRD) can be used to observe singles metal sites with host frameworks at different states such as defects [132]. Moreover, SEM/TEM and EDX are optical techniques that provide information about the morphology, chemical composition, and electrical behavior of a material. The chemical composition and the elemental distribution can be studied by EDX and mapping. Therefore, these techniques allow the extraction of the relative ratios for the linker and its distribution along the crystal [133, 134].

1.2.2.2 Premetalated Linker

On the other hand, premetalating the ligand and then introducing it in the MOF by direct synthesis is a more common practice. It involves a two-step reaction: first, a reaction to incorporate the metal into the ligand followed by another reaction to incorporate this into the framework. It is a longer work since it requires more steps but, most of the times, it is a more respectful procedure with the structure of the framework and the conditions for the formation of the same. There is a large number of reports that use this technique to introduce metals into the organic parts of the frameworks.

Chemical Reactivity. The premetalation of the ligands is a very used technique in the field of catalysis, as it allows to incorporate metals or metal complexes as metal sites for catalysis in a fast and safe way since it does not damage the original structure of the MOF. There are several examples of this kind of introduction of active metal sites like, for example, the work by Wu et al. who reported a Sn(IV) porphyrin MOF where they first metalled the porphyrin with Sn. They used this MOF for the photocatalytic reaction of oxygenation of phenols and sulfides, and they obtained quantitative yields for both reactions after four runs [135].

Another example of the same group was the one where they used the premetalated tetrakis(4-carboxyphenyl)porphyrin (TCPP) ligands with Mn and Ni and they combined these ligands with Mn(II) and Cd(II) ions to obtain the MOFs ZJU-18, ZJU-19 and ZJU-20. They used the MOFs for the oxidation of alkylbenzene and achieved the best results with ZJU-18 [136].

Characterization Techniques. Some of the characterization techniques that can help to know the nature of the ligand that has just been created by premetalation are the NMR and Mass spectroscopy. These are some techniques used in the work of Barron et al. where they use these to corroborate the ligand state [137]. In their work, they use in-situ NMR to monitor the composition of the liquid phase during the modification of the ligand of a Zr-MOF UiO-66.

Nuclear magnetic resonance techniques are divided into two blocks, those related to the liquid phase and those related to the solid phase (SS-NMR). The liquid phase is one of the most widely used technique to quantify the relative ratios in the structure of a complex, as well as to quantify the extent of the linker functionalization with another molecule [138]. On the other hand, SS-NMR is an effective tool to characterize the average distances and the distribution of linker. In this technique, it can be determined the chemical environment of a nuclei with a quantic number $1 \neq 0$ such as ^{1}H, ^{13}C, ^{15}N, ^{19}F, ^{31}P. It can provide a qualitative and quantitative analysis of the incorporated linker molecules and its functionalizations, and also, the linker distribution in the material.

Mass spectroscopy (MS) technique is used to obtain the mass-to-charge ratio of atoms or molecular fragments in the gas state. It is used to know the exact mass of any organic compound or complex mixtures by different methods such as ICP-MS, matrix-assisted laser desorption/ionization source with a time-of-flight mass analyzer (MALDI-TOF), accelerator mass spectrometry (AMS), thermal ionization-mass spectrometry (TIMS), or spark source mass spectrometry (SSMS). This technique is also commonly used in tandem with chromatography and other separation techniques to enhance the mass resolving/capabilities of the mass spectrometry. The only problem is that this is a destructive technique, so it is impossible to recover the samples [139].

1.2.2.3 Postgrafting Metal Complexes

In addition to the other two methods for the incorporation of metal complexes as active metal sites, another methodology to incorporate metal moieties is to synthesize a premetalated metal complex and then graft it into the organic ligands of the MOFs by covalent or coordination bonds [140, 141]. This approach has been used more recently than the others, but it holds great potential because it opens the door to the possibility to integrate well-defined metal complexes directly into the structure of the framework without any additional post-synthetical steps and helps to keep the structural integrity of the framework, avoiding degradation, and deactivation of the catalytic center as well [142].

Chemical Reactivity. The addition of metal complexes into the structure of a MOF by postgrafting is a common technique used to improve the catalytic behavior of the frameworks. There are several works published about this like the one by Ott et al. where they incorporate [Ru(bda)$(L)_2$] into a pyridine-decorated MIL-101(Cr) MOFs by the postgrafting method later exhibiting a tenfold increase in the turnover frequencies for water oxidation [143].

Characterization Techniques. For the characterization of ligands that have been modified by postgrafting metal complexes, one useful technique is X-Ray Photoelectron Spectroscopy (XPS). This is used to know if the metallic cluster created is introduced into the ligand retaining its properties, or if, on the contrary, it has been oxidized or reduced upon incorporated into the ligand. As reported by Shi et al., they use this technique to characterize the modification in the linker of the MIL-101-SO_3H-NHBOC and MIL-101-SO_3H-NH_2 and they saw their signals at a binding energy of 168.8eV which is consistent with the S^{6+} of the sulfonic acid group.

The fundamentals of XPS involve the detection of photoelectrons emitted from the sample as a consequence of the irradiation of the sample with X-ray photons. It can determine all elements through the detection of the binding energies of photoelectrons, with the exception of hydrogen and helium. As we have seen previously, this technique can give information about the chemical states of the species by small variations in the binding energies of the photoelectron lines as well as Auger lines satellite peaks. This way, XPS is a useful technique to obtain quantitative analysis of surface composition and to confirm the PSM of the linkers [144].

1.2.3 Introduction of Metals by Post-synthetic Modifications

Apart from the introduction of the metal by direct synthesis of the MOF or the premetalation of the ligand, there are other methods to introduce metals in ligands more commonly used by researchers, which are the PSM or post-synthetic metalations. In the last part of this section, we will detail the two post-synthetic methodologies most used for the introduction of metal active sites in the organic parts of the MOFs and the techniques of characterization usually used.

1.2.3.1 Post-synthetic Exchange or Solvent-Assisted Linker Exchange (SALE)

This efficient methodology is a ligand exchange strategy consisting of the replacement of certain ligands of the MOF by other metalloligands previously modified, which are in similar length, for the preparation of improved frameworks [145, 146]. These exchanges are normally carried out under milder conditions than direct synthesis. It is a proper route to control the spatial distribution of the building blocks, as during the process, the location of the linkers is determined by the synthesis conditions, so uniform and core–shell microstructures can be obtained, and the different spatial distributions of the building blocks lead to properties that cannot be achieved with the precursors of the pure MOF and can improve the application of these materials [147]. This methodology has been proven as a powerful tool to fabricate single-metal sites in MOFs for catalysis, due to the limitation of obtaining these kinds of materials by direct synthesis [121].

Chemical Reactivity. The SALE technique is a very used method to replace some ligands of a framework to other with different properties to enhance the catalytic properties of the MOFs, there are cases in which you can replace only some ligands of the structure and obtain a mixed-ligand MOF which combines the properties of both linkers, or you can replace all the ligands present in the structure. Some works as the one of Lillerud et al. explain how they do a linker exchange in an UiO-66 framework by changing benzoates by formate ligands, and it afforded a much more reactive material, which allows for the incorporation of different moieties important for the catalytic process [148] (Figure 1.33).

Characterization Techniques. Some useful characterization techniques to check if the ligand that has been replaced is in the right place and if it has not been damaged during the process are the Infrared (IR) and de UV–Vis. For example, Verpoort et al. [121] in their review show the use of IR and UV–Vis technique to characterize a MIL-101-Cr that has undergone a modification in its ligand. This modification involved a substitution of chlorine with imidazole,

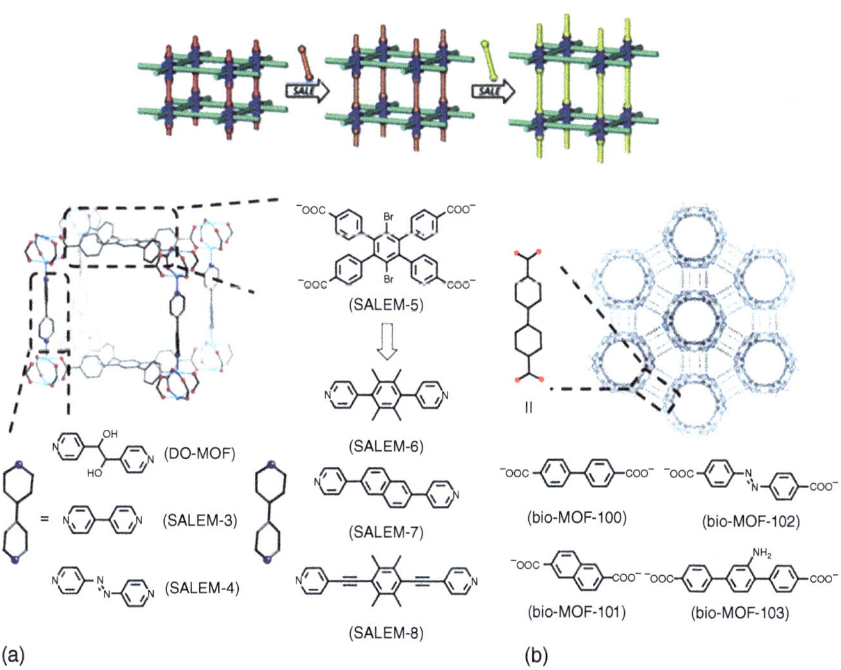

Figure 1.33 Top: Schematic representation of the incorporation of longer linkers into a MOF through SALE. Bottom: SALE in DO-MOF and SALEM-5 (a), and bio-MOF-101 (b). Source: Reproduced with permission from Karagiaridi et al. [149]/John Wiley & Sons.

used as a support for the immobilization of a catalysis moiety, a Mn(tcpp). They confirmed the presence of the complex by IR and UV–Vis. The resulting material, Mn(tpp)Cl@Im-MIL-101(Cr) MOF, is highly active, stable, and reusable for four consecutive runs without loss of catalytic activity.

Infrared spectroscopy is used to detect the adsorption bands of some molecules such as H_2, CO, NO, and CO_2. These molecules can be adsorbed on metal sites and present different vibrational frequencies and intensities when interacting with them, this is the way it can be demonstrated the presence of a single metal site in crystalline and amorphous materials [150]. UV–Vis technique is an absorption or reflectance spectroscopy that measures, as its name says, in the ultraviolet and the visible regions of the electromagnetic spectrum. It can be used to detect compounds that absorb in the range of the UV–Vis spectra, as for example chromophore ligands such as porphyrins. This technique is complementary to the fluorescence spectroscopy [151].

1.2.3.2 Post-synthetic Metalation

The precise incorporation of catalytic metal sites into MOFs by direct synthesis is a challenging thing, due to the difficulties in the synthesis of the MOFs, or the possible deactivation of the metal site during the direct synthesis, as mentioned in the introduction. That is why most people prefer to incorporate catalytic metal sites by PSM [152, 153]. This strategy normally involves mild conditions and has attracted people's attention lately [154]. This approach, consists of a two-step procedure, in which first a single metal bearing MOF assembly is prepared with the corresponding bifunctional ligand and then the introduction of a second metal ion occurs through a post-synthetic metalation resulting in a multi -metal MOF. This approach is usually chosen when the preformed complex would not handle the conditions required for the formation of the final MOF [155]. Usually, one of the organic ligands requirements to use the post-synthetic metalation is to have free chelating sites available at the functional groups, this is an important thing since these chelating sites would coordinate the different metal ions to form the active metal complexes for the catalysis. These single-metal sites can act as efficient heterogeneous catalysts in diverse organic reactions [156]. The post-synthetic metalation method has been broadly used in MOFs composed of different type of ligands such as pyridine, porphyrin, amine, thiocatechol, allyl, and phosphine groups as well as their derivative units and other organometallic species. An important thing to keep in mind is that monodentate functional groups are less stable for the chelation of metals than multidentate groups, the nature of metal binding in the first case is found to be difficult to elucidate and normally suffer from leaching of the metal species, but for the second case, these multidentate groups offer more stable chelating sites of the metal species. To date, a large number of metal precursors have been incorporated into the ligands of different MOFs using the post-synthetic metalation, to create diverse catalytic active metal sites [157, 158].

Chemical Reactivity. The post-synthetic modification is the most used method to incorporate metal moieties or other functionalization. There are a large number of works published using this methodology to add metal moieties for enhancing the catalytic properties of the MOF. One of them is, for example, the work of Cohen et al. where they use the PSM to introduce chelating sites in UMCM-1-NH_2 and then to metalate these sites with divalent (Cu^{2+}) and trivalent metals (Fe^{3+}), and the resultant MOFs UMCM-1-AMCupz and UMCM-1-AMFesal were used as robust catalysts for carbon–carbon bond forming reactions [159].

Characterization Techniques. To corroborate that the linker that has been modified post-synthetically has not suffered any damage during the transformation, there are some characterization techniques that can be helpful in seeing the composition and the chemical environment of the same, and they are the following: confocal microscopy coupled to Raman spectroscopy technique. This technique is used to determine the chemical composition along the crystal by measuring at different points and providing not only the chemical information, but also the microstructure. This way, the distribution of the linker along the crystal can be studied [160, 161]. Fluorescence lifetime imaging microscope (FLIM) technique can be complementary to the previous mentioned technique and supplies qualitative information about the spatial arrangement of the linker in the framework [162, 163]. Photo thermal induced resonance (PTIR) technique is the combination of two other techniques, IR-spectroscopy with lateral resolution of atomic force microscopy (AFM). This is another technique, complementary to the above mentioned, that can also provide information about the spatial ligand distribution. The technique provides this information by illuminating the sample at a fixed wavelength while the AFM tip scans the sample [164]. In an example of this, Kibria et al. published a work in which they use the Raman technique to corroborate the differences between two Zn-based MOF with different ligands, one with 1,2,4-triazole and other with 2-methylimidazole. In their work, they can see the bands of the in-plane bending modes and the bands of the C–H stretching of the azolate groups [160]. In another example published by Wuttke et al. they use the fluorescence imaging and lifetime analysis to determine the spatial arrangement of functionalities and the level of defects in a multivariable MOF UiO-67 [163]. Ramer et al. shows how the use of PTIR allows them to identify and obtain nanoscale images of the main components of each polymer they were working with, and also, their morphology when they have a mixture of them [164] (Figure 1.34).

Furthermore, there are some X-ray synchrotron radiation experiments that can shed some light in determining the coordination environment of the metal or metals

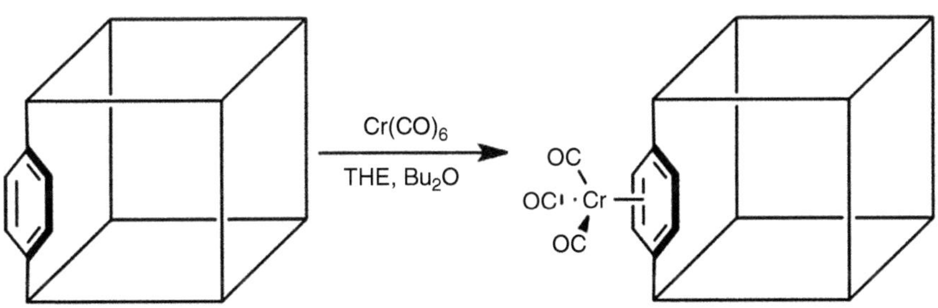

Figure 1.34 PSM addition by formation of organometallic chromium complex as demonstrated in MOF-5. Source: Reproduced with permission from Evans et al. [150]/Royal Society of Chemistry.

introduced in the framework and should be taken into account. These synchrotron techniques are useful for any type of metal introduced into the framework, either in the metal cluster (Section 1.1), in the organic ligand (Section 1.2), or in the cavities of the MOF itself (Section 1.3). The synchrotron technique referred to is the XAS, which can be further divided into EXAFS and XANES. On the one hand, Extended X-Ray Absorption Fine Structure (EXAFS) gives information about the distance and the number of the neighboring atoms as well as the chemical identity in the first coordination shell. This information can help in understanding the environment and chemical state of the metal in the organic ligand and material in general [165]. On the other hand, XANES provides information about the oxidation and the binding state of the metal, and also, the metal coordination geometry [166].

As it has been shown along the section, there are several ligands, metals, and techniques that can be used to introduce metal active sites in the organic ligands of the frameworks. As commented, there are a large number of characterization techniques to corroborate this. However, a comprehensive utilization of different techniques at a time for the structural understanding of the metal sites and their properties as catalysts is usually used. Moreover, more *in-situ* technologies to monitor the intermediates of the single-metal sites during the catalysis could be utilized, ultimately leading to a better understanding of the catalytic mechanism thus promoting the development of efficient single-site catalysts.

1.3 Metal-Based Guest Pore Engineering

According to the International Union of Pure and Applied Chemistry (IUPAC), porous materials can be classified as *microporous, mesoporous,* and *macroporous* depending on the pore size present within the materials. Microporous materials present a pore size smaller than 2 nm length, whereas mesoporous materials present a void width between 2 and 50 nm size [167–170]. Macroporous materials contain pore sizes bigger than 50 nm, so it is not relevant for the purpose of this chapter. MOFs usually present microporous or mesoporous porosity within their structure, which is important to bear in mind when encapsulating different molecules inside of them. The main challenge when immobilizing molecules in the pores is that some may not fit through; others may anchor themselves on the surface of the material or even aggregate before entering the cavities of the materials. For instance, just the porphyrin ring specie is estimated to be 0.84 nm in size not taking into account the delocalized electrons and the rigidity of the molecule [171], so the overall macromolecule would not fit through a micropore of the framework. In order to overcome these drawbacks, there have been many studies on the several approaches on how to immobilize different molecules within MOFs, not only after the material is already synthesized and the pore size is key, but also by growing the MOF around a guest that would not normally fit through the diffusion of the pores.

In this last part of the chapter, we will focus on the catalysis based on the different guests that can be encapsulated inside the MOF pores. As it has been previously mentioned, MOFs are organized crystalline structures with empty tuneable voids inside them that are able to immobilize and protect different catalytic species inside.

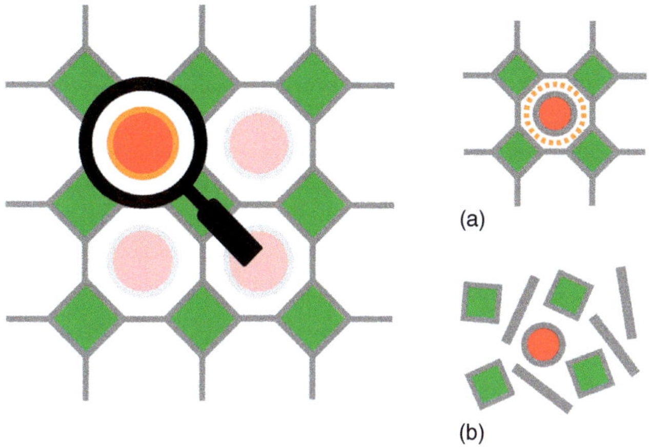

Figure 1.35 Schematic view of the two main general approaches to encapsulate molecular guests in MOF pores. (a) *MOFs as made*, by this approach we firstly synthesize the MOF and then we immobilize the guest; while (b) *In-situ encapsulation*, MOFs are grown over the guests by different strategies. Source: Figure produced by the authors of the chapter.

In this section, we will point out the distinct methodologies used either by *in-situ* encapsulation or by tailoring the pore size to store larger catalytic macromolecules. Furthermore, the various immobilization strategies of species, like metal NPs [172], polyoxometalates (POMS) [173], or large macromolecules (e.g. porphyrins) [171] will play a role in areas such as heterogeneous catalysis, photocatalysis, or electrocatalysis, among others. Overall, the summary of this chapter is represented in Figure 1.35.

1.3.1 Encapsulation Methodologies in As-Made Metal–Organic Frameworks

The large pore size distribution within MOFs makes them promising materials to capture guests with variable dimensions inside of them. In this approach, MOFs are made before the encapsulation takes place by conventional methodologies such as solvothermal synthesis where organic ligands and inorganic metal centers are mixed together in solution at room temperature or at high temperatures, depending on the desired MOF. Once we have obtained and washed the pristine materials, different methods are used to diffuse large molecules through the pores and immobilize them there. These methodologies could be considered as post-synthetical modifications, as the MOF is already formed when functionalized.

1.3.1.1 Incipient Wetness Impregnation

In this strategy, the already synthesized and activated MOF is submerged in a solution or gas flown containing an excess of the desired specie to be trapped by diffusion through the pores. The guest or precursor size is key as it has to fit across the cavities of the materials and stay there immobilized, thus it should not be too small or too big. One of the most common practical examples of this method is the encapsulation of large molecules such as POMS by the so-called incipient wetness impregnation. In this approach, the desired activated MOF with empty voids is submerged in an excess aqueous solution containing the POM specie [174]. Polyoxometalates are very interesting species due to their consideration as tuneable metal clusters which inside MOFs pores can be easily tailored by size, charge, or metal centers that will tune their properties making them more selective catalysts [175].

For instance, one of the first works of encapsulation of large molecules in porous materials is from Férey et al. in 2005 where they hydrothermally synthesized a Cr-based MIL-101 material with large pores sizes. After washing, activation, and characterization of the material, Keggin polyanions could fit through the voids and were incorporated by diffusion of the ion salts through the cavities [176]. The dried MOF was submerged in an aqueous solution of the Keggin salt for two hours. Moreover, the new composite material was successfully characterized to confirm the presence of the Keggin POM inside the MIL-101 (Cr). This study led to other impregnation of POMs reports for oxidative catalytic purposes such as the work from Kholdeeva et al. where they added an excess of M-POM (M=Co, Ti) to a Cr-based MIL-101 and successfully anchored the M-POM inside of the MIL-101 material [177].

On the other hand, not only POMs can be encapsulated within MOFs by incipient wetness impregnation. In a paper by Kaskel et al. in 2007, they described the immobilization of a palladium precursor inside the previously synthesized MOF-5 by this methodology (see Figure 1.37) [178]. The main difference among previous works is that in this case the guest precursor is subjected to the incipient wetness impregnation. First, they synthesized the material by solvothermal synthesis in autoclave at high temperatures, and then, once they had the MOF washed and activated, they slowly added the acetylacetonate palladium precursor dropwise with continuous stirring forming an orange paste. Subsequently, an evaporated solvent current was flown and evaporated for two hours in argon atmosphere, finally, letting the predried sample evaporate overnight to obtain the precursor MOF-5 material. After impregnation, the material was characterized to make sure

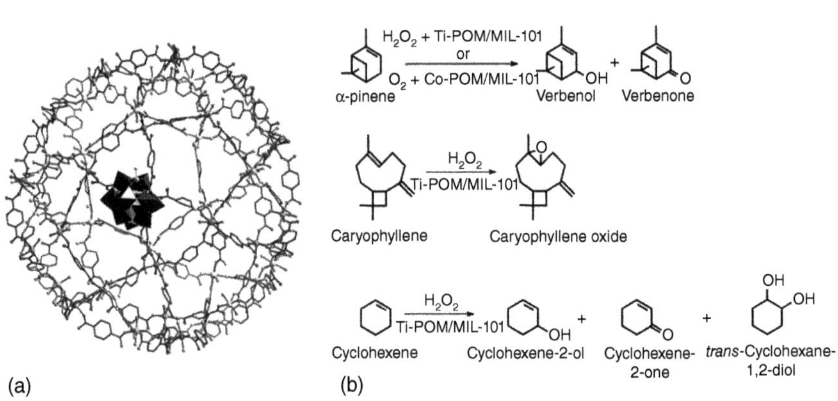

Figure 1.36 Schematic view of the M-POM@MIL-101 composite (a); and the cope of the oxidative catalytic reactions carried out in this study (b). Source: Reproduced with permission from Maksimchuk et al. [177]/Elsevier.

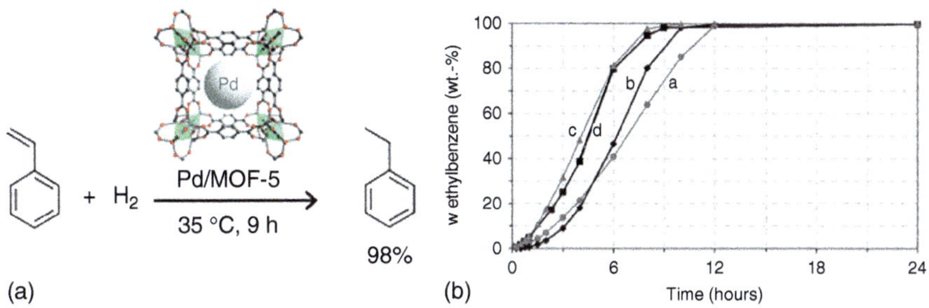

Figure 1.37 Hydrogenation reaction catalyzed by Pd@MOF-5 (a); and ethylene formation with time in the different Pd/MOF-5 studied in Kaskel's paper (b). The difference among the samples (a–d) is the drying method used for its obtaining. Source: Reproduced with permission from Sabo et al. [178]/Royal Society of Chemistry.

the precursor was diffused innards the micropores in order to thermally reduce the species precursor into Pd nanoparticles that fitted the cavities size.

A more complete study that followed Kaskel's work is a paper from Fischer et al. in 2009, where, by wetness impregnation and gas-phase infiltration, they diffused palladium precursors into Zinc-based MOF-5 followed by the thermally and photo-catalytically reduction of the species into Pd nanoparticles, respectively.

Characterization Techniques. In all the examples, Cr-based MIL-101 and MOF-5 were characterized by common techniques for MOFs such as PXRD, TGA, and N_2 isotherms. On MIL-101-Cr, large pore sizes of 29 to 34 Å and Brunauer Emmer Teller (BET) surface areas up to 5900 $m^2 g^{-1}$ were reported through the N_2 isotherms. The presence of the Keggin salt inside the MIL-101-Cr was determined by observing the differences between the initial and the final TGA weight loss, PXRD peak intensities, ^{31}P solid-state NMR, and N_2 sorption measurements. Herein, the decrease in BET surface area suggests a successful immobilization of the guests within the pores. IR was also used to confirm the integrity of the POM inside the MOF in both examples [176, 177]. By all these techniques, they could calculate the approximated number of 0.05 Keggin anions per chromium in the cluster in Férey's work [176].

On the nanoparticle examples, the Pd NPs inside MOF-5 were defined by the gradual decrease in the N_2 isotherms, the BET surface area, and the decrease in peak intensity in the PXRD patterns. Moreover, on the last work from Fischer, they do not only use the changes on the BET and PXRD to know if the nanoparticles are inside, but they also used Transmission Electron Microscopy (TEM) images to check the size and homogeneous distribution of the 2.4 nm Pd NPs within the MOF [179].

Chemical Reactivity. The composite formed by MIL-101 (Cr) and POMs was studied for selective alkene catalytic oxidations as can be seen in Figure 1.36b [177]. The stable material carried out the catalytic reaction successfully with high yields and good recyclability. In the other case mentioned, the Pd@MOF-5 was first utilized for hydrogen storage and then their catalytical properties were studied for the selective β-hydrogenation reaction of aromatic compounds as depicted in Figure 1.37. The reagents used were styrene, 1-octene, and *cys*-cyclooctene, but the optimization of the catalysis took place only with styrene. In Figure 1.37b, the four different composites formed by de Pd/MOF-5 are represented, where samples c and d show faster kinetic curves. Recyclability studies were also carried out showing a great stability of the material [178]. Lastly, on the work presented by Fischer, they performed the hydrogenation of cyclooctene but they do not focus on the catalytic properties of the composite rather on the characterization and optimization of the wetness impregnation technique itself by varying the loadings of palladium precursors into the Zn-MOF-5 [179].

1.3.1.2 Ship-in-a-Bottle

A more well-known strategy but similar to the incipient wetness impregnation would be the so-called *ship-in-a-bottle*. Herein, instead of impregnating the active specie directly into the material they diffuse the co-catalysts precursors through the pores. Once in the cavities, they would be transformed to form the immobilized catalytic species *in situ* inside the pores of the framework [180]. The name of the strategy comes from the guest being built inside the already-formed matrix material. The main challenge of this method would be the dispersion of the precursors inside the pores and not those getting stuck on the surface of the material, as in previous examples.

One early example of this would be the distribution of metal salts inside the pores in order to form metal nanoparticles by reduction inside the material. For instance, in 2012, Xu et al. used a previously reported double-solvent methodology to diffuse platinum-based salts inside MIL-101 (Cr) [181]. In this work, the previously synthesized Cr-based activated MOF, already mentioned, was dissolved in a dry non-polar solvent. The platinum salt was dissolved in an aqueous polar solvent and added dropwise into the stirring solution containing the MOF thus successfully achieving the inclusion of the platinum precursors in the voids. Afterward, the solution was left to reduce by increasing the temperature to finally obtain the Pt@MIL-101 composite with 2 nm Pt NPs inside of them.

Characterization Techniques. The MIL-101 (Cr) was characterized by PXRD, SEM, and N_2 adsorption. The key characterization technique carried out to certify the homogeneous distribution of the Pt NPs inside the MOF is TEM. As shown in Figure 1.38a,b, the Pt NPs size was also determined by TEM and fitted to a Gaussian distribution.

Chemical Reactivity. In order to prove the catalytic properties of this material, the catalytic performance was carried out for hydrogen generation reactions from the hydrolysis and later pyrolysis of ammonia borane. The studies show a high recyclability and efficiency not only in hydrogen generation reactions but also in gas-phase CO oxidation reactions. The main results from this work are summarized in Figure 1.38.

1.3.1.3 Metal–Organic Chemical Vapor Deposition (MOCVD)

Vapor deposition is a very well-studied synthetic methodology. In particular, metal–organic chemical vapor deposition (MOCVD), as its name recalls, consists of exposing a substrate to volatile precursors, being usually one of them

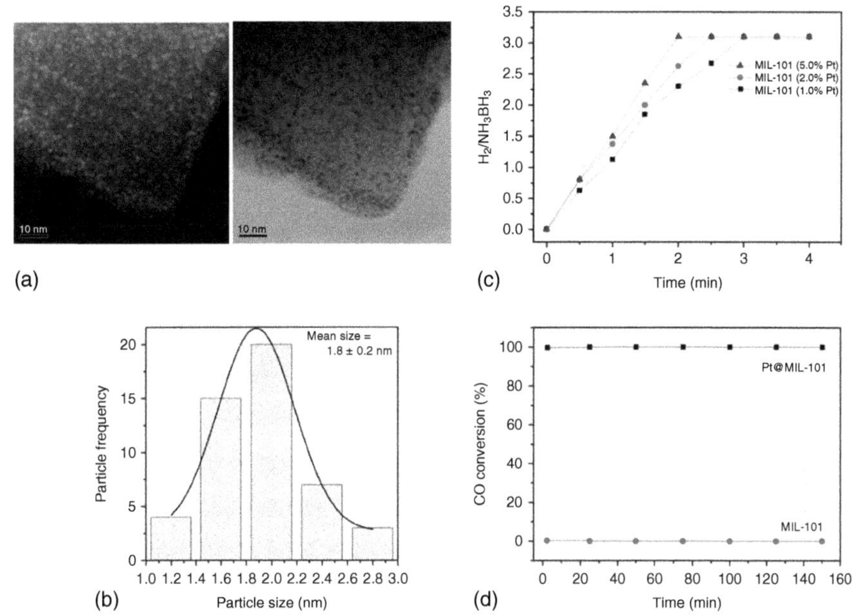

Figure 1.38 Summarize work of Xu et al. where (a) shows the TEM images of the Pt@MIL-101 composite, (b) histogram study on the Pt NP size distribution, and (c) and (d) are the catalytic studies on H_2 generation and CO oxidation, respectively.
Source: Reproduced with permission from Xu et al. [181], © 2012/American Chemical Society.

an organometallic compound, in order to be deposited onto the surface. These precursors usually react or decompose thus placed on the support. It can also be called MOCVD [182]. For the purpose of this chapter, our common substrate would be the porous frameworks while the organometallic volatile compounds are the desired guest to be embedded in the pores. Haruta et al. presented one of the earliest works on this methodology in 1997 for the preparation of gold catalysts anchored onto the surface of TiO_2 [183]. It was not until the following year where an example of CVD was reported for the deposition of gold nanoparticles in molecular sieves for catalytical applications [184].

However, examples of encapsulation of guest molecules inside MOFs by CVD came later in 2005 in a work by Fischer et al. where they propose to use MOF-5 as an absorption matrix suitable for the immobilization of metal–organic precursors by vapor deposition [185]. In this work they synthesized the previously mentioned along this chapter, Zn-based MOF-5 by solvothermal methodologies following the initially reported synthesis by Yaghi et al. Once the MOF-5 was dried and activated, it was exposed to the vapor of the organic palladium, copper, and gold precursors separately under vacuum and room temperature in a Schlenk tube. A color change was rapidly observed in the powder from colorless to dark red within five minutes in the case of Pd, and similar with the other two metal analogs. The metal–organic precursors are held inside the pores by weak interactions with the Zn cluster without

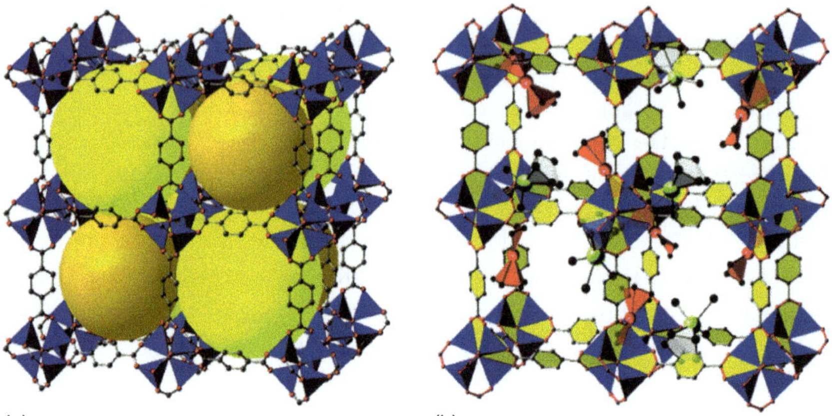

(a) (b)

Figure 1.39 MOF-5 unit cell (a); and $[L_nM]_a$@MOF-5 showing how the guest interacts inside the voids (b). Source: Reproduced with permission from Hermes et al. [182]/Royal Society of Chemistry.

altering the crystallinity or surface area of the whole material. Afterward, a hydrogen gas flow was introduced and therefore reducing the metal species to nanoparticles guests inside the cavities obtaining MNPs@MOF-5 materials (M=Pd, Cu, Au). A similar work was published in 2006 by the same authors but doing a greater scope of organic precursors in the CVD inclusion [182]. Fe, Pt, Pd, Au, Cu, Zn, and Sn precursors were utilized to acquire $[L_nM]_a$@MOF-5 compounds where they attributed their immobilization to weak Van der Waals forces between the Zn and them, seen schematically in Figure 1.39.

A more recent work by Kempe et al. uses the CVD strategy to embed Ni/Pd NPs inside MIL-101(Cr) [186]. Furthermore, they successfully achieved different loadings of Ni/Pd precursors inside the MOF cavities. Following a previously described CVD methodology, the MIL-101 powder was activated and a flowed of organic Ni/Pd organic precursors passed across in the gas phase under vacuum by simultaneous or successive loading of the precursors. Afterward, a hydrogen gas flow was introduced to obtain the metal NPs inside the voids. They observed that, depending on the loading approach, the metal contents and the NPs formation were different thus affecting the overall properties of the Pd/Ni@MIL-101 composite. Successive loading led to bigger NPs that were anchored on the surface rather than inside the material. On the other hand, simultaneous loading of the precursors guided the formation of Pd/Ni bimetallic NPs, which size was adapted to the MOF cavities. The materials obtained by this were Pd_4Ni_1@MIL-101 and Pd_3Ni_2@MIL-101.

Characterization Techniques. The M@MOF-5 composites formed in the work by Fischer et al. were characterized by conventional techniques such as PXRD, SEM, and their surface areas were calculated by Langmuir models in N_2 isotherms. The metal NPs size was determined by the width of the PXRDs where the main profile remains intact. TEM images estimated the Pd NPs size to be 1.4 nm, while the Cu ones were of approximately 3–4 nm. The Au NPs are the most polydisperse as their size goes from 5 to 20 nm. The calculation of guests per pore was made by elemental analysis measurements, which resulted to be between 2 and 4 species per cavity. The $[L_nM]_a$@MOF-5 from the next example from Fischer showed a more detailed characterization of the materials obtained using atomic absorption spectroscopy (AAS) to determine quantitatively the metal content inside the material. In addition, FT-IR, solid-state NMR spectroscopy, or PXRD were utilized as complementary techniques to prove the presence of the Van der Waals forces with the Zn centers (see Figure 1.40 below). In the last example, from Kempe et al., elemental analysis was carried out to get an estimated metal loading of approximately 20% in all the materials. Pd_4Ni_1@MIL-101 and Pd_3Ni_2@MIL-101 showed a decrease in the N_2 adsorption isotherms whereas the Pd@MIL-101 and Ni@MIL-101 remained similar in large surface areas.

Chemical Reactivity. The catalytic experiments shown in Fischer's work focused on methanol production from gas synthesis and hydrogenation of cyclooctene. Results exhibit a moderately active performance for Pd@MOF-5 and Cu@MOF-5. However, Au@MOF-5 did not display any catalytical activity due to the aggregation of the Au NPs inside the MOF. Following Fischer's work in 2008, another article was published where they immobilized Ruthenium NPs inside MOF-5 by CVD [187]. Although similar to previous publications, in this work a more in-depth study on the catalytical properties of the final materials was studied. They carried out the oxidation of benzyl alcohol following its evolution by Gas Adsorption/Mass Spectroscopy (GC/MS) and the hydrogenation of benzene using the Ru@MOF-5 material as catalyst. However, the sensitivity of MOF-5 to water gave them unsuccessful results for the oxidation of benzyl alcohol.

In the study of Pd/Ni@MIL-101, these materials demonstrated to have great catalytical applications. Hydrogenation reactions were carried out with phenol and cyclic ketones/dialkyl ketones. The composites obtained by 1:1 successive loading showed lower catalytic activity and conversion. Nevertheless, Pd_4Ni_1@MIL-101 and Pd_3Ni_2@MIL-101 exhibited a synergetic catalytic effect that gave them a significant conversion capacity, enhanced stability, and higher reusability under aggressive conditions compared with the Pd or Ni@MIL-101 materials. The hydrogenation of

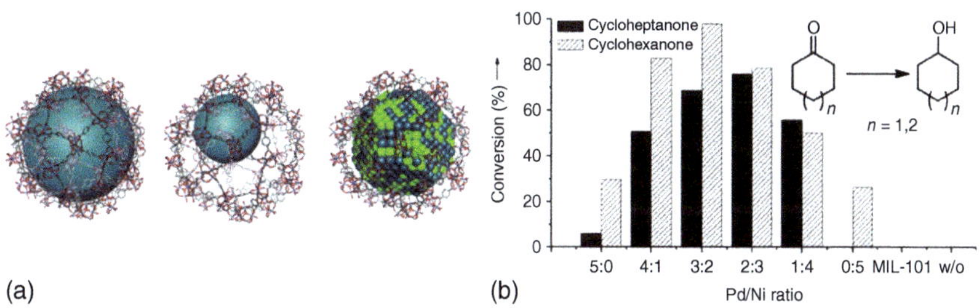

Figure 1.40 Simulation of the formation of the singles MNPs adapted to the MIL-101 cavities and the bimetallic NPS within the voids (a); and catalytic hydrogenation of phenol showing the % conversion with the different bimetallic NPs ratios (b). Source: Reproduced with permission from Hermannsdörfer et al. [186]/John Wiley & Sons.

dialkyl ketones was only feasible with the Pd_4Ni_1@MIL-101 and Pd_3Ni_2@MIL-101, which is attributed to their synergetic effect.

1.3.1.4 Metal-Ion Exchange

This encapsulation strategy is relatively new as the first example was published by Ma et al. in 2014 [188]. Their approach has similar characteristics that the one explained in Section 3.1.1: incipient wetness impregnation. However, the main variance is that the solvent from the impregnation is not evaporated. The previously synthesized MOF is submerged into the metal species they want to exchange without stirring and letting them decant at room temperature for some time. Afterward, small fragments of other components are anchored to the ionic centers.

In this specific example, Ma et al. synthesized bio-MOF-1 by reported methodologies consisting of mixing Zn salts with adenine and introducing an organic ligand such as biphenyldicarboxylic acid. The final material is a crystalline MOF where the SBUs are formed by zinc and adeninate species with organic ligands holding the cluster columns together [189]. This MOF-based material forms an anionic structure whose counterions are dimethylammonium cations originated from the decomposition of DMF and water molecules. Once bio-MOF-1 is acquired, the anionic material is submerged in a Co (II) acetate cation methanol solution for eight hours at room temperature. Therefore, a metal-ion exchange has taken place where the counterions have been replaced by positive charged metal centers such as Co, Ni, or Cu. The material is washed to avoid anchoring of metallic species onto the surface. Once the metallic centers are inside, they proceed to carry out a further functionalization by adding an organic linker in order to occupy the remaining place within the nanospace. The anchoring of the 1,2-dicyanobenzene (DCB), ligand is directed by the cationic metal center present. Moreover, they tried other encapsulation procedures to get the same materials, but they were unsuccessful (see Figure 1.41 below).

Characterization Techniques. In these works, conventional techniques are used to fully characterise the material such as PXRD, N_2 adsorption, XPS and MS. First, PXRD was used to check the integrity of the crystalline structure. ICP-MS, elemental analysis and MS confirmed the complete ion-metal exchange and the

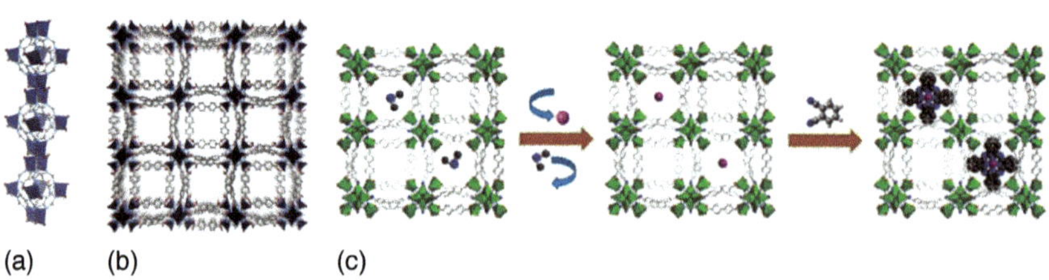

Figure 1.41 SBUs anion column formed by the Zn metal centers and the adeninates (a); crystalline structure of bio-MOF-1 (b), and schematic representation of the metal-ion exchange encapsulation process, with directional bonding of the organic linker at the end (c). Source: Adapted and reprinted with permission from An et al. [189]. Copyright 2009 American Chemical Society; Reproduced with permission from Li et al. [188]/American Chemical Society.

formation of the Co-Pc guest inside the nanospace. The BET surface area calculated through N_2 adsorption isotherms resulted in a decreased on the surface area due to the high occupancy of the pores. Complementary techniques such as XPS or UV–VIS were utilized to further verify the formation of the Co-Pc.

Chemical Reactivity. Catalytic styrene epoxidation was studied and the results were compared to ones obtained with the pristine material to see the enhancement of the process. Co-Pc@bio-MOF-1 material showed a 72% conversion and a 65% selectivity whereas bio-MOF-1 exhibited only an 8% conversion, under the same reaction conditions. No metallic leaching was observed in the brand-new composite material, which could be reused up to 3 times without decreasing its catalytic activity.

1.3.2 *In Situ* Guest Metal–Organic Framework Encapsulations

In this final section, called *in-situ* methodologies for encapsulation, the common characteristic would be that the precursors forming the MOFs and the chemical species desired to be immobilized are mixed and grown together at the same time to form the final composite *in situ*. In this approach, the guest size is not as significant as for the final material obtained. There have been different studies and classifications of the methodologies; however, we will summarize and give examples of the main approaches with the broad range of species used such as POMS, metal-NPs, enzymes, etc. [190, 191].

1.3.2.1 Solvothermal Encapsulation or One Pot

Amongst all the encapsulation techniques, solvothermal encapsulation or one-pot synthesis is the most commonly used out of them all. This strategy is very similar to the pristine synthesis of MOFs itself. The main characteristic is the mixing in solution of the metal precursors and the organic ligands of the desired MOF with the guest molecules, which are aimed to be embedded inside the pores.

As we have previously mentioned, porphyrins are large molecules, especially challenging to diffuse through the pores of already synthesized MOFs. Due to this, a lot of studies have been focused on these species. One early example of this inclusion method is a work by Vetromile et al. in 2011 where they successfully synthesized M-HKUST (M=Cu or Zn) with a metalloporphyrin embedded in its mesopores [192]. In this publication, they report the first metal–organic material enzyme (MOMzyme-1) by mixing the Hong Kong University of Science and Technology (HKUST) precursors with the porphyrin of choice and letting them react at 60 °C during seven hours. The porphyrins trapped inside the pores were functionalized in a second step with metal centers such as iron or manganese, mimicking the active site of some enzymes. Furthermore, these cationic centers are capable of exchanging themselves with other cationic metal centers which makes them able to tune their catalytical properties. The MOMzymne-1 still shows considerably enhanced porosity while protecting the porphyrin inside while small molecules can still pass through the small cages in HKUST in order to reach the porphyrin metallic center. Inspired by this methodology, a more recent work by Sun et al. shows the promising catalytic potential of these materials to convert CO_2 into more interesting products such as C_2H_4, which will be explained later in the chemical reactivity subsection below [193]. In this case, following the reported synthetic methodology, composites formed by M-TCPP porphyrins and Cu-MOFs were obtained by mixing BTC ligand with the metal-based porphyrin (Fe, Co, and Ni based) in an ethanol-DMF solution and added to the aqueous copper precursor solution. This mixture was left to react in an autoclave at 60 °C for 12 hours. Finally, five products were successfully acquired, one with the non-metalated porphyrin and the Cu-MOF and, the other four compounds with iron, cobalt, nickel, and copper porphyrin centers, respectively, inside the MOF cavities.

Another significant work to bear in mind in this approach is a publication from Hupp and coworkers where they selectively encapsulated different nanoparticles inside ZIF-8 by the one-pot methodology [194]. The main difference with previous examples is that in this work they do the encapsulation at room temperature. Herein, they achieve the synthesis of a big scope of nanoparticles with different sizes, shapes and compositions. Afterward, these nanoparticles were stabilized with organic components, such as polyvinylpirrolidone (PVP). Their goal was to use the stabilising organic compound to anchor the MOF precursors and start growing the MOFs over the metal-nanoparticles. This methodology could also be denominated build a "bottle-around-ship". Moreover, they successfully optimized the controlled encapsulation of NPs by adding different concentrations of them to the ZIF-8 precursors in methanol and letting them react at room temperature without stirring for 24 hours. In this work, they obtained Pt@ZIF8, Au@ZIF-8, Ag@ZIF-8, among other composites without altering the intrinsic porosity or crystallinity of the material, as you can see in Figure 1.43.

Characterization Techniques. As these are *in-situ* methodologies typical MOF characterisation techniques are utilised (PXRD, SEM, TEM, EDX, TGA, and N_2 isotherms). In contrast with previous approaches where the MOF can be characterized before forming the host-guest material, in this case the comparison is between the final functionalized material to the pristine MOF made in different synthetic batches. In the M-TCPP@Cu MOF work they use several techniques such as TGA to study the stability of the newly formed composite and, SEM and PXRD to check that the morphology and crystallinity of the initial porous material is maintained. But more importantly, to characterize the guest molecules, they use EDX-Mapping to check the porphyrin metal distribution within the material and, complementary, IR spectrum for the presence of characteristic porphyrin peaks such as C-H pyrrol bending at 1004 cm^{-1}. XPS studies were performed to check the chemical composition and their elemental state in the electrocatalysis. Results showed a reduction in the Cu centre from Cu^+ to Cu^0 which intervenes in the catalytic reaction, but the integrity of the material is maintained during the whole process. Furthermore, in the work from Sun et al. they mainly focused on the use of optical techniques such as HRTEM, SEM or EDX, to confirm: first, the formation of the nanoparticles previously synthesized and, consequentially, the homogeneous distribution of them inside ZIF-8, as you can see in Figure 1.43b. It is important to highlight that the presence of the different NPs does not alter the N_2 adsorption isotherms and neither the BET surface area obtained for the NP@ZIF-8, which are comparable to the pristine ZIF-8 itself.

Chemical Reactivity. In the work from Sun et al. once the materials were fully characterized, they were transferred to electrodes in order to carry out the electrochemical catalytic experiments. The proposed mechanisms for these consist of small molecules of carbon dioxide reaching the metal-porphyrin center through the pores where they were reduced to carbon monoxide in order to finally arrive at the copper MOF cluster to be transformed to ethane, as you can see in Figure 1.42. The enrichment of the Cu-MOF with copper metal centers in the porphyrins enhances the catalytic performance of the material compared to other previous composites obtained. Additionally, in Sun's paper, they studied the heterogeneous catalytic performance of the brand-new composites by the oxidation reaction of carbon monoxide and the hydrogenation reaction of alkenes such as *cis*-cyclooctene or *n*-hexene, see Figure 1.43.

1.3.2.2 Co-precipitation Methodologies

There are not many examples of this synthetic strategy as it is very similar to the one pot methodology which can difficult its differentiation. In previous methodologies, the main idea was to mix the precursors and the already synthesized guest together in order to obtain the final material. However, in a study by Kaskel et al. from 2007, they made this subtle difference [195]. Kaskel reported the preparation of palladium nanoparticles supported on MOF-5 composite, inspired by a previous work from Huang and coworkers consisting on the preparation of a Zn based metal-organic coordination polymer (MOCP) by what they called "direct mixing" strategy in a short period of time [196]. In order to do this, Kaskel and coworkers prepared a mixture of zinc nitrate hexahydrate salt, 1,4-benzenediccarboxylic acid and palladium nitrate trihydrate precursors by dissolving them in organic solvents at 60 °C and vigorous stirring. A precipitated was formed, dried and washed several times to obtain the pure Pd@MOF-5 material, without further modifications.

Characterization Techniques. The materials acquired were characterized by PXRD, nitrogen adsorption isotherms at 77K, and SEM. The highlighted technique from this work is the quantitative determination of the palladium content by AAS. Because of the fact that there were no significant differences in the isotherms, the AAS technique is the only probe that confirms the Pd supported in the MOF.

Chemical Reactivity. Catalytic hydrogenation reactions were performed in liquid–gas phase to see the applicability of this material. Hydrogen is in the gas phase while the reagents and the composite catalysts are in the liquid phase. Hydrogenation of ethyl cinnamate with the Pd@MOF-5 synthesized by this methodology showed an enhancement in activity compared to the Pd/C commercial catalyst with the same loading of Pd. However, it was observed after one cycle a complete loss in micropore volume without decreasing the catalytic activity.

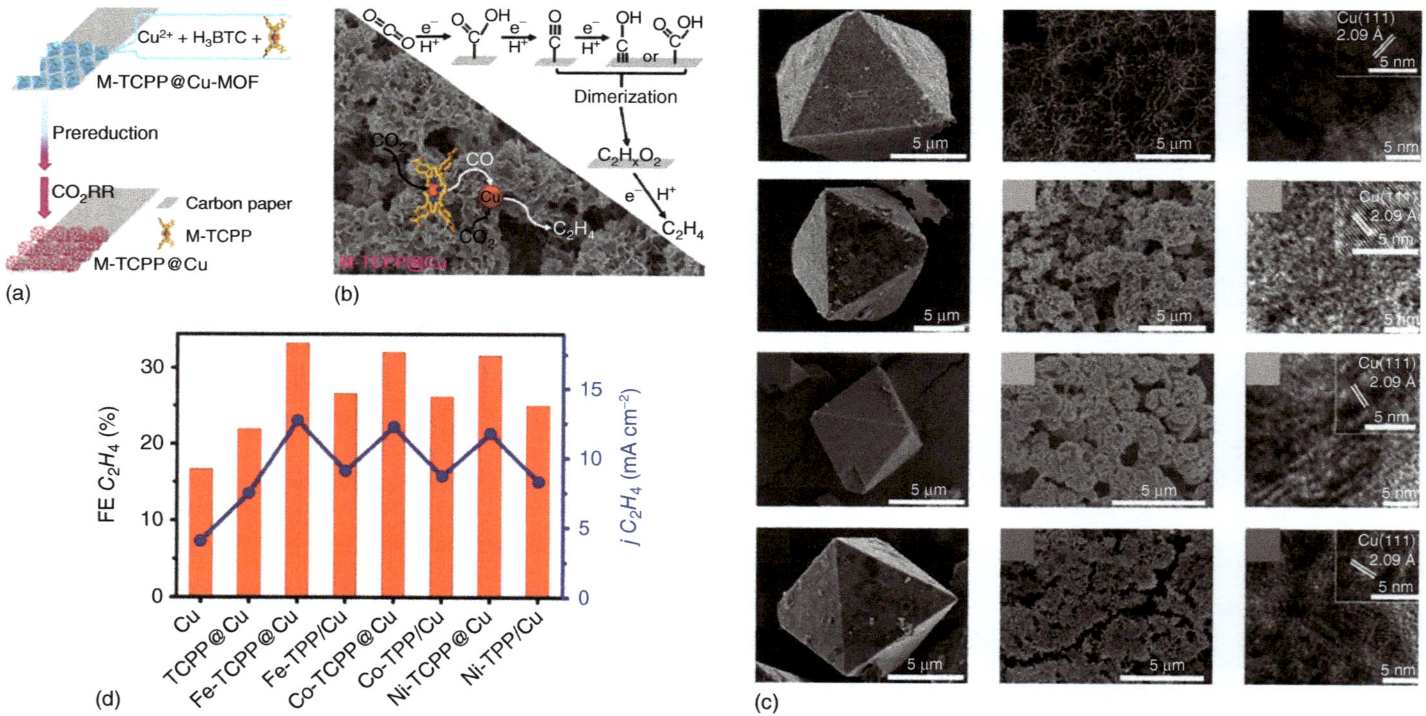

Figure 1.42 Schematic view of the synthetic mechanism proposed for the M-TCPP@Cu MOF (a); schematic view of the proposed catalytic mechanism with the enhancement provoked by the M-TCPP present (b); SEM and HRTEM images of the different composites obtained with the coral distribution inside the material (c); and C_2H_4 catalytic results from the different materials obtained (d). Source: Reproduced with permission from Sun et al. [193], © 2021/American Chemical Society.

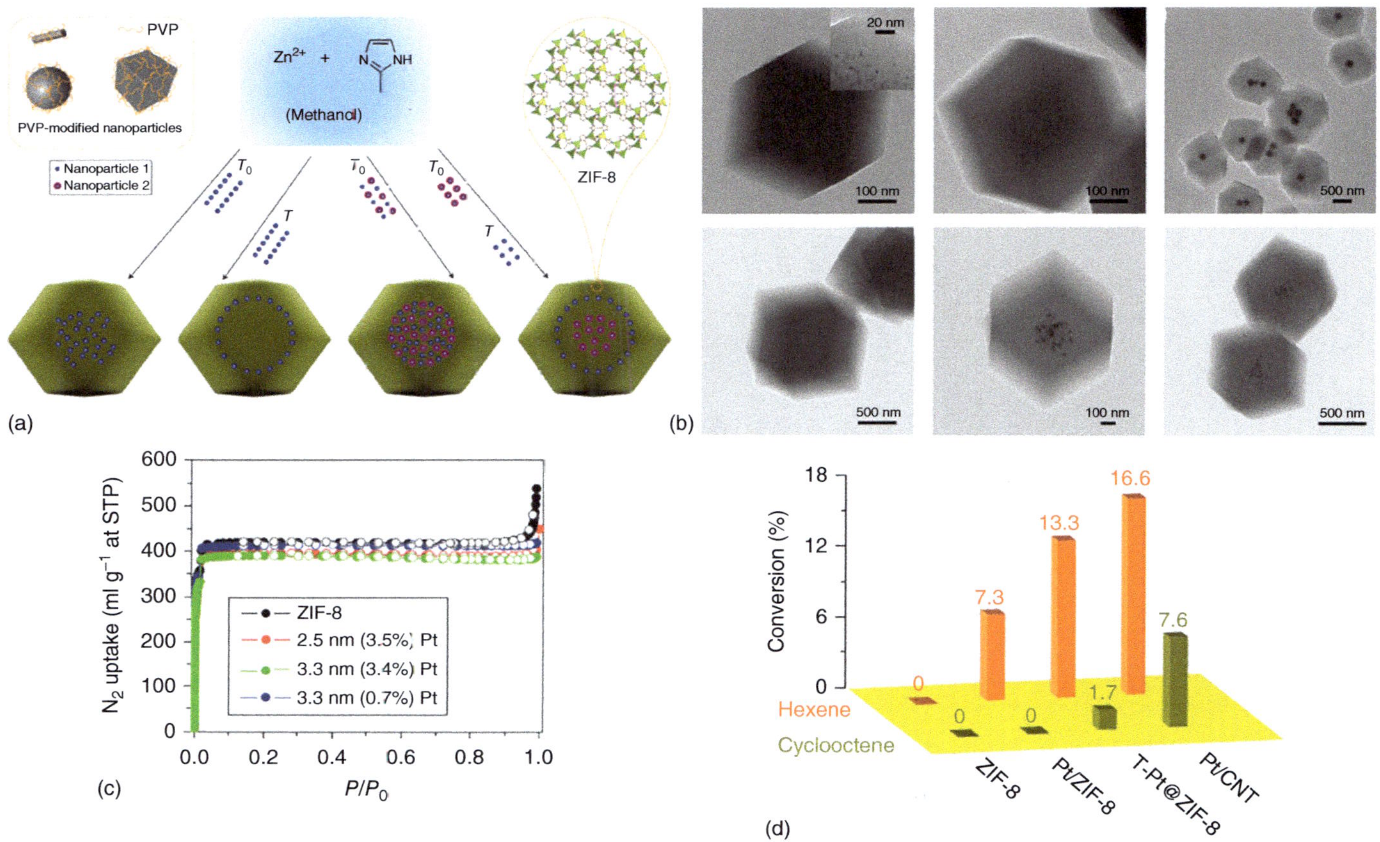

Figure 1.43 Schematic view of the controlled encapsulation of different nanoparticles in ZIF-8 by adding one type of NPs or two types of NPs (a); TEM images of the different NPs@ZIF-8 formed (b); N_2 isotherms of the Pt NPs@ZIF-8 composite where they observed no differences among them (c); catalytic performances of the Pt@ZIF-8 composites in the hydrogenation studies (d). Source: Reproduced with permission from Lu et al. [194], © 2012/Springer Nature.

List of Abbreviations

AAS	Atomic Absorption Spectroscopy
AFM	Atomic Force Microscopy
AIM	Atomic Layer Deposition in Metal–Organic Frameworks
ALD	Atomic Layer Deposition
AMS	Accelerator Mass Spectrometry
BET	Brunauer Emmer Teller
BTC	1,3,5-benzenetricarboxylic acid
DCB	1,2-dicyanobenzene
DFT	Density Functional Theory
EDX	Energy-Dispersive X-ray Spectroscopy
EXAFS	Extended X-Ray Absorption Fine Structure
FLIM	Fluorescence Lifetime Imaging Microscope
FTIR	Fourier-Transformed Infrared Spectroscopy
GC/MS	Gas Adsorption/ Mass Spectroscopy
HKUST	Hong Kong University of Science and Technology
HOMO	Highest Occupied Molecular Orbital
HRTEM	High Resolution Transmission Electron Microscopy
ICP-MS	Inductively Coupled Plasma-Mass Spectrometry
IR	Infrared Spectroscopy
LAC	Ligand Accelerated Catalysis
MALDI-TOF	Matrix-Assisted Laser Desorption/Ionization source with a Time-of-Flight
MCD	Missing Cluster Defects
MIL	Material of the Institute Lavoisier
MLD	Missing Linker Defects
MM-MOF	Mixed-Metal Metal–Organic Framework
MOCVD	Metal–Organic Chemical Vapor Deposition
MOCP	Metal–Organic Coordination Polymer
MOF	Metal–Organic Framework
MOM	Metal–Organic Material
NMR	Nuclear Magnetic Resonance
NPs	Nanoparticles
OMS	Open Metal Site
PDF	Pair Distribution Function
PIZA	Porphyrinic Illinois Zeolite Analogue
POMS	Polyoxometalates
PSM	Post-synthetic Modification
PTIR	Photo Thermal Induced Resonance
PVP	Polyvinylpirrolidone
PXRD	Powder X-ray Diffraction
SALE	Solvent Assisted Linker Exchange
SBU	Secondary Building Unit
SCXRD	Single-Crystal X-Ray Diffraction

SEM	Scanning Electron Microscopy
SIM	Solvothermal Deposition in Metal–Organic Frameworks
SOMC	Surface Organo-Metallic Chemistry
SSMS	Spark Source Mass Spectrometry
SS-NMR	Solid-State Nuclear Magnetic Resonance
TCPP	Tetrakis(4-carboxyphenyl) porphyrin
TEM	Transmission Electron Microscopy
TGA	Thermogravimetric Analysis
TIMS	Thermal Ionization-Mass Spectrometry
UNLPF	University of Nebraska-Lincoln Porous Framework
UV–Vis	Ultraviolet–Visible
XANES	X-Ray Absorption Near Edge Structure
XAS	X-Ray Absorption Spectroscopy
XPS	X-Ray Photoelectron Spectroscopy

References

1 Furukawa, H., Cordova, K.E., O'keeffe, M., and Yaghi, O.M. (2013). The chemistry and applications of metal-organic frameworks. *Science* 341: 1230444.

2 Zhou, S., Shekhah, O., Ramírez, A. et al. (2022). Asymmetric pore windows in MOF membranes for natural gas valorization. *Nature* 606: 706–712.

3 Liang, W., Xu, H., Carraro, F. et al. (2019). Enhanced activity of enzymes encapsulated in hydrophilic metal-organic frameworks. *J. Am. Chem. Soc.* 141 (6): 2348–2355.

4 Kim, E., Siegelman, R.L., Jiang, H.Z. et al. (2020). Cooperative carbon capture and steam regeneration with tetraamine-appended metal-organic frameworks. *Science* 369: 392–396.

5 Gascon, J., Corma, A., Kapteijn, F., and Llabrés I Xamena, F.X. (2014). Metal organic framework catalysis: quo vadis? *ACS Catal.* 4 (2): 361–378.

6 Bavykina, A., Kolobov, N., Khan, I.S. et al. (2020). Metal-organic frameworks in heterogeneous catalysis: recent progress, new trends, and future perspectives. *Chem. Rev.* 120 (16): 8468–8535.

7 McGuirk, C.M., Katz, M.K., Stern, C.L. et al. (2015). Turning on catalysis: Incorporation of a hydrogen-bond-donating squaramide moiety into a Zr metal-organic framework. *J. Am. Chem. Soc.* 137 (2): 919–925.

8 Corma, A., García, H., and Llabrés I Xamena, F.X. (2010). Engineering metal organic frameworks for heterogeneous catalysis. *Chem. Rev.* 110 (8): 4606–4655.

9 Yang, D., Babucci, M., Casey, W.H., and Gates, B.C. (2020). The surface chemistry of metal oxide qlusters: from metal-organic frameworks to minerals. *ACS Cent. Sci.* 6 (9): 1523–1533.

10 Trickett, C.A., Popp, T.M.O., Su, J. et al. (2019). Identification of the strong Brønsted acid site in a metal–organic framework solid acid catalyst. *Nat. Chem.* 11: 170–176.

11 Yoon, J.W., Seo, Y.-K., Hwang, Y.K. et al. (2010). Controlled reducibility of a metal-organic framework with coordinatively unsaturated sites for preferential gas sorption. *Angew. Chem. Int. Ed.* 49: 5949–5952.

12 Wei, Y.S., Zhang, M., Zou, R., and Xu, Q. (2020). Metal-organic framework-based catalysts with single metal sites. *Chem. Rev.* 120 (21): 12089–12174.

13 Vermoortele, F., Ameloot, A., Alaerts, L. et al. (2012). Tuning the catalytic performance of metal-organic frameworks in fine chemistry by active site engineering. *J. Mater. Chem.* 22: 10313–10321.

14 Ji, P., Feng, X., Oliveres, P. et al. (2019). Strongly Lewis acidic metal–organic frameworks for continuous flow catalysis. *J. Am. Chem. Soc.* 141 (37): 14878–14888.

15 Dhakshinamoorthy, A., Alvaro, M., Chevreau, H. et al. (2012). Iron(iii) metal-organic frameworks as solid Lewis acids for the isomerization of α-pinene oxide. *Catal. Sci. Technol.* 2: 324–330.

16 Alaerts, L., Séguin, E., Poelman, H. et al. (2006). Probing the lewis acidity and catalytic activity of the metal-organic framework $[Cu_3(btc)_2]$ (BTC = Benzene-1,3,5-tricarboxylate). *Chem. Eur. J.* 12: 7353–7363.

17 Nguyen, L.T.L., Nguyen, C.v., Dang, G.H. et al. (2011). Towards applications of metal-organic frameworks in catalysis: Friedel-Crafts acylation reaction over IRMOF-8 as an efficient heterogeneous catalyst. *J. Mol. Catal. A: Chem.* 349: 28–35.

18 Jiang, D., Mallat, T., Krumeich, F., and Baiker, A. (2008). Copper-based metal-organic framework for the facile ring-opening of epoxides. *J. Catal.* 257: 390–395.

19 Rojas-Buzo, S., Bohigues, B., Lopes, C.W. et al. (2021). Tailoring Lewis/Brønsted acid properties of MOF nodesviahydrothermal and solvothermal synthesis: simple approach with exceptional catalytic implications. *Chem. Sci.* 12: 10106–10115.

20 Zhai, Q.G., Bu, X., Mao, C. et al. (2016). Systematic and dramatic tuning on gas sorption performance in heterometallic metal-organic frameworks. *J. Am. Chem. Soc.* 138 (8): 2524–2527.

21 Syzgantseva, M.A., Ireland, C.P., Ebrahim, F.M. et al. (2019). Metal substitution as the method of modifying electronic structure of metal-organic frameworks. *J. Am. Chem. Soc.* 141 (15): 6271–6278.

22 Castells-Gil, J., Padial, N.M., Almora-Barrios, N. et al. (2018). Chemical engineering of photoactivity in heterometallic titanium–organic frameworks by metal doping. *Angew. Chem. Int. Ed.* 57: 8453–8457.

23 Gaillac, R., Pullumbi, P., Beyer, K.A. et al. (2017). Liquid metal–organic frameworks. *Nat. Mater.* 16: 1149–1155.

24 Terban, M.W. and Billinge, S.J.L. (2022). Structural analysis of molecular materials using the pair distribution function. *Chem. Rev.* 122 (1): 1208–1272.

25 Castillo-Blas, C., Moreno, J.M., Romero-Muñiz, I., and Platero-Prats, A.E. (2020). Applications of pair distribution function analyses to the emerging field of: non-ideal metal-organic framework materials. *Nanoscale* 12: 15577–15587.

26 Castells-Gil, J., Padial, N.M., Almora-Barrios, N. et al. (2020). Heterometallic titanium-organic frameworks as dual metal catalysts for synergistic non-buffered hydrolysis of nerve agent simulants. *Chem* 6 (11): 3118–3131.

27 Wang, L.J., Deng, H., Furukawa, H. et al. (2014). Synthesis and characterization of metal-organic framework-74 containing 2, 4, 6, 8 and 10 different metals. *Inorg. Chem.* 53 (12): 5881–5883.

28 Viciano-Chumillas, M., Liu, X., Leyva-Pérez, A. et al. (2022). Mixed component metal-organic frameworks: heterogeneity and complexity at the service of application performances. *Coord. Chem. Rev.* 451: 214273.

29 Syzgantseva, M.A., Ireland, C.P., Ebrahim, F.M. et al. (2019). Metal substitution as the method of modifying electronic structure of metal-organic frameworks. *J. Am. Chem. Soc.* 141 (15): 6271–6278.

30 Trousselet, F., Archereau, A., Boutin, A., and Coudert, F.X. (2016). Heterometallic metal-organic frameworks of MOF-5 and UiO-66 families: insight from computational chemistry. *J. Phys. Chem. C* 120 (43): 24885–24894.

31 Nouar, F., Devic, T., Chevreau, H. et al. (2012). Tuning the breathing behaviour of MIL-53 by cation mixing. *Chem. Commun.* 48: 10237–10239.

32 Depauw, H., Nevjestić, I., Wang, G. et al. (2017). Discovery of a novel, large pore phase in a bimetallic Al/V metal-organic framework. *J. Mater. Chem. A* 5: 24580–24584.

33 Horcajada, P., Surblé, S., Serre, C. et al. (2007). Synthesis and catalytic properties of MIL-100(Fe), an iron(III) carboxylate with large pores. *Chem. Commun.* 2820–2822.

34 Castells-Gil, J., Padial, N.M., Almora-Barrios, N. et al. (2019). Novo synthesis of mesoporous photoactive titanium(iv)-organic frameworks with MIL-100 topology. *Chem. Sci.* 10: 4313–4321.

35 Jia, H., Han, Q., Luo, W. et al. (2022). Sequence control of metals in MOF by coordination number precoding for electrocatalytic oxygen evolution. *Chem Catal.* 2 (84): 84–101.

36 Dissegna, S., Vervoorts, P., Hobday, C.L. et al. (2018). Tuning the mechanical response of metal-organic frameworks by defect engineering. *J. Am. Chem. Soc.* 140 (37): 11581–11584.

37 Shearer, G.C., Chavan, S., Bordiga, S. et al. (2016). Defect engineering: tuning the porosity and composition of the metal-organic framework UiO-66 via modulated synthesis. *Chem. Mater.* 28 (11): 3749–3761.

38 Wu, H., Chua, Y.S., Krungleviciute, V. et al. (2013). Unusual and highly tunable missing-linker defects in zirconium metal-organic framework UiO-66 and their important effects on gas adsorption. *J. Am. Chem. Soc.* 135 (28): 10525–10532.

39 Rodríguez-Albelo, L.M., López-Maya, E., Hamad, S. et al. (2017). Selective sulfur dioxide adsorption on crystal defect sites on an isoreticular metal organic framework series. *Nat. Commun.* 8: 14457.

40 Liu, Y., Klet, R.C., Hupp, J.T., and Farha, O. (2015). Probing the correlations between the defects in metal-organic frameworks and their catalytic activity by an epoxide ring-opening reaction. *Chem. Commun.* 52: 7806–7809.

41 Taddei, M. (2017). When defects turn into virtues: the curious case of zirconium-based metal-organic frameworks. *Coord. Chem. Rev.* 343: 1–24.

42 Feng, X., Jena, H.S., Krishnaraj, C. et al. (2021). Generating catalytic sites in UiO-66 through defect engineering. *ACS Appl. Mater. Interfaces* 13 (51): 60715–60735.

43 Lázaro, I.A., Almora-Barrios, N., Tatay, S., and Martí-Gastaldo, C. (2021). Effect of modulator connectivity on promoting defectivity in titanium-organic frameworks. *Chem. Sci.* 12: 2586–2593.

44 Gutov, O.v., Hevia, M.G., Escudero-Adán, E.C., and Shafir, A. (2015). Metal-organic framework (MOF) defects under control: insights into the missing linker sites and their implication in the reactivity of zirconium-based frameworks. *Inorg. Chem.* 54 (17): 8396–8400.

45 Valenzano, L., Civalleri, B., Chavan, S. et al. (2011). Disclosing the comp lex structure of UiO-66 metal organic framework: a synergic combination of experiment and theory. *Chem. Mater.* 23 (7): 1700–1718.

46 Ravon, U., Savonnet, M., Aguado, S. et al. (2010). Engineering of coordination polymers for shape selective alkylation of large aromatics and the role of defects. *Microporous Mesoporous Mater.* 129: 319–329.

47 Vermoortele, F., Bueken, B., Le Bars, G. et al. (2013). Synthesis modulation as a tool to increase the catalytic activity of metal-organic frameworks: the unique case of UiO-66(Zr). *J. Am. Chem. Soc.* 135 (31): 11465–11468.

48 Liu, L., Chen, Z., Wang, J. et al. (2019). Imaging defects and their evolution in a metal–organic framework at sub-unit-cell resolution. *Nat. Chem.* 11: 622–628.

49 Katz, M.J., Klet, R.C., Moon, S.-Y. et al. (2015). One step backward is two steps forward: enhancing the hydrolysis rate of UiO-66 by decreasing [OH-]. *ACS Catal.* 5 (8): 4637–4642.

50 Drake, T., Ji, P., and Lin, W. (2018). Site isolation in metal-organic frameworks enables novel transition metal catalysis. *Acc. Chem. Res.* 51 (9): 2129–2138.

51 Cui, X., Li, W., Ryabchuk, P. et al. (2018). Bridging homogeneous and heterogeneous catalysis by heterogeneous single-metal-site catalysts. *Nat. Catal.* 1: 385–397.

52 Gao, Z., Lai, Y., Gong, L. et al. (2022). Robust Th-MOF-supported semirigid single-metal-site catalyst for an efficient acidic oxygen evolution reaction. *ACS Catal.* 12 (15): 9101–9113.

53 Mandal, S., Natarajan, S., Mani, P., and Pankajakshan, A. (2021). Post-synthetic modification of metal–organic frameworks toward applications. *Adv. Funct. Mater.* 31 (4): 2006291.

54 Baranowski, C.J., Brandon, J., Bahmanpour, A.M., and Kröcher, O. (2021). Grafting of alkali metals on fumed silica for the catalytic dehydrogenation of methanol to formaldehyde. *ChemCatChem* 13 (17): 3864–3877.

55 Brozek, C.K. and Dincă, M. (2013). Ti^{3+}-, $V^{2+/3+}$-, $Cr^{2+/3+}$-, Mn^{2+}-, and Fe^{2+}-substituted MOF-5 and redox reactivity in Cr- and Fe-MOF-5. *J. Am. Chem. Soc.* 135 (34): 12886–12891.

56 Mondloch, J.E., Bury, W., Fairen-Jimenez, D. et al. (2013). Vapor-phase metalation by atomic layer deposition in a metal-organic framework. *J. Am. Chem. Soc.* 135 (28): 10294–10297.

57 Justin Kunene, T., Kwanda Tartibu, L., Ukoba, K., and Jen, T.C. (2022). Review of atomic layer deposition process, application and modeling tools. *Mater. Today Proc.* 62: S95–S109.

58 George, S.M. (2010). Atomic layer deposition: an overview. *Chem. Rev.* 110 (1): 111–131.

59 Yang, D., Odoh, S.O., Wang, T.C. et al. (2015). Metal-organic framework nodes as nearly ideal supports for molecular catalysts: NU-1000- and UiO-66-supported iridium complexes. *J. Am. Chem. Soc.* 137 (23): 7391–7396.

60 Noh, H., Cui, Y., Peters, A.W. et al. (2016). An exceptionally stable metal-organic framework supported molybdenum(VI) oxide catalyst for cyclohexene epoxidation. *J. Am. Chem. Soc.* 138 (44): 14720–14726.

61 Kim, I.S., Borycz, J., Platero-Prats, A.E. et al. (2015). Targeted single-site MOF node modification: trivalent metal loading via atomic layer deposition. *Chem. Mater.* 27 (13): 4772–4778.

62 Hackler, R.A., Pandharkar, R., Ferrandon, M.S.M. et al. (2020). Isomerization and selective hydrogenation of propyne: screening of metal-organic frameworks modified by atomic layer deposition. *J. Am. Chem. Soc.* 142 (48): 20380–20389.

63 Berijani, K., Morsali, A., and Hupp, J.T. (2019). An effective strategy for creating asymmetric MOFs for chirality induction: a chiral Zr-based MOF for enantioselective epoxidation. *Catal. Sci. Technol.* 9: 3388–3397.

64 Copéret, C., Comas-Vives, A., Conley, M.P. et al. (2016). Surface organometallic and coordination chemistry toward single-site heterogeneous catalysts: strategies, methods, structures, and activities. *Chem. Rev.* 116 (2): 323–421.

65 Samantaray, M.K., Pump, E., Bendjeriou-Sedjerari, A. et al. (2018). Surface organometallic chemistry in heterogeneous catalysis. *Chem. Soc. Rev.* 47: 8403–8437.

66 Basset, J.M. and Pelletier, J.D.A. (2018). Predictive approach of heterogeneous catalysis. *Natl. Sci. Rev.* 5: 633–635.

67 Manna, K., Ji, P., Greene, F.X., and Lin, W. (2016). Metal-organic framework nodes support single-site magnesium-alkyl catalysts for hydroboration and hydroamination reactions. *J. Am. Chem. Soc.* 138 (24): 7488–7491.

68 Thiam, Z., Abou-Hamad, E., Dereli, B. et al. (2020). Extension of surface organometallic chemistry to metal–organic frameworks: development of a well-defined single site $[(\equiv Zr-O-)W(=O)(CH2^tBu)_3]$ olefin metathesis catalyst. *J. Am. Chem. Soc.* 142 (39): 16690–16703.

69 Yang, Q., Xu, Q., and Jiang, H.L. (2017). Metal–organic frameworks meet metal nanoparticles: synergistic effect for enhanced catalysis. *Chem. Soc. Rev.* 46 (15): 4774–4808.

70 Sui, J., Liu, H., Hu, S. et al. (2022). A general strategy to immobilize single-atom catalysts in metal–organic frameworks for enhanced photocatalysis. *Adv. Mater.* 34 (6): 2109203.

71 Du, Q., Rao, R., Bi, F. et al. (2022). Preparation of modified zirconium-based metal-organic frameworks (Zr-MOFs) supported metals and recent application in environment: a review and perspectives. *Surf. Interfaces* 28: 101647.

72 Zhuang, Z., Mai, Z., Wang, T., and Liu, D. (2020). Strategies for conversion between metal–organic frameworks and gels. *Coord. Chem. Rev.* 421: 213461.

73 Abednatanzi, S., Najafi, M., Gohari Derakhshandeh, P., and Van Der Voort, P. (2022). Metal- and covalent organic frameworks as catalyst for organic transformation: comparative overview and future perspectives. *Coord. Chem. Rev.* 451: 214259.

74 Wei, Y.-S., Zhang, M., Zou, R., and Xu, Q. (2020). Metal–organic framework-based catalysts with single metal sites. *Chem. Rev.* 120: 12089–12174.

75 Hwang, Y.K., Hong, D.-Y., Chang, J.-S. et al. (2008). Amine grafting on coordinatively unsaturated metal centers of MOFs: consequences for catalysis and metal encapsulation. *Angew. Chem.* 120 (22): 4212–4216.

76 Cohen, S.M., Zhang, Z., and Boissonnault, J.A. (2016). Toward "metallo-MOFzymes": metal-organic frameworks with single-site metal catalysts for small-molecule transformations. *Inorg. Chem.* 55 (15): 7281–7290.

77 Wen, Y., Zhang, J., Xu, Q. et al. (2018). Pore surface engineering of metal–organic frameworks for heterogeneous catalysis. *Coord. Chem. Rev.* 376: 248–276.

78 Yao, Y., Ma, Z., Dou, Y. et al. (2022). Random occupation of multimetal sites in transition metal-organic frameworks for boosting the oxygen evolution reaction. *Chem. A Eur. J.* 28 (14): e202104288.

79 Liberman, I., Shimoni, R., Ifraemov, R. et al. (2020). Active-site modulation in an Fe-porphyrin-based metal-organic framework through ligand axial coordination: accelerating electrocatalysis and charge-transport kinetics. *J. Am. Chem. Soc.* 142 (4): 1933–1940.

80 Song, F., Zhang, T., Wang, C., and Lin, W. (2012). Chiral porous metal-organic frameworks with dual active sites for sequential asymmetric catalysis. *Proc. R Soc.* 468 (2143): 2035–2052.

81 Isaeva, V.I. and Kustov, L.M. (2010). The application of metal-organic frameworks in catalysis (review). *Pet. Chem.* 50 (3): 167–180.

82 Li, D., Xu, H.Q., Jiao, L., and Jiang, H.L. (2019). Metal-organic frameworks for catalysis: state of the art, challenges, and opportunities. *EnergyChem* 1 (1): 100005.

83 Wu, C.-D., Zhao, M., Wu, C.-D., and Zhao, M. (2017). Incorporation of molecular catalysts in metal–organic frameworks for highly efficient heterogeneous catalysis. *Adv. Mater.* 29 (14): 1605446.

84 Tu, W., Xu, Y., Yin, S., and Xu, R. (2018). Rational design of catalytic centers in crystalline frameworks. *Adv. Mater.* 30 (33): 1707582.

85 Shi, K., Lin, Y., Xiong, Z. et al. (2022). Synergistic effects of porphyrin-ring catalytic center and metal catalytic site from crosslinked porphyrin-based porous polyimides cathode host for lithium polysulfides conversion in lithium-sulfur batteries. *Chem. Eng. J.* 430: 132692.

86 Manna, K., Zhang, T., and Lin, W. (2014). Postsynthetic metalation of bipyridyl-containing metal-organic frameworks for highly efficient catalytic organic transformations. *J. Am. Chem. Soc.* 136 (18): 6566–6569.

87 Ranjan Jena, S., Mandal, T., and Choudhury, J. (2022). Metal-terpyridine assembled functional materials for electrochromic, catalytic and environmental applications. *Chem. Rec.* 22 (12): e202200165.

88 Chen, L., Jiang, Y., Huo, H. et al. (2020). Metal-organic framework-based composite Ni@MOF as heterogenous catalyst for ethylene trimerization. *Appl. Catal., A* (594): 117457.

89 Xu, Z., Shen, J., Li, L. et al. (2022). (Salen)Titanium-catalyzed asymmetric hydrogen atom transfer for epoxides reduction. *Angew. Chem.* 61 (50): e202214111.

90 Drake, T., Ji, P., and Lin, W. (2018). Site isolation in metal-organic frameworks enables novel transition metal catalysis. *Acc. Chem. Res.* 51 (9): 2129–2138.

91 Ma, L., Abney, C., and Lin, W. (2009). Enantioselective catalysis with homochiral metal–organic frameworks. *Chem. Soc. Rev.* 38 (5): 1248–1256.

92 Senge, M.O., Sergeeva, N.N., and Hale, K.J. (2021). Classic highlights in porphyrin and porphyrinoid total synthesis and biosynthesis. *Chem. Soc. Rev.* 50 (7): 4730–4789.

93 Barona-Castaño, J.C., Carmona-Vargas, C.C., Brocksom, T.J. et al. (2016). Porphyrins as catalysts in scalable organic reactions. *Molecules* 21, 21 (3): 310.

94 Usov, P.M., Ahrenholtz, S.R., Maza, W.A. et al. (2016). Cooperative electrochemical water oxidation by Zr nodes and Ni–porphyrin linkers of a PCN-224 MOF thin film. *J. Mater. Chem. A* 4 (43): 16818–16823.

95 Shultz, A.M., Farha, O.K., Hupp, J.T., and Nguyen, S.T. (2009). A catalytically active, permanently microporous MOF with metalloporphyrin struts. *J. Am. Chem. Soc.* 131 (12): 4204–4205.

96 Fateeva, A., Chater, P.A., Ireland, C.P. et al. (2012). A water-stable porphyrin-based metal–organic framework active for visible light photocatalysis. *Angew. Chem. Int. Ed.* 51 (30): 7440–7444.

97 Bai, Y., Dou, Y., Xie, L.H. et al. (2016). Zr-based metal–organic frameworks: design, synthesis, structure, and applications. *Chem. Soc. Rev.* 45 (8): 2327–2367.

98 Feng, D., Gu, Z.Y., Li, J.R. et al. (2012). Zirconium-metalloporphyrin PCN-222: mesoporous metal-organic frameworks with ultrahigh stability as biomimetic catalysts. *Angew. Chem. Int. Ed.* 51 (41): 10307–10310.

99 Jiang, H.L., Feng, D., Wang, K. et al. (2013). An exceptionally stable, porphyrinic Zr metal-organic framework exhibiting pH-dependent fluorescence. *J. Am. Chem. Soc.* 135 (37): 13934–13938.

100 Huang, N., Wang, K., Drake, H. et al. (2018). Tailor-made pyrazolide-based metal-organic frameworks for selective catalysis. *J. Am. Chem. Soc.* 140 (20): 6383–6390.

101 Ishimizu, Y., Ma, Z., Hada, M., and Fujii, H. (2021). Rate-limiting step of epoxidation reaction of the oxoiron(IV) porphyrin π-cation radical complex: electron transfer coupled bond formation mechanism. *Inorg. Chem.* 60: 17687–17698.

102 Nakagaki, S., Ferreira, G.K.B., Ucoski, G.M., and KAD, d.F.C. (2013). Chemical reactions catalyzed by metalloporphyrin-based metal-organic frameworks. *Molecules* 18: 7279–7308.

103 Hod, I., Sampson, M.D., Deria, P. et al. (2015). Fe-porphyrin-based metal-organic framework films as high-surface concentration, heterogeneous catalysts for electrochemical reduction of CO_2. *ACS Catal.* 5 (11): 6302–6309.

104 Bipyridine. (2006) Wikipedia. https://en.wikipedia.org/wiki/Bipyridine.

105 Canivet, J., Aguado, S., Schuurman, Y., and Farrusseng, D. (2013). MOF-supported selective ethylene dimerization single-site catalysts through one-pot postsynthetic modification. *J. Am. Chem. Soc.* 135 (11): 4195–4198.

106 Song, F., Wang, C., Falkowski, J.M. et al. (2010). Isoreticular chiral metal-organic frameworks for asymmetric alkene epoxidation: tuning catalytic activity by controlling framework catenation and varying open channel sizes. *J. Am. Chem. Soc.* 132 (43): 15390–15398.

107 Shimizu, S., Watanabe, N., Kataoka, T. et al. (2012). Pyridine and pyridine derivatives. *Ullmann's Encycl. Ind. Chem.* 30: 557–586.

108 Wang, C., Xie, Z., Dekrafft, K.E., and Lin, W. (2011). Doping metal-organic frameworks for water oxidation, carbon dioxide reduction, and organic photocatalysis. *J. Am. Chem. Soc.* 133 (34): 13445–13454.

109 Hou, C.C., Li, T.T., Cao, S. et al. (2015). Incorporation of a [Ru(dcbpy)(bpy) 2] 2+ photosensitizer and a Pt(dcbpy)Cl 2 catalyst into metal–organic frameworks for photocatalytic hydrogen evolution from aqueous solution. *J. Mater. Chem. A* 3 (19): 10386–10394.

110 Chen, G.H., Leu, W.J., Guh, J.H. et al. (2018). Synthesis, characterization and cancer cell growth inhibition activity of ruthenium(II) complexes bearing bidentate pyrrole-imine ligands. *J. Organomet. Chem.* 868: 122–130.

111 Jin, Y., Zhu, Y., and Zhang, W. (2013). Development of organic porous materials through Schiff-base chemistry. *CrystEngComm* 15 (8): 1484–1499.

112 Schenkel, L.B. and Ellman, J.A. (2003). Novel sulfinyl imine ligands for asymmetric catalysis. *Org. Lett.* 5 (4): 545–548.

113 Cozzi, P.G. (2004). Metal–Salen Schiff base complexes in catalysis: practical aspects. *Chem. Soc. Rev.* 33 (7): 410–421.

114 Kitaura, R., Onoyama, G., Sakamoto, H. et al. (2004). Immobilization of a metallo schiff base into a microporous coordination polymer. *Angew. Chem. Int. Ed.* 43 (20): 2684–2687.

115 Yuan, G., Jiang, H., Zhang, L. et al. (2019). Metallosalen-based crystalline porous materials: synthesis and property. *Coord. Chem. Rev.* (378): 483–499.

116 Li, C., Ma, Z., Li, S. et al. (2018). Syntheses, structures and catalytic activities of molybdenum carbonyl complexes based on pyridine-imine ligands. *Transition Met. Chem.* 43 (2): 193–199.

117 Yoon, M., Srirambalaji, R., and Kim, K. (2012). Homochiral metal-organic frameworks for asymmetric heterogeneous catalysis. *Chem. Rev.* 112 (2): 1196–1231.

118 Su, D.S., Zhang, J., Frank, B. et al. (2010). Metal-free heterogeneous catalysis for sustainable chemistry. *ChemSusChem* 3 (2): 169–180.

119 Palucki, M., Finney, N.S., Pospisil, P.J. et al. (1998). The mechanistic basis for electronic effects on enantioselectivity in the (salen)Mn(III)-catalyzed epoxidation reaction. *J. Am. Chem. Soc.* 120 (5): 948–954.

120 Phan, N.T.S., Nguyen, C.K., Nguyen, T.T., and Truong, T. (2014). Towards applications of metal–organic frameworks in catalysis: C–H direct activation of benzoxazole with aryl boronic acids using Ni 2 (BDC) 2 (DABCO) as an efficient heterogeneous catalyst. *Catal. Sci. Technol.* 4 (2): 369–377.

121 Chughtai, A.H., Ahmad, N., Younus, H.A. et al. (2015). Metal–organic frameworks: versatile heterogeneous catalysts for efficient catalytic organic transformations. *Chem. Soc. Rev.* 44 (19): 6804–6849.

122 Wei, Y.S., Zhang, M., Zou, R., and Xu, Q. (2020). Metal-organic framework-based catalysts with single metal sites. *Chem. Rev.* 120 (21): 12089–12174.

123 Mackay, L.G., Stephen Wylie, R., Sanders, M., and JK. (1994). Catalytic acyl transfer by a cyclic porphyrin trimer: efficient turnover without product inhibition. *J. Am. Chem. Soc.* 116 (4): 8980.

124 Huang, Y.B., Liang, J., Wang, X.S., and Cao, R. (2017). Multifunctional metal–organic framework catalysts: synergistic catalysis and tandem reactions. *Chem. Soc. Rev.* 46 (1): 126–157.

125 Chakrabarty, R., Mukherjee, P.S., and Stang, P.J. (2011). Supramolecular coordination: Self-assembly of finite two- and three-dimensional ensembles. *Chem. Rev.* 111 (11): 6810–6918.

126 Yang, X.L., Xie, M.H., Zou, C. et al. (2012). Porous metalloporphyrinic frameworks constructed from metal 5,10,15,20-tetrakis(3,5-biscarboxylphenyl) porphyrin for highly efficient and selective catalytic oxidation of alkylbenzenes. *J. Am. Chem. Soc.* 134 (25): 10638–10645.

127 Grützmacher, H. (2008). Cooperating ligands in catalysis. *Angew. Chem. Int. Ed.* 47 (10): 1814–1818.

128 Walter, C.J., Anderson, H.L., and Sanders, J.K.M. (1993). exo -Selective acceleration of an intermolecular Diels–Alder reaction by a trimeric porphyrin host. *J. Chem. Soc., Chem. Commun.* 0 (5): 458–460.

129 Berrisford, D.J., Bolm, C., and Sharpless, K.B. (1995). Ligand-accelerated catalysis. *Angew. Chem. Int. Ed.* 34 (10): 1059–1070.

130 Jacobsen, E.N., Markó, I., Mungall, W.S. et al. (1988). Asymmetric dihydroxylation via ligand-accelerated catalysis. *J. Am. Chem. Soc.* 110 (6): 1968–1970.

131 Carson, F., Martínez-Castro, E., Marcos, R. et al. (2015). Effect of the functionalisation route on a Zr-MOF with an Ir–NHC complex for catalysis. *Chem. Commun.* 51 (54): 10864–10867.

132 Song, L.F., Jiang, C.H., Jiao, C.L. et al. (2010). Two new metal-organic frameworks with mixed ligands of carboxylate and bipyridine: synthesis, crystal structure, and sensing for methanol. *Cryst Growth Des.* 10 (12): 5020–5023.

133 Masoomi, M.Y., Morsali, A., Dhakshinamoorthy, A., and Garcia, H. (2019). Mixed-metal MOFs: unique opportunities in metal-organic framework functionality and design. *Angew. Chem. Int. Ed. Engl.* 141 (43): 15330–15347.

134 Wang, C., Wang, J.L., and Lin, W. (2012). Elucidating molecular iridium water oxidation catalysts using metal-organic frameworks: a comprehensive structural, catalytic, spectroscopic, and kinetic study. *J. Am. Chem. Soc.* 134 (48): 19895–19908.

135 Xie, M.H., Yang, X.L., Zou, C., and De, W.C. (2011). A SnIV-porphyrin-based metal-organic framework for the selective photo-oxygenation of phenol and sulfides. *Inorg. Chem.* 50 (12): 5318–5320.

136 Zhao, M., Ou, S., and De, W.C. (2014). Porous metal-organic frameworks for heterogeneous biomimetic catalysis. *Acc. Chem. Res.* 47 (4): 1199–1207.

137 Taddei, M., Wakeham, R.J., Koutsianos, A. et al. (2018). Post-synthetic ligand exchange in zirconium-based metal-organic frameworks: beware of the defects! *Angew. Chemie.* 130 (36): 11880–11884.

138 Hoffmann, H.C., Debowski, M., Müller, P. et al. (2012). Solid-state NMR spectroscopy of metal–organic framework compounds (MOFs). *Materials* 5: 2537–2572.

139 Kim, M., Cahill, J.F., Fei, H. et al. (2012). Postsynthetic ligand and cation exchange in robust metal-organic frameworks. *J. Am. Chem. Soc.* 134 (43): 18082–18088.

140 Yu, X. and Cohen, S.M. (2016). Photocatalytic metal-organic frameworks for selective 2,2,2-trifluoroethylation of styrenes. *J. Am. Chem. Soc.* 138 (38): 12320–12323.

141 Roy, S., Bhunia, A., Schuth, N. et al. (2018). Light-driven hydrogen evolution catalyzed by a cobaloxime catalyst incorporated in a MIL-101(Cr) metal–organic framework. *Sustainable Energy Fuels* 2 (6): 1148–1152.

142 García-Simón, C., Gramage-Doria, R., Raoufmoghaddam, S. et al. (2015). Enantioselective hydroformylation by a Rh-catalyst entrapped in a supramolecular metallocage. *J. Am. Chem. Soc.* 137 (7): 2680–2687.

143 Bhunia, A., Johnson, B.A., Czapla-Masztafiak, J. et al. (2018). Formal water oxidation turnover frequencies from MIL-101(Cr) anchored Ru(bda) depend on oxidant concentration. *Chem. Commun.* 54 (56): 7770–7773.

144 Li, B., Zhang, Y., Ma, D. et al. (2012). A strategy toward constructing a bifunctionalized MOF catalyst : post-synthetic modification of MOFs on organic ligands and coordinatively unsaturated metal sites. *Chem. Commun.* 48 (49): 6151–6153.

145 Maza, W.A. and Morris, A.J. (2014). Photophysical characterization of a ruthenium(II) tris(2,2′- bipyridine)-doped zirconium UiO-67 metal-organic framework. *J. Phys. Chem. C* 118 (17): 8803–8817.

146 Pullen, S., Roy, S., and Ott, S. (2017). [FeFe] Hydrogenase active site model chemistry in a UiO-66 metal–organic framework. *Chem. Commun.* 53 (37): 5227–5230.

147 Kajiwara, T., Fujii, M., Tsujimoto, M. et al. (2016). Photochemical reduction of low concentrations of CO_2 in a porous coordination polymer with a ruthenium(II)–CO complex. *Angew. Chem. Int. Ed.* 55 (8): 2697–2700.

148 Shearer, G.C., Vitillo, J.G., Bordiga, S. et al. (2016). Functionalizing the defects: postsynthetic ligand exchange in the metal organic framework UiO-66. *Chem. Mater.* 28 (20): 7190–7193.

149 Karagiaridi, O., Bury, W., Mondloch, J.E. et al. (2014). Solvent-assisted linker exchange: an alternative to the De Novo synthesis of unattainable metal–organic frameworks. *Angew. Chem. Int. Ed.* 53 (18): 4530–4540.

150 Evans, J.D., Sumby, C.J., and Doonan, C.J. (2014). Post-synthetic metalation of metal–organic frameworks. *Chem. Soc. Rev.* 43 (16): 5933–5951.

151 Zhao, W., Li, G., and Tang, Z. (2019). Metal-organic frameworks as emerging platform for supporting isolated single-site catalysts. *Nano Today* 27: 178–197.

152 Gao, W.Y., Chrzanowski, M., and Ma, S. (2014). Metal–metalloporphyrin frameworks: a resurging class of functional materials. *Chem. Soc. Rev.* 43 (16): 5841–5866.

153 Luo, S., Li, X., Gao, W. et al. (2019). An MOF-derived C@NiO@Ni electrocatalyst for N2 conversion to NH3 in alkaline electrolytes. *Sustainable Energy Fuels* 4 (1): 164–170.

154 Wu, S., Chen, L., Yin, B., and Li, Y. (2015). "Click" post-functionalization of a metal–organic framework for engineering active single-site heterogeneous Ru(III) catalysts. *Chem. Commun.* 51 (48): 9884–9887.

155 Mon, M., Ferrando-Soria, J., Grancha, T. et al. (2016). Selective gold recovery and catalysis in a highly flexible methionine-decorated metal-organic framework. *J. Am. Chem. Soc.* 138 (25): 7864–7867.

156 Mon, M., Rivero-Crespo, M.A., Esfflsf Errando-Soria, J. et al. (2018). Synthesis of densely packaged, ultrasmall Pt02 clusters within a thioether-functionalized MOF: catalytic activity in industrial reactions at low temperature. *Angew. Chem.* 130 (21): 6294–6299.

157 Shi, W., Cao, L., Zhang, H. et al. (2017). surface modification of two-dimensional metal–organic layers creates biomimetic catalytic microenvironments for selective oxidation. *Angew. Chem.* 129 (33): 9836–9841.

158 Zuo, Q., Tingting Liu, T., Chen, C. et al. (2019). Ultrathin metal–organic framework nanosheets with ultrahigh loading of single Pt atoms for efficient visible-light-driven photocatalytic H2 evolution. *Angew. Chem. Int. Ed.* 58 (30): 10198–10203.

159 Tanabe, K.K. and Cohen, S.M. (2009). Engineering a metal–organic framework catalyst by using postsynthetic modification. *Angew. Chem.* 121 (40): 7560–7563.

160 Al-Attas, T.A., Marei, N.N., Yong, X. et al. (2021). Ligand-engineered metal-organic frameworks for electrochemical reduction of carbon dioxide to carbon monoxide. *ACS Catal.* 11 (12): 7350–7357.

161 Boissonnault, J.A., Wong-Foy, A.G., and Matzger, A.J. (2017). Core-shell structures arise naturally during ligand exchange in metal-organic frameworks. *J. Am. Chem. Soc.* 139 (42): 14841–14844.

162 Becker, W. (2012). Fluorescence lifetime imaging – techniques and applications. *J. Microsc.* 247 (2): 119–136.

163 Schrimpf, W., Jiang, J., Ji, Z. et al. (2018). Chemical diversity in a metal–organic framework revealed by fluorescence lifetime imaging. *Nat. Commun.* 9 (1): 1–10.

164 dos Santos, A.C.V.D., Lendl, B., and Ramer, G. (2022). Systematic analysis and nanoscale chemical imaging of polymers using photothermal-induced resonance (AFM-IR) infrared spectroscopy. *Polym. Test.* 106: 107443.

165 Thacker, N.C., Lin, Z., Zhang, T. et al. (2016). Robust and porous β-diketiminate-functionalized metal-organic frameworks for earth-abundant-metal-catalyzed C-H amination and hydrogenation. *J. Am. Chem. Soc.* 138 (10): 3501–3509.

166 He, C., Liang, J., Zou, Y.H. et al. (2022). Metal-organic frameworks bonded with metal N-heterocyclic carbenes for efficient catalysis. *Natl. Sci. Rev.* 9 (6).

167 Sing, K.S.W., Everett, D.H., Haul, R.A.W. et al. (1985). Reporting physisorption data for gas/solid systems with special reference to the determination of surface area and porosity. *Pure Appl. Chem.* 57 (4): 603–619.

168 Rouquerol, J., Avnir, D., Fairbridge, C.W. et al. (1994). Recommendations for the characterization of porous solids. *Pure Appl. Chem.* 66 (8): 1739–1758.

169 Thommes, M., Kaneko, K., Nimark, A.V. et al. (2015). Physisorption of gases, with special reference to the evaluation of Surface area and pore size distribution. *Pure Appl. Chem.* 87 (9–10): 1051–1069.

170 Farha, O.K., Zhang, X., Chen, Z. et al. (2020). A historical overview of activation and porosity of metal-organic frameworks. *Chem. Soc. Rev.* 49: 7406–7427.

171 Suendo, V., Nurhayati, Alni, A. et al. (2020). Revealing the real size of a porphyrin molecule with quantum confinement probing via temperature-dependent photoluminescence spectroscopy. *J. Phys. Chem. A* 124 (13): 2672–2682.

172 Xiang, W., Zhang, Y., Lin, H., and Liu, C. (2017). Nanoparticles/metal-organic framework composites for catalytic applications: current status and perspective. *Molecules* 22 (12): 2103–2036.

173 Farha, O.K. and Buru, C.T. (2020). Strategies for incorporating catalytically active polyoxometalates in metal-organic frameworks for organic transformations. *ACS Appl. Mater. Interfaces* 12 (5): 5345–5360.

174 Sun, J., Abednatanzi, S., Van Der Voort, P. et al. (2020). POM@MOF hybrids: synthesis and applications. *Catalysts* 10 (5): 578–608.

175 Farha, O.K., Hupp, J.T., Liu, J. et al. (2022). MOF-enabled confinement and related effects for chemical catalyst presentation and utilization. *Chem. Soc. Rev.* 51: 1045–1097.

176 Férey, G., Mellot-Drazmieks, C., Serre, C. et al. (2005). A chromium terephthalate-based solid with unusually large pore volumes and surface area. *Science* 309 (5743): 2040–2042.

177 Maksimchuk, N.V., Timofeeva, M.N., Melgunov, M.S. et al. (2008). Heterogeneous selective oxidation catalysts based on coordination polymer MIL-101 and transition metal-substituted polyoxometalates. *J. Catal.* 257 (2): 315–323.

178 Sabo, M., Henschel, A., Fröde, H. et al. (2007). Solution infiltration of palladium into MOF-5: synthesis, physisorption and catalytic properties. *J. Mater. Chem.* 17: 3827–3832.

179 Esken, D., Zhang, X., Lebedev, O.I. et al. (2009). Pd@MOF-5: limitations of gas-phase infiltration and solution impregnation of $[Zn_4O(bdc)_3]$ (MOF-5) with metal-organic palladium precursors for loading with Pd nanoparticles. *J. Mater. Chem.* 19: 1314–1319.

180 Kuang, Q., Chen, L., Zhang, X. et al. (2020). The function of metal-organic frameworks in the application of MOF-based composites. *Nanoscale Adv.* 2: 2628–2647.

181 Xu, Q., Aijaz, A., Karkamkar, A. et al. (2012). Immobilizing highly catalytically active Pt nanoparticles inside the pores of metal-organic framework: a double solvents approach. *J. Am. Chem. Soc.* 134 (34): 13926–13929.

182 Hermes, S., Schröder, F., Amirjalayer, S. et al. (2006). Loading of porous metal-organic open frameworks with organometallic CVD precursors: inclusion compounds of the type $[L_nM]_a$@MOF-5. *J. Mater. Chem.* 16: 2464–2472.

183 Okumura, M., Tanaka, K., Ueda, A., and Haruta, M. (1997). The reactivities of dimethylgold (III) β-diketone on the surface of TiO_2: a novel preparation method for Au catalysts. *Solid State Ionics* 95 (1–2): 143–149.

184 Okumura, M., Tsubota, S., Iwamoto, M., and Haruta, M. (1998). Chemical vapor deposition of gold nanoparticles on MCM-41 and their catalytic activities for the low-temperature oxidation of CO and of H_2. *Chem. Lett.* 27 (4): 315–316.

185 Hermes, S., Schröter, M.-K., Schmid, R. et al. (2005). Metal@MOF: loading of highly porous coordination polymers host lattices by metal organic chemical vapor deposition. *Angew. Chem. Int. Ed.* 44 (38): 6237–6241.

186 Hermannsdörfer, J., Friedrich, M., Miyajima, N. et al. (2012). Ni/Pd@MIL-101: synergistic catalysis with cavity-conform Ni/Pd nanoparticles. *Angew. Chem. Int. Ed.* 51 (46): 11473–11477.

187 Schröder, F., Esken, D., Cokoja, M. et al. (2008). Ruthenium nanoparticles inside porous $[Z_{n4}O(bdc)_3]$ by hydrogenolysis of adsorbed [Ru(cod)(cot)]: a solid-state reference system for surfactant-stabilized ruthenium colloids. *J. Am. Chem. Soc.* 130 (19): 6119–6130.

188 Li, B., Zhang, Y., Ma, D. et al. (2014). Metal-cation-directed *de Novo* assembly of a functionalized guest molecule in the nanospace of a metal-organic framework. *J. Am. Chem. Soc.* 136 (4): 1202–1205.

189 An, J., Geib, S.J., and Rosi, N.L. (2009). Cation-triggered drug release from porous zin-adeninate metal-organic framework. *J. Am. Chem. Soc.* 131 (24): 8376–8377.

190 Feng, D., Liu, T.F., Su, J. et al. (2015). Stable metal-organic frameworks containing single-molecule traps for enzyme encapsulation. *Nat. Commun.* 6: 5979–5986.

191 Chen, L., Zhang, X., Cheng, X. et al. (2020). The function of metal-organic frameworks in the application of MOF-based composites. *Nanoscale Adv.* 2: 2628–2647.

192 Larsen, R.W., Wojtas, L., Perman, J. et al. (2011). Mimicking heme enzymes in the solid state: metal-organic materials with selectively encapsulated heme. *J. Am. Chem. Soc.* 133 (27): 13056–10359.

193 Sun, W.Y., Liu, Z.Q., Yan, T., and Guo, J.H. (2021). Metalloporphyrin encapsulation for enhanced conversion of CO_2 to C_2H_4. *ACS Appl. Mater. Interfaces* 13 (22): 25937–25945.

194 Lu, G., Li, S., Guo, Z. et al. (2012). Imparting functionality to a metal-organic framework material by controlled nanoparticles encapsulation. *Nat. Chem.* 4: 310–316.

195 Opelt, S., Turk, S., Dietzsch, E. et al. (2008). Preparation of palladium supported on MOF-5 and its use as hydrogenation catalyst. *Catal. Commun.* 9 (6): 1286–1290.

196 Huang, L., Wang, H., Chen, J. et al. (2003). Synthesis, morphology control and properties of porous metal-organic coordination polymers. *Microporous Mesoporous Mater.* 58 (2): 105–114.

2

Engineering the Porosity and Active Sites in Metal–Organic Framework

Ashish K. Kar, Ganesh S. More, and Rajendra Srivastava

Indian Institute of Technology Ropar, Catalysis Research Laboratory, Department of Chemistry, Rupnagar 140001, Punjab, India

2.1 Introduction

Metal–organic framework (MOF), also known as porous coordination polymers, are emerging as a new class of multifunctional nanoporous materials containing inorganic building units and organic linkers [1]. The MOF chemistry received significant research interest after Omar M. Yaghi discovered a robust and highly porous Zn-containing MOF-5 [2]. Because of their structural variability in the potential geometry, size, novel three-dimensional porous structures, and adjustable pore size, MOF can be designated as one of the classified modern-day efficient materials in several catalytic applications. A wide range of MOFs has been developed in the literature. A very high degree of uniformly distributed metal contents is present in MOF; hence, they are a potential candidate for catalysis.

MOF can be designated based on their properties and framework, such as IRMOF corresponds to isoreticular MOFs [3]. The isoreticular concept in the MOF design opens up several opportunities in the development of a wide range of MOFs with varying functionalities. During the rationale coordination-driven self-assembly of organic and inorganic building units, a cage-like repeating network structure is obtained, which has the periodic arrangements of the rigid organic linker and metal nodes classified as catalytically active sites [3–5]. These cage network structures impart unambiguous porosity with a large surface area, special tunability features, and designable topology. Moreover, due to the high surface area and porosity, the MOF structure often serves as the structural podium for the various chemical guest species to make the MOF material more promising. It is also noted that the functional surface of the MOF structure provides numerous opportunities to further modify its structure through integrating some structural functionalities, and the phenomenon is known as post-synthetic modifications (PSMs). The PSM strategy has been hugely adopted to extend the scope and versatility of the MOF structure to generate novel scaffolds exhibiting pronounced characteristics compared to the parent frameworks. The various techniques, such as framework modulation with coordination bonds [6–9], modification with covalent approach [10–13],

Catalysis in Confined Frameworks: Synthesis, Characterization, and Applications, First Edition.
Edited by Hermenegildo Garcia and Amarajothi Dhakshinamoorthy.

noncovalent approach (incorporation of metals or metal ions) [14–16], and systematic tandem modification [17–19], associated in PSM techniques to gain the chemical modulation in the MOF structure. The auxiliary ligand in the MOF can be removed under some specific experimental conditions to achieve the coordinatively unsaturated metal sites with a more pronounced electronic environment and reactivity than the parent MOF. Such unsaturated metallic sites are available for desired secondary ligating moieties for further functionalization in the MOF. The covalent modification deals with the making or breaking of covalent bonds, which facilitates the impactful structural ascertain and is believed to be the most powerful tool for the structural modulation in the MOF. In the covalent modification approach, the appropriate undisturbed ligating sites in the MOF are available for the post-synthetic operations via several covalent transformations, such as imine condensation, *N*-alkylation/arylation, halogenation, condensation, coupling reaction, hydrazone formation, protection/deprotection, and click reactions. The MOF modifications through noncovalent approaches include guest removal and exchange, ion exchange in metal nodes, and incorporation of metallic units such as metal nanoparticles (MNPs) in the MOF (either inside the pores or on the external surface of the MOF). The systematic tandem modification of MOF structure is a highly efficient and controllable approach for the performance of more than one chemical transformation. In this process, the partial modification was first carried out using a suitable reagent at the reactive sites of the MOF, and then the subsequent tandem modification was further achieved by another reagent over the remaining reactive sites in the MOF. It is also possible that the first MOF gets complete modification near the saturated active sites using the desired reagent, which has latent functional sites. These latent functional sites of the incorporated reagent in the MOF allow further modification with another reagent that makes it ideal for various cascade catalytic reactions [20–23]. The objective of the PSM strategies mainly focuses on the engineering and integration of active sites in the MOF to make them more committed to their catalytic application. The active site engineering of MOF with functional groups imparts huge opportunities for a plethora of applications, including catalysis in modern-day science. For instance, the imperative condition is the retention of the structural integrity of the PSM process in the MOF structure. Moreover, the rational choice of PSM precursors must be meticulously chosen in order to derivatization of the MOF into multifunctional units, which enables an actual way to methodically fine-tune and optimize MOF structure. Unlike the pure MOF, the PSM MOF exhibited better chemical stability and sometimes provided a hydrophobic engineered surface, which has a pronounced effect in catalysis and other applications.

With these considerations, this chapter stresses the brief discussion about the active sites in MOF and the research progress on the systematic engineering of the active sites and porosity in MOF to enhance the catalytic applications in various organic transformations. The synthesis methodologies and the associated characterization techniques are discussed to understand the structure–activity relationship and the extent of the engineering process in the MOF.

2.2 Active Sites in MOF

The MOF distinguish themself from other porous nanomaterials because of the rational combination of building units of variable geometry and multivariate organic linker, which provides an opportunistic framework structure with highly dispersed active sites. The isolated homogeneous distribution of these active sites is the main structural feature of the MOF. As discussed, the MOF having the unambiguous porosity, numerous metal nodes, and ligating junctions, which are functional. These constructing moieties are believed to be the epitome of MOF's catalytic activity. The schematic illustration of the distribution of the active sites in the MOF structure is given in Scheme 2.1.

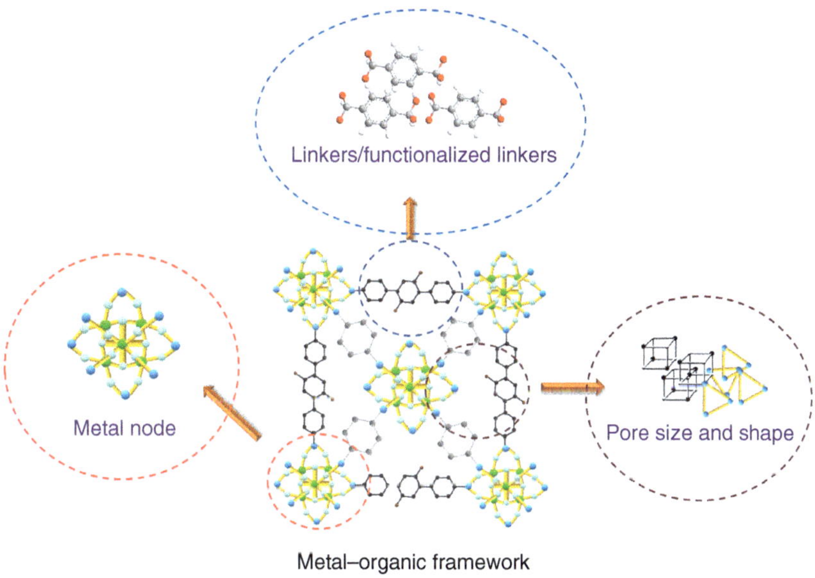

Scheme 2.1 The distribution of active sites in the MOF.

2.2.1 Active Sites Near Pores in MOF

The pores channel in the MOF triggers the diffusion and mass transfer process and, therefore, plays a bold and decisive role in controlling the reactivity in some catalytic reactions. For instance, the pore surface directly dictates MOF's activity, making them an ideal molecular catalyst for numerous catalytic applications [24, 25]. The multivariate porosity of the MOF is known to exhibit shape-selective catalysis, which is one of the viable and most desirable chemical transformations to be achieved. The careful choice of MOF precursors for constructing variable porous arrays in the MOF structure is one of the significant research topics in MOF.

2.2.2 Active Sites Near Metallic Nodes in MOF

The metallic nodes in the MOF coordinate with the organic linkers and, to some extent, solvent molecules. Of note, the MOF contains a high density of dispersedly metallic nodes, which stimulates the reactivity of MOF, thus responsible for the activity of the MOF during some catalytic transformations [26, 27]. However, in many MOFs, these metallic nodes are completely engaged in the framework construction and not completely available for the incoming reactants. On the contrary, unsaturated metallic centers can be created during the synthesis of some MOF, which imparts high reactivity toward the catalytic activity of some organic transformations.

2.2.3 Active Sites Near Ligand Center in MOF

The other active sites located in the MOF structure are due to the organic linker in the MOF. Both the auxiliary ligand and the primary ligand centers not only provide the structural tenacity to the MOF but, in some instances, also regulate the reactivity of certain chemical transformations [28, 29]. The variety of functional moieties, such as amino, amide, pyridyl, sulfoxy, and bipyridyl, are associated with the organic linker in the MOF and termed as active centers for various catalytic applications.

2.3 Synthesis and Characterization

The engineering of active sites in the MOF extensively depends upon the adopted synthesis route, which controls the possibility and extent of the active site engineering of a MOF. Since the MOF is sensitive to thermal and chemical stability, it is essential to follow the appropriate synthesis and modification methodologies for engineering the active sites in MOF. The synthesis methodologies are broadly classified into two methods known as (i) de novo synthesis method and (ii) PSM method. During the de novo or PSM method, it is necessary to consider the degree of crystallization and structural integrity of the MOF. The de novo synthesis method is mainly achieved during the MOF synthesis by using appropriate chemical moieties, modulators, auxiliary ligands, mixed ligands, or multivariate ligands. The addition of such additives regulates the rate of the crystallization process and, therefore, manipulates the electronic and spatial nature of the MOF structure. In addition, during the MOF synthesis, the alteration of metal to linker molar ratio modifies the electronic properties near the metal node in the MOF structure. The PSM process can be achieved after the synthesis of the parent MOF. The synthesized MOF is allowed for various activation and modification processes, which results in the induced electronic and some sort of structural modification in the MOF. Several modifications include covalent and coordinate modification, mechanical and thermal activation followed by modification, surface integration, and encapsulation, leading to the formation of optimized and electronically and structurally modulated MOF, provided the employing moieties must not disturb the framework integrity or stability of the

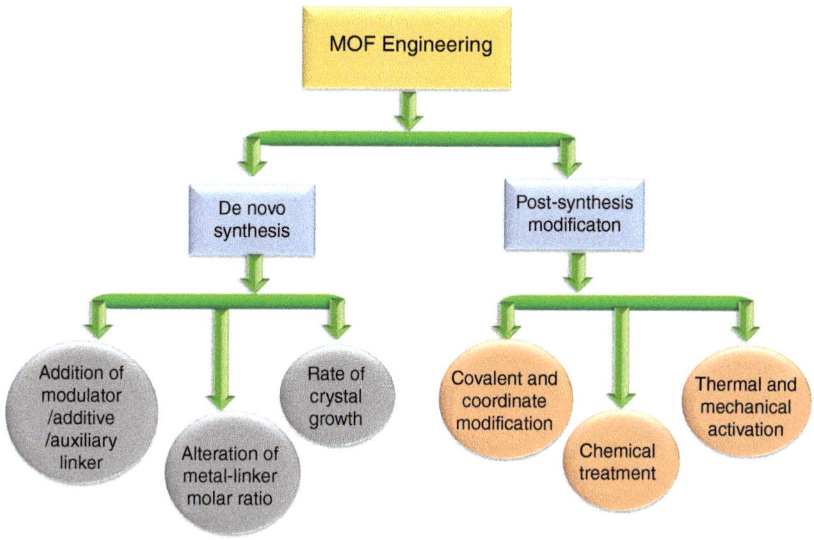

Scheme 2.2 Strategies involved in the active site engineering of MOF.

MOF structure. A schematic presentation of the possible engineering of active sites in the MOF structure is provided in Scheme 2.2.

It is challenging to understand and monitor the effect of active site engineering of MOF due to the incomprehensible and polymeric functional framework of MOF. The main challenge associated with the active site engineering of MOF is its physical characterization methodologies. However, the collective contribution of various physical characterizations gives the obligatory idea about the extent and impact of active site engineering in MOF. The primary characterization processes employed to understand the active site engineering of MOF structures are X-ray diffraction (XRD), Fourier transform infrared (FT-IR), X-ray photoelectron spectroscopy (XPS), thermogravimetric analysis (TGA), Brunauer-Emmett-Teller (BET), HR-TEM, XANES, EXFAS, etc. MOF is a crystalline material; thus, by measuring the extent of phase purity and peak shifting, the extent of structural modifications can be qualitatively estimated. The powder XRD analysis is a potential analytical tool to identify metal node engineering by missing linker defects in MOF. The N_2 adsorption–desorption isotherm and Barrett-Joyner-Halenda (BJH) methods are used to characterize the effect of surface area, pore size, and pore tunability of an active site-engineered MOF. The organic linkers and PSM strategies that tune the surface area and pore size of MOF (micropores and mesopores) are easily analyzed by the adsorption–desorption technique. The PSM at ligand centers, and the integration of new organic functionalities near the linker sites, can be easily understood from the FT-IR analysis. Scanning electron microscopy (SEM) is an effective method for determining morphological information of surface embedded or modified MOF, while high-resolution transmission electron microscopy (HR-TEM) offers excellent insight into the information of nanostructures of the modified MOF and also provides additional information about the incorporated crystalline nanoscopic species

Table 2.1 Various characterization methods employed to characterize the modified MOF structure.

Sr. no	MOF engineering	Characterization	References
1	MIL-101Cr-NH_2	XRD, N_2-sorption analysis	[30]
2	Zr-MOF (PCN-700)	Single-crystal X-ray diffraction (SC-XRD), N_2-sorption analysis	[31]
3	Zr-MOF (UIO-66)	Single-crystal X-ray diffraction (SC-XRD), H^1-NMR, TGA-MS	[32]
4	MOF-808-SO_4	Powder XRD, N_2 adsorption, and TGA-DSC	[33]
5	Cu-BTC MOF	FT-IR, N_2-adsorption, and PXRD	[34]
6	Zn-MOF (DABCO)	Powder XRD, N_2-adsorption, and FT-IR	[35]
7	Cr(III)-MOF (Cr-SXU-1)	Powder XRD, N_2-adsorption	[36]
8	Co/Ni-MOF-74	Powder XRD and EXAFS	[37]
9	Zn-BP-BTC MOF	FT-IR, powder XRD	[38]
10	I_2@Cd-MOF	FT-IR, powder XRD, and XPS	[39]
11	Zr_6 (UIO-66)	Extensive I.R. study	[40]

by reflecting its corresponding lattice fringes. Using X-ray photoelectron spectroscopy, the change in the chemical and electronic environment of a modified MOF structure and its composition and oxidation state can be easily identified. The shift in the binding energies of metal and organic linkers provides information about defects and the binding of metal and organic linkers in engineered MOF. X-ray absorption near-edge structure (XANES) and extended X-ray absorption fine structure (EXAFS) spectroscopy give information about the change in coordination structure and chemical state after the MOF modification. The examples of various characterization methods employed to elucidate the structural and chemical modulation upon the MOF modification through active site engineering are provided in Table 2.1.

2.4 Engineering of Active Sites in MOF Structure for Catalytic Transformations

The functional nature of MOF provides ample opportunities to integrate the catalytically enhanced moieties with the MOF to make it a robust and highly demandable material with broad catalytic scope. The active sites in MOF are highly tailorable and modulable to improve their overall structural and electronic properties to become a stable catalytic system. A wide scope of numerous methodologies has been reported in the literature to modify and engineer the active sites in the MOF. As discussed in the above section, the active sites associated with the MOF are pores, metal nodes, and ligand centers which actively participate in the various catalytic application. Therefore, in this section, the detailed methodologies involved in engineering active sites in MOF are summarized.

2.4.1 Pore Tunability

As discussed above, the basic concept of MOF structure is assembling metal and linkers in a repetitive phenomenon to create a large volume of porous network structure. It is known that, during the catalytic process, the respective reactant molecule tends to interact with the internal and external surfaces of the MOF. The internal surface of the MOF contributes to the total pore wall which occurs inside the pore. The reactivity for the pore-mediated catalytic processes can be enhanced by fine-tuning the porous channel through the pore engineering concept, which provides optimum contact between the reactant molecules and the internal pore surface. The porous channel in the MOF significantly controls the mass-diffusion limitation during catalysis. Hence, it is necessary to architect the appropriate porous channel through pore engineering to avoid the mass-diffusion limitation. The pore channel in the MOF also does the immobilization and encapsulation of the nanoscopic catalytically active species such as MNPs and ultra-small size metal oxides (MOs). Moreover, the proper tunability of porous architecture with functionalized ligand body provides a suitable hydrophilic and hydrophobic surface at the internal pore surface, which is one of the desirable requirements for the easy binding and interaction of the reactant species in heterogeneous catalysis [41]. This section discusses the methodologies associated with the active engineering site of MOF and the resultant catalytic applications based on pore engineering.

The most significant properties of the porous network in the MOF structure are the ability to encapsulate the nanoscopic guest species, such as MNPs or MOs, inside the porous channel of the MOF through noncovalent interaction. The functional nature of the MOF pore wall and the geometrical dimensions provide optimum chemical surrounding that helps in the prevention of the agglomeration of MNPs or MOs, which is a common problem in stabilizing the MNPs or MOs when supported on a support material. The confinement effect of the MOF pore wall plays a crucial role in the stabilization of several MNPs, such as Pd, Pt, Au, and Cu, inside the pore of MOF to refrain them from leaching activity. The resultant MNPs@MOF systems have wide scope applications in several heterogeneous catalyses [42, 43]. Pascanu et al. have reported different loading of Pd-embedded MIL-101Cr-NH_2 Suzuki–Miyaura cross-coupling reaction between aryl halide and pinacol phenylboronate (Scheme 2.3) [30].

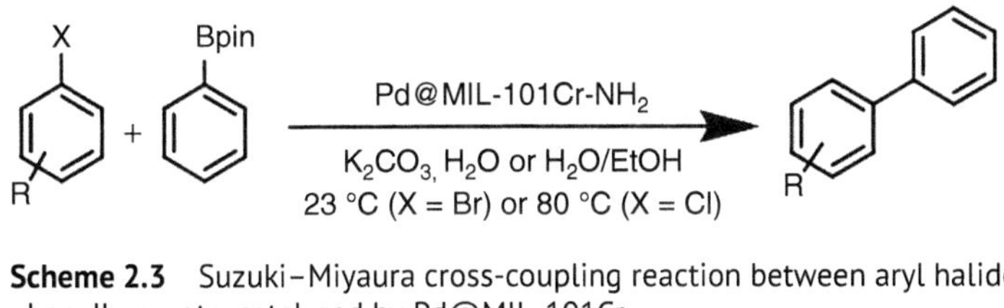

Scheme 2.3 Suzuki–Miyaura cross-coupling reaction between aryl halide and pinacol phenylboronate, catalyzed by Pd@MIL-101Cr.

The retention of the Pd NPs inside the pore was confirmed by the N_2-sorption analysis and electron tomography. Both the surface area and the pore volume significantly decreased when the Pd NPs encapsulated inside the pore of the

MIL-101Cr-NH_2. The Pd NPs homogeneously distributed inside the pores of MIL-101Cr-NH_2 were visualized by the HR-TEM analysis and 3D tomographic images. However, on increasing the Pd NPs loading concentration, the Pd NPs started to accumulate on the surface of the MOF and became larger in size. Pore tunability via PSM strategy is a valuable technique to achieve the MOF material for size-selective catalysis. The PCN-700 MOF was installed with BPYDC(Cu) (BPYDC = 2,2′-bipyridine-5,5′-dicarboxylate) and TPDC-R_2 (TPDC = terphenyl-4,4″-dicarboxylate, R = Me, Ph, or Hex) ligands by soaking PCN-700 in DMF solution of these two ligands using linker installation method via PSM strategy for the size-selective alcohol oxidation reaction [31]. The successful incorporation of these ligands was confirmed by single-crystal X-ray diffraction (SCXRD) analysis and N_2-sorption analysis. The N_2-uptake capacity and the cell volume were enhanced by 33.4% and 17.4%, respectively, suggesting the successful incorporation of the ligand moieties in the PCN-700 framework via the ligand exchange method. Figure 2.1 demonstrates the active catalytic site is Cu^+, chelated by a bipyridine group of BPYDC, and its accessibility was controlled by TPDC-R_2 moiety; hence, TPDC-R_2 regulates the selectivity.

The catalytic activity in the oxidation of benzyl alcohol, 1-naphthalenemethanol, and 9-anthracenemethanol was achieved over the homogeneous (bpy) Cu catalyst, which has no experience in size-selective catalysis. However, PCN-700-BPYDC, PCN-700-BPYDC(Cu)-TPDC-Me, PCN-700-BPYDC(Cu)-TPDC-Ph, and PCN-700-BPYDC(Cu)-TPDC-Hex exhibited the catalytic activity based on their pore size (Figure 2.2). Since the size of benzyl alcohol (~4.3 Å) is much smaller than the pore size of PCN-700-BPYDC(Cu) (~14 Å), it exhibited the highest activity, whereas the

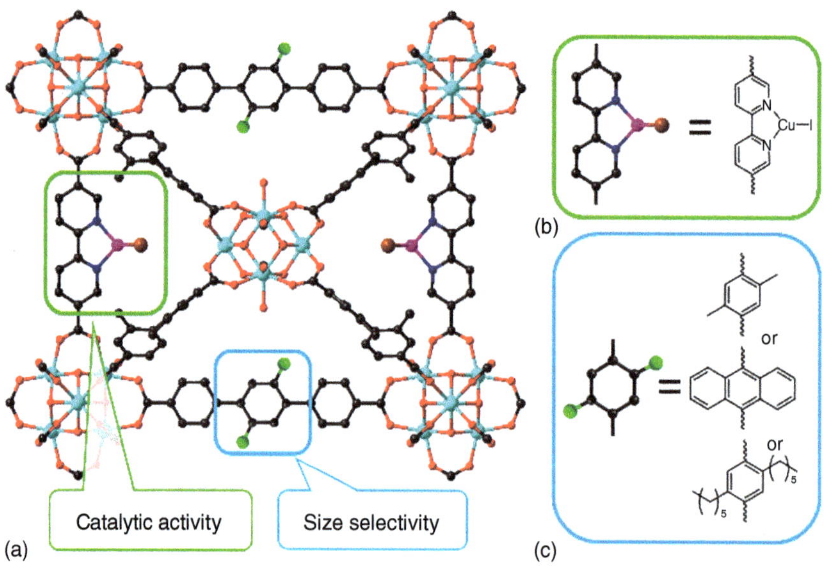

Figure 2.1 (a) Size-selective catalytic system in modified PCN-700. (b, c) Structure of catalytic center and size-selective moiety. Source: Reproduced from Yuan et al. [31] with permission from American Chemical Society, Copyright 2016.

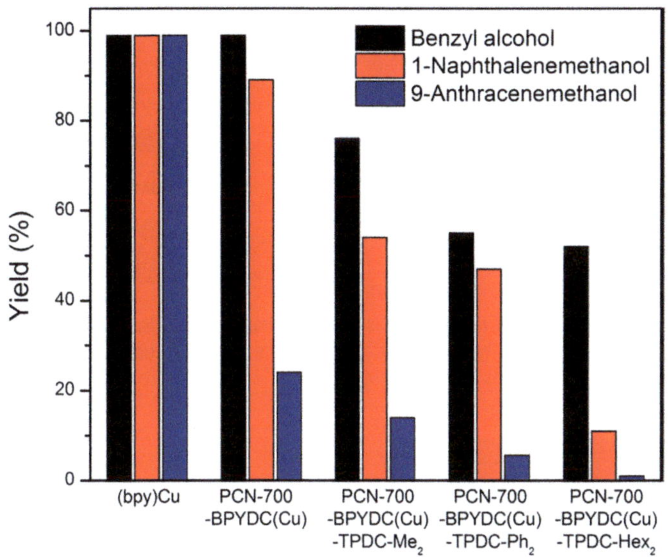

Figure 2.2 Size-selective alcohol aerobic oxidation using PCN-700 modified with different linkers. Source: Reproduced from Yuan et al. [31] with permission from American Chemical Society, Copyright 2016.

comparatively larger molecular size of 9-anthracenemethanol (~9.2 Å) exhibited lower catalytic activity due to the diffusion and transport limitation (Figure 2.2). On further increasing the size of the substituent in TPDC-R_2, the shielding effect of Cu^+ was observed, and its accessibility toward the catalytic activity was diminished (Figure 2.2).

The tuning of the mixed linker approaches can also create a multivariate porous channel which significantly influences the catalytic activity. Liu et al. reported the modulation of the pore channel in the range of 1.4 to 2.25 nm in MOF-808-SO_4 through a defect engineering approach by varying the amount of isophthalic acid ligand [33]. The resultant pore size modulation significantly controlled the activity of the addition reaction of isobutylene with ethylene glycol. It was found that the large pore-containing catalyst having a high concentration of isophthalic acid exhibited the highest conversion of ethylene glycol, while the optimum pore size (1.99 nm) responsible for the high selectivity toward the target product. The multimodal porous channel containing MOF can be obtained by adopting an appropriate templating method. A dicationic geminal surfactant modified Cu-BTC MOF designated as Cu-BTC-MOF(C18-6-18) was synthesized, having the hierarchically interconnected micropore–mesopore channel for the oxidation of various aromatic alcohol (Scheme 2.4) [34].

For comparison study, one conventional mono-cationic surfactant cetyltrimethylammonium bromide (CTMABr) modified Cu-BTC (designated as Cu-BTC CTMABr) and a TMB as an auxiliary pore-forming agent with dicationic gemini surfactant modified Cu-BTC MOF (Cu-BTC-MOF(C18-6-18-TMB) were synthesized. The TEM analysis, N_2-sorption analysis, and FE-SEM analysis suggested

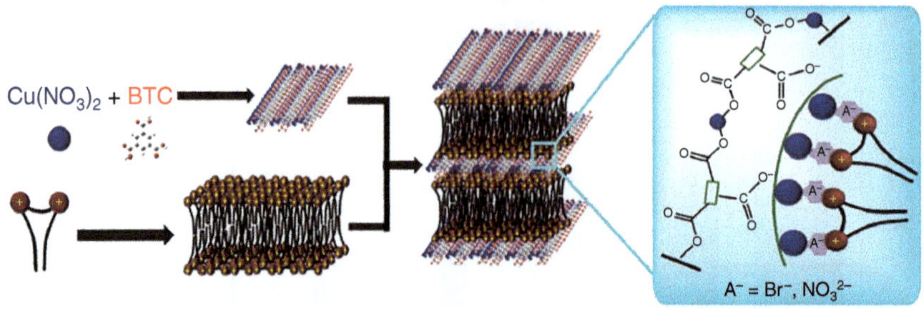

Scheme 2.4 Schematic presentation of the synthesis of micro-mesoporous interconnected Cu-BTC-MOF(C18-6-18) using dicationic gemini surfactant. Source: Reproduced from Rani and Srivastava [34] with permission from RSC, Copyright 2018.

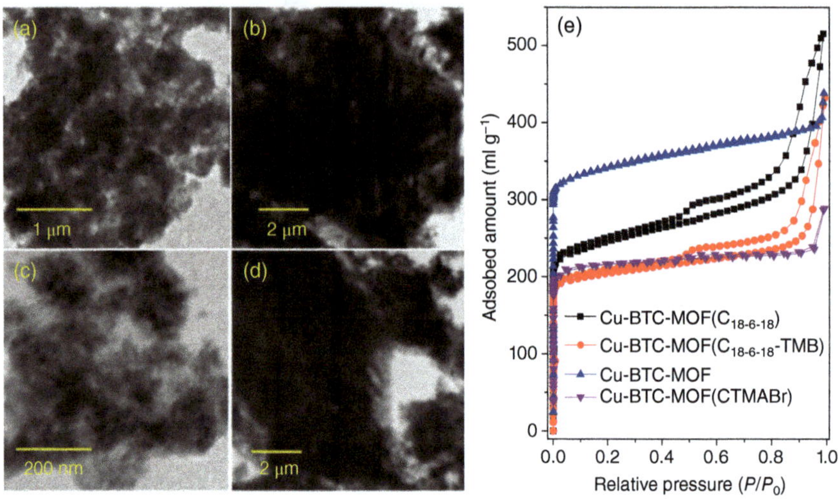

Figure 2.3 (a–c) TEM images of Cu-BTC-MOF(C18-6-18), (d) Cu-BTC-MOF(C18-6-18-TMB), and (e) N_2-adsorption isotherms of different materials prepared in this study. Source: Reproduced with permission from Rani and Srivastava [34], ©2018/Royal Society of Chemistry.

the Cu-BTC-MOF(C18-6-18) exhibited large porous architecture compared to the pristine Cu-BTC MOF and other modified Cu-BTC MOF (Figures 2.3 and 2.4).

When the cinnamyl alcohol oxidation reaction was carried out, the effect of the porosity in the catalytic activity was understandable. The pristine Cu-BTC MOF having a smaller pore size exhibited lower activity; however, the hierarchical mesopore-induced Cu-BTC MOF was able to catalyze this reaction because of the ease of accessibility of active sites and diffusion of reactant/product molecules due to the large porous channel. Similarly, Gu et al. have reported a DABCO-modified 3D Zn-MOF for the size-selective catalysis of Henry reaction between 4-n itrobenzaldehyde and different dimensions of nitroalkanes (Scheme 2.5) [35]. The reactivity purely depended on the pore size of the modified Zn-MOF as the pore wall was modified and functionalized with DABCO modification that provided the active basic sites inside microporous 1D channels to catalyze the reaction.

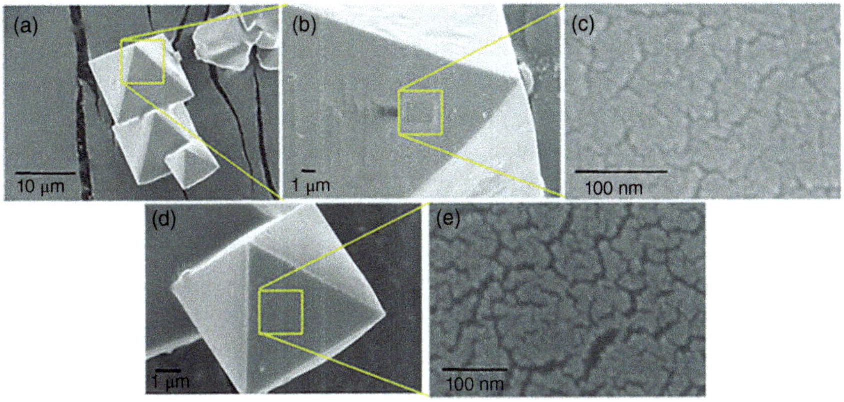

Figure 2.4 FE-SEM images of (a–c) Cu-BTC-MOF(C18-6-18) and (d, e) Cu-BTC-MOF (C18-6-18-TMB). Source: Reproduced with permission from Rani and Srivastava [34], ©2018/Royal Society of Chemistry.

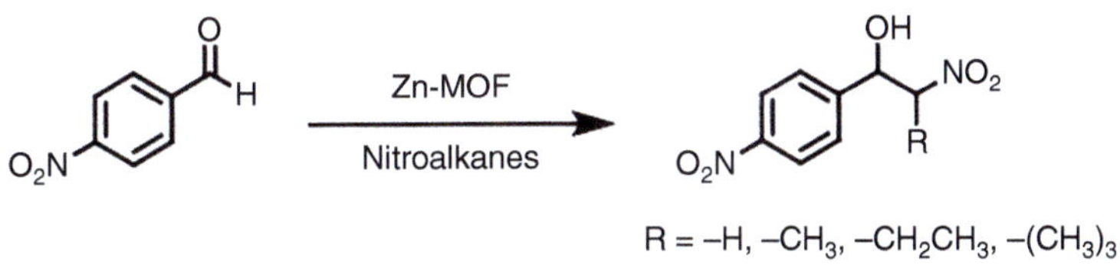

Scheme 2.5 Henry reaction between 4-nitrobenzaldehyde and nitroalkanes over DABCO-modified Zn-MOF.

2.4.2 Metal Nodes

The MOF contains a high density of metallic concentrations, which are spatially present across the MOF matrix and impart a viable contribution to the catalytic process. The few metal nodes in the MOF are partially coordinated with solvent molecules or labile ligands and are loosely connected with the framework integrity. Moreover, these metal sites are comparatively labile in nature and available for the metal exchange phenomena for the multivariate metallic nodes in the MOF. The metal exchange process or the trans-metalation process involves the simultaneous coordinate bond breaking of the framework metal and formation with the desirable incoming metal ions. However, during metal exchange reactions, special attention should be paid to choose the appropriate incoming metal ion to maintain the framework compatibility. A general hindrance to the trans-metalation process is the MOF stability; hence, it is important to consider several factors like the crystal field stabilization energy (CFSE) and the ionic radius of both the metal ions. It is always advisable to choose the metal ion with comparable ionic radii. Similarly, for Cu-based MOF, during the trans-metalation process, the influence of the John–Teller effect has to be considered [44]. Wang et al. reported a post-synthetic trans-metalation process for the solvent-assisted synthesis of the Cr(III)-MOF from an analog Fe(III)-MOF as it was difficult in the direct synthesis of a Cr(III)-MOF [36]. Acetone, as an appropriate solvent, provides the optimum electronic environment and plays a decisive role in the successful trans-metalation process. The crystallinity and porosity were retained after the trans-metalation process, which was confirmed using the PXRD and N_2-sorption analysis [36]. The extent of the trans-metalation process depends upon the liability of exchangeable framework metal ion, choosing of appropriate solvent, coordination liability of the incoming metal ion, and the structural as well as the chemical stability of the MOF. However, in common practice, partial trans-metalation occurs, and it is also possible for the complete metal ion exchange in a MOF. The metal exchange occurs in the solution state, where the choice of solvent selection has the dominant factor that regulates the trans-metalation process. Dutta et al. reported two Zn-porphyrinic MOF with chemical structure of $[Zn_3(C_{40}H_{24}N_8)(C_8H_4O_4)_2(DMF)](DMF)_5(H_2O)_{12}$ designated as (2) and $[Zn_3(C_{40}H_{24}N_8)(C_{12}H_6O_4)_2(DMA)_2](H_2O)_7$ designated as (3) with porphyrin 5,10,15,20-tetrakis(4-pyridyl)porphyrin. Two trans-metalated Cu-porphyrinic MOFs were employed with 100% Zn exchange, followed by a room temperature metalation approach. The metalation reaction was conducted at room temperature in the DMF solution of the Cu salt. The solution was decanted every alternate day for 12 hours. The successful incorporation of the Cu sites was investigated using energy-dispersive X-ray spectroscopy (EDAX) and atomic absorption spectroscopy (AAS). The resultant trans-metalated Cu-porphyrinic framework was enhanced the nucleophilic reaction to achieve the chemo- and regioselective enamination of β-ketoesters, production of propargylamine, and click reactions between regioselective cycloadditions of alkyne and azide (Scheme 2.6)a–c [45].

Similarly, a different amount of Co-substituted Ni-MOF-74 was prepared via a post-synthetic metal exchange approach carried out in DMF solvent was reported by Sun et al [37]. The Co-modified Ni-MOF-74 was employed for the oxidation of cyclohexene and exhibited boosted catalytic activity compared to the pristine Ni-MOF-74 and the Co-MOF-74 (Scheme 2.7). The extent of the metal exchange was depend upon the temperature and incubation time. The successful Co incorporation was estimated using inductively coupled plasmonic mass spectroscopy (ICPMS), powder X-ray diffraction (PXRD), and EXAFS analysis.

An interesting achievement of the trans-metalation process is the development of MOF on MOF via seed-mediated core-shell growth. After the trans-metalation, the exchanged metal around the surface of the MOF can be responsible for the growth of another MOF to develop MOF@MOF, which resembles the noble nano reactor-based system and could be a potential catalytic system. Liu et al. reported a seed-mediated MOF on the MOF system of the core-shell ZIF-8@ZIF-67 (Figure 2.5) [46].

Another way of metal node engineering is the doping of metal ions to increase the total metallic concentration in the MOF structure. The resultant metal-rich MOF can be potentially applied to enhance the catalytic activity in various catalytic applications [47]. Recently, Kar et al. prepared an Nb-modified UIO-66 MOF (Nb@UIO-66) to improve the glucose isomerization reaction. The successful Nb incorporation was proved by XPS and HR-TEM analysis (Figure 2.6). The TEM images demonstrated the surface accumulation of Nb species resulting in its epitaxial growth stabilized by the Zr-O cluster and forming Zr-O-Nb in the Nb@UIO-66 MOF, while the pristine UIO-66 possessed a smooth surface (Figure 2.6a–d). The XPS analysis also confirmed

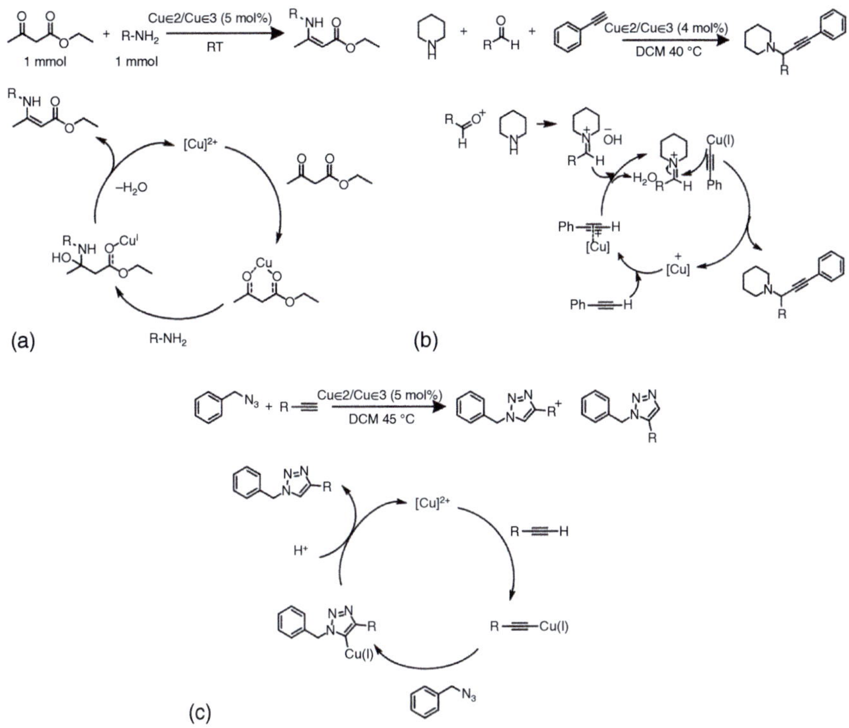

Scheme 2.6 Cu-porphyrinic framework mediated chemo- and regioselective (a) enamination of β-ketoesters, (b) production of propargylamine, and (c) click reactions between regioselective cycloadditions of alkyne and azide.

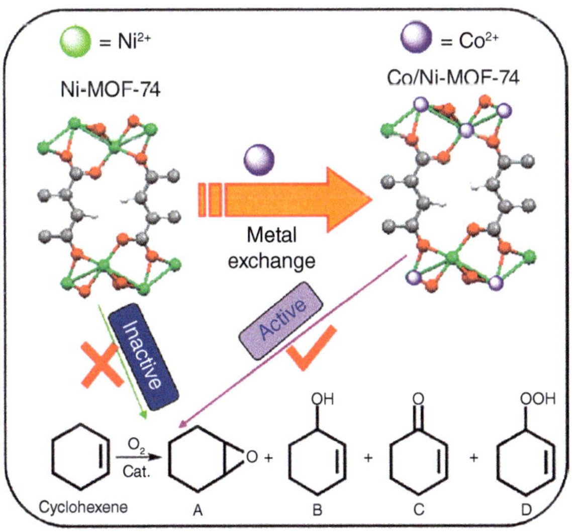

Scheme 2.7 Schematic presentation of Co substitution by post-synthetic metal exchange process and catalytic activity in the oxidation of cyclohexene. Source: Reproduced from Sun et al. [37] with permission from ACS inorganic chemistry, Copyright 2015.

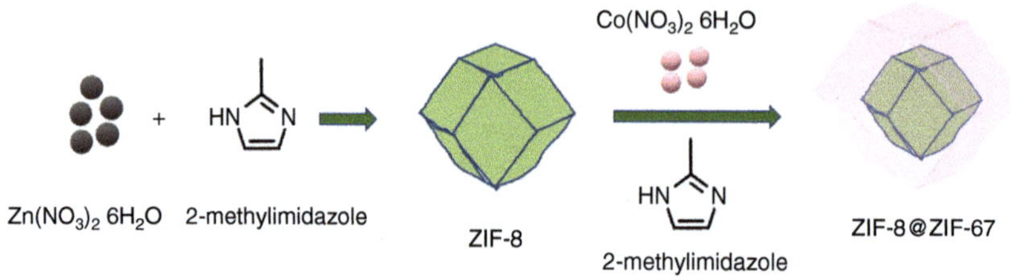

Figure 2.5 Schematic representation of ZIF-8@ZIF-67 synthesis.

Nb in Nb@UIO-66 and the modifications of the electronic surrounding of Zr and O after the Nb modification (Figure 2.6). Moreover, the Nb incorporation was proven to exhibit modulated and pronounced effects in the surface acidity, which was the cause for controlling the isomerization activity and selectivity of glucose to fructose transformation [48].

The external metal nodes (metal-oxo) in the MOF resembled the MOs kind support system for the unambiguous growth of the metal complex and other catalytic stimulating moieties. A Rh(I) complex was grafted on the nodes of the UIO-67 MOF for the ethylene hydrogenation and dimerization reaction in a flow reactor [49]. The Rh(I) complex was grafted using PSM strategy by employing dried n-pentane as a solvent at room temperature. It can be concluded that the metal exchange, metal doping, and metal complex-supported MOF synthesis phenomena for the metal node engineering are conducted via the PSM process.

The labile ligand and solvent molecules can be removed upon the activation (heating or evacuation) of MOF. It results in the coordinatively unsaturated sites in the framework, which resemble the potential Lewis acid and are highly recommended in several acid-mediated heterogeneous catalytic reactions [50]. These activated metallic nodes are catalytically active metal nodes in the MOF. For example, simple removal of coordinated H_2O molecule from the crystal structure of Cu-BTC MOF ($Cu_3(BTC)_2(H_2O)_3$) by adopting a careful MOF synthesis route to generate coordinatively unsaturated Cu sites which behave as weak Lewis acid and are employed in the cyanosilylation of benzaldehyde. The coordinatively unsaturated metal sites were prepared by de novo synthesis route by using an appropriate modulator, structure directing agent, or by varying the ligand-to-metal molar ratio. In these cases, the coordinatively unsaturated active metal sites are generated, which are highly prone to high catalytic activity in various organic transformations. The defect engineering concept can progress the development, characterization, and catalytic applications of such coordinatively unsaturated metal sites in a more holistic approach. The formation of coordinatively unsaturated metal sites can be effectively achieved by missing linker phenomena where the framework linker was missing and left out with coordinatively unsaturated metal centers. The appropriate extent of the missing linker can be obtained by a different method such as (i) using suitable modulators such as mineral acid or organic acids including HCl, HCOOH, CH_3COOH, and CF_3COOH[51, 52], (ii) structure directing group such as CTAB and amino acid can effectively produce linker deficiency in the MOF [53, 54], and (iii) by taking the low molar ratio of the linker to the high molar ratio of metal during the synthesis [55].

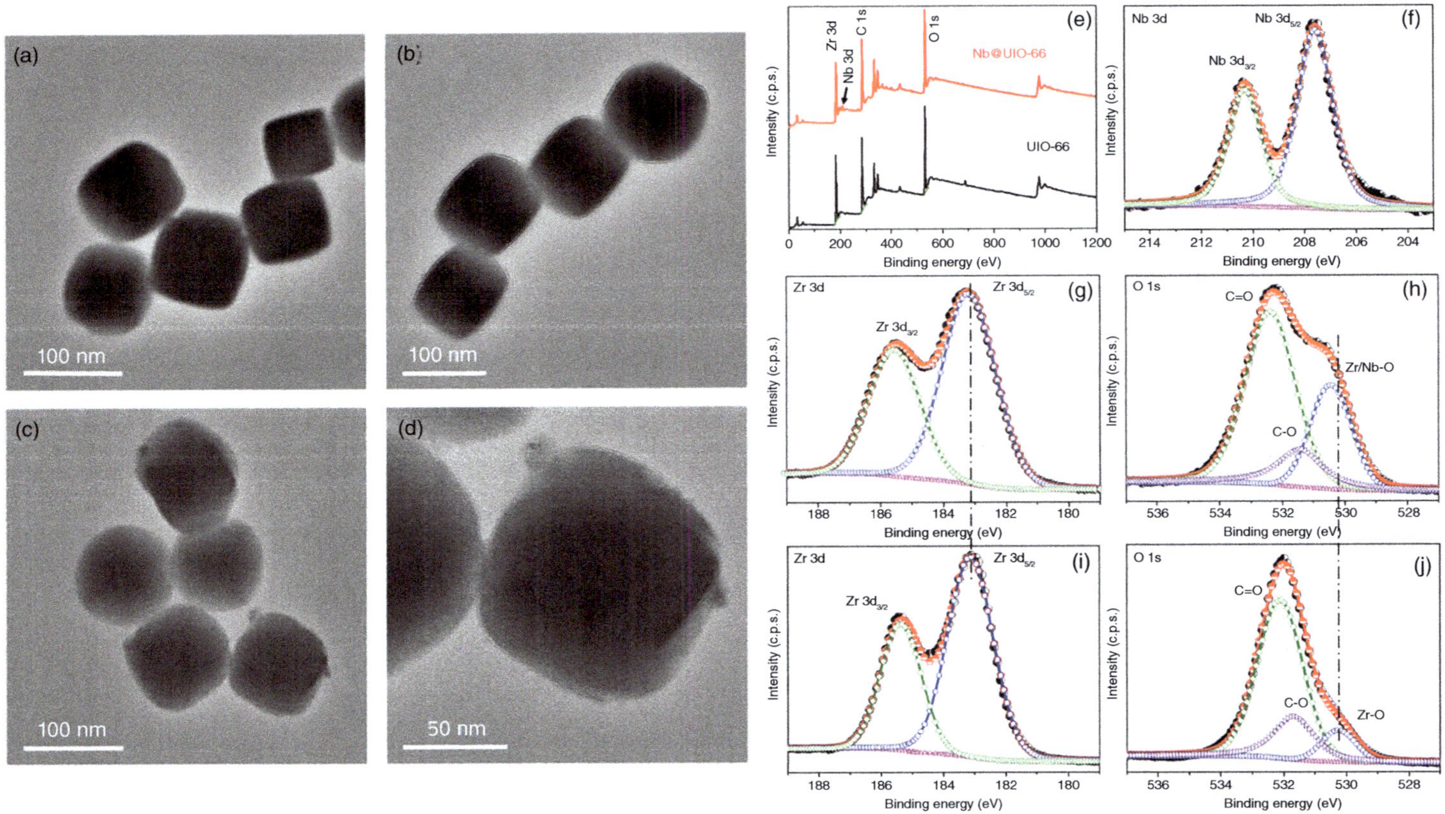

Figure 2.6 TEM images of pristine (a–b) UIO-66, (c–d), Nb@UIO-66 (e) XPS surface-survey spectra of UIO-66 and Nb@UIO-66, (f) high-resolution Nb 3d spectrum of Nb@UIO-66, (g, i) high-resolution Zr 3d spectrum of Nb@UIO-66 and UIO-66, (h, j) high-resolution O 1s spectrum of Nb@UIO-66 and UIO-66. Source: Reproduced with permission from Kar and Srivastava [48], © 2022/John Wiley & Sons.

All these methods fall under de Novo synthesis approach. It is crucial and challenging to track the extent of missing linker efficiency in the MOF. Moreover, till now, it has been very difficult to identify the number of missing linkers and the associated number of unsaturated metal sites. However, few characterization techniques have been extensively used to understand the missing linker phenomena and coordinatively unsaturated metal sites. The coordinatively unsaturated metal sites upon the missing linker possess a high degree of Lewis acidity, which imparts high activity in isomerization and biomass transformation reactions. Valekar et al. reported Zr-MOFs with different metal-linker connectivity for the defect-induced catalytic transfer hydrogenation (CTH) of furfural to furfuryl alcohol using isopropanol [56]. The defect-induced unsaturated Zr centers of MOF-808 by simple methanol activation potentially activate the furfural and IPA molecule to facilitate the CTH process, which was proven both experimentally and theoretically (Scheme 2.8) [56].

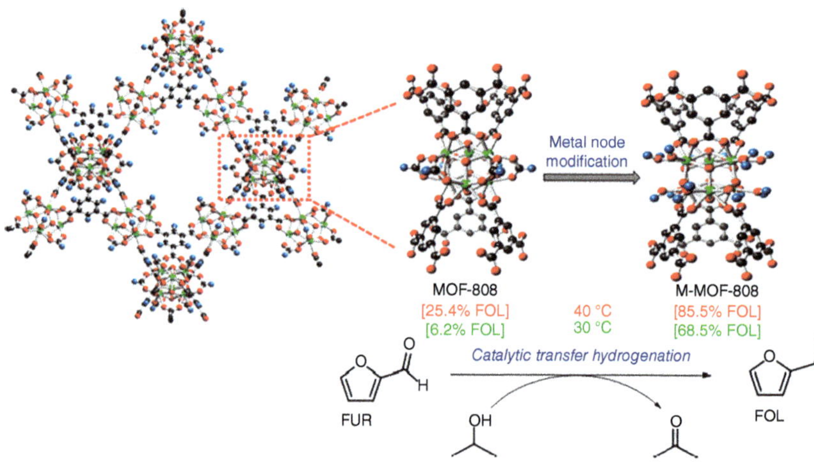

Scheme 2.8 Schematic presentation of the MOF-808 and defect-induced MOF-808 Zr metal node due to methanol activation (designated as M-MOF-808) in the CTH of furfural. Source: Reproduced from Valekar et al. [56] with permission from ACS, Copyright 2020.

Table 2.2 summarizes the engineering of the metal node by missing linker approach in various catalytic applications in which the engineered structure displayed more activity than pristine MOF. The post-synthetically modified coordinatively unsaturated metal sites containing MOF can also be obtained when MOF is heated at relatively lower temperatures (less than the framework decomposition temperature). The low-temperature calcined MOF through de-ligandation process, known as quasi-MOF, has modulated framework structure because, during the low-temperature calcination process, the volatile solvent molecules and the auxiliary ligand are removed from the framework, which left out coordinatively unsaturated metal centers. Moreover, during this heating process, the partial decomposition of the framework also occurs, which eventually generates the unsaturated sites. The major drawback is the loosing of framework identity and stability during the calcination process.

Table 2.2 Various metal node modifications and their catalytic applications.

Sr. no	Engineering of MOF	Active sites	Applications	References
1	UIO-66 (Zr)	Zr(IV)	Meerwein–Ponndorf–Verley (MPV) reduction	[57, 58]
2	UIO-66-NO_2 (Zr)	Zr(IV)	Citronellal cyclization	[59]
3	Co(bpb)-3DMF	Co(II)	Oxidation reaction	[60]
4	Zr/Hf-UiO-66	Zr/Hf	Ring-opening reaction	[61]
5	Co/Ni-MOF-74	Co/Ni	Oxidation of cyclohexene	[37]
6	UIO-66 (Zr)	Zr	Conversion of cyclohexanone to cyclohexanol	[62]
7	Zn/Cu-MOF	Zn/Cu	Regioselective cycloadditions of alkyne and azide	[45]
8	$[Eu_2(fum)_3(H_2O)_4](3H_2O)$	Eu(III)	Allylation	[63]
9	UiO-66(Zr)-V	Zr-V	Allyl alcohol epoxidation	[64]
10	$Cu_2(bpdc)_2(bpy)$	Cu(II)	C–H arylation	[65]
11	$Cu_3(btc)_2(H_2O)_3\ xH_2O$	Cu(II)	Cynosilylation	[66]
12	Ni-MOF-74	Ni(II)	Michael addition	[67]
13	Fe(btc)-MOF	Fe(III)	Aza-Michael reaction	[68]

2.4.3 Ligand Centers

Organic linkers are connected to metal centers to form MOF network porous channels, which are widely utilized as active sites in several catalytic transformations. The advantages of organic linkers resulted in variations in functionalities, surface area, and pore size without altering the architecture. Moreover, modified organic linkers with varied functional groups and PSM can design and tailor the interior of the MOF. Additionally, linkers bind to metallic sites by enabling metalation or chelation and providing additional active sites in MOF [69]. The functionalized organic linker in the MOF allows the possibility of anchoring the various catalytic enhancing moieties through covalent PSM [70]. The functional group of the organic linker interprets the functionality, porosity, and active site distribution in the MOF. The organic linkers-modified MOFs have been widely used for organic transformations and succeeded in synthesizing various platform chemicals.

The C–C bond formation gives a significant attention to organic synthesis; therefore, the MOF was modified with different ligands to facilitate organic transformations. M. Kathiresan et al. synthesized Zn-Bp-BTC MOF using two different organic linkers (Bp-4,4′-bipyridine; BTC-1,3,5-benzene tricarboxylic acid). The synthesized MOF was characterized by PXRD, while the functionality was confirmed by FT-IR [38]. The stretching frequency of –COOH (1,3,5-benzene tricarboxylic acid [BTC]) was slightly shifted to a lower frequency (1708 to 1639 cm^{-1}), suggesting that the –COO– moiety was coordinated with Zn metal while stretching frequency 3430 cm^{-1}

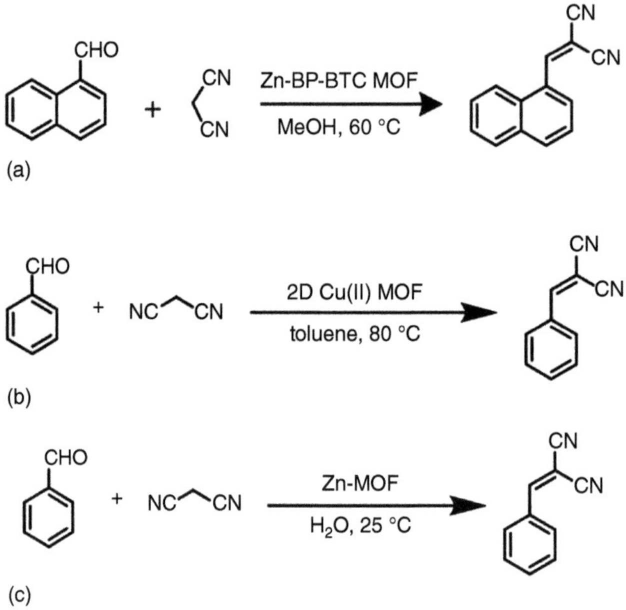

Scheme 2.9 Knoevenagel condensation reaction by using various ligand-modified metal–organic frameworks.

showed –OH stretching, attributed to free –COOH groups. In addition, the 4,4′-bipyridine ligand's C=N stretching frequency was observed at 1597 cm^{-1}; however, it changed to 1569 cm^{-1} after coordination with the metal ion, demonstrating the 4,4′-bipyridine ligand was coordinated to the metal center. The basicity of the 4,4′-bipyridine was enhanced the catalytic activity of the Knoevenagel condensation reaction and found ~99% yield (Scheme 2.9a). The role of basicity was confirmed by the analogous Zn-BTC MOF and $Zn(NO_3)_2$ $6H_2O$ and obtained a relatively lower yield. A novel 2D Cu(II) MOF has been developed by F. Guo [71]. The MOF was synthesized by self-assembly of Cu(II) with 4-(5-methyl-3-pyridine)-1,2,4-triazole and characterized by SCXRD and FT-IR spectroscopy, while the elemental composition was analyzed by elemental analysis. The nitrogen content in 4-(5-methyl-3-pyridine)-1,2,4-triazole linker acted as a Bronsted base and facilitated the Knoevenagel condensation reaction and obtained a 96% yield of 2-benzylidenemalononitrile (Scheme 2.9b). S. M. Cohen et al. described the synthesis of two separate catalytic domains in the MOF [72]. Organocatalytic amine and organometallic Ir(I) groups were added to Zn(II)-based IRMOF-9-Irdcppy-NH_2 by both pre-synthesis and post-synthetic functionalization techniques. It was demonstrated that the segregated amine and Ir(I) sites of IRMOF-9-Irdcppy-NH_2 were active sites for Knoevenagel condensation and an allylic *N*-alkylation. The basic amino group catalyzed the Knoevenagel condensation and found a 95% yield, while the Ir(I) was responsible for the subsequent alkylation. Furthermore, L. Zhao et al. reported a triazole-carboxylate-based mixed ligand (5-(4H-1,2,4-triazol-4-yl)isophthalic acid), synthesized a water-stable Zn-MOF, and characterized by FT-IR spectroscopy [73].

Hydrothermal synthesis produced a 2D MOF that was then developed into a 3D supramolecular network by strong hydrogen bonds. The Lewis acidic site increased the reactivity of benzaldehyde, whereas the Bronsted basic site aided in activating the methylene group of malononitrile (Scheme 2.9c).

The PSM of MOF enhances the various organic transformation. A straightforward one-step PSM process to produce a number of covalently modified MOF catalysts with a basic alkyl amino group that may function as a Bronsted base and is extensively utilized for the production of C–C bonds. Lah et al. reported that amine-tagged fragmented ligand was installed for modification of MOF-74 [70]. They have successfully shown that amine-tagged fragment installation was used to synthesize functional derivatives of DEMOF-I and DEMOF-II. To create iminopyridine functionality, amine-tagged DEMOFs were covalently modified with an aldehyde functional group. The functionality was then changed further using post-synthetic metalation. The synthesized MOF was thoroughly characterized by PXRD and N_2-sorption analysis to confirm the modification and utilized for Suzuki–Miyaura cross-coupling reaction. P. Rani et al. reported that amine-grafted Zr-BDC-MOF synthesized by PSM strategy and is employed for various organic reactions [74]. Amines, including *N*-butyl amine, diethyl amine, 4-dimethylaminopyridine, and ionic liquids (ILs) with –OH, and butyl methyl imidazole were grafted on Zr-BDC MOF and thoroughly characterized by P-XRD, FT-IR, XPS, CO_2-TPD, and TGA analysis. The FT-IR spectra showed the stretching frequencies of the free $-NH_2$ group at 3300 and 3250 cm^{-1}, while the bending and stretching vibrations of N–H and C–N appeared in 1637–1107 cm^{-1}, confirming that the primary amine was successfully grafted over the Zr-BDC MOF (Figure 2.7). Also, the stretching and bending vibrations at 1648 and 1220 cm^{-1} in DMAP-Zr-BDC-MOF were attributed to C=N and C–N stretching, respectively. Meanwhile, the CO_2-TPD was conducted to identify the basicity of Zr-BDC and amine-grafted Zr-BDC MOF. It was found that DMAP-Zr-BDC-MOF possessed the highest basicity than *N*-butylamine-Zr-BDC-MOF and diethylamine-Zr-BDC-MOF, respectively, while [BMIM][OH]-Zr-BDC-MOF exhibited even lower basicity than Zr-BDC-MOF.

The amine-grafted Zr-BDC MOF was employed for the Knoevenagel condensation reaction (Scheme 2.10). Among them, DMAP-Zr-BDC-MOF showed superior activity. The higher basicity of DMAP-Zr-BDC MOF and Lewis acidic sites of Zr metal facilitated the reaction and found >98% benzaldehyde conversion in Knoevenagel condensation. The amine-grafted Zr-MOF was used for the cross-aldol condensation and cycloaddition of CO_2 (Scheme 2.10). In the case of cross-aldol condensation, *N*-butylamine-Zr-BDC-MOF exhibited the highest jasminaldehyde yield. The selectivity of jasminaldehyde followed the following order *N*-butylamine-Zr-BDC-MOF > diethylamine-Zr-BDC-MOF > DMAP-Zr-BDC-MOF. The obtained order of reactivity showed that the primary amine-grafted Zr-BDC MOF enhanced the catalytic activity. The basic sites of the catalyst promoted the synthesis of the carbanion of 1-heptanal according to the mechanistic route, whereas the acid sites of the catalyst activate the carbonyl group of benzaldehyde. Hence, the primary amine-grafted Zr-BDC MOF facilitated cross-aldol condensation. The amine-grafted Zr-MOF was also employed for the cycloaddition of CO_2.

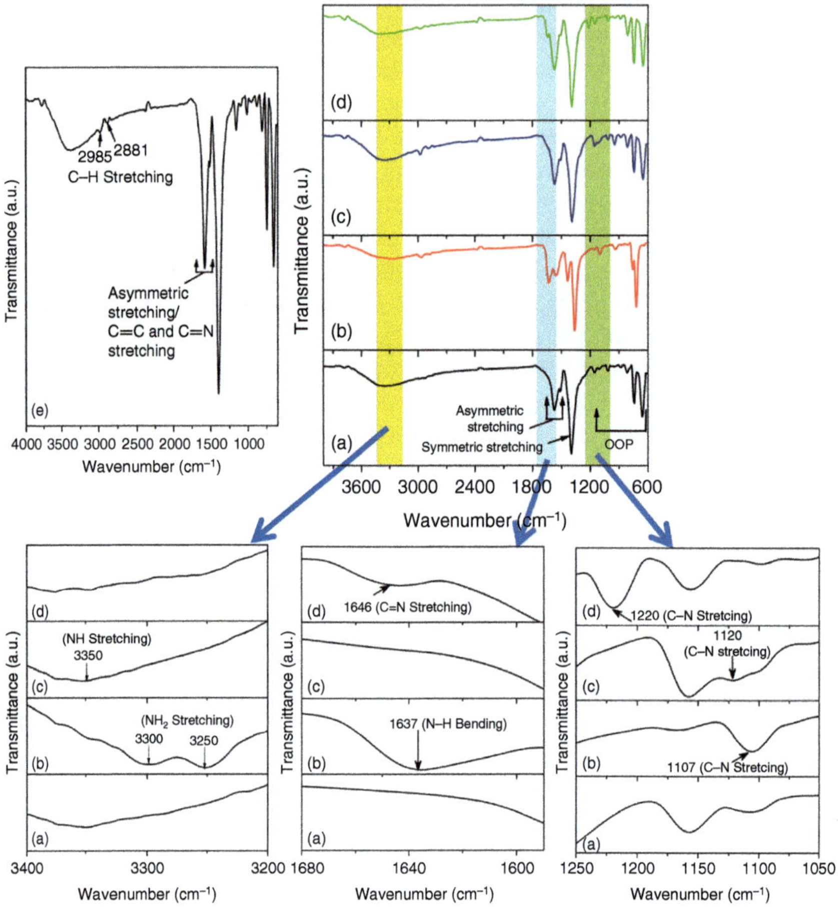

Figure 2.7 FT-IR spectra of (a) Zr-BDC-MOF and (b–e) *N*-butyl amine, diethylamine, DMAP, and [BMIM][OH] grafted Zr-BDC-MOF, respectively. Source: Reproduced from Rani and Srivastava [74] with permission from RSC, Copyright 2017.

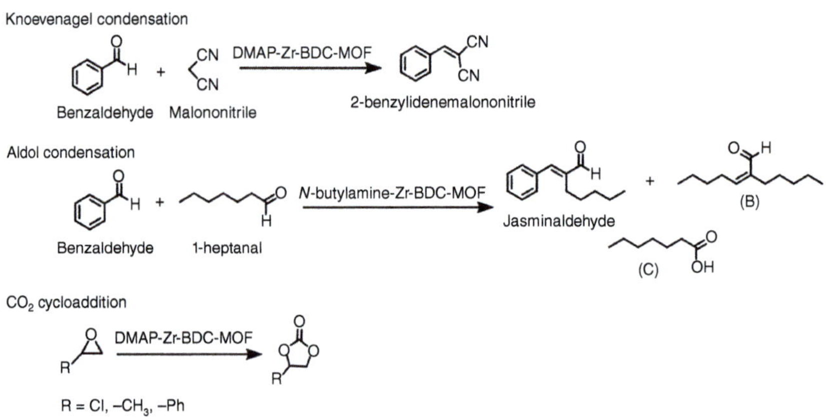

Scheme 2.10 Knoevenagel condensation reaction, aldol condensation reaction, and cycloaddition of CO_2.

Among the various amine-grafted MOFs, DMAP-Zr-BDC-MOF exhibited very high activity for the cycloaddition of CO_2 to epoxide and found >92% epichlorohydrin conversion and 96% cyclic carbonate selectivity.

Several bifunctional MOFs have been developed, and the different functionalities were used to catalyze the tandem one-pot reactions. Z. Shi et al. developed a new synthesis strategy for the development of bifunctional MOFs by PSM [75]. The bifunctional MOF was synthesized by the protection of amino group in MIL-101 to yield MIL-101-NHBOC. Furthermore, MIL-101-NHBOC was sulfonated with chlorosulfonic acid giving MIL-101-SO_3H-NHBOC. Thereafter, deprotection of –NHBOC by thermal treatment provided MIL-101-SO_3H-NH_2, and the obtained bifunctional MOF was thoroughly characterized by FT-IR, XPS, PXRD, and N_2-adsorption–desorption measurements. The MIL-101 and the dehydrated MIL-101-NHBOC's FT-IR displayed the n(NH) and n(CH) characteristic peaks, demonstrating the effective grafting of mono-BOC-ethylenediamine. The absorption peak for n(CH) and n(NH) at 2864, 2934, and 3263 cm^{-1} for MIL-101-SO_3H-NHBOC and MIL-101-SO_3H-NH_2 showed that Lewis base sites preserved after the modification procedure. The IR adsorption bands at 1207 cm^{-1} were obtained for the –SO_3H unit in MIL-101-SO_3H-NHBOC and MIL-101-SO_3H-NH_2, strongly indicating that the Lewis acid sites were modified. The peak at 168.9 eV in XPS spectra confirmed sulfur in MIL-101-SO_3H-NHBOC and MIL-101-SO_3H-NH_2. Thereafter, the bifunctional MOF was utilized for the hydrolysis of acetals, followed by the Henry reaction. The hydrolysis of acetals was facilitated by –SO_3H group, and the subsequent Henry reaction was influenced by the –NH_2 group (Scheme 2.11). The monofunctional MIL-101-SO_3H facilitated the hydrolysis of benzylacetals to form benzaldehyde and a negligible yield of 2-nitrovinyl benzene, while MIL-101-NH_2 did not facilitate the hydrolysis of acetals suggesting the one-pot tandem reaction was only feasible by MIL-101-SO_3H-NH_2. Furthermore, W. S. Ahn et al. proposed an alternative synthetic route for the bifunctional MIL-101 MOF [76]. MIL-101-SO_3H-NH_2 was synthesized using two organic linkers, such as 2-sulfoterephthlate (H_2BDC–SO_3Na) and 2-nitrobenzene-1,4-dicarboxylic acid (H_2BDC–NO_2). The synthesized MIL-101-SO_3H-NO_2 was subsequently reduced by $SnCl_2$ to obtain MIL-101-SO_3H-NH_2. The MIL-101-SO_3H-NH_2 having Bronsted acid and base sites to the organic linkers was a highly effective design for isolating antagonistic activities in a single system and facilitating the one-pot tandem acetal hydrolysis, followed by Henry reaction with nitromethane.

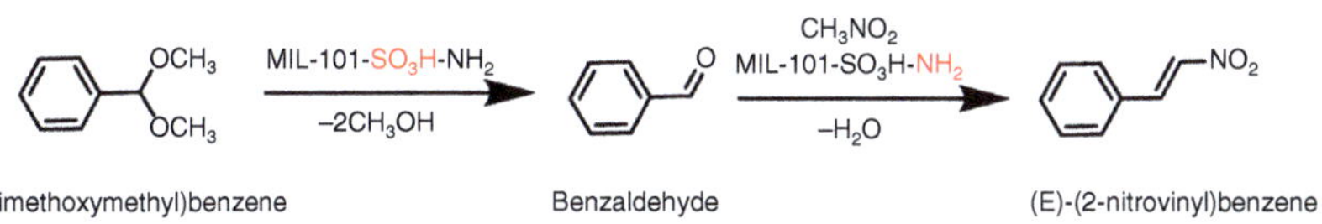

Scheme 2.11 Tandem one-pot reaction of acetal hydrolysis followed by Henry reaction over acid–base-bifunctional MOF.

Linker engineering is a powerful method to change the chemical environment or introduce new active sites in MOF with inherent catalytic sites because it enables the exact and systematic construction of the active sites in the molecular dimension. Therefore, various linkers have been used to modify the MOF's active sites and introduce bifunctionality/multifunctionality. The bifunctionality/multifunctionality of the MOF facilitates various organic reactions, C–C bond formations, and asymmetric synthesis. Su et al. reported integrated multifunctional linkers into a 3D MOF $[(Zr_6(\mu 3\text{-}O)_8(H_2O)_8(L1)_4]$ (LIFM-28, L1=2,2′-bis (trifluoromethyl)-4,4′-biphenyldicarboxylate) [77]. Then integration of 2,2′-bipyridine-4,4′-dicarboxylate (dcbpy) and the amino-functionalized dicarboxylate ligand yielded LIFM-80, which has both chelation and covalent sites for further modification. The amino group was modified with 2-sulfobenzoic anhydride (SBA) to obtain LIFM-80-$ArSO_3H$. The further modification with Cu produced multipurpose catalysts having modulated active sites and was utilized in Baylis–Hillman reactions, Click reactions, acetal reactions, and Knoevenagel condensation reactions and afforded good to excellent yield (Scheme 2.12).

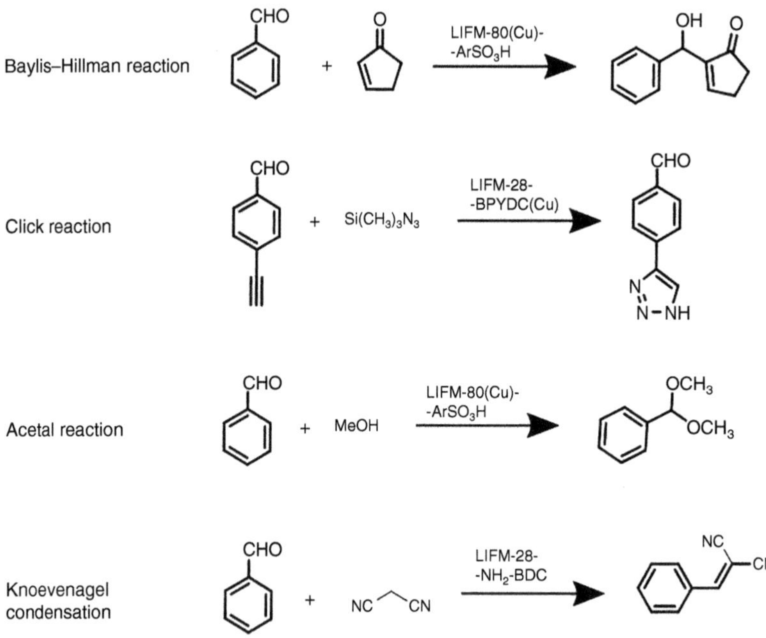

Scheme 2.12 Catalysis through multifunctionality in MOF.

Furthermore, the functional linker also provides an opportunity for the surface modifications of the MOF by catalytically active nanoscopic species like MNPs. The size of the MNPs decides their retention ability either inside the pores or on the pore wall away from the internal pores. The heteroatoms in the linker provide the functionality nature to stabilize the MNPs that act as catalytic entities for organic transformations. Kar et al. reported Pd NPs supported Ce-BTC MOF (Pd/Ce-BTC) for the catalytic hydrogenolysis of aryl ethers under mild reaction

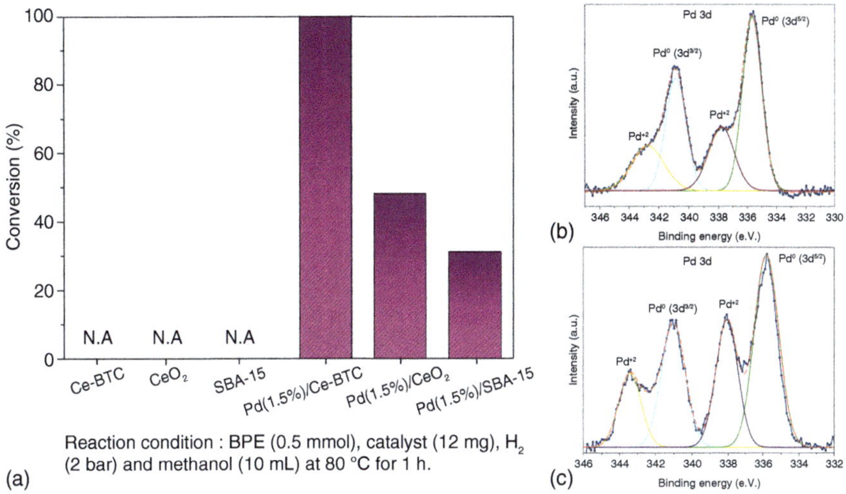

Figure 2.8 (a) BPE hydrogenolysis over various synthesized catalysts in this study, (b) XPS spectra of Pd in Pd/Ce-BTC MOF, and (c) XPS spectra of Pd in Pd/CeO_2. Source: Reproduced from Kar et al. [78] with permission from RSC, Copyright 2020.

conditions [78]. The catalytic activity was compared with the Pd/CeO_2 and Pd/SBA-15 to justify the advantageous role of Ce-BTC MOF as a support material. The catalytic activity followed the order Pd/Ce-BTC MOF > Pd/CeO_2 > Pd/SBA-15 (Figure 2.8a). The high catalytic activity of Pd/Ce-BTC MOF over Pd/CeO_2 was due to the protagonist nature of the Ce-BTC MOF that provided a suitable podium for stabilizing higher Pd^0 content in Pd/Ce-BTC MOF than Pd/CeO_2. It was confirmed by the XPS analysis (Figure 2.8b,c) and Hirshfeld charge analysis during theoretical calculation. The higher adsorption of the reactant on Ce-BTC-MOF and higher contents of Pd^0 species in Pd/Ce-BTC-MOF dissociatively chemisorbed H_2 to form abundant Pd-H active species than Pd/CeO_2 were the key reasons for the high hydrogenolysis activity of Pd/Ce-BTC-MOF over Pd/CeO_2.

Similarly, Cu NPs-supported MOFs were synthesized for the catalytic transfer reduction of styrene as a model compound utilizing hydrazine hydrate (N_2H_4 H_2O) as a hydrogen donor (Figure 2.9) [79]. Cu-BTC MOF was more active than other transition metal-based MOF. Moreover, the catalytic activity was improved upon Cu NPs incorporation on the external surface of the Cu-BTC MOF. The carboxylate linker stabilized the colloidal Cu NPs and accommodated them on the external surface, which enhanced the reduction reaction activity by activating N_2H_4 H_2O.

Peng et al. demonstrated an IL surface-modified Cu-BTC in stabilizing Pd NPs for the selective hydrogenation of alkyne [80]. The IL-modified Cu-BTC surface stabilized the Pd NPs by the surface interaction between "N" of the IL functioned in Cu-BTC MOF (Scheme 2.13). The resultant IL-modified MOF stabilized the small-sized Pd NPs, prevented their agglomeration, and enhanced the catalytic activity.

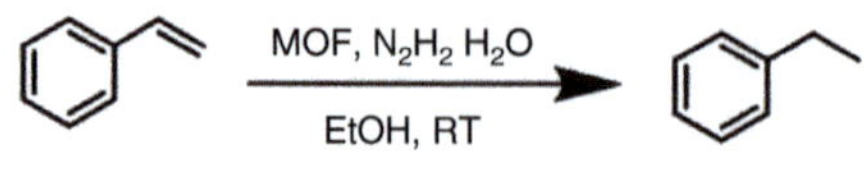

Entry no.	MOF	Amount (mmol)	Ethyl benzene yield (%)	TON
1	Zr-BDC-MOF	0.0122	15.6	7.6
2	Mn-BDC-MOF	0.0301	24.2	4.8
3	Cu-BDC-MOF	0.0664	61.1	5.5
4	Cu-BTC-MOF	0.0330	66.3	12
5	Ni-BTC-MOF	0.0335	22.2	4
6	Cu(10%)@Cu-BTC-MOF	0.0358	96.4	16.2
7	Cu(5%)@Cu-BTC-MOF	0.0355	74.2	12.6
8	Cu(15%)@Cu-BTC-MOF	0.0360	86.9	14.5

Figure 2.9 Catalytic activity of styrene reduction using N_2H_4 H_2O over various transition metal-based MOF. Source: Reproduced from Kar and Srivastava [79] with permission from RSC, Copyright 2018.

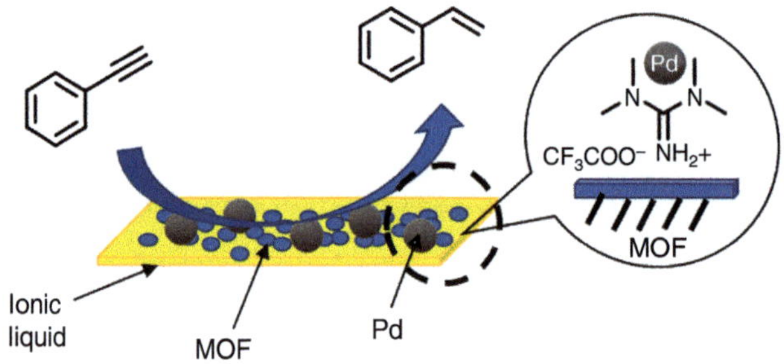

Scheme 2.13 Schematic presentation of the Pd/IL/MOF and its application in the selective hydrogenation of phenylacetylene.

2.5 Conclusion

In summary, a brief discussion of the various isolated active sites in MOF structure and the possible associated applications in heterogeneous catalysis were discussed. The functional nature of MOF allows for the engineering of the active sites to form a wide range of high-scope catalytic materials useful in several heterogeneous catalytic transformations. The high catalytic activities of these engineered or modified MOFs can be directly correlated with the induced structure–activity relationship due to the active site engineering phenomena in MOF. The tailorable properties of the MOF structure by means of modulating the pore size and chemical environment of the pore wall, altering the chemical connectivity and tuning the electronic surrounding of the metal node, and finally modifying the functionalities, size, and ligating sites of the organic linker. The book chapter also emphasized the various engineering strategies and necessary precautions needed to take care during the modification process. The overall electronic properties of the engineered MOF dictate the catalytic efficacy; provided, the stability issue is taken into consideration. One of the critical issues attributed to the MOF is its thermal and chemical stability; this issue is even more pronounced after the active site engineering of the MOF structure. Therefore, following the meticulous reaction procedure while modifying the MOF structure. In several pieces of literature, it has been reported that the active site of the engineered MOF surprisingly differs from its pristine MOF structure and exhibits different chemical and physical properties, which need to be understood by several advanced analytical techniques.

In this book chapter, we have attempted to summarize various heterogeneous catalytic applications based on the exposed active sites during the functional site modifications of MOF. However, based on the recent literature reports, the modified or electronically modulated MOF is highly appreciated in various biomass transformations, as such transformations require spatial and optimum acidic and basic sites to achieve. The fine tune adjustment of the active sites in the MOF structure needs further research for the apparent understanding and execution in the development of more committed catalytic systems. This book chapter is expected to be a useful guide in further developing the necessary engineering of the active sites of MOF for challenging and unique catalytic applications.

References

1 Kim, J., Chen, B., Reineke, T.M. et al. (2001). Assembly of metal-organic frameworks from large organic and inorganic secondary building units: new examples and simplifying principles for complex structures. *J. Am. Chem. Soc.* 123: 8239–8247.

2 Li, H., Eddaoudi, M., Groy, T.L., and Yaghi, O.M. (1998). Establishing microporosity in open metal-organic frameworks: gas sorption isotherms for Zn(BDC) (BDC = 1,4-Benzenedicarboxylate). *J. Am. Chem. Soc.* 120: 8571–8572.

3 Mai, Z. and Liu, D. (2019). Synthesis and applications of isoreticular metal–organic frameworks IRMOFs-n (n = 1, 3, 6, 8). *Cryst. Growth Des.* 19: 7439–7462.

4 Allen, C.A., Boissonnault, J.A., Cirera, J. et al. (2013). Chemically crosslinked isoreticular metal–organic frameworks. *Chem. Commun.* 49: 3200–3202.

5 Yang, L.M., Ravindran, P., Vajeeston, P., and Tilset, M. (2012). Formation of an intermediate band in isoreticular metal–organic framework-993 (IRMOF-993) and metal-substituted analogues M-IRMOF-993. *J. Mater. Chem.* 22: 16324–16335.

6 Song, Y.F. and Cronin, L. (2008). Postsynthetic covalent modification of metal–organic framework (MOF) materials. *Angew. Chem. Int. Ed.* 47: 4635–4637.

7 Ding, M., Cai, X., and Jiang, H.L. (2019). Improving MOF stability: approaches and applications. *Chem. Sci.* 10: 10209–10230.

8 Alqadami, A.A., Khan, M.A., Siddiqui, M.R., and Alothman, Z.A. (2018). Development of citric anhydride anchored mesoporous MOF through post synthesis modification to sequester potentially toxic lead (II) from water. *Microporous Mesoporous Mater.* 261: 198–206.

9 Liu, M., Wu, J., and Hou, H. (2019). Metal–organic framework (MOF)-based materials as Heterogeneous Catalysts for C–H Bond activation. *Chem. Eur. J.* 25: 2935–2948.
10 Tanabe, K.K. and Cohen, S.M. (2011). Postsynthetic modification of metal-organic frameworks a progress report. *Chem. Soc. Rev.* 40: 498–519.
11 Wang, Z. and Cohen, S.M. (2009). Postsynthetic modification of metal-organic frameworks. *Chem. Soc. Rev.* 38: 1315–1329.
12 Jin, Y., Lu, H., and Yan, B. (2021). A multivariate luminescent MOF based on dye covalently modification serving as a sensitive sensor for $Cr_2O_7^{2-}$, CrO_4^{2-} anions and its applications. *Dyes and Pigments* 194: 109588.
13 Jiang, W.L., Ding, L.G., Yao, B.J. et al. (2016). A MOF-membrane based on the covalent bonding driven assembly of a NMOF with an organic oligomer and its application in membrane reactors. *Chem. Commun.* 52: 13564–13567.
14 Singh, M., Kumar, D., Thomas, J., and Ramanan, A. (2010). Crystallization of copper(II) sulfate based minerals and MOF from solution: chemical insights into the supramolecular interactions. *J. Chem. Sci.* 122: 757–769.
15 Uflyand, I.E., Zhinzhilo, V.A., Nikolaevskaya, V.O. et al. (2021). Recent strategies to improve MOF performance in solid phase extraction of organic dyes. *Microchem. J.* 168: 106387.
16 Agostoni, V., Horcajada, P., Noiray, M. et al. (2015). A "green" strategy to construct noncovalent, stable and bioactive coatings on porous MOF nanoparticles. *Sci Rep.* 5: 7925.
17 Chen, C., Allen, C.A., and Cohen, S.M. (2011). Tandem postsynthetic modification of metal-organic frameworks using an inverse-electron-demand diels–Alder reaction. *Inorg. Chem.* 50: 10534–10536.
18 Yu, Q., Li, Z., Cao, Q. et al. (2020). Advances in luminescent metal-organic framework sensors based on post-synthetic modification. *Trends Anal. Chem.* 129: 115939.
19 Wang, Z. and Cohen, S.M. (2008). Tandem modification of metal-organic frameworks by a postsynthetic approach. *Angew. Chem. Int. Ed.* 47: 4699–4702.
20 Zhang, X., Chen, L., Li, Y. et al. (2019). Palladium N.P.s supported on sulfonic acid functionalized metal-organic frameworks as catalysts for biomass cascade reactions. *Dalton Trans.* 48: 5515–5519.
21 Huang, Y., Zhao, M., Han, S. et al. (2017). Growth of Au nanoparticles on 2D metalloporphyrinic metal-organic framework nanosheets used as biomimetic catalysts for cascade reactions. *Adv. Mater.* 29: 1700102.
22 Wang, Y., Zhang, N., Zhang, E. et al. (2019). Heterogeneous metal–organic-framework-based biohybrid catalysts for cascade reactions in organic solvent. *Chem. Eur. J.* 25: 1716–1721.
23 Dhakshinamoorthy, A. and Garcia, H. (2014). Cascade reactions catalyzed by metal organic frameworks. *ChemSusChem* 7: 2392–2410.
24 Xu, C., Fang, R., Luque, R. et al. (2019). Functional metal-organic frameworks for catalytic applications. *Coord. Chem. Rev.* 388: 268–292.
25 Wena, Y., Zhang, J., Xu, Q. et al. (2018). Pore surface engineering of metal-organic frameworks for heterogeneous catalysis. *Coord. Chem. Rev.* 376: 248–276.

26 Liu, T.F., Zou, L., Feng, D. et al. (2014). Stepwise synthesis of robust metal–organic frameworks via postsynthetic metathesis and oxidation of metal nodes in a single-crystal to single-crystal transformation. *J. Am. Chem. Soc.* 136: 7813–7816.

27 Jiao, L., Wang, Y., Jiang, H.L., and Xu, Q. (2018). Metal-organic frameworks as platforms for catalytic applications. *Adv. Mater.* 30: 1703663.

28 Islamoglu, T., Goswami, S., Li, Z. et al. (2017). Postsynthetic tuning of metal–organic frameworks for targeted applications. *Acc. Chem. Res.* 50: 805–813.

29 Kaur, M., Kumar, S., Younis, S.A. et al. (2021). Post-Synthesis modification of metal-organic frameworks using Schiff base complexes for various catalytic applications. *Chem. Eng. J.* 423: 130230.

30 Pascanu, V., Yao, Q., Gomez, A.B. et al. (2013). Sustainable catalysis: rational Pd loading on MIL-101Cr-NH_2 for more efficient and recyclable suzuki–miyaura reactions. *Chem. Eur. J.* 19: 17483–17493.

31 Yuan, S., Chen, Y.P., Qin, J.S. et al. (2016). Linker installation: engineering pore environment with Precisely placed functionalities in zirconium MOFs. *J. Am. Chem. Soc.* 138: 8912–8919.

32 Trickett, C.A., Gagnon, K.J., Lee, S. et al. (2015). Definitive molecular level characterization of defects in UiO-66 crystals. *Angew. Chem. Int. Ed.* 54: 11162–11167.

33 Liu, H., Zhang, Z., Tang, J. et al. (2019). Quest for pore size effect on the catalytic property of defect-engineered MOF-808-SO_4 in the addition reaction of isobutylene with ethylene glycol. *J. Solid State Chem.* 269: 9–15.

34 Rani, P. and Srivastava, R. (2018). Exploring the dicationic gemini surfactant for the generation of mesopores: a step towards the construction of a hierarchical metal-organic framework. *Inorg. Chem. Front.* 5: 2856–2867.

35 Gu, J.M., Kim, W.S., and Huh, S. (2011). Size-dependent catalysis by DABCO-functionalized Zn-MOF with one-dimensional channels. *Dalton Trans.* 40: 10826–10829.

36 Mandal, S., Natarajan, S., Mani, P., and Pankajakshan, A. (2021). Post-synthetic modification of metal–organic frameworks toward applications. *Adv. Funct. Mater* 31: 2006291 (1–22).

37 Sun, D., Sun, F., Deng, X., and Li, Z. (2015). Mixed-metal strategy on metal–organic frameworks (MOFs) for functionalities expansion: co substitution induces aerobic oxidation of cyclohexene over inactive Ni-MOF-74. *Inorg. Chem.* 54: 8639–8643.

38 Madasamy, K., Kumaraguru, S., Sankar, V. et al. (2019). A Zn based metal organic framework as a heterogeneous catalyst for C–C bond formation reactions. *New J. Chem.* 43: 3793–3800.

39 Kaur, H., Venkateswarulu, M., Kumar, S. et al. (2018). A metal-organic framework based multifunctional catalytic platform for organic transformation and environmental remediation. *Dalton Trans.* 47: 1488–1497.

40 Yang, D., Bernales, V., Islamoglu, T. et al. (2016). Tuning the surface chemistry of metal organic framework nodes: proton topology of the metal-oxide-like Zr_6 nodes of UiO-66 and NU-1000. *J. Am. Chem. Soc.* 138: 15189–15196.

41 Chen, K. and Wu, C.D. (2019). Designed fabrication of biomimetic metal–organic frameworks for catalytic applications. *Coord. Chem. Rev.* 378: 445–465.

42 Habib, N.R., Nieto, E.A., Taddesse, A.M., and Diaz, I. (2021). Non-noble MNPS@MOF materials: synthesis and applications in heterogeneous catalysis. *Dalton Trans.* 50: 10340–10353.

43 Guo, C., Liang, C., Qin, X. et al. (2020). Zeolitic imidazolate framework cores decorated with Pd nanoparticles and coated further with metal–organic framework shells (ZIF-8@Pd@MOF-74) as nanocatalysts for chemoselective hydrogenation reactions. *ACS Appl. Nano Mater.* 3: 7242–7251.

44 Lalonde, M., Bury, W., Karagiaridi, O. et al. (2013). Transmetalation: routes to metal exchange within metal–organic frameworks. *J. Mater. Chem. A* 1: 5453–5468.

45 Dutta, G., Jana, A.K., and Natarajan, S. (2017). Assembling Porphyrins into extended network structures by employing aromatic dicarboxylates: synthesis, metal exchange, and heterogeneous catalytic studies. *Chem. Eur. J.* 23: 8932–8940.

46 Pan, Y., Sun, K.A., Liu, S.J. et al. (2018). Core–shell ZIF-8@ZIF-67-derived CoP nanoparticle-embedded N-Doped carbon nanotube hollow polyhedron for efficient overall water splitting. *J. Am. Chem. Soc.* 140: 2610–2618.

47 Botas, J.A., Calleja, G., Saanchez, M.S., and Orcajo, M.G. (2010). Cobalt doping of the MOF-5 framework and its effect on gas-adsorption properties. *Langmuir* 26: 5300–5303.

48 Kar, A.K. and Srivastava, R. (2022). Improving the glucose to fructose isomerization via epitaxial-grafting of niobium in UIO-66 framework. *ChemCatChem* 14: e202200721.

49 Bernales, V., Yang, D., Yu, J. et al. (2017). Molecular rhodium complexes supported on the metal-oxide-like nodes of metal organic frameworks and on zeolite HY: catalysts for ethylene hydrogenation and dimerization. *ACS Appl. Mater. Interfaces* 9: 33511–33520.

50 Ren, H.Y., Yao, R.X., and Zhang, X.M. (2015). Enhanced catalysis activity in a coordinatively unsaturated cobalt-MOF generated via single-crystal-to-single-crystal dehydration. *Inorg. Chem.* 54: 6312–6318.

51 Kondo, Y., Kuwahara, Y., Mori, K., and Yamashita, H. (2021). Dual role of missing-linker defects terminated by acetate ligands in a zirconium-based MOF in promoting photocatalytic hydrogen peroxide production. *J. Phys. Chem. C* 125: 27909–27918.

52 Zhao, Y., Zhang, Q., Li, Y. et al. (2017). Large-scale synthesis of monodisperse UiO-66 crystals with tunable sizes and missing linker defects via acid/base co-modulation. *ACS Appl. Mater. Interfaces* 9: 15079–15085.

53 Suárez, C.A., Daniel, R., Molina, V., Alem, V.H. et al. (2020). Synthesis of ordered microporous/microporous MOF-808 through modulator-induced defect formation, and surfactant self-assembly strategies. *Phys. Chem. Chem. Phys.* 22: 12591–12604.

54 Kirchon, A., Feng, L., Drake, H.F. et al. (2018). From fundamentals to applications: a toolbox for robust and multifunctional MOF materials. *Chem. Soc. Rev.* 47: 8611–8638.

55 Hardian, R., Dissegna, S., Ullrich, A. et al. (2021). Tuning the properties of MOF-808 via defect engineering and metal nanoparticle encapsulation. *Chem. Eur. J.* 27: 6804–6814.

56 Valekar, A.H., Lee, M., Yoon, J.W. et al. (2020). Catalytic transfer hydrogenation of furfural to furfuryl alcohol under mild conditions over Zr-MOFs: exploring the role of metal node coordination and modification. *ACS Catal.* 10: 3720–3732.

57 Vermoortele, F., Bueken, B., Le Bars, G. et al. (2013). Synthesis modulation as a tool to increase the catalytic activity of metal–organic frameworks: the unique case of UiO-66(Zr). *J. Am. Chem. Soc.* 135: 11465–11468.

58 Vandichel, M., Hajek, J., Vermoortele, F. et al. (2015). Active site engineering in UiO-66 type metal–organic frameworks by intentional creation of defects: a theoretical rationalization. *CrystEngComm.* 17: 395–406.

59 Vermoortele, F., Vandichel, M., Van de Voorde, B. et al. (2012). Electronic effects of linker substitution on lewis acid catalysis with metal–organic frameworks. *Angew. Chem. Int. Ed.* 51: 4887–4890.

60 Lua, Y., Tonigold, M., Bredenkötter, B. et al. (2008). A Cobalt(II)-containing metal-organic framework showing catalytic activity in oxidation reactions. *Z. Anorg. Allg. Chem.* 634: 2411–2417.

61 Liu, Y., Klet, R.C., Hupp, J.T., and Farha, O. (2016). Probing the correlations between the defects in metal–organic frameworks and their catalytic activity by an epoxide ring-opening reaction. *Chem. Commun.* 52: 7806–7809.

62 Wang, J., Liu, L., Chen, C. et al. (2020). Engineering effective structural defects of metal-organic frameworks to enhance their catalytic performances. *J. Mater. Chem. A* 8: 4464–4472.

63 Freitas, J.C.R., de Oliveira, C.K., Cunha, E.C. et al. (2013). Allylation of aldehydes with potassium allyltrifluoroborate catalyzed by lanthanide-based metal-organic framework. *Tetrahedron Lett.* 54: 1558–1561.

64 Geravand, E., Farzaneh, F., and Ghiasi, M. (2019). Metalation and DFT studies of metal organic frameworks UiO-66(Zr) with vanadium chloride as allyl alcohol epoxidation catalyst. *J. Mol. Struct.* 1198: 126940.

65 Le, H.T.N., Nguyen, T.T., Vu, P.H.L. et al. (2014). Ligand-free direct C-arylation of heterocycles with aryl halides over a metal-organic framework $Cu_2(BPDC)_2$ (BPY) as an efficient and robust heterogeneous catalyst. *J. Mol. Catal. A Chem.* 391: 74–82.

66 Schlichte, K., Kratzke, T., and Kaskel, S. (2004). Improved synthesis, thermal stability and catalytic properties of the metal-organic framework compound $Cu_3(BTC)_2$. *Microporous Mesoporous Mater.* 73: 81–88.

67 Valvekens, P., Vandichel, M., Waroquier, M. et al. (2014). Metal-dioxidoterephthalate MOFs of the MOF-74 type: microporous basic catalysts with well-defined active sites. *J. Catal.* 317: 1–10.

68 Nguyen, L.T.L., Nguyen, T.T., Nguyen, K.D., and Phan, N.T.S. (2012). Metal-organic framework MOF-199 as an efficient heterogeneous catalyst for the aza-michael reaction. *Appl. Catal. A: Gen.* 426: 44–52.

69 Wang, Y., Chen, L., Hou, C.C. et al. (2020). Multiple catalytic sites in MOF-based hybrid catalysts for organic reactions. *Org. Biomol. Chem.* 18: 8508–8525.

70 Lim, J., Lee, S., Ha, H. et al. (2021). Amine-tagged fragmented ligand installation for covalent modification of MOF-74. *Angew. Chem. Int. Ed.* 60: 9296–9300.

71 Guo, F. (2019). A mononuclear Cu(II)-based metal-organic framework as an efficient heterogeneous catalyst for chemical transformation of CO_2 and knoevenagel condensation reaction. *Inorg. Chem. Commun.* 101: 87–92.

72 Dau, P.V., Cohen, S.M., and Bifunctional, A. (2015). Site-isolated metal–organic framework-based tandem catalyst. *Inorg. Chem.* 54: 3134–3138.

73 Liu, F., Kumar, S., Li, S. et al. (2020). Bifunctional design of stable metal-organic framework bearing triazole–carboxylate mixed ligand: highly efficient heterogeneous catalyst for Knoevenagel condensation reaction under mild conditions. *Catal. Commun.* 142: 106032.

74 Rani, P. and Srivastava, R. (2017). Tailoring the catalytic activity of metal organic frameworks by tuning the metal center and basic functional sites. *New J. Chem.* 41: 8166–8177.

75 Li, B., Zhang, Y., Ma, D. et al. (2012). A strategy toward constructing a bifunctionalized MOF catalyst: post-synthetic modification of MOFs on organic ligands and coordinatively unsaturated metal sites. *Chem. Commun.* 48: 6151–6153.

76 Lee, Y.R., Chung, Y.M., and Ahn, W.S. (2014). A new site-isolated acid-base bifunctional metal-organic framework for one-pot tandem reaction. *RSC Adv.* 4: 23064–23067.

77 Cao, C.C., Chen, C.X., Wei, Z.W. et al. (2019). Catalysis through dynamic spacer installation of multivariate functionalities in metal–organic frameworks. *J. Am. Chem. Soc.* 141: 2589–2593.

78 Kar, A.K., Kaur, S.P., Dhilip Kumar, T.J., and Srivastava, R. (2020). Efficient hydrogenolysis of aryl ethers over Ce-MOF supported Pd NPs under mild conditions: mechanistic insight using density functional theoretical calculations. *Catal. Sci. Technol.* 10: 6892–6690.

79 Kar, A.K. and Srivastava, R. (2018). An efficient and sustainable catalytic reduction of carbon-carbon multiple bonds, aldehydes, and ketones using a Cu nanoparticle decorated metal-organic framework. *New J. Chem.* 42: 9557–9567.

80 Peng, L., Zhang, J., Yang, S. et al. (2015). The ionic liquid microphase enhances the catalytic activity of Pd nanoparticles supported by a metal-organic framework. *Green Chem.* 17: 4178–4182.

3

Characterization of Organic Linker-Containing Porous Materials as New Emerging Heterogeneous Catalysts

Ali R. Oveisi[1], *Saba Daliran*[1], *and Yong Peng*[2]

[1] *Department of Chemistry, Faculty of Sciences, University of Zabol, P.O. Box: 98615-538, Zabol, Iran*
[2] *Consejo Superior de Investigaciones Científicas, Universitat Politècnica de València, Instituto de Tecnología Química CSIC-UPV, Av. de los Naranjos s/n, Valencia 46022, Spain*

3.1 Introduction

Metal–organic frameworks (MOFs), also known as porous coordination networks (PCNs) or porous coordination polymers (PCPs), are porous and crystalline materials which are self-assembled from metal ions or clusters with organic linkers [1–3]. Porous organic polymers (POPs) are porous materials made in one, two, or three dimensions from the combination of organic building blocks through covalent bonds [4, 5]. They are commonly classified into crystalline covalent organic frameworks (COFs) [6, 7] and noncrystalline materials. Amorphous POPs have also been labeled by many different names, including containing conjugated microporous polymers (CMPs) [8], polymers of intrinsic microporosity (PIMs) [9], porous aromatic frameworks (PAFs) [10], hypercross-linked polymers (HCPs) [11], and most covalent triazine frameworks (CTFs) [12]. Owning to their advantages of low density, high intrinsic porosity, excellent stability (POPs rather than MOFs), and predictable and tunable structures and functions, these porous materials have received increasing attention and interest for their remarkable potential applications in catalysis [13–16], photocatalysis [17–22], gas storage/separation [23–25], etc. [1, 5, 7, 23, 26].

In both heterogeneous catalysis and photocatalysis, parameters, such as catalyst structure and porosity, morphology (shape and size) [27, 28], active site dispersion onto a support [29, 30], and spatial location on external surface or within pores [31, 32], as well as coordination environment (connectivity and geometry) [33, 34], influence the activity for specific reactions. To better correlate the catalyst properties and the reaction mechanisms, and thus to establish clear guidance for catalyst design and optimization, exhaustive characterization of catalysts under *ex/in-situ* and even *operando* conditions is a powerful strategy [35–37]. In addition, theoretical calculations (normally based on density functional theory; DFT) are very important and frequently applied to complement experimental characterization

Catalysis in Confined Frameworks: Synthesis, Characterization, and Applications, First Edition.
Edited by Hermenegildo Garcia and Amarajothi Dhakshinamoorthy.

and provide understanding of the catalytic data measured experimentally. Validated theoretical calculations can serve to anticipate novel catalysts and determine reaction mechanisms [37–40]. In the following, several advanced microscopies, spectroscopies, and macroscopic testing techniques, as well as theoretical calculations will be also introduced to exemplify the type of information provided and how this information has been applied to correlate the properties of porous materials with their catalytic activity. This chapter has focused mainly on characterizations of crystalline and amorphous porous frameworks including MOFs, COFs, and POPs, which have recently become a frontier research topic in chemistry.

Techniques considered are scanning electron microscopy (SEM), energy-dispersive X-ray spectroscopy (EDS/EDX), transmission electron microscopy (TEM), atomic force microscopy (AFM), X-ray diffraction (XRD), X-ray photoelectron spectroscopy (XPS), X-ray absorption fine structure (XAFS) methods, nuclear magnetic resonance (NMR), electron paramagnetic resonance (EPR), ultraviolet-visible diffuse reflectance spectroscopy (UV–Vis DRS), inductively coupled plasma (ICP) spectroscopy, thermogravimetric analysis (TGA), N_2 adsorption–desorption measurements, and DFT calculations. Researchers usually use couple of these techniques to support their claim on the structure.

3.2 Microscopy Techniques

3.2.1 Scanning Electron Microscopy (SEM)

The most extensively used microscopy techniques in nanoscience and catalysis are SEM, TEM, and AFM. Among them, SEM acquires images by scanning in a raster fashion the specimen with a focused electron beam that typically possesses energy below 30 keV. By selectively detecting the electrons emitted by the interaction of the source electron beam with the specimen, a bidimensional image can be produced. Operating the SEM at higher accelerating voltages allows for larger penetration depth and better resolution images but leads to less surface structural information. In addition, the organic-containing solids can possibly be damaged in the high-voltage electron microscope. SEM has advantages in determining three-dimensional particle morphology compared to TEM and combining it with other techniques, such as XRD, SEM, EDX, and XPS can provide additional information including preferential exposed crystal facet, element mapping, and crystal structures.

SEM samples need to be stabilized on a specimen holder to withstand the high vacuum and high intensity of electron beam and operational conditions. Samples must be enough electrically conductive to avoid the accumulation of electrostatic charge, and thus nonconductive samples normally need an ultrathin coating of either gold or platinum or osmium by low vacuum sputter or high vacuum evaporation. However, the proper choice of the coating material is required to prevent overlapping of elemental peaks (or lines) with peaks from coating materials (or their contaminants) for a quantitative analysis. For example, iridium and platinum show similar EDS peaks, and iridium quantification by EDS is not accurate [41]. Therefore, it is needed either to change the coating material or to use ICP to quantify the Ir loading.

The most commonly detected signals are the secondary electrons (SE) and backscattered electrons(BSE), of which the SE signals are electrons emitted from atoms of the sample that are excited by the incident electron beam, while BSE signals are the incident beam electrons that are reflected from the specimen by elastic scattering. The images obtained plotting SE are used for evaluating the surface topography, while the BSE signals normally provide information about the distribution of different elements in the sample and crystallinity. In this way, coupling with energy-dispersive X-ray spectroscopy (EDS or EDX or EDAX) detector, the BSE signal can be used as an analytical technique to determine the composition of the surveyed particle. EDS is a simple and relatively rapid elemental analysis technique. However, EDS cannot provide reliable signals for light elements with atomic number below 11 (Na) and detect trace elements (concentrations below 0.01 wt%). The elements such as carbon, oxygen, nitrogen, and sulfur can only be detected by EDS for their existence. Wavelength dispersive spectroscopy (WDS) is more accurate than EDS for detecting light elements but is a less well-known related technique that is usually added to an existing EDS system. EDS is also capable of providing elemental distribution mapping analysis in a sample.

Elsaidi et al. [42] reported the controlled synthesis of core@shell MOF@MOF structures (SIFSIX-3-Ni@HKUST-1 and SIFSIX-3-Ni@SIFSIX-1-Cu, HKUST-1 stands for Hong Kong University of Science and Technology) in 2020. SIFSIX-M (M=Zn, Cu, and Ni) family is a class of isoreticular MOFs based on metal coordination networks with inorganic hexafluorosilicate (SiF_6^{2-}, SIFSIX) pillars and organic linkers. The synthesis of the core–shell architectures was examined by SEM/EDX and other techniques. The surface morphology and growth of the shell layer, SIFSIX-3-Ni, were well observed by SEM (Figure 3.1a–d, SEM images of the top view). The compositions of solids were assessed using EDX elemental mapping (Figure 3.1e–g) and EDX line scan across a single particle (Figure 3.2h). EDX elemental mapping indicated the homogeneous distribution of Cu and Ni in the samples. In addition, SEM was also applied to determine the quality and cross section morphologies of the core–shell particles (Figures 3.2i–j).

Thomas et al. [44] reported a macro/microporous COF and the honeycomb-like matrix with pore size around 200 nm was clearly observed by SEM. Hu et al. [43] reported the synthesis of MOF-containing polyoxometalate ($H_5PV_2Mo_{10}O_{40}$) (NENU-3a) with controllable facet exposure of NENU-3a. SEM images showed that octahedral NENU-3a crystals are formed without morphology modulator in the synthesis (see Figure 3.2a). In contrast, the addition of *p*-toluic acid (pTA) caused changes in the MOF crystal morphology that evolved to cubic morphology when the pTA to H_3BTC (trimesic acid and the organic linker) ratio equals to 20 (Figure 3.2b). The changes in the morphology observed by SEM were caused by the enhanced growth of (100) facet with respect to the (111) facet, as confirmed by PXRD (Figure 3.2c). Catalytic activity for esterification reaction was tested with different faceted crystals, and the results showed that the cubic crystals with (100) facet domain exhibited higher activity than the octahedral crystals with (111) facet domain. This enhanced catalytic activity was rationalized based on experimental studies and theoretical calculations that indicate that the cavity window of (100) facet is larger than that of (111) facet. Therefore, the larger dimensions of the 100 cavities exposed to the external surface facilitate the mass diffusion of reactants and products, and as a consequence, resulted in a higher reaction activity.

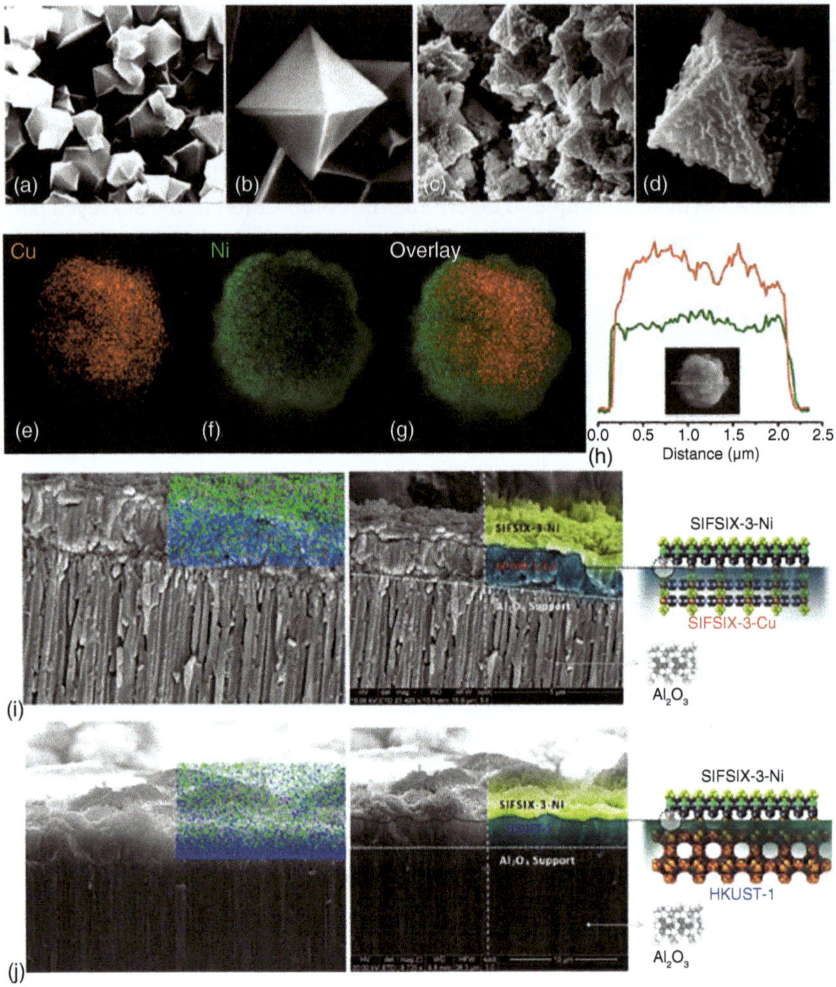

Figure 3.1 SEM images for different materials; (a and b) HKUST-1, (c and d) SIFSIX-3-Ni@HKUST-1 combined with (e-g) EDX mapping and (h) EDX line scan, and (i-j) SEM cross-sectional images combined with EDX analysis for SIFSIX-3-Ni@SIFSIX-1-Cu and SIFSIX-3-Ni@HKUST-1 (the top of the Al_2O_3 disc). Source: Reproduced with permission from Elsaidi et al. [42], Elsevier.

3.2.2 Transmission Electron Microscopy (TEM)

Owing to the maturity of aberration-corrected technique, TEM has become one of the most powerful techniques at the nanoscale, reaching structure, and composition at atomic resolution. Aberration-corrected TEM is now a mainstream technique to preliminary discern single atoms/sub-nanometric clusters. There are two operation modes in TEM, one being the parallel electron beam transmission mode (TEM mode), and the other one is the focused electron beam raster scanning mode (STEM mode). For TEM mode, the information of the specimen is acquired by detecting the transmitted electron beam after crossing the sample. In STEM

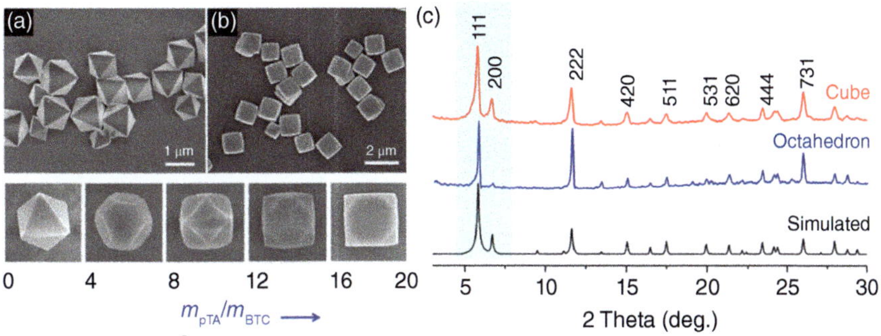

Figure 3.2 SEM images of NENU-3a in the absence (a) or in the presence (b) of pTA morphology modulator at a pTA:H_3BTC mass ratio of 20. SEM images of a single particle below with labels from 0 to 20 represent the morphology of NENU-3a prepared at different mass ratio of pTA and H_3BTC. (c) PXRD patterns of simulated, octahedral, and cubic NENU-3a crystals, respectively. Source: Reproduced with permission from Liu et.al. [43], American Chemical Society.

mode, focalized electron beams scan the specimen surface point-by-point in a raster fashion, and the scattered electron signals are collected by an annular dark-field (ADF) detector, forming an image composed by different contrast areas depending on the Z value of the elements at each point. Especially, if the ADF detector collects electrons from the high-angle direction, only incoherent scattered electrons are detected, and the acquired images are termed as high-angle annular dark-field (HAADF) images or Z contrast imaging, which is the image of direct projection of the atoms in the specimen. In this case, the resolution only depends on the size of the focused electron beam. Therefore, aberration correcting is always coupled with HAADF-STEM to correct the electron beam focus point to get a high-quality atomic-level resolution, and thus it is able to discern the existence of single atoms/sub-nanometric clusters.

Porous materials are extensively employed as the support to confine single atom/sub-nanometric clusters as active sites due to their high surface area and confinement effects disfavoring agglomeration. In most of these researches, the active site dispersion is conventionally evaluated by TEM. Zhang et al. [45] designed a POPs with aminopyridine functional group, and iridium active sites stabilized by pyridinyl and carboxyl groups of the POPs scaffold. The homogeneous distribution of Ir on the POPs support was established by HAADF-STEM images, wherein highly contrast atomic dispersity of Ir can be easily discerned due to its high Z value that is much higher than those of the elements of the support. Li et al. [46] synthesized ZIF-8 confining atomically dispersed ruthenium species. By taking advantages of ruthenium precursors and confinement effects of ZIF-8 cavities characterized by its small windows (0.4 nm diameter), either single Ru atoms or Ru_3 clusters were both successfully synthesized and confined in ZIF-8. Specifically, $Ru_3(CO)_{12}$ was encapsulated inside the cavities of the ZIF-8 by thermal adsorption. In this sample, Ru_3 clusters were separately confined inside the cages of ZIF-8, as confirmed by aberration-corrected HAADF-STEM. As can be observed in Figure 3.3a, dispersed bright dots corresponding to Ru atoms can be clearly distinguished from the

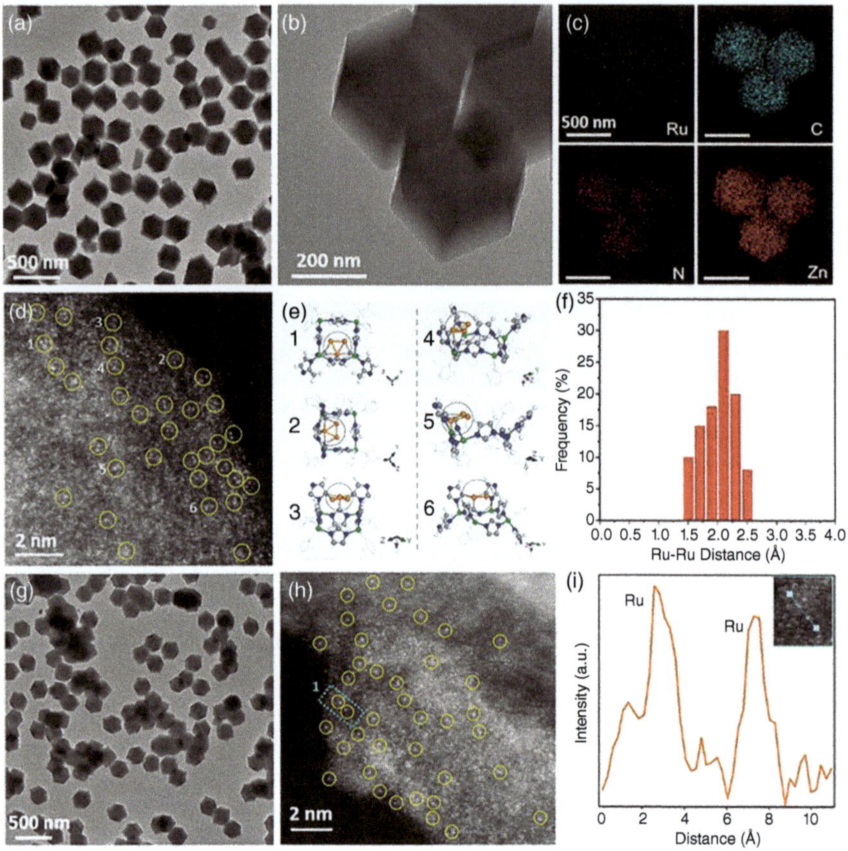

Figure 3.3 (a) TEM, (b) enlarged TEM, (c) EDS maps, (d) AC HAADF-STEM images of Ru_3@ZIF-8 highlighting Ru_3 clusters in yellow, (e) Illustrations of Ru_3 clusters from top and side view. (f) Ru–Ru distance histogram of Ru–Ru distance in the observed Ru_3 clusters. (g) TEM and (h) AC HAADF-STEM images of Ru_1@ZIF-8. (i) Intensity file of Ru_1@ZIF-8 in the cyan-dotted rectangle area of (h). Source: Reproduced with permission from Ji et al. [46], John Wiley & Sons.

ZIF-8 matrix. For a better identification, Figure 3.3d shows groups of three Ru atoms marked by yellow cycles, confirming the existence of Ru_3 clusters in the MOF support. In addition, the alignment of adjacent Ru atoms coincides with the expected projection for one of the possible crystal orientations simulated by DFT (Figure 3.3b). The average neighbor Ru distance obtained by statistical analyzing of more than 100 sites from aberration-corrected HAADF-STEM images matches well with the Ru–Ru bond length, further confirming the bright dots are Ru_3 clusters. Interestingly, if the $Ru(acac)_3$ was used as the precursor, aberration-corrected HAADF-STEM images indicate the presence of single Ru atoms confined in ZIF-8 (Figure 3.3g,h). In another example, Chen and co-workers used TEM to evaluate the progress of production of biomacromolecule@COF capsules through the MOF etching of core–shell structure of biomacromolecule@ZIF-90@COF-42-B [47].

In 2020, Farha et al. also found and monitored a phase transition from a microporous MOF, NU-906 (NU, Northwestern University), to a mesoporous MOF, NU-1008, using *in-situ* TEM techniques as the morphology and lattice space of the crystals were changed over time [48].

3.2.3 Atomic Force Microscopy (AFM)

AFM is also one of the most important tools for surface imaging and measuring thickness of films and materials on a (sub)nanometric scale. The information is collected by detecting the force or distance changes of the cantilever when the probe tip of the cantilever interacts with the sample surface. One advantage of AFM compared to SEM is that it can characterize both conductive and nonconductive samples without requiring the deposition of a conductive coating that may shade some of information of the pristine sample surface.

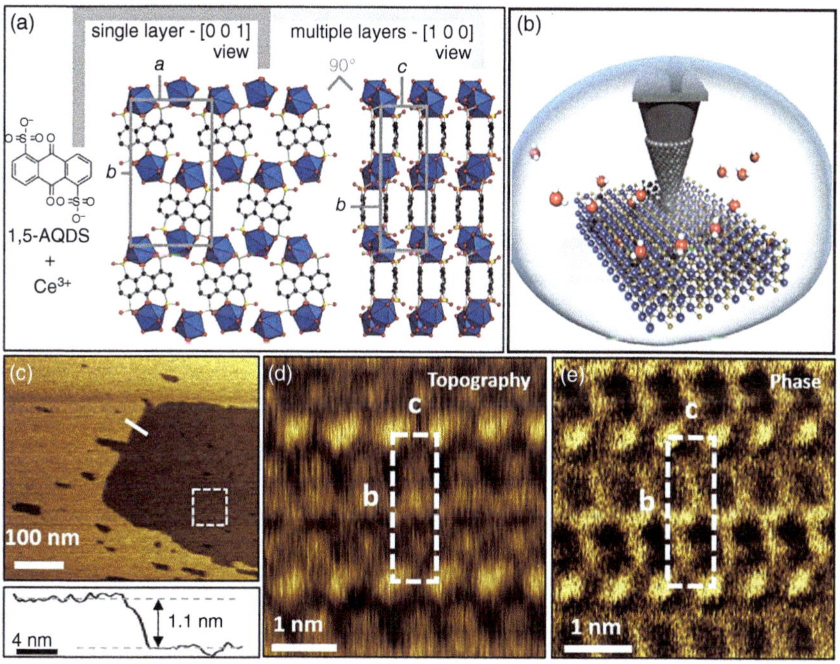

Figure 3.4 Structure of the MOF and AFM images. (a) The structure of Ce-RPF-8 involves double layers disposed perpendicular to the ab plane. Top and side views of the layers are presented in the figure, displaying unit cell dimensions. Cerium atoms are denoted as blue polyhedra, carbon, sulfur, and oxygen atoms are black, yellow, and red balls, respectively. (b) Schematic representation of the liquid-phase AFM experimental set-up. (c) AFM topography image of a region of a Ce-RPF-8 bc plane and particle thickness. The image has been gained in glycerol. (d) Molecular resolution image of the region marked in (c) (topography). (e) Phase contrast image of the region marked in (c). Imaging parameters for panels d,e: A 0 = 0.5 nm, A sp = 0.14 nm, fast axis scanning frequency = 20 Hz. Source: Reproduced from Chiodini et al. [49], Springer Nature/CC BY 4.0.

Garcia et al. [49] monitored the surface changes of MOF in the MOF-liquid interface by AFM characterization. In that case, using photothermal excitation the amplitude of the microcantilever modulated in a range from free amplitude to a given set-point. These measurements allowed acquiring angstrom-level resolution images. By means of the AFM characterization, they observed a surface reconstruction of MOF when immersed in water and glycerol (Figure 3.4). Besides the surface topography, another extensive application of AFM is to determine the thickness of 2D layered materials and films, such as graphene [50], 2D MOF [51, 52], LDH [53, 54], C_3N_4 [55, 56], and Mxenes [57, 58].

3.3 Spectroscopy Techniques

3.3.1 X-ray Spectroscopy

3.3.1.1 X-ray Diffraction (XRD)

Powder X-ray diffraction (PXRD) patterns are produced by the interaction of incident X-ray and power crystal samples. The crystal planes in crystalline materials have distances commensurate with X-ray wavelength and produce diffraction of X-rays with regions with increased or decreased X-ray intensity. PXRD is one of the most powerful methods to determine the crystal structure of material. The position of diffraction peaks correlates with the interplanar distance according to the Bragg's law:

$$n\lambda = 2d \sin \theta$$

where n is an integer number, λ is the wavelength of the X-ray source, d is the distance of adjacent diffracting planes, and θ is the incent angle of the X-ray source.

From the PXRD pattern, massive information, including crystal structures and orientations [59, 60], particle size (Scherrer equation), and defects [61–64], can be acquired. The catalytic performance of crystalline porous materials can be tuned *via* defect engineering [65, 66]. As an example, Farha's group synthesized a UiO(Zr/Hf) MOF series with various degrees of defects. The MOFs with more defects, having more available Brønsted-acid sites, indicated higher catalytic activity in the ring-opening reaction [67].

Research has shown that the two peaks in low angle of PXRD pattern (Miller index fcc planes (100) and (110), symmetry forbidden reflections) are disappeared for defective UiO-66 MOFs (the missing cluster defects) [68, 69].

Also, *in-situ* PXRD measurement can provide information on the phase evolution versus pressure, temperature, or reaction progress, allowing researchers to get a deeper understanding of the properties of materials under reaction conditions and their catalytic performance [70–72].

PXRD measurements are carried out in polycrystalline samples, and therefore, defects and random orientations are unavoidable. This makes it difficult or impossible to calculate the accurate unit cell dimension, especially for unknown materials without reported patterns in the Cambridge Crystallographic Data Center database (CCDC database). Therefore, to precisely obtain the diffraction patterns of certain materials, single crystals are usually grown and subjected to XRD analysis (similar instrument to that for PXRD), and the technique is called single-crystal x-ray diffraction (SC-XRD). Actually, due to the vast number of possible organic building blocks and metal nodes, MOFs and COFs can have tens of thousands of structure variants. Therefore, the foremost task for a newly discovered crystalline material is to obtain its single crystal to acquire its SC-XRD for the determination of its crystal structure. Crystal structure and phase purity of the solids can be assessed by comparing the experimental PXRD to simulated ones generated from SC-XRD or through the usage of computational approaches [73–76]. However, POPs (except crystalline COFs) are not formed as crystal structures to study by SCXRD. In addition, their PXRD measurements of the POPs show no or broad diffraction peaks representing their amorphous nature as expected [17, 77].

3.3.1.2 X-ray Photoelectron Spectroscopy (XPS)

XPS is conventionally employed to analyze element compositions, valence states, and chemical bonds of a thin shallow of material surface. For a given element, the electron energy in an electronic state depends on its valence state and coordination environment. When an incident beam of high energy X-ray (for example, Al K_{α}, $E_{photon} = 1486.7$ eV) interacts with the atoms in a sample, the energy can be absorbed by orbit electrons, and as a consequence, electrons are emitted with certain kinetic energy. Under high vacuum conditions, these electrons can be detected and converted into a current signal. Thus, the binding energy (BE) can be deduced based on the following equation:

$$E_{binding} = E_{photon} - (E_{kinetic} + \Phi),$$

where $E_{binding}$ is the binding energy of the detected electrons in an atom, $E_{kinetic}$ is the kinetic energy of the emitted electrons as measured by the spectrometer, and Φ is the work function of the spectrometer.

Each XPS signal measured with high-resolution mode has typically a few eV in width and can be analyzed and deconvoluted as resulting from the simultaneous contribution of individual components of much narrow binding energy (typically 0.5 eV or less) in various proportions. By comparing the obtained binding energy of the individual components and the values from databases and literature, rich information about the distribution of the element on the surface in various species, including their proportion, their valence state, possible effects of the interaction of the active site and the support, as well as the chemical bond, can be obtained [78–81]. In addition, XPS measurements can be performed *ex/in-situ* conditions, and recent instrument developments make it possible to work under certain pressure, thus allowing monitoring the properties of the active site closer to reaction conditions [82, 83].

Zhang et al. [84] proved the photoinduced electron transfer process within the MOF by means of XPS and other techniques. In this work, Eu-based MOF with a linker-containing pyridinium moiety was successfully synthesized (organic linker, $H_3TTTPCBr_3$ =1,1′,1″-(2,4,6-trimethylbenzene-1,3,5-triyl)-trimethylene-*tris*-(4-carboxypyridinium) tribromide) and its crystal structure solved by SC-XRD.

Upon UV light (365 nm) irradiation, an eye-detectable color change from colorless to pale yellow occurred, which was speculated to be caused by the generation of pyridinium radicals. This pale-yellow color is stable for months, thus allowing XPS to study the mechanism of the photoinduced electron transfer. From the XPS N1s spectrum, a new component at 399.8 eV was observed in the colored sample, besides the two contributions at 402.3 and 407.2 eV corresponding to pyridinium N and nitrate N atoms. This new N type was assigned to the pyridinium radicals. Further evidence of the electron transfer from Br^-/carboxylate $-CO_2^-$ to pyridinium N atoms was obtained from Br3d and O1s spectra, wherein new components at higher binding energies were observed in the colored sample compared to the parent MOF (Figure 3.5).

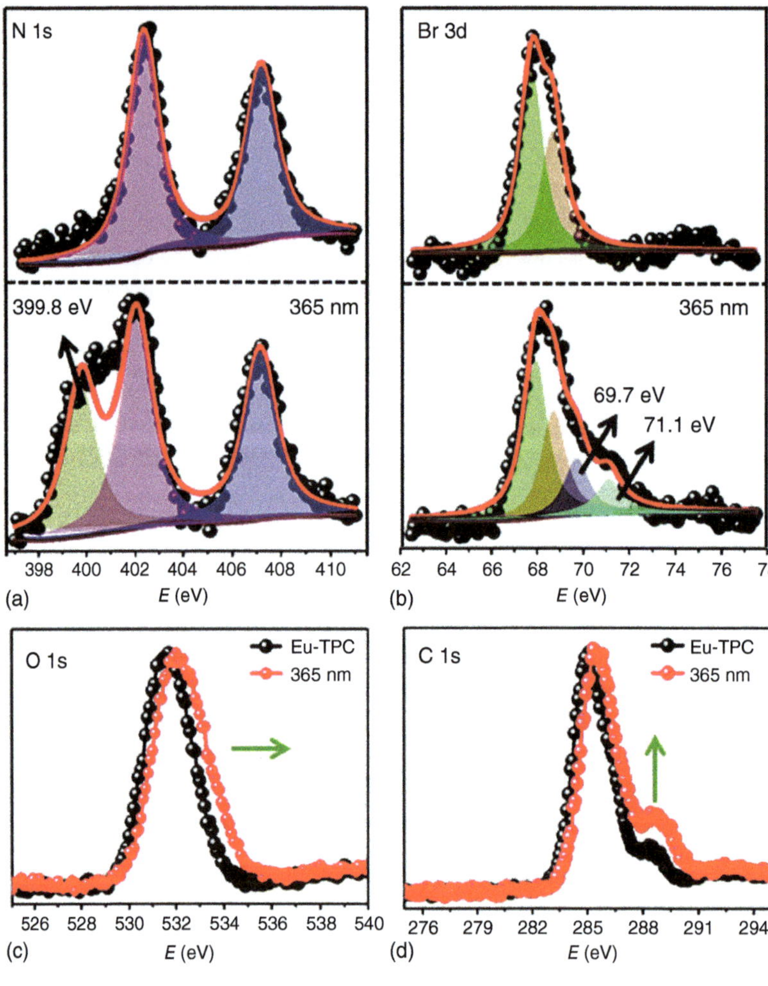

Figure 3.5 XPS core-level spectra of N1s (a), Br3d (b), O1s (c), and C1s (d) of Eu-TPC before and after 365 nm light irradiation. Source: Yang 2019, figure 3, p. 12831. Reproduced with permission of the Royal Society Chemistry. Yang et al. [84].

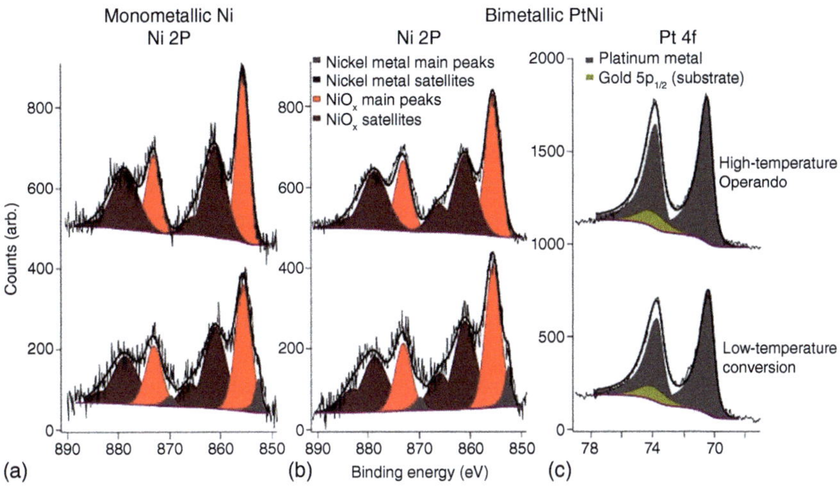

Figure 3.6 Temperature programmed CO oxidation over the Ni@UiO-67 and Pt/Ni@UiO-67 catalysts under NAP-XPS conditions (CO/O_2 ratio = 2, total pressure = 3 mbar): (a) Ni2p spectra of Ni@UiO-67, (b) Ni2p spectra of PtNi@UiO-67, (c) Pt4f spectra of PtNi@UiO-67. Source: Vakili 2019, figure 3, p. 526. Reproduced with permission of Elsevier. Vakili et al. [85].

In another study, Fan et al. [85] revealed the evolution of Pt/Ni bimetallic structure supported on UiO-67 during CO oxidation reaction by means of *operando* near-ambient-pressure X-ray photoelectron spectroscopy (NAP-XPS). By collecting the spectra of the monometallic nanoparticles and bimetallic nanoparticles (Ni@UiO-67, Pt@UiO-67, and Pt/Ni@UiO-67) after the H_2 activation, they observed a full reduction of Ni in Ni@UiO-67, while a partial reduction takes place in the case of Pt/Ni@UiO-67. In addition, a lower binding energy of Pt was also noted in Pt/Ni@UiO-67 compared to the value measured for Pt@UiO-67, which together with the partial reduction of Ni, demonstrates the electron transfer from Ni to Pt atoms (Figure 3.6). Furthermore, *operando* XPS revealed the formation of a NiOx shell on the surface of Pt/Ni nanoparticles, and this NiOx/Pt core–shell structure was proposed to be the active sites in the CO oxidation reaction.

3.3.1.3 X-ray Absorption Fine Structure (XAFS) Techniques

The energy of X-rays absorbed by core electrons is scattered by the surrounding coordination atoms. This makes that depending on the coordination number and interatomic distance, the absorption intensity of X-rays oscillates as a function of X-ray energy. Thus, the local electronic and geometric structure of atoms can be studied in a system by analysis of these oscillation signals. This specific shape of the fluctuation of X-ray absorption intensity vs. X-ray energy is termed XAFS. The availability of synchrotron X-ray sources enables the practical application of XAFS, since it has greatly reduced the time for XAFS measurement and increased the signal-to-noise ratio.

There are basically two types of XAFS measurements, depending on whether the analysis is applied to the shape of the edge of X-ray absorption (X-ray absorption near-edge structure, XANES) or to the shape at higher X-ray energies (extended X-ray absorption fine structure, EXAFS). The difference between XANES and EXAFS is, therefore, the absorption energy range. It is generally accepted that XANES is in the range of 50 eV above the absorption edge (multiple scattering region), while EXAFS is within 50–1000 eV above the absorption edge (single scattering region). From XANES, information about the element's oxidation state, unoccupied electronic states, spin state, and bonding environment can be attained, since XANES depend on the local molecular environment. By means of Fourier transform (FT) analysis of EXAFS signals, key information for longer distances reaching the surrounding atoms, specific bond distances, and coordination numbers can be obtained [86–88].

Single-atom dispersed catalysts are currently a frontier hot research topic and it is deemed as the bridge to connect homogeneous and heterogeneous catalysis [89]. Porous materials have been extensively employed as supports of single atoms due to their high surface area and micro/meso porous structure that are suited to immobilize a large density of single-atom active sites inside the cages. To confirm the atomic dispersity of the active sites, XAFS is an indispensable technique, together with aberration-corrected HAADF-STEM. Taking POPs as an example, Zhang et al [90] incorporated M-N_4 and M-N_2O_2 (M=Co and Ni) into the POPs based on phthalocyanine and salophen subunits. XANES of Co and Ni revealed that both metals are in +2 valence state. Co K-edge FT-XAFS spectra revealed an average bond distance of 1.38 Å, which is lying between the values of Co–N and Co–O, supporting Co coordination as Co-N_2O_2 configuration in the case of salophen (Figure 3.7). In addition, no values that correspond to Co–Co or Ni–Ni bond distance demonstrate the single atom dispersion of Co and Ni.

XAF spectroscopies are very suited for *in-situ* and *operando* studies, increasing even further the importance of these techniques. Thus, *in-situ/operando* EXAFS provides researchers opportunities to monitor the changes in the chemical environment under reaction conditions [91]. For instance, it has been observed *in situ* the proportion difference and coordination difference of oxidized CoNi supported on

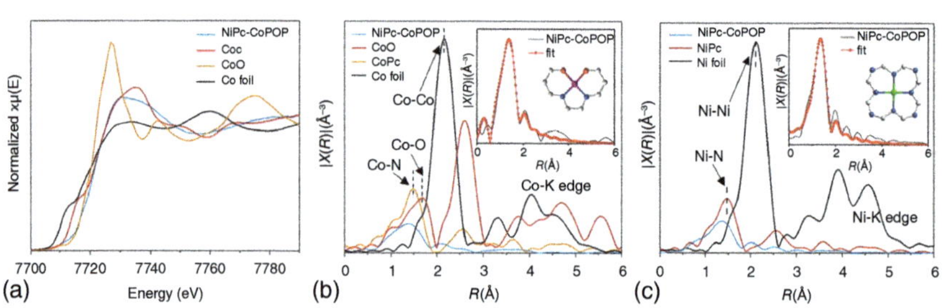

Figure 3.7 (a) XANES and (b) FT-EXAFS spectra at the Co K-edge of NiPc–CoPOP, CoPc, CoO, and Co foil sample. Inset: EXAFS fitting for NiPc–CoPOP and proposed schematic model of Co coordination environment in NiPc-CoPOP. (c) FT-EXAFS spectra at the Ni K-edge of NiPc–CoPOP, NiPc, and Ni foil sample. Inset: EXAFS fitting for NiPc-CoPOP and proposed schematic model of Ni coordination environment in NiPc-CoPOP. The red, blue, gray, pink, and green spheres represent O, N, C, Co, and Ni atoms, respectively. Source: Dong 2021, figure 2, p. 3. Reproduced with permission of Wiley-VCH. Dong et al. [90].

either bulk or nanosheet MOFs during oxygen evolution reaction. These differences observed during the reaction have been correlated with differences in catalyst performance [92]. Yan et al. employed *operando* synchrotron radiation XAS and revealed that $Ni^{3+/4+}$ as the active sites in oxygen evolution reaction with NiFe-MOF nanosheets as the electrocatalyst [93].

3.3.2 Nuclear Magnetic Resonance (NMR)

When a nucleus that possesses a ½ magnetic moment is placed in a strong magnetic field, the spin state parallel (denoted as α spin state) to the external magnetic field will possess lower energy compared to the spin state (denoted as β spin state) antiparallel to the magnetic field. Absorption of an electromagnetic wave in the radio frequency range can flip the α to the β spin state, which is called resonance. The NMR spectrum is a plot of the absorption intensity as a function of the radiation wavelength expressed in ppm with respect to a standard value. The NMR spectroscopy is an extremely powerful technique to determine the atomic level structure of a material, especially organic molecules-containing materials, both in solution or in solid state.

Since MOFs are insoluble in conventional NMR solvents, the solid digestion by using an acid or base (i.e. HCl, DCl, HF, H_2SO_4, D_2SO_4, NaOH, NaOD, NH_4HCO_3, and $NaHCO_3$) prior to its solution NMR measurement is required [91, 94–103]. The dissolution NMR experiments are commonly used to determine the ratios of linkers, modulator to linker, post-linker exchange, functionalization of organic linkers, and missing-linker defects, as well as solvent removal in porous materials.

In recent years, solid-state NMR (SS-NMR) spectroscopy has already proven its unique role to characterize the structure of porous compounds [104, 105], offering a useful technique for the description of chemical structure, the location of binding sites, understanding about the host–guest interactions, and dynamics of guests inside a porous solid [104, 106, 107].

For quadrupolar nuclei with integer spin, for example, $Al,^{27}Zn,^{67}$ $Mg,^{25}$ $Ni,^{63}$ etc., which are common as the node of MOFs, the NMR measurements can be performed in solid state by magic angle spinning (MAS) with 2D heteronuclear NMR technique. By the combination of ^{1}H, ^{13}C, ^{15}N, ^{17}O, and metal node MAS 2D NMR, structure–property relationships can be established [108]. An extended reading regarding the solid-state NMR spectroscopy for microporous materials can be accessed in a recent review from Deng et al. [109]. However, ss NMR, generally, suffers from poor spectral resolution and sensitivity owning to broad peaks as a result of strong anisotropic interactions.

The unit building blocks of COFs and POPs are organic molecules, and thus determination of their structure can be done by ^{1}H and ^{13}C NMR spectroscopies. For instance, the parent TpPa-1 COF, consisting of 1,3,5-triformylphloroglucinol (Tp) and p-phenylenediamine (Pa-1) components, exhibited three different H environments in the ratio of 1 : 1 : 2 in its ^{1}H NMR spectrum that corresponds to N-H (a), C=C-H (b), and Ar-H (c) types of hydrogens, as shown in Figure 3.8a. The proportion of these type of hydrogens changed to 0.75 : 1 : 2 after the incorporation of Zn^{2+}ions, indicating the partial exchange of H atoms in environment (N-H) by Zn^{2+} (Figure 3.8) [110].

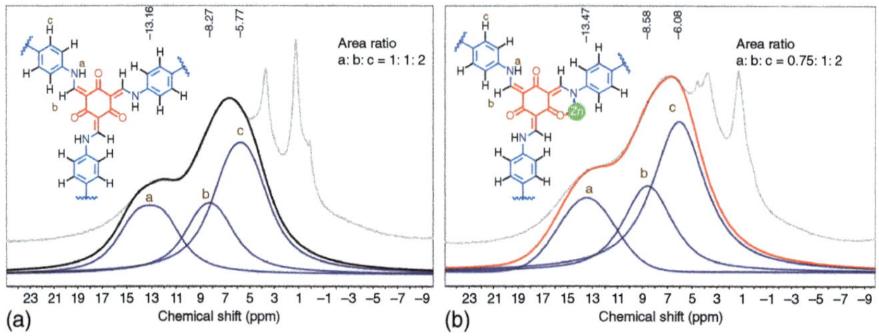

Figure 3.8 ^{1}H NMR spectra with deconvoluted individual peaks of pristine TpPa-1 (a) and Zn (10)-TpPa (b). Source: Cao 2021, figure 1, p. 3. Reproduced with permission of Elsevier. Cao et al. [110].

3.3.3 Electron Paramagnetic Resonance (EPR)

The basic concepts of EPR (electron spin resonance, ESR, or electron magnetic resonance, EMR, spectroscopy) are similar to NMR, except the spin excited are that of unpaired electrons instead of nuclei. In addition, instead of fixing the magnetic field like in NMR, the EPR instrument emits a microwave with constant frequency and scans by changing the magnetic field strength.

EPR spectroscopy has been extensively applied to detect and identify radical species intermediates (such as singlet oxygen, superoxide ions, hydroxyl and alkyl radicals) formed during the reactions, as well as surface atom vacancies (V, Zn^{+}, Ti^{3+}, etc.), and transition metal valence states [19, 20, 111–118]. However, direct EPR detection of many free radicals is almost impossible in solution at room temperature due to their high reactivity and very short half-life. Spin trapping, a technique in which a diamagnetic molecule (spin trap), usually an organic nitrone or nitroso compound, is employed to react with the reactive free radical to produce a more stable free radical (so-called radical/spin adduct) that can be detected by EPR. Some of spin traps are 5,5-dimethyl-1-pyrroline *N*-oxide (DMPO, a spin trap for oxygen-centered radicals mostly superoxide and hydroxyl species), 5-diethoxyphosphoryl-5-methyl-1-pyrroline *N*-oxide (DEPMPO, a spin trap for hydroxyl and superoxide radicals), and 2,2,6,6-tetramethyl-4-piperidinol (TEMP, a spin trap for singlet oxygen) [20, 113, 119], which are used to detect reactive oxygen species (ROS) generated by the irradiation of light on porous photocatalysts.

For instance, a carbazolyl porphyrin-based conjugated microporous polymer (TCPP-CMP) was successfully synthesized and used as a metal-free heterogeneous photocatalyst for several aerobic oxidation reactions under visible light irradiation [119]. To gain insight into the detailed photocatalytic mechanism, EPR experiments were performed in the presence of DMPO and TEMP as well-known trapping agents for $O_2^{\cdot-}$ and $^{1}O_2$, respectively. After adding the trapping agents and visible light irradiation under O_2 atmosphere, the characteristic EPR signals of superoxide (a sextet pattern (or a quartet in some cases) [120, 121] assigned to DMPO-$O_2^{\cdot-}$ or DMPO/OOH·) (Figure 3.9a) and singlet oxygen radicals

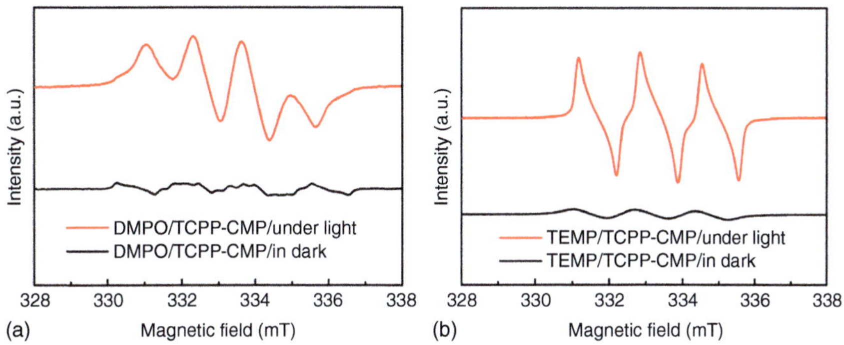

Figure 3.9 EPR spectra of the samples: (a) DMPO or (b) TEMP solution with TCPP-CMP and O_2 under visible light illumination ($\lambda > 420$ nm) and in dark. Source: Jiang 2020, figure 7, p. 3528. Reproduced with permission of Wiley-VCH. Jiang et al. [119]

(a 1 : 1 : 1 triplet pattern assigned to 2,2,6,6-tetramethylpiperidine-1-oxyl (TEMPO)) (Figure 3.9b) were obviously observed, meaning that TCPP-CMP is capable of producing $O_2^{\cdot-}$ and $^1O_2$ under photocatalytic conditions. Under other experimental conditions, such as in the absence of light or oxygen, no EPR signal is detected. It should be noted that the EPR detection of DMPO/OOH· is not without its difficulties such as interference of transition metals, short lifetime of DMPO/OOH·, reaction of $O_2^{\cdot-}$ with DMPO/OOH· and DMPO/OH·, and the possibility that DMPO/OOH· spontaneously breakdowns to form DMPO/OH.

In another instance, Zhong et al. [117] observed photoinduced electron transfer from $Ru(bpy)_3^{2+}$ excited state to Zr-oxo nodes upon light irradiation based on the EPR peak at g = 2.002. Introduction of CO_2 results in a decrease of this signal, indicating the transfer of electrons from the Zr-oxo nodes to adsorbed CO_2 as reactant. In addition, the intensity of Cu(II) signal decreases and the intensity increases in Ni(II) signal in the EPR spectrum under CO_2 atmosphere compared to that in inert atmosphere, together with the evidence of Cu(I) peak in the XPS Auger spectrum, further demonstrate the electron transfer from Zr-oxo to CO_2 via Cu/Ni cocatalyst [117].

3.3.4 Ultraviolet-Visible Diffuse Reflectance Spectroscopy (UV–Vis DRS)

UV–Vis DRS is widely used to study the electronic properties and to calculate the optical band gap energy (E_g) of solids [122, 123]. To calculate the bandgap of photocatalysts (i.e. MOFs and POPs) from DRS measurements, the Kubelka–Munk (K–M) function and Tauc plot are most commonly used [20, 123–127]. The Tauc method is based on the hypothesis that the energy-dependent absorption coefficient α is related to the following equation:

$$(\alpha h\nu)^{1/n} = B(h\nu - E_g)$$

where h is the Planck's constant, ν is the frequency of incident photon, E_g is the band gap energy, and B is a constant. The n factor relies on the nature of the electron transition and is equal to 1/2 or 2 for the direct and indirect transition band gaps, respectively [123]. Band gap is estimated from diffuse reflectance UV–Vis spectrum according to Tauc plots by drawing $(\alpha h\nu)^{1/2}$ (for indirect E_g value) or $(\alpha h\nu)^2$ (for direct E_g value) against energy $h\nu$ (eV). Based on the most often used method, the E_g is also calculated by applying the Kubelka–Munk function by the intercept of the tangents to the plots of $(F(R)\ h\nu)^{1/2}$ (for indirect E_g value) or $(F(R)\ h\nu)^2$ (for direct E_g value) vs photon energy ($h\nu$): where $h\nu$ is the photon energy ($h\nu$ (eV) $= E = hc/\lambda = 1240$/incident wavelength), and $F(R)$ is Kubelka–Munk function ($F(R) = (1-R)^2/2R$) (R is the reflectance) [123]. Placing $F(R)$ instead of α into the previous eq offers the following equation:

$$(F(R)\ h\nu)^{1/n} = B(h\nu - E_g)$$

However, it is needed to know the property of a new material (whether or not conductor or semi-conductor or insulator is that) and the accurate method for determination of the band-gap energy [123].

For instance, the UV–Vis DRS of some acridine-containing COFs (Tp-Acr, DHTA-Acr, HTA-Acr, and Tp-DAA; Acr: 2,6-diaminoacridine and DAA: 2,6-diaminoanthracene) is shown in Figure 3.10a [128]. Accordingly, the spectra reveal that the acridine COFs can absorb light in the visible region, with absorption edges at about 680 nm tailing up to more than 800 nm. Based on the Kubelka–Munk formula, the optical band gaps of the acridine COFs were calculated to be 1.82–1.83 eV (Figure 3.10b) [128]. In another example, the calculated optical bandgap of UiO-66-NH_2 was obtained as 2.75 eV as shown in Figure 3.10c [129]. Also, by applying the direct Tauc method, optical band gaps were calculated to be 2.62 and 2.60 eV for NU-1000 MOF [130] and NU-1102 [130], respectively (Figure 3.10d,e), which announce the samples for visible-light photocatalysis.

3.3.5 Inductively Coupled Plasma (ICP) Analysis

ICP can generate a torch with temperature over 5000 K, which results in the degradation of the samples to individual elements and thus, this technique is applied for element concentration quantification. Depending on the detection method, it can be classified into inductively coupled plasma optical emission spectroscopy (ICP-OES), of which the quantitation is based on measurements of the intensity of characteristic wavelength of the target element, and inductively coupled mass spectrometry (ICP-MS), which measures an atom's mass. The detection limit of ICP-MS is as low as parts per trillion (ppt), while for ICP-OES, concentrations over parts per billion (ppb) can be quantified [131]. For both techniques, the samples for quantification must be accurately weight, and the targeted elements must be completely digested by acid or piranha solution [41] (and usually under microwave irradiation) before injection into the instruments. If the atomic ratios from ICP are needed, the exact weight of the sample is not necessary. This is extremely important when dealing with compounds that are resistant to acids (even aqua regia), for example, ruthenium species.

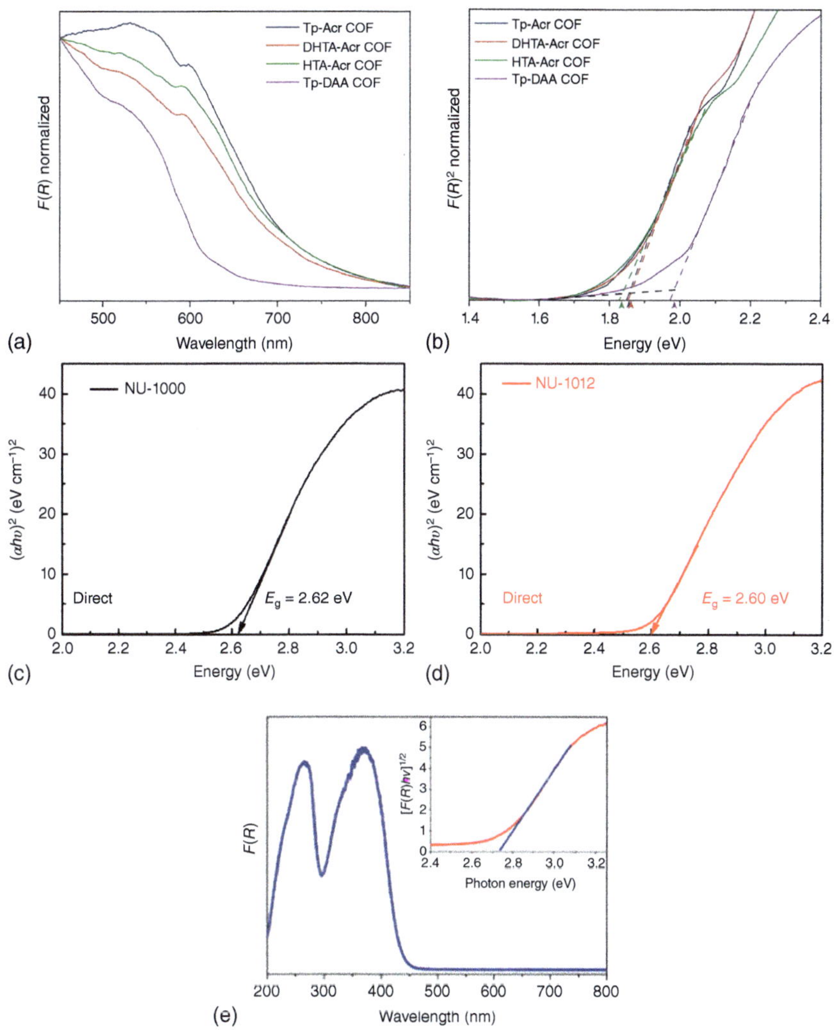

Figure 3.10 Solid-state UV–Vis diffuse reflectance spectra for Tp-Acr, DHTA-Acr, HTA-Acr, and Tp-DAA (a), and Tauc plots for absorption spectra obtained with the Kubelka–Munk function and the linear fit for the COFs (Source: Traxler 2022, figure 3, p. 4. Reproduced with permission of Wiley-VCH. Traxler et al. [128]). (b) and UiO-66-NH_2 (Source: Long 2012, figure 1, p. 11565. Reproduced with permission of the Royal Society Chemistry. Long et al. [129]). (c). Tauc plots for NU-1000 (d) and NU-1012 (e) (Source: Wong 2022, figure S16, p. 29. Reproduced with permission of American Chemistry Society. Wang et al. [130]).

Compared to energy-dispersive X-ray spectroscopy (EDS), which heavily depends on the distribution of the elements on a given samples, ICP-MS and ICP-OES offer the average concentration of each element throughout the sample. Therefore, the combination of ICP and EDS can present a comprehensive information of the element distribution, as well as the accurate concentration. For instance, Farha et al. incorporated polyoxometalate ($[PW_{12}O_{40}]_3^-$) in NU-1000 MOF by

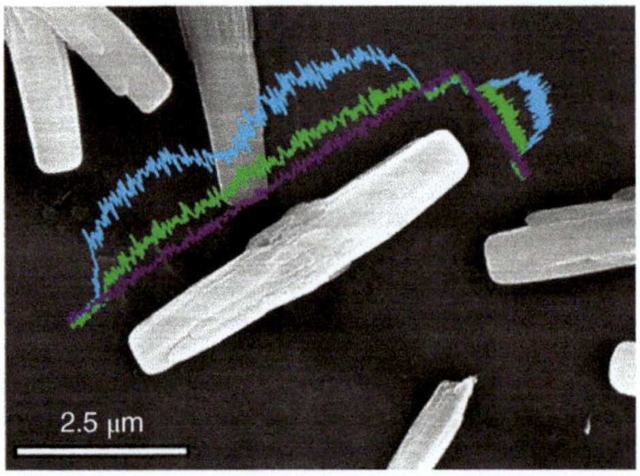

Figure 3.11 SEM-EDS of NU-1000 after $[PW_{12}O_{40}]^{3-}$ incorporation. EDS legend: light blue = W, green = Zr, purple = Zn (baseline). Source: Reproduced with permission from Buru et al. [132], American Chemical Society.

impregnation method, and the amount of $[PW_{12}O_{40}]_3^-$ per Zr_6 node was estimated by ICP-OES [132]. To further confirm the distribution of $[PW_{12}O_{40}]_3^-$ inside the MOF, EDS mapping was conducted, and the results (Figure 3.11) clearly indicated that the $[PW_{12}O_{40}]_3^-$ located uniformly within the MOF. For bimetallic node MOFs [41, 133, 134], or the MOFs obtained from metathesis [135], ICP is the most straightforward way to determine the atomic ratio of each metal component.

3.4 Other Techniques

3.4.1 Thermogravimetric Analysis (TGA)

TGA monitors the mass loss of a sample exposed to a given atmosphere (air or O_2 or inert gases) as the temperature increases, which can provide a variety of information, including phase transition, thermal decomposition, desorption, etc. One of the important properties of porous materials that can be evaluated by TGA is thermo-stability that directly determines the working temperature and recyclability constraints and is porosity that defines through potentially solvent-accessible volume [96, 131, 136, 137]. It should be noted that the decomposition pathway of a porous organic linker-containing material is different when switching the atmosphere from inert to air (or to oxygen). In addition, the mass loss of a sample in a defined atmosphere is changed with the type and content of solvent [138, 139]. Figure 3.12 presents a representative TGA curve of two MOFs. As it can be seen there, the first mass loss occurs from 25 to 100 °C and corresponds to water desorption (dehydration), followed by a slightly decreasing mass from 100 to ~300 °C [141], which is due to the removal of the solvent (DMF) (solvent removal). The subsequent mass loss over 300 °C is attributable to dehydration of the $Zr_6O_4(OH)_4$ nodes into Zr_6O_6 (dehydroxylation) associated with initial degradation of the organic linker around 500 °C [126]. The next stage at temperatures higher than 500 °C corresponds to complete linker and framework decomposition (T_{dec}) [140]. At the end of TGA experiment (heating up to 800 °C), about 40% of the starting weight remained and related to the formation of ZrO_2 with a very low degree of crystallinity [126].

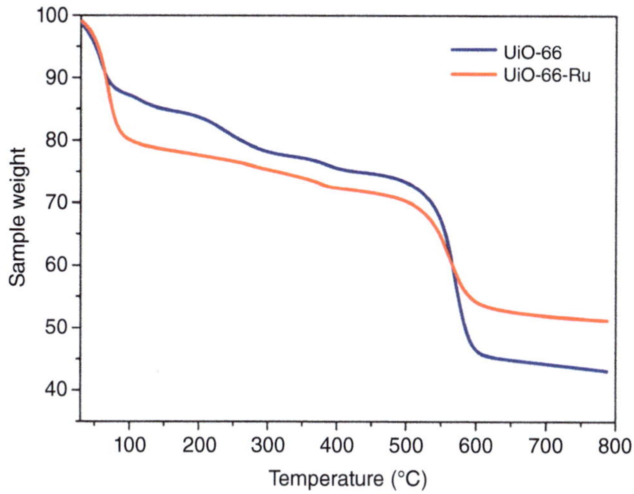

Figure 3.12 TGA plots of UiO-66 and UiO-66-Ru. Source: Yang 2018, figure 8, p. 4204. Reproduced with permission of American Chemistry Society. Yang et al. [140].

Notably, the TGA data is commonly used to indicate the fraction of missing-linker defects in MOFs [63, 142]. For example, based on the TGA measurements, a quantitative amount of the defects in UiO-66 MOF has been estimated according to the method reported by Lillerud and co-workers [64]. So, in an aerobic TGA curve of the UiO-66 samples, the end weight of the product ($6ZrO_2$) is supposed as 100%. The molecular weight of the dehydroxylated defect-free MOF, $Zr_6O_6(BDC)_6$, is ideally presumed 220% (2.2 times the mass larger than of $6ZrO_2$). In the perfect MOF, each Zr_6 cluster is connected to 12 carboxylates from 1,4-benzenedicarboxylate (BDC) linkers (12 linkers per Zr_6 node). Then, to calculate the number of missing BDC linker defects per Zr_6 node, the actual weight of defective UiO-66 prior to its thermal decomposition is compared with these principles. As an example, this number was calculated to be 4.4 for HI-UiO-66 and 6 (maximum of defects) for HI-UiO-66-SO_4 (HI = hemilabile, 4-sulfonatobenzoate (PSBA) used as the hemilabile linker) which means degree loss of connectivity of 4.4 and 6, respectively, out of 12 linkers per each Zr_6 node (Figure 3.13) [63]. The presence of defects is also assessed by a variety of techniques such as dissolution ^1H NMR, PXRD (the disappearance of symmetry forbidden reflections) [64], and N_2 adsorption (mostly, not all times, the increase of surface area and pore size distribution compared to theoretically expected and the ideal structure) [64, 143, 144].

3.4.2 N_2 Adsorption

The Brunauer–Emmet–Teller (BET) surface area and pore size distribution (PSD) measured by N_2 physisorption are two important properties for porous materials, considering their two main featured properties, namely, high surface area and regular porosity. As already commented, porosity makes this type of materials

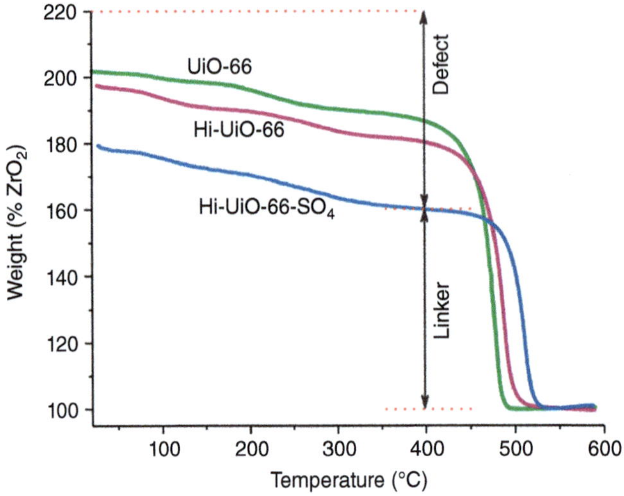

Figure 3.13 TGA profiles of defective UiO-66, HI-UiO-66, and HI-UiO-66 after soaking the samples in an H_2SO_4 solution. Source: Feng et al. [63]/Public Domain.

suitable for a variety of applications including gas storage, adsorption separation, and catalysis. In addition, porous materials with different cavity size and window diameter can discriminate reactants according to their dimensions for their access to the active sites located in the cavity. This size selectivity can be used to implement shape-selective reactions with a mixture of substrates [43]. For porous materials, sample pre-treatment with vacuum and temperature (most commonly method) or solvent exchange [145] or supercritical carbon dioxide [146], removing accurately the adsorbed solvents, so-called activation process, is crucial prior to running further experiments [131, 146, 147].

Proper choice of the gas adsorption (usually N_2 gas at 77 K) and the region of the isotherm selected (linear regions), make it possible to correctly report BET surface area and the pore size distribution for such porous materials [131, 145]. Carbon dioxide is frequently used over nitrogen for a material with high-density hydrogen bonding in its small pores [148].

There are six types of adsorption isotherms based on the International Union of Pure and Applied Chemistry (IUPAC) classification for gas–solid systems as presented in Figure 3.14 [149]. Briefly, microporous materials (Type I isotherms; Type I(a) and Type I(b) are microspores <~1 and <~2.5 nm, respectively); non-porous or macroporous adsorbents (Type II isotherms); Type III and Type V isotherms for samples in which there are relatively weak adsorbent-adsorbate interactions; mesoporous materials (Type IV isotherms; Type IV(a) is for larger mesopores (>4 nm) which lead to hysteresis loops and Type IV(b) is used for smaller mesopores); and Type VI isotherms indicate that the adsorption happens layer-by-layer on a smooth nonporous surface. In addition, the hysteresis patterns can also provide important information about the pore structure (pore geometry and pore connectivity) of the porous network, which is crucial for guest storage and transport-related applications [150–152].

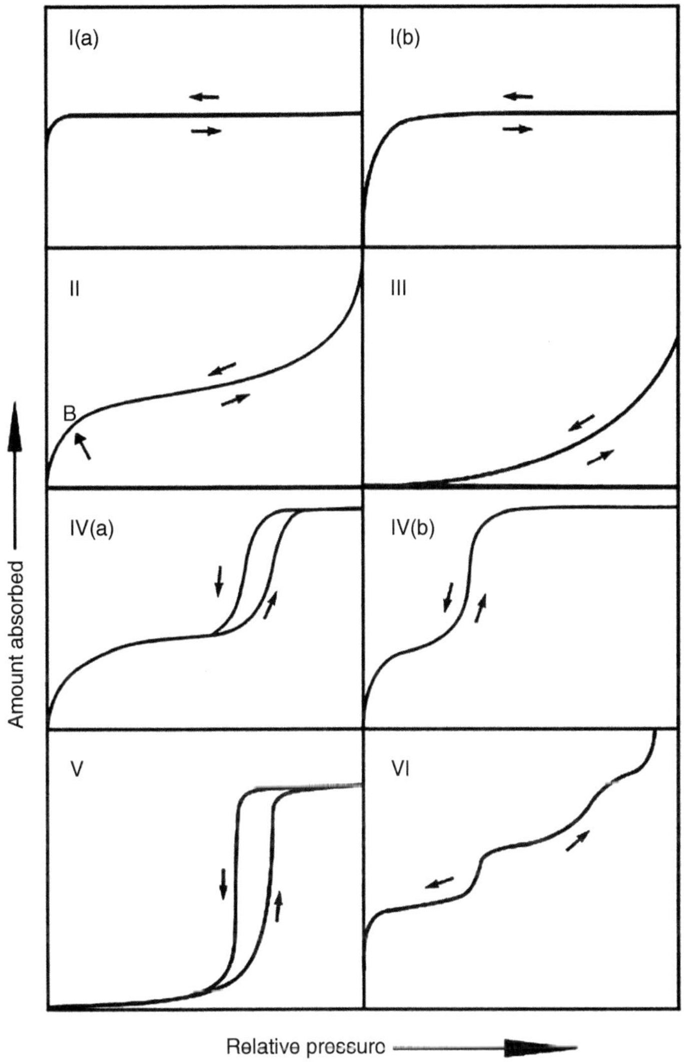

Figure 3.14 The IUPAC classification of physisorption isotherms. Source: Thommes 2015, figure 2, p. 1058. Reproduced with permission of Walter de Gruyter. Thommes et al. [149].

McGrier et al. reported the synthesis of a mesoporous 3D COF-containing dehydrobenzoannulene (DBA), namely, DBA-3D-COF 1 [153]. The COF was then post-metalated with $Ni(COD)_2$ affording Ni-DBA-3D-COF. The porosity of COFs was confirmed by nitrogen adsorption experiments at 77 K (Figure 3.15). A type IV isotherm of DBA-3D-COF 1 shows a steep increase at $P/P_0 < 0.1$, followed by a sharp step between $P/P_0 = 0.05$ and 0.13, indicating its mesoporosity. The BET surface area of $5083\,m^2\,g^{-1}$ was obtained for PCN-600(Fe) over the low-pressure region ($0.07 < P/P_0 < 0.11$). The calculated total pore volume at $P/P_0 = 0.974$ was $1.85\,cm^3\,g^{-1}$. A nonlocal density functional theory (NLDFT) simulation from the

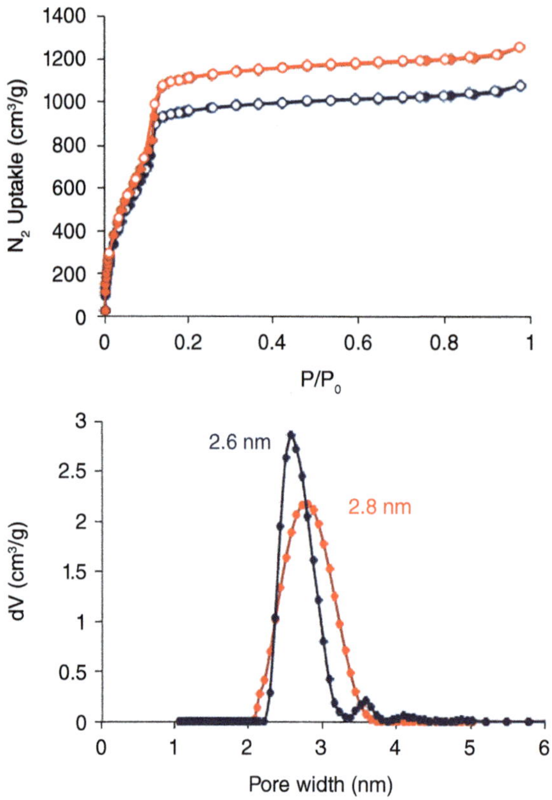

Figure 3.15 N_2 adsorption–desorption isotherms (top) and NLDFT pore size distributions (bottom) of DBA-3D-COF 1 (red) and Ni-DBA-3D-COF (blue) at 77 K. Source: Baldwin 2016, figure 3, p. 15134. Reproduced with permission of American Chemistry Society. Baldwin et al. [153]

N_2 sorption curve offers that there is an average pore size of 2.8 nm. In addition, the BET surface area, total pore volume, and pore size of the pristine COF were clearly decreased after the metalation.

3.4.3 Density Functional Theory (DFT) Calculations

DFT calculations have been extensity used to complement and rationalize the experimental data, such as those provided by XAFS, FTIR, ESR, and XPS, and to predict structure–property relationships for specific reactions, thereby providing guidance for further catalyst design [154–157]. Taking the methane oxidation to methanol on Cu-oxo catalyst as an example [37], the configuration of Cu ion immobilized in NU-1000 has according to EXAFS four-coordinate square-planar $Cu(OH)_2$-like structure. To further study the detailed structure, DFT calculations were performed to compute a series of mononuclear (M-A) and dinuclear (D-A) Cu coordination structures, and two optimized structures, namely, M-A and D-A structures, were proposed (Figure 3.16). The simulated EXAF spectra of the model structures are coincidence with the experimental results (Figure 3.16), illustrating the conversion of mononuclear and dinuclear Cu configuration in the case of low and high Cu loading, respectively. For both dinuclear and mononuclear Cu sites, it was predicted from the computation that the Cu(II) hydroxide and Cu(II) oxide were the most

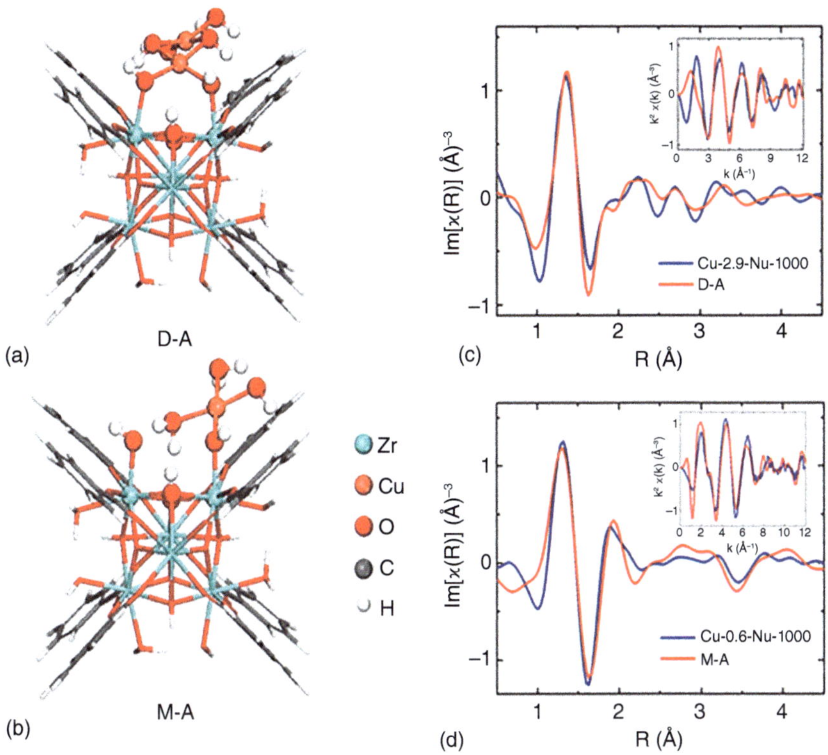

Figure 3.16 Possible DFT-optimized structure of dinuclear D-A complex $[Cu^{II}_2(OH)_4\ (H_2O)]$ on the MOF (a) and mononuclear M-A complex $[Cu^{II}\ (OH)_2\ (H_2O)_2]$ on the MOF (b); k^2-weighted Cu-EXAFS Im[χ(R)] spectra for (c) Cu-2.9- NU-1000 and (d) Cu-0.6-NU-1000 and simulated spectra from the DFT optimized D-A and M-A models. Insets display the respective *x*(*k*) plots. Source: Zheng 2019, figures 6 and 7, p. 9297. Reproduced with permission of American Chemistry Society. Zheng et al. [37].

stable structures, while a small fraction of Cu(II) was transformed to oxyl species $((Cu^{II}(O^{\cdot})(OH)$ and $Cu^{II}_2(O^{\cdot})(OH)_3)$ in the case of low and high Cu loading, respectively. By computing the energy barrier of each intermediate, it was found that in both cases (dinuclear and mononuclear), Cu oxyl species are the active sites, instead of Cu(II) hydroxide or Cu(II) oxide, responsible for C-H activation and functionalization (Figure 3.17).

Besides validation of experimental results, another powerful use of DFT calculations is prediction of structure/performance correlations. In this way, DFT can guide the catalyst screening/design to address specific reactions. For instance, Snurr et al. [158] employed DFT calculation to study the correlations of a Fe_2M variant in PCN-250 MOF regarding its activity for light alkane oxidative dehydrogenation to olefins. The energy barrier of two uphill reaction steps, namely, N_2O (oxidants) activation and C-H activation were computed varying the Fe_2M composition (M=Ti, V, Mn, Fe, Co, Cu, and Zn). From the oxygen-binding energy values, it was calculated that oxygen has preference to bind early transition metals, including Ti, V, and Cr,

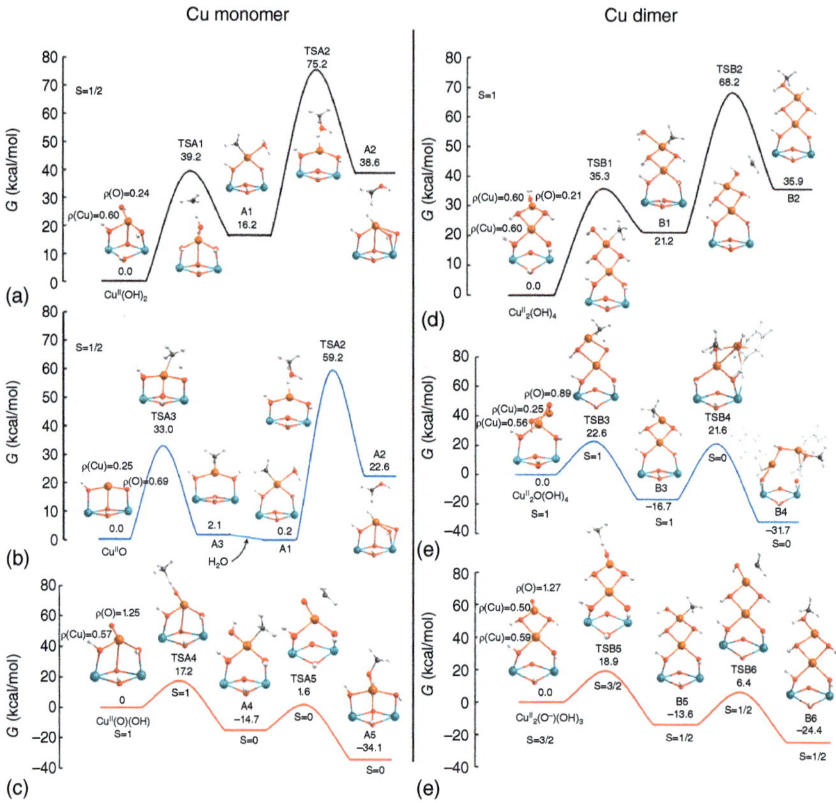

Figure 3.17 Reaction pathways of methane oxidation in the presence of possible active copper sites for (a) $Cu^{II}(OH)_2$, (b) $Cu^{II}O$, (c) $Cu^{II}(O\cdot)(OH)$, (d) $Cu^{II}_2(OH)_4$, (e) $Cu^{II}_2O(OH)_2$, and (f) $Cu^{II}_2(O\cdot)(OH)_3$. The quantities are free energies of equilibrium species and free energies of activation of transition states, both relative to the left-most species in the panel. Source: Zheng 2019, figure 10, p. 9297. Reproduced with permission of American Chemistry Society. Zheng et al. [37].

and thus those metals form the M=O active species during the N_2O activation, whilst in the case of Ni, Cu, and Zn, Fe=O will be the active site due to higher oxygen binding energy of Fe (Figure 3.18a). Further computation indicated that the energy barriers of N_2O activation are positively correlated to the oxygen binding energy, and the energy barrier of C-H activation exhibited an inverse trend (Figure 3.18b). Based on those two plots, the authors predicted that Fe_2M clusters closer to the cross point (cross point in Figure 3.18b) possess higher activity. Four different MOF catalyst, including PCN-250(Fe_2Mn), PCN-250(Fe_3), PCN-250(Fe_2Co), and PCN-250(Fe_2Ni), were prepared to validate the predicted activity. As revealed from the product (propene) concentration, the catalytic activity follows the trend PCN-250(Fe_2Mn) ~ PCN-250(Fe_3) > PCN-250(Fe_2Co) > PCN-250(Fe_2Ni), which is in coincidence with the N_2O activation energy barrier as depicted in Figure 3.18b, demonstrating how useful has been the prediction ability of DFT calculation for such reaction.

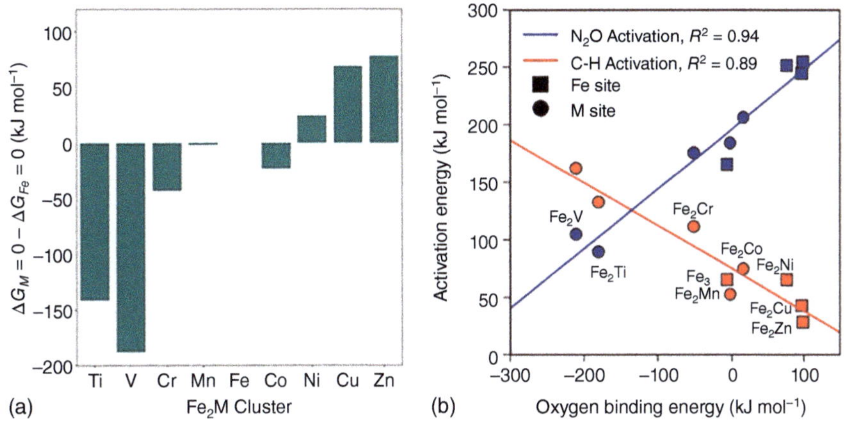

Figure 3.18 (a) Predilection of oxygen binding for the Fe or M atom, suggested by the difference between the binding free energies of oxygen on the M site ($\Delta G_{M=O}$) and on the Fe site ($\Delta G_{Fe=O}$). A negative value is related to a favorable binding to the M site and a positive value is related to a favorable binding on the Fe site. (b) Correlation between the oxygen-binding energy and the N_2O activation barrier (blue) and the C-H activation barrier (red). Source: Barona 2020, figure 5, p. 1463. Reproduced with permission of American Chemistry Society. Barona et al. [158].

3.5 Conclusions

In this chapter, we summarize the common and advanced techniques used to identify porous organic linker-containing materials such as MOFs, COFs, and POPs. The chemistry of porous materials is a cutting-edge research field and with new progresses come new techniques. Various measurements should commonly be done to approve successful synthesis of these materials. We hope that this chapter will contribute to the greater scientific knowledge base related to the field of crystalline and amorphous materials. It is also expected to be valuable for the students and researchers to get more motivation in studying areas we described.

Acknowledgments

This work was supported by the University of Zabol (UOZ) and is gratefully acknowledged (Grant Number: IR-UOZ-GR-9381). We also acknowledge the support of the Iran Science Elites Federation (ISEF).

References

1 Sakamaki, Y., Tsuji, M., Heidrick, Z. et al. (2020). *J. Chem. Educ.* 97: 1109–1116.
2 Zhou, H.-C., Long, J.R., and Yaghi, O.M. (2012). *Chem. Rev.* 112: 673–674.
3 Redfern, L.R. and Farha, O.K. (2019). *Chem. Sci.* 10: 10666–10679.

4 McKeown, N.B. and Budd, P.M. (2006). *Chem. Soc. Rev.* 35: 675–683.
5 Mohamed, M.G., El-Mahdy, A.F.M., Kotp, M.G., and Kuo, S.-W. (2022). *Mater. Adv.* 3: 707–733.
6 Côté, A.P., Benin, A.I., Ockwig, N.W. et al. (2005). *Science* 310: 1166–1170.
7 Ding, S.-Y. and Wang, W. (2013). *Chem. Soc. Rev.* 42: 548–568.
8 Lee, J.-S.M. and Cooper, A.I. (2020). *Chem. Rev.* 120: 2171–2214.
9 McKeown, N.B. (2020). *Polymer* 202: 122736.
10 Tian, Y. and Zhu, G. (2020). *Chem. Rev.* 120: 8934–8986.
11 Tan, L. and Tan, B. (2017). *Chem. Soc. Rev.* 46: 3322–3356.
12 Liu, M., Guo, L., Jin, S., and Tan, B. (2019). *J. Mater. Chem. A* 7: 5153–5172.
13 Fang, R., Dhakshinamoorthy, A., Li, Y., and Garcia, H. (2020). *Chem. Soc. Rev.* 49: 3638–3687.
14 Yang, D. and Gates, B.C. (2019). *ACS Catal.* 9: 1779–1798.
15 Guo, J. and Jiang, D. (2020). *ACS Cent. Sci.* 6: 869–879.
16 Kaur, P., Hupp, J.T., and Nguyen, S.T. (2011). *ACS Catal.* 1: 819–835.
17 Zhang, T., Xing, G., Chen, W., and Chen, L. (2020). *Mater. Chem. Front.* 4: 332–353.
18 Dhakshinamoorthy, A., Li, Z., and Garcia, H. (2018). *Chem. Soc. Rev.* 47: 8134–8172.
19 Karamzadeh, S., Sanchooli, E., Oveisi, A.R. et al. (2022). *Appl. Catal. B Environ.* 303: 120815.
20 Daliran, S., Khajeh, M., Oveisi, A.R. et al. (2022). *ACS Sustain. Chem. Eng.* 10: 5315–5322.
21 Wang, H., Wang, H., Wang, Z. et al. (2020). *Chem. Soc. Rev.* 49: 4135–4165.
22 Wang, T.-X., Liang, H.-P., Anito, D.A. et al. (2020). *J. Mater. Chem. A* 8: 7003–7034.
23 Wu, M.-X. and Yang, Y.-W. (2017). *Chin. Chem. Lett.* 28: 1135–1143.
24 Li, H., Li, L., Lin, R.-B. et al. (2019). *EnergyChem* 1: 100006.
25 Bhanja, P., Modak, A., and Bhaumik, A. (2019). *ChemCatChem* 11: 244–257.
26 Mal, A., Ding, H., Li, M. et al. (2022). *ACS Appl. Nano. Mater.* 5, 10, 13972–13984. https://doi.org/10.1021/acsanm.2c01517.
27 Wei, X., Yang, X.-F., Wang, A.-Q. et al. (2012). *J. Phys. Chem. C* 116: 6222–6232.
28 Zheng, X., Li, Y., Zhang, L. et al. (2019). *Appl. Catal. B Environ.* 252: 98–110.
29 Kang, Y.-H., Liu, X.-D., Yan, N. et al. (2016). *J. Am. Chem. Soc.* 138: 6099–6102.
30 Li, S.-J., Zhou, Y.-T., Kang, X. et al. (2019). *Adv. Mater.* 31: 1806781.
31 Yang, Q., Liu, W., Wang, B. et al. (2017). *Nat. Commun.* 8: 14429.
32 Khan, M.A., Nadeem, M.A., and Idriss, H. (2016). *Surf. Sci. Rep.* 71: 1–31.
33 Li, X., Rong, H., Zhang, J. et al. (2020). *Nano Res.* 13: 1842–1855.
34 Wu, W., Tran, D.T., Cheng, S. et al. (2021). *Microporous Mesoporous Mater.* 311: 110710.
35 Hülsey, M.J., Zhang, B., Ma, Z. et al. (2019). *Nat. Commun.* 10: 1330.
36 Foster, A.J. and Lobo, R.F. (2010). *Chem. Soc. Rev.* 39: 4783–4793.
37 Zheng, J., Ye, J., Ortuño, M.A. et al. (2019). *J. Am. Chem. Soc.* 141: 9292–9304.
38 Yang, C., Liu, S., Wang, Y. et al. (2019). *Angew. Chem. Int. Ed. Eng.* 58: 11242–11247.

39 Brogaard, R.Y., Kømurcu, M., Dyballa, M.M. et al. (2019). *ACS Catal.* 9: 5645–5650.

40 Chen, C., Zhu, X., Wen, X. et al. (2020). *Nat. Chem.* 12: 717–724.

41 Chang, T.-E., Chuang, C.-H., Chen, Y.-H. et al. (2022). *ChemCatChem* 14, e202200199. https://doi.org/10.1002/cctc.202200199, e202200199.

42 Elsaidi, S.K., Venna, S.R., Mohamed, M.H. et al. (2020). *Cell Rep. Phys. Sci.* 1: 100059.

43 Liu, Y., Liu, S., He, D. et al. (2015). *J. Am. Chem. Soc.* 137: 12697–12703.

44 Zhao, X., Pachfule, P., Li, S. et al. (2019). *J. Am. Chem. Soc.* 141: 6623–6630.

45 Shao, X., Yang, X., Xu, J. et al. (2019). *Chem* 5: 693–705.

46 Ji, S., Chen, Y., Zhao, S. et al. (2019). *Angew. Chem. Int. Ed.* 58: 4271–4275.

47 Li, M., Qiao, S., Zheng, Y. et al. (2020). *J. Am. Chem. Soc.* 142: 6675–6681.

48 Lyu, J., Gong, X., Lee, S.-J. et al. (2020). *J. Am. Chem. Soc.* 142: 4609–4615.

49 Chiodini, S., Reinares-Fisac, D., Espinosa, F.M. et al. (2017). *Sci. Rep.* 7: 11088.

50 Lui, C.H., Liu, L., Mak, K.F. et al. (2009). *Nature* 462: 339–341.

51 Zhang, X., Zhang, P., Chen, C. et al. (2019). *Green Chem.* 21: 54–58.

52 Wu, S., Qin, L., Zhang, K. et al. (2019). *RSC Adv.* 9: 9386–9391.

53 Song, F. and Hu, X. (2014). *Nat. Commun.* 5: 4477.

54 Zhao, Y., Zhang, X., Jia, X. et al. (2018). *Adv. Energy Mater.* 8: 1703585.

55 Gao, D., Liu, Y., Liu, P. et al. (2016). *Sci. Rep.* 6: 35768.

56 Chen, J., Xu, X., Li, T. et al. (2016). *Sci. Rep.* 6: 37318.

57 Sheng, X., Li, S., Huang, H. et al. (2021). *J. Mater. Sci.* 56: 4212–4224.

58 Zhou, J., Zha, X., Zhou, X. et al. (2017). *ACS Nano* 11: 3841–3850.

59 Peng, Y., Shang, L., Cao, Y. et al. (2015). *Appl. Surf. Sci.* 358: 485–490.

60 Mateo, D., Esteve-Adell, I., Albero, J. et al. (2017). *Appl. Catal. B Environ.* 201: 582–590.

61 Kim, B.-K., Gwak, G.-H., Okada, T., and Oh, J.-M. (2018). *J. Solid State Chem.* 263: 60–64.

62 Zhang, W., Ming, J., Zhao, W. et al. (2019). *Adv. Funct. Mater.* 29: 1903641.

63 Feng, X., Hajek, J., Jena, H.S. et al. (2020). *J. Am. Chem. Soc.* 142: 3174–3183.

64 Shearer, G.C., Chavan, S., Ethiraj, J. et al. (2014). *Chem. Mater.* 26: 4068–4071.

65 Fang, Z., Bueken, B., De Vos, D.E., and Fischer, R.A. (2015). *Angew. Chem. Int. Ed. Eng.* 54: 7234–7254.

66 Feng, Y., Chen, Q., Jiang, M., and Yao, J. (2019). *Ind. Eng. Chem. Res.* 58: 17646–17659.

67 Liu, Y., Klet, R.C., Hupp, J.T., and Farha, O. (2016). *Chem. Commun.* 52: 7806–7809.

68 Shearer, G.C., Chavan, S., Bordiga, S. et al. (2016). *Chem. Mater.* 28: 3749–3761.

69 Cliffe, M.J., Wan, W., Zou, X. et al. (2014). *Nat. Commun.* 5: 4176.

70 Peng, Y., Albero, J., Álvarez, E., and García, H. (2019). *Sustain. Energy Fuel* 3: 2356–2360.

71 Tsoukalou, A., Abdala, P.M., Stoian, D. et al. (2019). *J. Am. Chem. Soc.* 141: 13497–13505.

72 Zhang, F., Liu, Z., Zhang, S. et al. (2018). *ACS Catal.* 8: 3550–3560.

73 Kleemiss, F., Dolomanov, O.V., Bodensteiner, M. et al. (2021). *Chem. Sci.* 12: 1675–1692.

74 Wilmer, C.E., Leaf, M., Lee, C.Y. et al. (2012). *Nat. Chem.* 4: 83–89.

75 Zhang, Y.-B., Su, J., Furukawa, H. et al. (2013). *J. Am. Chem. Soc.* 135: 16336–16339.

76 Ma, T., Kapustin, E.A., Yin, S.X. et al. (2018). *Science* 361: 48–52.

77 Oveisi, A.R., Zhang, K., Khorramabadi-zad, A. et al. (2015). *Sci. Rep.* 5: 10621.

78 Peng, Y., Albero, J., and García, H. (2019). *ChemCatChem* 11: 6384–6390.

79 Lv, C., Zhong, L., Liu, H. et al. (2021). *Nat. Sustain.* 4: 868–876.

80 Zhang, L., Cong, M., Ding, X. et al. (2020). *Angew. Chem. Int. Ed.* 132: 10980–10985.

81 Yang, J., Li, W., Wang, D., and Li, Y. (2020). *Adv. Mater.* 32: 2003300.

82 Bukhtiyarov, A., Prosvirin, I., Saraev, A. et al. (2018). *Faraday Discuss.* 208: 255–268.

83 Vuong, T.H., Radnik, J.R., Rabeah, J. et al. (2017). *ACS Catal.* 7: 1693–1705.

84 Yang, X.-D., Gao, N., Ma, S. et al. (2019). *Chem. Commun.* 55: 12829–12832.

85 Vakili, R., Slater, T.J.A., Hardacre, C. et al. (2020). *J. Catal.* 391: 522–529.

86 Quartieri, S. (2018). *Encycl. Archaeol. Sci.* S.L. López Varela (Ed.). https://doi.org/10.1002/9781119188230.saseas0613.

87 Ye, X., Yang, C., Pan, X. et al. (2020). *J. Am. Chem. Soc.* 142: 19001–19005.

88 Øien, S., Agostini, G., Svelle, S. et al. (2015). *Chem. Mater.* 27: 1042–1056.

89 L. Liu and A. Corma, *Chem*, 2021, 7, 2347-2384.

90 Dong, X.-Y., Si, Y.-N., Wang, Q.-Y. et al. (2021). *Adv. Mater.* 33: 2101568.

91 Ahn, S., Nauert, S.L., Buru, C.T. et al. (2018). *J. Am. Chem. Soc.* 140: 8535–8543.

92 Zhao, S., Wang, Y., Dong, J. et al. (2016). *Nat. Energy* 1: 16184.

93 Ji, Q., Kong, Y., Wang, C. et al. (2020). *ACS Catal.* 10: 5691–5697.

94 Chu, J., Ke, F.-S., Wang, Y. et al. (2020). *Commun. Chem.* 3: 5.

95 Wang, T.C., Vermeulen, N.A., Kim, I.S. et al. (2016). *Nat. Protoc.* 11: 149–162.

96 Daliran, S., Ghazagh-Miri, M., Oveisi, A.R. et al. (2020). *ACS Appl. Mater. Interfaces* 12: 25221–25232.

97 Oudi, S., Oveisi, A.R., Daliran, S. et al. (2020). *J. Colloid Interface Sci.* 561: 782–792.

98 Garibay, S.J. and Cohen, S.M. (2010). *Chem. Commun.* 46: 7700–7702.

99 Liu, H., Xi, F.-G., Sun, W. et al. (2016). *Inorg. Chem.* 55: 5753–5755.

100 Marshall, R.J. and Forgan, R.S. (2016). *Eur. J. Inorg. Chem.* 2016: 4310–4331.

101 Fei, H., Pullen, S., Wagner, A. et al. (2015). *Chem. Commun.* 51: 66–69.

102 Lammert, M., Wharmby, M.T., Smolders, S. et al. (2015). *Chem. Commun.* 51: 12578–12581.

103 Zhang, Z.-H., Peng, S.-Q., Chi, S. et al. (2021). *Polym. Chem.* 12: 6745–6754.

104 Lucier, B.E.G., Chen, S., and Huang, Y. (2018). *Acc. Chem. Res.* 51: 319–330.

105 Reif, B., Ashbrook, S.E., Emsley, L., and Hong, M. (2021). *Nat. Rev. Methods Primers* 1: 2.

106 Brunner, E. and Rauche, M. (2020). *Chem. Sci.* 11: 4297–4304.

107 Chierotti, M.R., Amin, M., Hassan, Y.S. et al. (2017). *J. Phys. Chem. C* 121: 8850–8856.

108 Deng, H., Doonan, C.J., Furukawa, H. et al. (2010). *Science* 327: 846–850.
109 Li, S., Lafon, O., Wang, W. et al. (2020). *Adv. Mater.* 32: 2002879.
110 Cao, Q., Zhang, L.-L., Zhou, C. et al. (2021). *Appl. Catal. B Environ.* 294: 120238.
111 Xu, C., Pan, Y., Wan, G. et al. (2019). *J. Am. Chem. Soc.* 141: 19110–19117.
112 Li, Z., Pan, X., and Yi, Z. (2019). *J. Mater. Chem. A* 7: 469–475.
113 Davies, M.J. (2016). *Methods* 109: 21–30.
114 Ai, M., Zhang, J.W., Wu, Y.W. et al. (2020). *Chem.- Asian J.* 15: 3599–3619.
115 Ghaleno, M.R., Ghaffari-Moghaddam, M., Khajeh, M. et al. (2019). *J. Colloid Interface Sci.* 535: 214–226.
116 Xu, C., Liu, H., Li, D. et al. (2018). *Chem. Sci.* 9: 3152–3158.
117 Li, J., Huang, H., Xue, W. et al. (2021). *Nat. Catal.* 4: 719–729.
118 Šimėnas, M., Kobalz, M., Mendt, M. et al. (2015). *J. Phys. Chem. C* 119: 4898–4907.
119 Jiang, J., Liang, Z., Xiong, X. et al. (2020). *ChemCatChem* 12: 3523–3529.
120 Li, P., Li, J., Feng, X. et al. (2019). *Nat. Commun.* 10: 2177.
121 Clément, J.-L., Ferré, N., Siri, D. et al. (2005). *J. Org. Chem.* 70: 1198–1203.
122 George, P. and Chowdhury, P. (2019). *Analyst* 144: 3005–3012.
123 Makuła, P., Pacia, M., and Macyk, W. (2018). *J. Phys. Chem. Lett.* 9: 6814–6817.
124 Landi, S., Segundo, I.R., Freitas, E. et al. (2022). *Solid State Commun.* 341: 114573.
125 Patel, M., Chavda, A., Mukhopadhyay, I. et al. (2016). *Nanoscale* 8: 2293–2303.
126 Valenzano, L., Civalleri, B., Chavan, S. et al. (2011). *Chem. Mater.* 23: 1700–1718.
127 Wang, J., Tian, X.-X., Yu, L. et al. (2021). *J. Mater. Chem. A* 9: 25474–25479.
128 Traxler, M., Gisbertz, S., Pachfule, P. et al. (2022). *Angew. Chem. Int. Ed. Eng.* 61: e202117738.
129 Long, J., Wang, S., Ding, Z. et al. (2012). *Chem. Commun.* 48: 11656–11658.
130 Wang, X., Ma, K., Goh, T. et al. (2022). *J. Am. Chem. Soc.* 144: 12192–12201.
131 Howarth, A.J., Peters, A.W., Vermeulen, N.A. et al. (2017). *Chem. Mater.* 29: 26–39.
132 Buru, C.T., Li, P., Mehdi, B.L. et al. (2017). *Chem. Mater.* 29: 5174–5181.
133 Zhang, X., Luo, J., Wan, K. et al. (2019). *J. Mater. Chem. A* 7: 1616–1628.
134 Guo, S.-H., Qi, X.-J., Zhou, H.-M. et al. (2020). *J. Mater. Chem. A* 8: 11712–11718.
135 Zou, L., Feng, D., Liu, T.-F. et al. (2016). *Chem. Sci.* 7: 1063–1069.
136 Ragon, F., Campo, B., Yang, Q. et al. (2015). *J. Mater. Chem. A* 3: 3294–3309.
137 Shultz, A.M., Farha, O.K., Hupp, J.T., and Nguyen, S.T. (2011). *Chem. Sci.* 2: 686–689.
138 Mu, L., Liu, B., Liu, H. et al. (2012). *J. Mater. Chem.* 22: 12246–12252.
139 Holst, J.R., Stöckel, E., Adams, D.J., and Cooper, A.I. (2010). *Macromolecules* 43: 8531–8538.
140 Yang, Q., Zhang, H.-Y., Wang, L. et al. (2018). *ACS Omega* 3: 4199–4212.
141 Kandiah, M., Nilsen, M.H., Usseglio, S. et al. (2010). *Chem. Mater.* 22: 6632–6640.

142 Athar, M., Rzepka, P., Thoeny, D. et al. (2021). *RSC Adv.* 11: 38849–38855.

143 Xiang, W., Zhang, Y., Chen, Y. et al. (2020). *J. Mater. Chem. A* 8: 21526–21546.

144 Wu, H., Chua, Y.S., Krungleviciute, V. et al. (2013). *J. Am. Chem. Soc.* 135: 10525–10532.

145 Wang, T.C., Bury, W., Gómez-Gualdrón, D.A. et al. (2015). *J. Am. Chem. Soc.* 137: 3585–3591.

146 Wang, K., Feng, D., Liu, T.-F. et al. (2014). *J. Am. Chem. Soc.* 136: 13983–13986.

147 Ma, J., Kalenak, A.P., Wong-Foy, A.G., and Matzger, A.J. (2017). *Angew. Chem. Int. Ed. Eng.* 56: 14618–14621.

148 Moon, S.-Y., Bae, J.-S., Jeon, E., and Park, J.-W. (2010). *Angew. Chem., Int. Ed. Engl.* 49: 9504–9508.

149 Thommes, M., Kaneko, K., Neimark, A.V. et al. (2015). *Pure Appl. Chem.* 87: 1051–1069.

150 Cychosz, K.A. and Thommes, M. (2018). *Engineering* 4: 559–566.

151 Comesaña-Gándara, B., Chen, J., Bezzu, C.G. et al. (2019). *Energy Environ. Sci.* 12: 2733–2740.

152 Furukawa, H. and Yaghi, O.M. (2009). *J. Am. Chem. Soc.* 131: 8875–8883.

153 Baldwin, L.A., Crowe, J.W., Pyles, D.A., and McGrier, P.L. (2016). *J. Am. Chem. Soc.* 138: 15134–15137.

154 Sun, X., Sun, S., Gu, S. et al. (2019). *Nano Energy* 61: 245–250.

155 Liu, D., He, Q., Ding, S., and Song, L. (2020). *Adv. Energy Mater.* 10: 2001482.

156 Wan, J., Zhao, Z., Shang, H. et al. (2020). *J. Am. Chem. Soc.* 142: 8431–8439.

157 Tian, S., Peng, C., Dong, J. et al. (2021). *ACS Catal.* 11: 4946–4954.

158 Barona, M., Ahn, S., Morris, W. et al. (2019). *ACS Catal.* 10: 1460–1469.

4

Mixed Linker MOFs in Catalysis

Mohammad Y. Masoomi and Lida Hashemi

Arak University, Department of Chemistry, Faculty of Sciences, Karballa Blvd, Sardasht, Arak 3848177584, Iran

4.1 Introduction

4.1.1 Introduction to Mixed Linker MOFs

Framework materials are a class of polymers whose structures impart unique properties distinct from other macromolecular architectures. These materials are defined by their periodicity and permanent porosity, which results in exceptionally large surface areas and makes them privileged for applications in separations, payload storage and release, and catalysis. Metal–organic frameworks (MOFs) in particular have attracted intense interest for these applications and othersas a consequence of their periodicity, high specific surface area, and rational synthesis. The diverse set of metal coordination bonds and geometries provides an infinite number of potential network topologies, with thousands of MOFs described thus far [1]. With progressively lengthening the organic moieties, great success has been achieved in obtaining a series of isoreticular MOFs from microporosity to mesoporosity [2]. Integrating multiple ligands with identical linking chemistry represents an efficient route to access multivariate metal–organic frameworks (MTV-MOFs) with increased complexity and functions [3]. When two or more organic ligands of distinct sizes, shapes, and coordination features are used, the synergetic coordination of different ligands with metals is involved [4–6]. Moreover, the cooperation among different ligands allows one to obtain multifunctional MOFs with designable topology, adjustable porosity, tunable functionality, and variable surface environments within a single material. In some cases, the complementary or contradictory properties can be combined to enable functionality of a single MOF with mixed-linker feature [7]. In this context, it is highly desirable to employ mixed-linker strategy for the construction of MOFs owing to the following advantages: (i) to place different functional groups into the frameworks at the desired positions and thus modify the pore environments; (ii) to fabricate multifunctional porous materials that synergistically

Catalysis in Confined Frameworks: Synthesis, Characterization, and Applications, First Edition.
Edited by Hermenegildo Garcia and Amarajothi Dhakshinamoorthy.

exhibit two complementary or contradictory properties; and (iii) to tolerate partial absence of the second linker and thus create structural defects while preserving the integrity of the whole framework, which influences the properties of MOFs.

4.2 Strategies for Synthesizing Mixed-Linker MOFs

Frameworks might be functionalized to improve their gas adsorption behavior, enable covalent attachment of molecular payloads, or include active catalysts. The first and most straightforward method is an isostructural mixed linker (IML) approach, which involves mixing two or more monomers with identical linking geometry [8]. The IML approach is often employed to incorporate reactive functionality along the walls of a framework. Alternatively, monomers bear different linkage geometries in a heterostructural mixed linker (HML) strategy. At some monomer feed ratios, the resulting framework distributes the minority component throughout a defective lattice of the majority component. However, certain ratios provide lower-symmetry topologies that are less prone to forming interpenetrated structures. The last strategy employs a second monomer with a reduced number of functional groups, denoted as a truncated mixed linker (TML) approach. The second monomer acts as a capping agent that, depending on the kinetics of framework crystallization, either directs crystallite morphology and surface chemistry or enables functionalization of the framework interior (Figure 4.1) [9].

4.2.1 IML Frameworks

Perhaps the most simple and versatile way to incorporate multiple monomers into framework materials is to use two or more organic building blocks with identical size and linkage chemistry, denoted as an IML approach [8, 10]. For example, combining two (or more) terephthalic acid derivatives yields the mixed-composition MOF-5

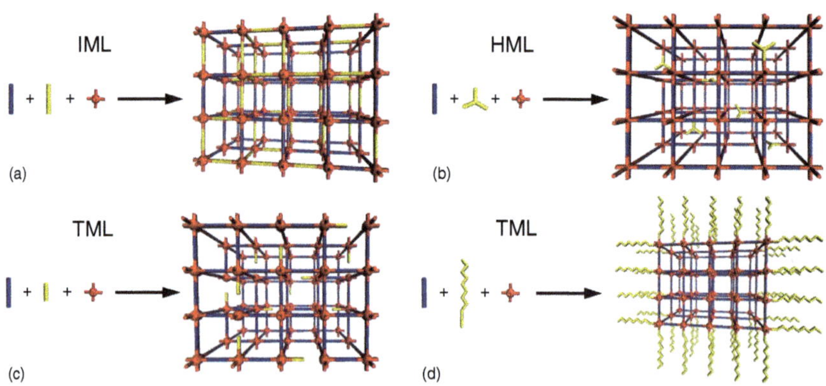

Figure 4.1 Illustrations of the (a) isostructural mixed linker (IML), (b) heterostructural mixed linker (HML), and (c, d) two outcomes of truncated mixed linker (TML) approaches to framework functionalization. Source: Bunck and Dichtel [9]/John Wiley & Sons.

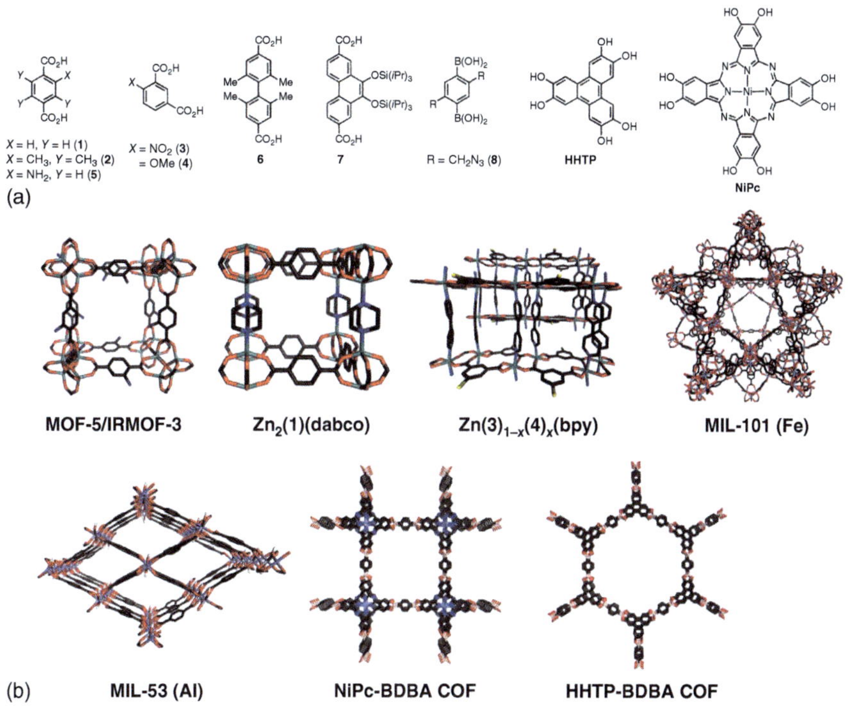

Figure 4.2 (a) Monomers employed in the IML strategy with (b) depictions of their corresponding frameworks Source: Bunck and Dichtel [9]/John Wiley & Sons.

cubic framework (Figure 4.2). To the best of our knowledge, Kim and coworkers were the first to employ the IML strategy when they incorporated a mixture of terephthalic acid (**1**) and tetramethylterephthalic acid (**2**) into the [$Zn_2(1)_2$dabco] framework (Figure 4.2b) [11]. Kitagawa and coworkers subsequently tuned pore flexibility across a series of MOFs containing varying ratios of 5-nitroisophthalic acid (**3**) and 5-methoxyisophthalic acid (**4**) within a [$Zn(\mathbf{3})_{1-x}(\mathbf{4})_x$(bpy)] network (Figure 4.2b) [12]. Several studies have employed mixtures of terephthalic acids to load specific functionality into the pores of various frameworks, including MOF-5 (based on Zn) [13], MIL-53 (Al) [14, 15], and MIL-101 (Fe) [16] (Figure 4.2b). Baiker and coworkers characterized the thermal stability of amine-functionalized MOF-5 derivatives and used these amines both as nucleophilic catalysts [17] and as ligands for Pd-catalyzed cross-coupling reactions [18].

Yaghi and coworkers demonstrated the versatility of the IML strategy by preparing 18 mixed-composition MOF-5 derivatives using nine substituted terephthalic acid building blocks too [3].

4.2.2 HML Frameworks

Though less intuitive than the IML strategy, mixing building blocks with different coordination geometries, denoted as a HML approach, provides opportunities

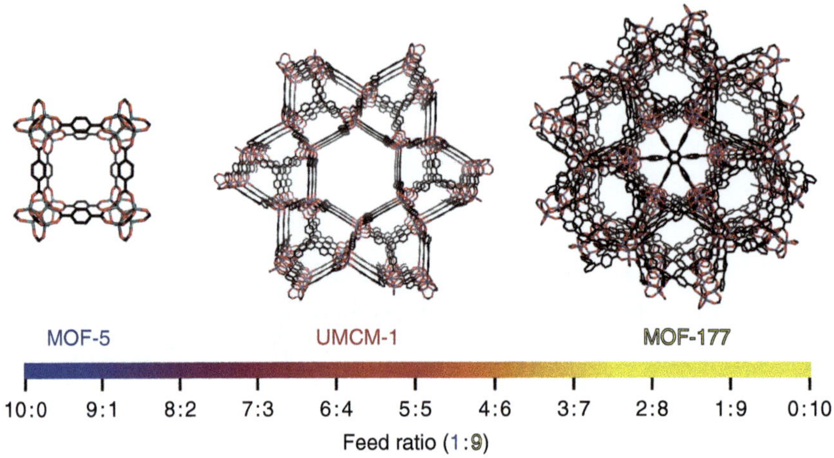

Figure 4.3 Phase diagram for UMCM-1 synthesis. Source: Bunck and Dichtel [9]/John Wiley & Sons.

to functionalize existing frameworks and to access lower symmetry topologies with large pore sizes. Matzger and coworkers performed the first rigorous study of a MOF derived from the HML strategy, in which terephthalic acid and a trifunctional carboxylic acid were cocrystallized in the presence of $Zn(NO_3)_2$ [4]. On their own, terephthalic acid produces MOF-5, and trifunctional carboxylic acid produces MOF-177 (Figure 4.3). Cocrystallization of the two linkers produces a new MOF (UMCM-1) containing 3.1 nm mesoporous hexagonal channels surrounded by 1.4 nm microporous cages. The HML approach, along with other strategies to desymmetrize MOF building blocks [19], shows great promise in this area. UMCM-1 is noninterpenetrated and retains its high surface area ($4160\ m^2\ g^{-1}$) even after heating at 300 8 °C for three hours [9].

Small variations in the relative linker lengths of mixtures of di- and trifunctional linkers dramatically affect the topologies of HML MOFs. These topological differences have important consequences for guest uptake and release, as illustrated by the vastly different diffusion behaviors observed in a single-molecule study of diffusion within the pores of UMCM-1, -2, and -4 [20].

4.2.3 TML Frameworks

A third strategy to obtain multicomponent frameworks employs a TML approach, in which a polyfunctional monomer is condensed with a monomer bearing fewer reactive groups. The relative rates of bond formation and exchange dictate the role of truncated monomer in the crystallization. If growth is faster than exchange, the truncated monomer is incorporated throughout the network, which grows around these defect sites. If exchange is rapid relative to framework growth, the truncated monomer will preferentially reside at the faces of the growing crystal, providing a means to control its size, shape, and surface functionality [9]. The TML approach operates across a broad continuum of exchange and error correction rates, providing outcomes ranging from internal functionalization to anisotropic crystallization. These results also highlight the unanswered question of what reaction and exchange rates are required for framework formation [9].

4.3 Types of Mixed-Linker MOFs

Here we highlight four kinds of mixed-linker MOFs according to their structural features (Figure 4.4). The first typical subgroup is pillared-layer mixed-linker MOFs, which were formed by the introduction of a secondary linker to support the pre-designed metal-organic layer. The second one is cage-directed mixed-linker MOFs, which are usually constructed by the introduction of a secondary linker to bridge an existing metal-organic cage-like structure. A secondary ligand was employed to extend a metal-organic cluster, resulting in the formation of the third kind of mixed-linker MOFs, cluster-based mixed-linker MOFs. The fourth kind is structure-template mixed-linker MOFs, which were built from the introduction of a secondary organic linker or metal-organic fragment into a porous framework [21].

4.3.1 Pillared-Layer Mixed-Linker MOFs

Pillared-layer MOFs are one of the most important and classical branches of porous MOFs. However, the accessible tunability of porosity is limited in pillared-layer MOFs that contain only one single organic linker. To get more possibility to tune the porosity featurc, pillared-layer mixed-linker MOFs caught more and more attention in recent years. Kitagawa and coworkers demonstrated a series of outstanding works on pillared-layer coordination polymers. The typical mixed-linker one is on the basis of 2D sheet [M(pzdc)$_n$] (M = Cu or Cd, pzdc = pyrazine-2,3-dicarboxylate) pillared by pyrazine (pyz) and pyridine derivatives [22].

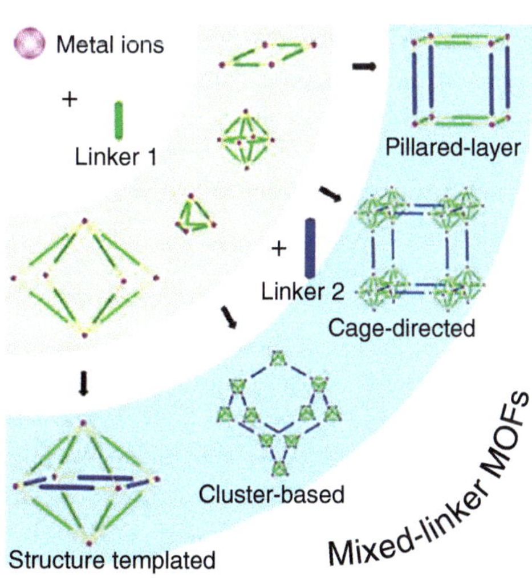

Figure 4.4 Four types of mixed-linker MOFs: (i) pillared-layer mixed-linker MOFs, (ii) cage-directed mixed-linker MOFs, (iii) cluster-based mixed-linker MOFs, and (iv) structure-template mixed-linker MOFs. Source: Qin et al. [21]/Royal Society of Chemistry.

4.3.2 Cage-Directed Mixed-Linker MOFs

Discrete molecular cage structures constructed by the coordination of metal ions and organic linkers have recently attracted considerable attention because of their intriguing structures, potential applications, and relevance to biological self-assembly. Metal-organic cage-like frameworks possess individual cages carrying large voids interconnected by relatively small windows, which are different from typical MOFs with open-channel type pores. The use of cages as building blocks for the construction of porous frameworks is an efficient approach, since the inherent cavities of the fused cages are maintained in tessellation in space [23, 24]. In addition, the cage feature is promising for the storage and release of small molecules because the adsorbed guests may remain kinetically trapped inside the cages [21].

4.3.3 Cluster-Based Mixed-Linker MOFs

Some high-symmetry metal-carboxylate clusters were frequently encountered to act as secondary building units (SBUs), for instance, $[Cr_3(\mu_3\text{-}O)(COO)_6]$ in MIL-101 and $[Zn_4(\mu_4\text{-}O)(COO)_6]$ in MOF-5. These metal-oxygen clusters can be connected by a single-organic linker to generate various three-dimensional (3D) metal cluster-based MOFs, except for a few mixed-linker examples [25–28]. In light of this, a series of metal-organic polynuclear clusters with small multidentate ligands were designed, which were further extended by the secondary ligands to construct multifunctional mixed-linker MOF materials [21].

4.3.4 Structure Templated Mixed-Linker MOFs

MOFs are shown to be good and promising platforms for a new class of crystalline porous materials for guest encapsulation. It is required for advanced applications to place multiple functional molecules in highly ordered MOFs through coordination bonds that allow the tailoring of the pore environment. However, it is still a challenge to encapsulate multiple functional molecules simultaneously into a MOF pore with a high level of control over the distribution of the incorporated functionalities [21]. In recent years, several strategies to modify the pore environments of the MOFs by introducing size-matching organic ligands by deliberately anchoring them onto the open metal sites in the pores have been proposed [29–32].

Besides the above commented strategies based on the preparation of mixed-linker MOFs in a single step, a different strategy that has been also commonly used consists in the post-synthetic modification (PSM) of a preformed MOF containing a single metal or a single linker (Figure 4.5). PSM can also be used to partially replace the linker present in the original MOF with a different one. In a strategy different from exchange, PSM could be based on performing organic transformation at the organic linker. Typically, in these PSM reactions, only a part of the original linkers react, probably as consequence of steric encumbrance for the reaction inside MOF voids as the degree of functionalization increases.

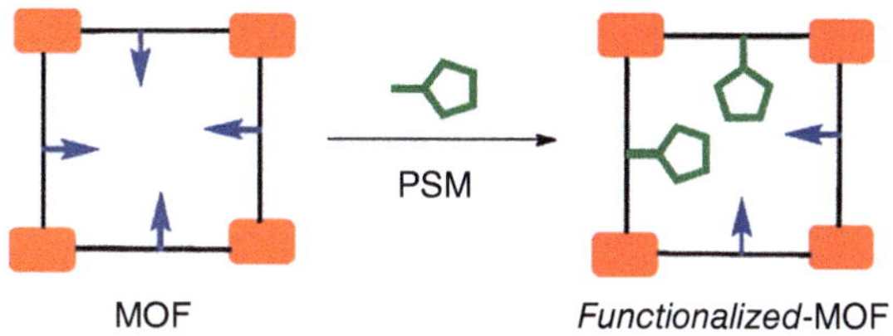

Figure 4.5 Post-synthetic modification of MOF by linker exchange or linker derivatization.

4.4 Introduction to Catalysis with MOFs

Since MOFs have high crystallinity and do not have theoretical pore size limitations, they provide unique opportunities for catalysis. This means that it is possible to have a homogeneous distribution of one or more active sites due to the high crystallinity of the material and, at the same time, to overcome diffusion and pore size limitations. Moreover, the fine structure and nature of the active site can be controlled. The known strategies for the building of specific catalytic sites in crystalline MOFs can be categorized into three classes: framework activity, encapsulation of active species, and PSM [33].

MOF structures with a high surface area and pore volume, especially those with coordination unsaturated open metal sites, showed efficient Lewis acid-catalyzed and oxidation reactions for fine chemical synthesis. MOFs containing coordinatively unsaturated metal sites are also useful for the incorporation of a range of organic catalytic groups on the metal sites via PSM, leading to new transition metal complex-immobilized catalysts for organic transformation [34].

4.5 Mixed-Linker MOFs as Heterogeneous Catalysts

MOFs are among the most studied heterogeneous catalysts that have been applied to promote a wide range of reactions. Most of the initial studies on the catalytic activity of MOFs were based on the use of materials containing a single metal and a single linker. However, the most recent trend in the field is to exploit the synthetic flexibility offered by MOFs to obtain new MOFs having in their structure two different linkers (*mixed linkers*) that result in materials having superior catalytic activity over the corresponding single metal or single linker MOFs [35]. Functional groups present on the organic linker can be used as heterogeneous catalysts and using of mixed linker is a strategy to increase the activity of the MOFs in some selected reactions.

Due to the presence of active and selectivity transition metals in MOFs, the huge potential can be exploited in heterogeneous catalysis [33]. As selectivity is the most important feature in catalysis, it can be maximized if the selective site is known and only this site of catalyst is synthesized [36]. Nevertheless, the biggest problem in supported catalysis processes is instability of introduction of complexes to support material because of pore blockage and framework collapse, as well as their thermal stability. To overcome this instability issue, introduction of mixed-linker metal–organic frameworks (MIXMOFs) as palladium-supported catalysts has been

Table 4.1 Mixed-linker MOFs with similar size/directionality (isomorphous substitution) employed as heterogeneous catalysts.

Catalyst	Mixed linkers		Reaction	Reference
MOF-5-NH_2	BDC	ABDC NH_2-BDC	Cycloaddition of CO_2 to propylene oxide	[17]
Pd/MOF-5-NH_2	BDC	ABDC	CO Oxidation	[18]
Pd/MIL-53(Al)-NH_2	BDC	ABDC	Heck coupling	[38]
$[M_3^{II,II}(BTC)_2]$ (M = Cu, Mo, Cr, Ni, Zn)	BTC	PYDC	Reduction of 1-octene	[39]
$Cu_3(BTC)_2$-PYDC	BTC	PYDC	Hydroxylation of toluene	[40]
$[Zn_4O(NH_2$-BPDC)1.5 (DPP-BPDC)1.5] or $[Zn_4O(NH_2\text{-BPDC})_{1.2}(\text{MeBPDC})_{1.8}]$	NH_2-BPDC NH_2-BPDC	DPP-BPDC MeBPDC	Knoevenagel condensation	[41]
NH_2-UiO-66(Zr)-DTA	DTA	ABDC	CO_2 reduction	[42]
Zr-MOF	ABDC	X-BDC, X = H, F, Cl, Br	Benzyl alcohol oxidation	[43]

Abbreviations: BDC:1,4-benzenedicarboxylic acid; ABDC: 2-amino-1,4-benzenedicarboxylic acid; BTC: 1,3,5-benzenetricarboxylic acid; PYDC: pyridinedicarboxylic acid; BPDC: (1,1′-biphenyl)-4,4′-dicarboxylic acid; NH_2-BPDC: 2-amino-(1,1′-biphenyl)-4,4′-dicarboxylic acid; DPP-BPDC: diphenylphosphoryl-(1,1′-biphenyl)-4,4′-dicarboxylic acid; MeBPDC: methyl-(1,1′-biphenyl)-4,4′-dicarboxylic acid; DTA: 2,5-diaminoterephthalic acid.

taken into account [37]. The general idea of MIXMOFs is the formation of two isorecticular linker molecules that are randomly combined into a mixed structure.

4.5.1 Mixed-Linker MOFs with Similar Size/Directionality Linkers

The first part will briefly state use of mixed linker MOFs, which were reported in the literature as heterogeneous catalysts. Table 4.1 lists those mixed linker MOFs in which the linkers have similar size/directionality through isomorphous substitution.

A series of mixed-linker MOF-5 was synthesized by partial replacement of BDC by ABDC, and their activity was tested for the synthesis of propylene carbonate (Scheme 4.1). The amount of amino groups was tuned by choosing the appropriate BDC/ABDC linker ratio, which in turn exhibited different activity [17].

The resulting mixed linker MOFs are still stable enough to be used as heterogeneous catalysts within the temperature limit of 300 °C. Under optimized reaction conditions, MOF-5 as catalyst resulted in 44% yield of propylene carbonate. The yield of propylene carbonate is further enhanced to 63% with mixed linker MOF (40% using ABDC) (Figure 4.6). Although the product yield decreased from 63% to 45% and 41% in the second and third runs, respectively.

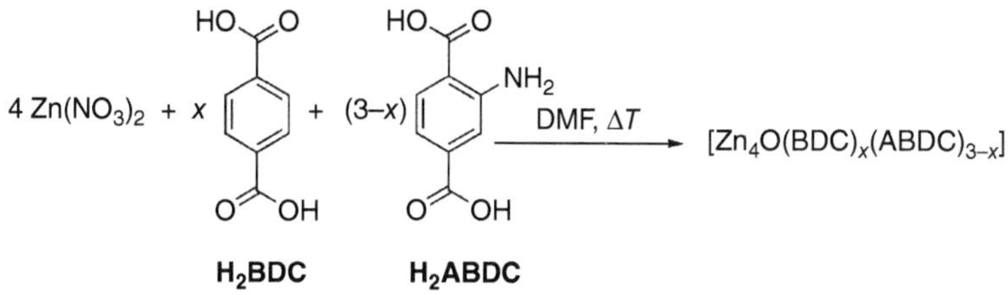

Scheme 4.1 Synthesis of mixed-linker MOFs containing variable ratios of benzene-1,4-dicarboxylate (BDC) and 2-aminobenzene-1,4-dicarboxylate (ABDC) linkers.

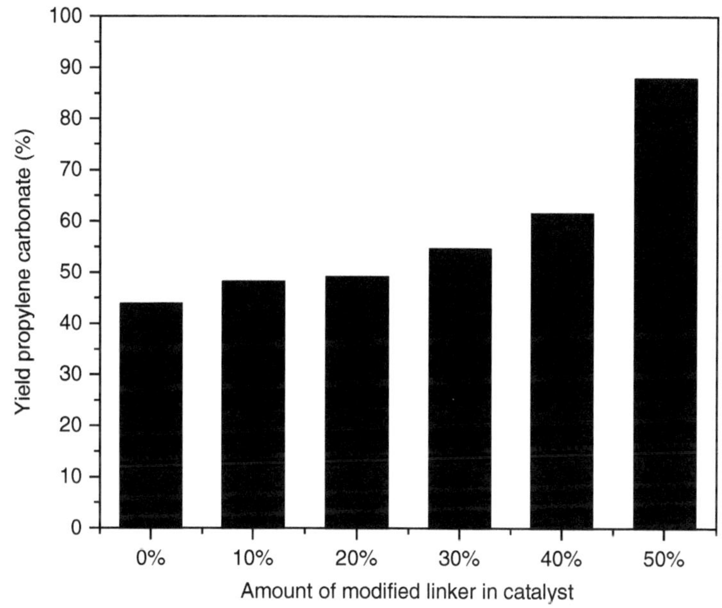

Figure 4.6 Influence of ABDC content on catalytic performance. Source: Kleist et al. [17]/John Wiley & Sons.

A comparison of the different materials in the mix-linker MOF series clearly indicates a dependence of activity on the amount of amino groups in the material. Due to the fact that pure MOF-5 (without NH_2 groups) also catalyzes the reaction, the activation of the substrates takes place most probably both at the basic amino groups and at accessible Zn centers at the surface (free coordination sites) [17]. The presence of amino groups is also beneficial for the synthesis of Pd/mixed-linker MOF catalysts, which can be obtained by equilibrium adsorption. The interaction between the amino groups and the Pd leads to highly dispersed Pd species that do not form particles big enough to block pores or channels in the microporous MOF structure. Although the presence of Pd inside the pores of the material affects the thermal stability to some extent, the Pd/MIXMOF catalysts can be used as effective catalysts in the oxidation of CO at elevated temperatures [18]. In similar work for

MIL-53(Al)-NH_2 in Heck coupling reaction this process has occurred, and these data suggest the presence of amino group on the functionalized linker is beneficial for the stabilization of Pd NPs [38].

Incorporation of PYDC, which is the same size as BTC but carries lower charge, as a second, defective linker in [$M_3^{II,II}(BTC)_2$] (M = Cu, Mo, Cr, Ni, Zn) resulted in the formation of mixed-linker isoreticular derivatives of Ru-MOF, which display specific characteristics different from those of the defect-free framework. Along with the creation of additional coordinatively unsaturated sites, the incorporation of PYDC induces the partial reduction of ruthenium. Hence, the MOFs with the modified Ru sites were highly active in the reduction of 1-octene using molecular hydrogen as reducing agent [39]. The parent [$Ru_3(BTC)_2Cl_{1.5}$] MOF resulted in low activity by producing only 12% conversion of 1-octene after 20 hours. On the other hand, the use of [$Ru_3(BTC)_{1.4}(PYDC)_{0.6}Cl$] (c. 30 mol% of PYDC) exhibited 50% conversion under identical conditions.

Baiker and coworkers reported a mixed-linker MOF based on the $Cu_3(BTC)_2$ structure in which BTC linkers were partially replaced by PYDC [40]. The pyridine unit, similarly to the previous structure, can be seen as a defect site in the local coordination environment of the dimeric copper units, leading to a significant alteration of their electronic structure with remarkable consequences for their catalytic properties. $Cu_3(BTC)_2$ and the mixed $Cu_3(BTC)_2$-PYDC MOFs were tested for the hydroxylation of toluene both in acetonitrile and under neat conditions.

Several research groups have successfully prepared individual MOFs containing different functional groups as MIXMOFs or MTV-MOFs [3, 14, 44]. Xu and coworkers have described activity of two bifunctional MOFs with IRMOF-9 topology that contain amino, phosphine oxide, and methyl groups. Bifunctional MOFs, LSK-6 [$Zn_4O(NH_2$-BPDC$)_{1.5}$(DPP-BPDC$)_{1.5}$] and LSK-9 [$Zn_4O(NH_2$-BPDC$)_{1.2}$(Me-BPDC$)_{1.8}$] BPDC = (1,1′-biphenyl)-4,4′-dicarboxylic acid) were synthesized in DMF with an equal amount of two linkers and $Zn(NO_3)_2$ $4H_2O$ under solvothermal conditions. The amino group acts as an active site for the Knoevenagel condensation of ortho, meta, or para nitrobenzaldehyde (Scheme 4.2) and malononitrile, while the non-active site phosphine oxide or methyl group moderates the spatial characteristics inside the MOF pores. The functional groups induce a unique catalytic response on bulky substrates. The catalytic activity of LSK-6 and LSK-9 was higher in the Knoevenagel condensation of para, meta, or ortho nitrobenzaldehyde and malononitrile than that of aniline [45, 46]. DPP functionalized amino MOF LSK-6 slowed the reactivity of ortho nitrobenzaldehyde reaction more than the methyl functionalized amino MOF,

Scheme 4.2 Bifunctional MOFs (LSK-6 and LSK-9) catalyze Knoevenagel condensation reactions of nitrobenzaldehyde (ortho, meta, and para nitrobenzaldehyde regioisomers) with malononitrile [41].

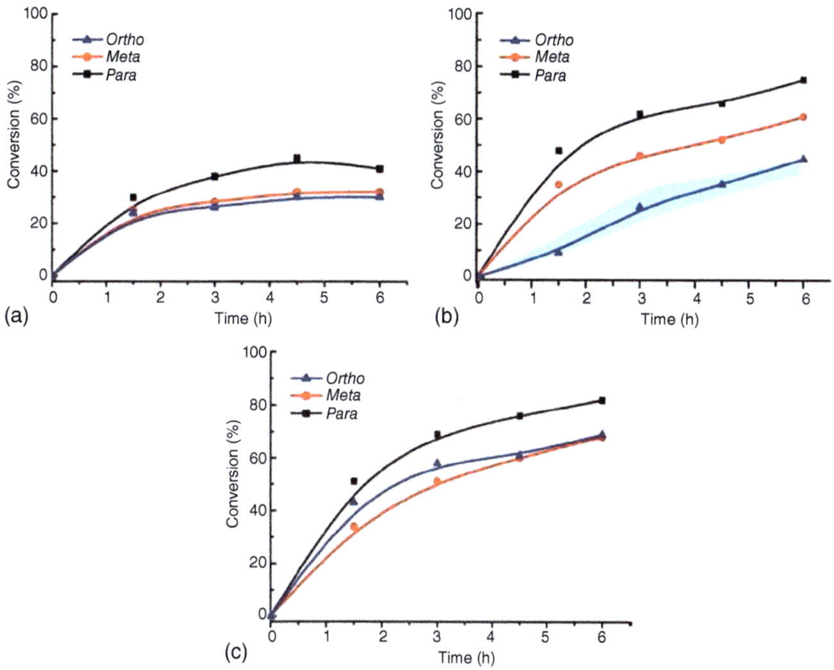

Figure 4.7 Nitrobenzaldehyde conversion in the Knoevenagel condensation with malononitrile catalyzed by aniline (a); Nitrobenzaldehyde conversion in the Knoevenagel condensation with malononitrile catalyzed by LSK-6 (b) and LSK-9 (c). Source: Xu et al. [41]/John Wiley & Sons.

LSK-9 (Figure 4.7). Size analysis of open channels and guest molecules revealed that the accessibility of the product of ortho isomer was the reason for such reactivity [41].

Photocatalytic activity of DTA-modified NH_2-UiO-66(Zr) and NH_2-UiO-66(Zr) was evaluated for CO_2 reduction using TEOA as sacrificial agent under visible light irradiation. Higher photocatalytic activity of the mixed-linker MOF can be correlated with its high light absorption and increased CO_2 adsorption due to the presence of additional –NH_2 groups in mixed-linker NH_2-UiO-66(Zr) [42]. In other work, A series of mixed-linker zirconium-based MOFs were synthesized containing ABDC as the primary linker and 2-X-1,4-benzenedicarboxylate (X-BDC, X = H, F, Cl, Br) as a secondary linker, and their activity was evaluated for the visible light photocatalytic oxidation of benzyl alcohol (Scheme 4.3) [43]. Replacement of H-BDC by halogenated linkers (X-BDC, X = F, Cl, Br) led to four- to five-fold enhancement of the activity of the corresponding Zr-MOFs for the benzyl alcohol oxidation in comparison to UiO-66–NH_2–H. This enhancement in catalytic activity could also be related to the preferential interaction of the halogen groups with benzyl alcohol molecules that drives the equilibrium of the reaction to the formation of benzaldehyde (Table 4.2) [43].

Other reports have confirmed that the presence of amino-functionalized linker in the MOFs improves the photocatalytic activity [47–50].

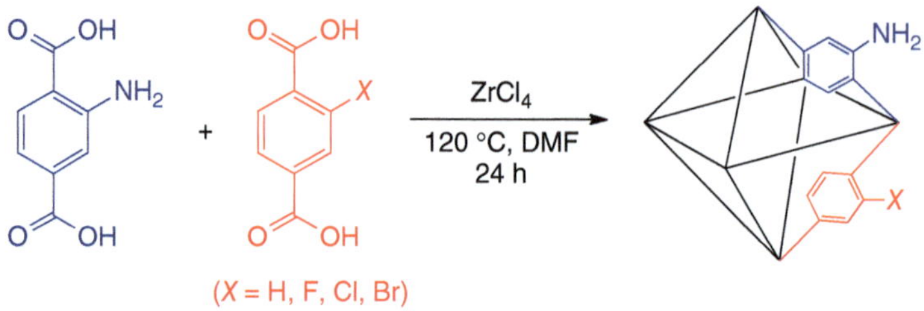

Scheme 4.3 Synthesis of the mixed-linker Zr-MOFs. Equimolar quantities of primary linker 2-amino 1,4-benzenedicarboxylicacid and secondary linker 2-X-1,4-benzenedicarboxylicacid (X = H, F,Cl, Br).

Table 4.2 Benzyl alcohol oxidation catalyzed by Zr-MOF.

Entry	Catalyst	MOF (mmol)	Conv. (%)	TOF (lmol min^{-1} g^{-1})
1	Blank	—	1	—
2	UiO-66	0.10	1	
3	UiO-66–F	0.10	1	—
4	UiO-66–NH_2	0.1	18.4	140
5	UiO-66–NH_2–H	0.10	11.7	89
6	UiO-66–NH_2–F	0.1	53.9	410
7	UiO-66–NH_2–Cl	0.10	38.2	290
8	UiO-66–NH_2–Br	0.10	43.4	330
9	UiO-66–NH_2–Br	0.10	1	—

In another piece research, it has been reported that a direct mixing synthesis strategy of two different ligands, i.e. ligands containing nitrogen groups and terepthalic acid, into solution containing salt. Their interest lies in synthesized MIXMOFs containing different nitrogen ligands as palladium-supported catalysts. In this study, three types of nitrogen groups, such as NH_3, NO_2, and pyridine, have been selected as functionalized organic ligands for MIXMOFs. The prepared heterogeneous palladium-supported catalyst, Pd/MIXMOF-NH_2 (M1), Pd/MIXMOF-NO_2 (M2), and Pd/MIXMOF-pyridine (M3) have been applied in the Suzuki–Miyaura cross coupling reaction (Schemes 4.4 and 4.5) [51].

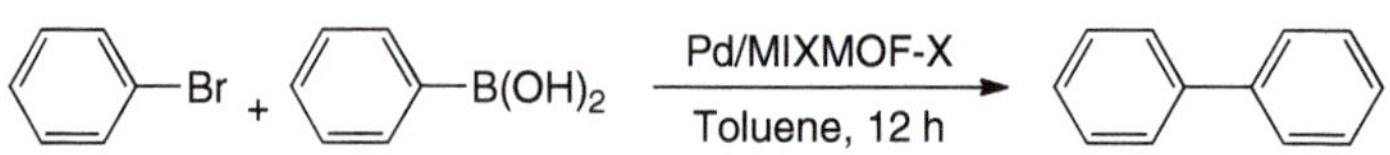

Scheme 4.4 Suzuki–Miyaura coupling reaction of bromobenzene and phenylboronic acid over Pd/MIXMOF catalyst.

Results in this study have shown that the Pd-MIXMOFs catalyst exhibits typical behavior of heterogeneous catalysts. These palladium-supported catalysts

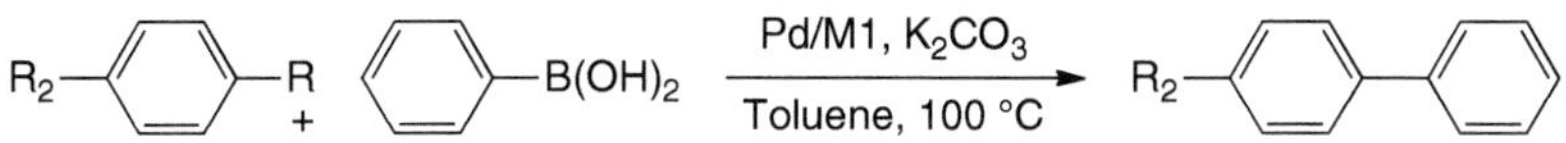

Scheme 4.5 Suzuki–Miyaura coupling reactions of aryl halides and phenylboronic acid over Pd/M1 catalyst.

demonstrated high catalytic activities in Suzuki–Miyaura cross-coupling reaction and can be reused several times without any visible loss of activity. The presence of amino groups on the linker complexes proved to be beneficial for the stabilization of Pd species by electrostatic interaction [51].

In similar work, two series of amine-functionalized Zr-based mixed-linker metal–organic frameworks (MOFs; UiO-66-Mix and UiO-67-Mix) have been synthesized with different ratios of 1,4-benzenedicarboxylate/2-amino-1,4-benzenedicarboxylate and 4,4′-biphenyldicarboxylic acid/2-amino-biphenyl-4,4′-dicarboxylic acid incorporated into their structures. The pendant amino groups were postmodified by the condensation reaction with pyridine-2-carboxaldehyde, and Pd metal centers were anchored onto the MOFs by coordination with the pyridylimine moieties. The two series of heterogeneous Pd catalysts (UiO-66-Mix-PIPd and UiO-67-Mix-PI-Pd) were utilized for the Suzuki and Heck cross-coupling reactions. The porosity of the MOFs as used with mix ligands and the density of the metal binding sites of the Pd catalysts had great influence on the Suzuki and Heck cross-coupling reactions. The best results were achieved with a very low Pd loading [52].

In contrast, in another study, a series of functionalized mixed-linker bipyridyl MOF-supported palladium catalysts were developed. These mixed-linker bipyridyl MOF-supported palladium(II) catalysts were used to elucidate the electronic and steric effects of linker substitution on the activity of these catalysts in the context of Suzuki–Miyaura cross-coupling reactions. *m*-6,6′-Me_2bpy-MOF-$PdCl_2$ exhibited a 110-fold and 496-fold enhancement in the activity compared to non-functionalized m-bpy-MOF-$PdCl_2$ and *m*-4,4′-Me_2bpy MOF-$PdCl_2$ (Figure 4.8). This result clearly demonstrates that the stereoelectronic properties of metal-binding linker units are critical to the activity of single-site organometallic catalysts in MOFs and highlights the importance of linker engineering in the design and development of efficient MOF catalysts [53].

Recently, Huang and coworkers reported selective selenization of mixed-linker Ni-MOFs. These mixed-linker Ni-MOFs are composed of coordinating anionic carboxylate linkers and N-coordinating neutral ligands. Initially, the coordinating anionic carboxylate linkers could be easily replaced by Se_2^{2-} anions during selective selenization process, while the neutral N-coordinating ligands were maintained and embedded in the $NiSe_2$ nanocrystals. During the carbonization process at 450 °C, the N-ligands are converted into ultrathin N-doped carbon and segregated onto the $NiSe_2$ surface. The N concentration and pyridinic-N contents were controlled by varying the N-ligands in MOFs (N-ligands = 4,4′-bipyridine, 1,4-diazabicyclooctane, pyrazine, and aminopyrazine), thus providing a highly controllable platform for mechanistic studies [54].

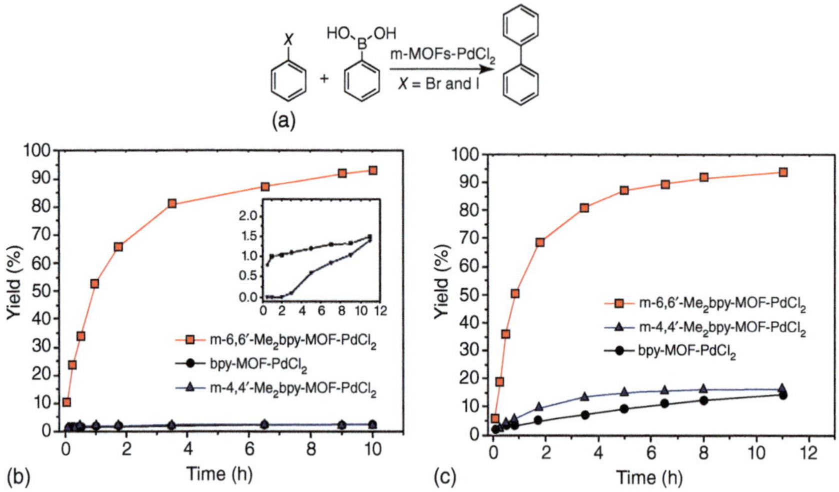

Figure 4.8 (a) Suzuki–Miyaura cross-coupling reaction catalyzed by m-MOF-$PdCl_2$; (b) Catalytic difference of m-MOF-$PdCl_2$ in coupling reaction of iodobenzene. The inset shows the reaction yield by m-bpy-MOF-$PdCl_2$ and m-4,4′-Me_2bpy-MOF-$PdCl_2$; (c) Catalytic difference of m-MOF-$PdCl_2$ in coupling reaction of bromobenzene. Reaction conditions: Aryl halide (0.1 mmol), phenylboronic acid (0.15 mmol), K_2CO_3 (0.2 mmol), toluene (0.5 ml), m-MOF-$PdCl_2$ (1.0 mol% Pd), Ar atmosphere, 85 °C. Source: Li et al. [53]/American Chemistry Society.

4.5.2 Mixed-Linker MOFs with Structurally Independent Linkers

The second part will briefly state the use of mixed-linker MOFs as heterogeneous catalysts that are constructed with linkers having different sizes or shapes (Table 4.3).

Zr-based MOFs were synthesized by using mixtures of NDC and NDC–NH_2 or BPDC and BPDC–NH_2 as organic linkers. The materials with low loadings of functionalized linkers (–NH_2) exhibited higher long-term chemical stability and similar high thermal stability relative to those with 100% amino linkers. Finally, data clearly demonstrate the effect of the percentage of amino functionalized linker on crystal structure, MOF stability, and, consequently, catalytic activity of the resulting material, which is lower for amorphous solids [55].

UiO-67 MOFs were synthesized by incorporating some urea derivatives, such as H_2-urea, with the BPDC linker, leading to a MOF with the molecular formula of $Zr_6O_4(OH)_4(urea)_{1.44}(BPDC)_{2.56}$ and having two pore sizes with diameters of 23 and 12 Å. The catalytic activity of this mixed-linker MOF was tested in the Henry reaction of benzaldehyde and nitromethane (Scheme 4.6) [56].

$Zr_6O_4(OH)_4(urea)_{1.44}(BPDC)_{2.56}$ exhibited after 24 hours more than a threefold increase in the formation of the desired product 2-nitro-1-phenylethanol (67% yield), while UiO-67-Urea catalyst showed 14% yield in the control experiment. Under identical reaction conditions, UiO-67 showed 11% yield after 24 hours (Figure 4.9). The enhanced catalytic activity of $Zr_6O_4(OH)_4(urea)_{1.44}(BPDC)_{2.56}$ is due to its high surface area (1550 $m^2 g^{-1}$) relative to that of UiO-67-urea (390 $m^2 g^{-1}$). Furthermore, the pore size distribution indicates that the mixed-linker MOF (21.5 and 12 Å) did

Table 4.3 Mixed-linker MOFs with structurally different linkers as heterogeneous catalysts.

Catalyst	Mixed linkers		Reaction	Reference
UiO-67 or UiO-67-NH_2	BPDC	NH_2-BPDC	Cascade reaction	[55]
$Zr_6O_4(OH)_4(urea)_{1.44}(BPDC)_{2.56}$	Urea	BPDC	Henry reaction	[56]
$([Cd(ABDC)(BPHZ)_{0.5}]\cdot DMF\cdot H_2O\}_n$	ABDC	BPHZ	Knoevenagel condensation	[57]
Zn_2(ABDC)(L-lac)(HCOO)	ABDC	L-lactic acid	A^3 coupling	[58]
$Cu_2(BDC)_2(BPY)$	BDC	BPY	Coupling reaction	[59]
$[Co_3(OBA)_3(O)(PY)_{0.5}]_n$ 4DMF·Py (TMU-12)	OBA	PY	Desulfurization of DBT	[60]
$[H_2NMe_2]_2Cd_3\{TiO_6(TiL)_3\}(BPDC)_3(H_2O)_3]$ $16H_2O$	chiral Ti(salen)	BPDC	Oxidation of thioethers to sulfoxides	[61]
Vanadium-salen Cd-BPDC	chiral V(salen)	BPDC	Asymmetric cyanosilylation of benzaldehyde	[62]
$[Cu_2(BTEC)(BTX)_{1.5}]_n$	BTEC	BTX	Degradation of MO	[63]
$[Co(L)_{0.5}(HIP)]_n$	L	HIP	Degradation of MO	[64]
$[Co(L1)(BPDC)]_n$	L1	BPDC	Degradation of Congo red	[65]
$[Cu^{II}(BPAH)_2(OBA)(H_2O)]$	BPAH	OBA	Degradation of MB, MO and RhB	[66]
$[Zn(1,3\text{-}BDC)(bmimb)]_n$	1,3-BDC	bmimb	Degradation of RhB	[67]

Abbreviations: BPY: bipyridine; PYDC: pyridinedicarboxylic acid; BPDC: (1,1′-biphenyl)-4,4′-dicarboxylic acid; ABDC: 2-amino-1,4-benzenedicarboxylic acid; BPHZ: 1,2-bis(4-pyridylmethylene)hydrazine); OBA: 4,4′-oxydibenzoic acid; PY: pyrazine; DBT: dibenzothiophene; TMU: Tarbiat Modares University; BTEC: 1,2,4,5- benzenetetracarboxylate; BTX: 1,4-bis(1,2, 4-triazol-1-ylmethyl)benzene; L: 1,4-bis(5,6- dimethylbenzimidazole)butane; HIP: 5-hydroxyisophthalic acid; L1: 1,4-bis(5,6-dimethylbenzimidazol-1-ylmethyl)benzene; MO: methyl orange; BPAH: N,N′-bis(4-pyridinecarboxamide)-1,2-cyclohexane); MB: methylene blue; RhB: Rhodamine B; 1,3-BDC: 1,3-benzenedicarboxylic acid; bmimb: 4,4′-bis(4-methyl-1-imidazolyl) biphenyl); NDC: 2,6-naphthalenedicarboxylic acid; NDC-NH_2: 4-amino-2,6-naphthalenedicarboxylic acid.

not show significant variations in the pore diameters in comparison to the parent UiO-67 (23 and 11.5 Å) and notably, both are significantly larger than that of the pure strut UiO-67-Urea (12 and 9 Å) [56].

A novel homochiral Zn-containing MOF referred to as CUP-1 with the formula Zn_2(atpt)(L-lac)(HCOO) was successfully prepared based on the reaction of the mixed linkers of H_2atpt and L-lactic acids in an OP synthesis. By virtue of free NH_2 group in the frameworks of CUP-1 and IRMOF-3, the catalytically active chiral L-proline or gold was chosen to modify the CUP-1 and/or IRMOF-3 catalysts by

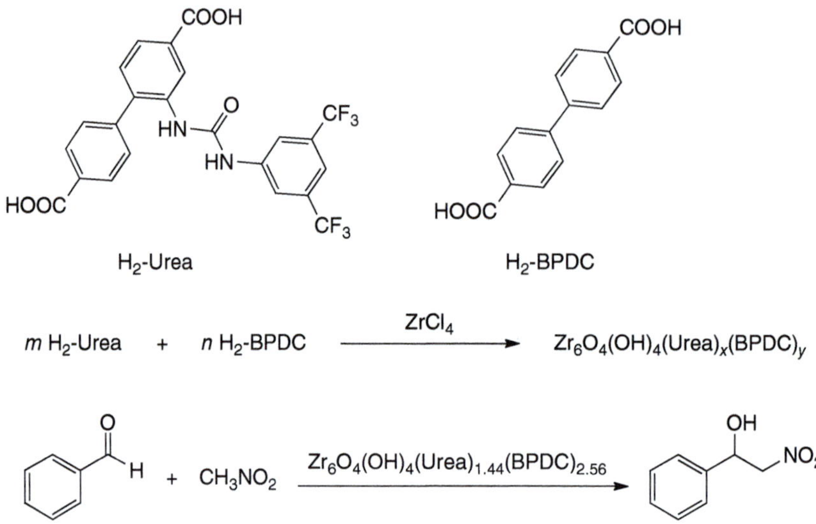

Scheme 4.6 Synthesis of $Zr_6O_4(OH)_4(urea)_x(BPDC)_y$ and its catalytic activity in Henry reaction of benzaldehyde with nitromethane.

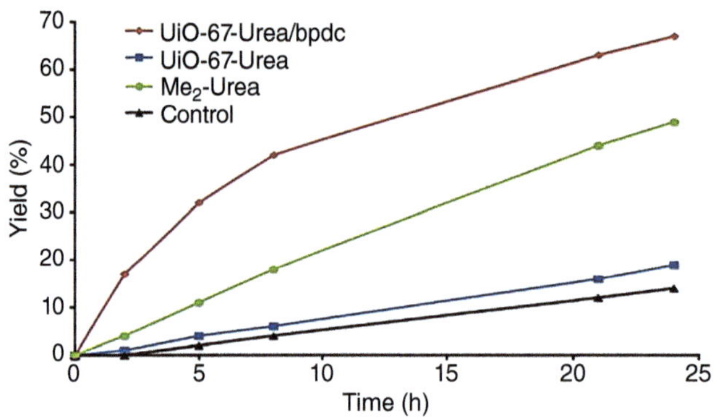

Figure 4.9 Catalytic activities of the mixed-linker MOF UiO-67-Urea/BPDC, pure MOF UiO-67- Urea and Me_2-Urea strut. Source: Siu et al. [56]/Royal Society of Chemistry.

OP and PM strategies. The L-proline functionalized IRMOF-3 catalysts showed fair to excellent enantioselectivity (up to 98%) in asymmetrical aldol reactions with a higher TON and catalytic stabilities than the homogeneous counterpart of L-proline [58].

In other work, the role of bipyridine ligand as a secondary ligand in mixed-linker MOFs has been investigated. $Cu_2(BDC)_2(BPY)$ has been synthesized and used as an efficient heterogeneous catalyst for oxidative cross coupling reactions (Scheme 4.7). The $Cu_2(BDC)_2(BPY)$ catalyst offered high activity and selectivity since only trace amount of diyens by-products were observed. Studies on catalyst modification indicated that bipyridine ligands enabled the stability of catalysts, while BDC was the optimal linker. $Cu_2(BDC)_2(BPY)$ resulted in quantitative yield of the

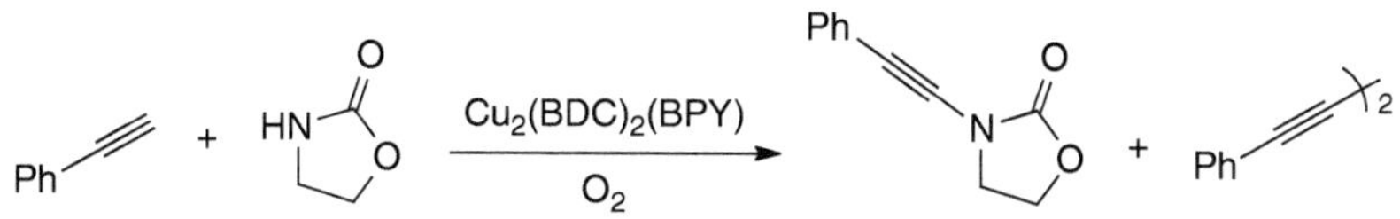

Scheme 4.7 Oxidative cross-coupling reactions between activated N–H amine and terminal alkyne.

Table 4.4 Catalyst activity of other mixed-linker MOFs.

Entry	Type	Catalyst	GC yield (%)	Selectivity
1	Cu-MOFs	$Cu_2(BDC)_2(BPY)$	100 (TOF = 41.3)	95
2		$Cu_2(BDC)_2(DABCO)$	100	85
3		Cu(BDC)	92	84
4		Cu(BDC) 2nd run	56	51
5		$Cu_2(OBA)_2(BPY)$	58	75
6		$Cu_3(BTC)_2$	39	47
13	Other MOFs	$Ni_2(BDC)_2(BPY)$	0	—
14		$Co_2(BDC)_2(DABCO)$	0	—
15		$Mn_2(BDC)_2(DMF)_2$	0	—
16		$Fe_3O(BDC)_3$	0	—
17		ZIF-67	0	—

Source: Le et al. [59]/Royal Society of Chemistry.

coupling product with 95% selectivity, while in contrast Cu(BDC) gave 92% yield with 84% selectivity. On the other hand, $Cu_3(BTC)_2$ exhibited 39% yield with 47% selectivity [59].

On the other hand, mixed linker MOFs with other metal ions like $Ni_2(BDC)_2(BPY)$, $Co_2(BDC)_2(DABCO)$, and $Mn_2(BDC)_2(DMF)_2$ failed to promote this oxidative cross-coupling reaction (Table 4.4). These data clearly indicate the specific role played by the metal ion and the synergy that can be achieved by the appropriate selection of the mixed linkers. This effect is apparently achieved by taking advantage of the Cu^{2+} coordination sphere in the mixed-linker MOF [59].

In new study, two cobalt-based MOFs were identified $[Co_6(OBA)_5(OH)_2(H_2O)_2(DMF)_4]_n$ 5DMF (TMU-10) and $[Co_3(OBA)_3(O)(PY)_{0.5}]_n$ 4DMF.PY (TMU-12) has been synthesized, and their catalytic activity in desulfurization of DBT has been examined. TMU-12 is constructed from two OBA and pyrazine ligands, while the structure of TMU-10 is built from only OBA ligand. Desulfurization performance of TMU-12 is nearly two times larger than that of TMU-10 for removal of DBT under optimized conditions after eight hours [60]. This enhanced activity of TMU-12 may be related to the unsaturated coordination number around Co, which is responsible for more adsorption and oxidation of DBT in the presence of TBHP, while in the case of TMU-10, there are no unoccupied positions around Co centers. This is a very nice example of where the activity of mixed-linker MOF shows higher activity than its pure counterpart due to the different structural features around the metal center [60].

$[Cd_3(BTEC)(BTX)_{0.5}(\mu_3\text{-}OH)(H_2O)]\ H_2O\}_n$ and $[Cu_2(BTEC)(BTX)_{1.5}]_n$ MOFs were synthesized with BTEC and BTX as mixed-linker MOFs. BTX as secondary linker cannot only extend the structural dimension of its complexes through the pillared-layered method but can also modify the structures to obtain highly connected frameworks. The catalytic activity of $[Cu_2(BTEC)(BTX)_{1.5}]_n$ was studied in the degradation of MO by Fenton-like reaction, but no catalytic data regarding the activity of Cd-based MOF was provided [63].

A new 3D supramolecular framework, namely, $[Co(L)_{0.5}(HIP)]_n$, has been synthesized using 1,4-bis(5,6-dimethylbenzimidazole)butane (L) and HIP and catalytic activity in the degradation of MO by sodium persulfate has been examined. Results have shown that the degradation efficiency of MO is about 49.1% within 105 minutes, and there is no data regarding the catalyst's stability in terms of leaching or reusability [64].

In other work, Wang and his coworker's has been introduced four mixed-linker MOFs based on the flexible bis(5,6-dimethybenzimidazole) and aromatic dicarboxylic acids, namely, $[Co(L1)(BPDC)]_n$, $[Co(L1)(NPHT)]\ 0.5H_2O\}_n$, $[Co(L2)(BPDC)]_n$, and $[Co(L3)(BPDC)(H_2O)]_n$ (L1 = 1,4-bis(5,6-dimethylbenzimidazol-1-ylmethyl)benzene, L2 = 1,3-bis(5,6-dimethylbenzimidazol-1-ylmethyl)benzene, H_2npht = 3-nitrophthalic acid, L3 = 1,10-bis(5,6-dimethylbenzimidazole)methane) were synthesized and their activity tested in the degradation of Congo red dye. Degradation of Congo red in the presence of $[Co(L1)(BPDC)]_n$ or $[Co(L1)(NPHT)]\ 0.5H_2O\}_n$ resulted in 89% or 98% after 130 minutes, respectively. However, when MOFs such as $[Co(L2)(BPDC)]_n$ or $[Co(L3)(BPDC)(H_2O)]_n$ were used as photocatalysts, only about 56% or 52% of the dye was decolorized, respectively. The difference in activity between these MOFs arises from their structural integrity under photocatalytic conditions and their ability to produce more hydroxyl radicals upon irradiation [65].

The new Cu^{II}-mixed-linker MOF $[Cu^{II}(BPAH)_2(OBA)(H_2O)]$ has been synthesized and catalytic activity has been investigated in the degradation of MB, MO, and RhB. The degradation percentage after 240 minutes reached about 60%, 28%, and 22%, for MB, MO, and RhB, respectively [66].

Similary, a series of mixed-linker MOFs, namely, $[Zn(1,3\text{-}BDC)(bmimb)]_n$, $[Zn(1,4\text{-}BDC)(bmimb)]_n$, $[Cd(1,3\text{-}BDC)(bmimb)]_n$, and $[Cd(1,4\text{-}BDC)(bmimb)]_n$ has been reported to be active for the photochemical degradation of RhB too (Scheme 4.8 and Figure 4.10) [67].

L-Proline-functionalized zirconium-based sulfonated MOF catalysts were successfully synthesized using a solvothermal method under aprotic polar solvent conditions. The use of two dicarboxylic acids (H_2bdc and H_2bdc-SO_3Na) in the synthesis led to mixed-linker MOFs, and use of 30% H_2bdc-SO_3Na in the synthesis led to optimum values for the surface area, pore volume, and crystallinity compared to the other mixed-linker ratios. L-Proline immobilization was carried out by adding L-proline directly into the reaction mixture (Scheme 4.9). The aldol reaction of

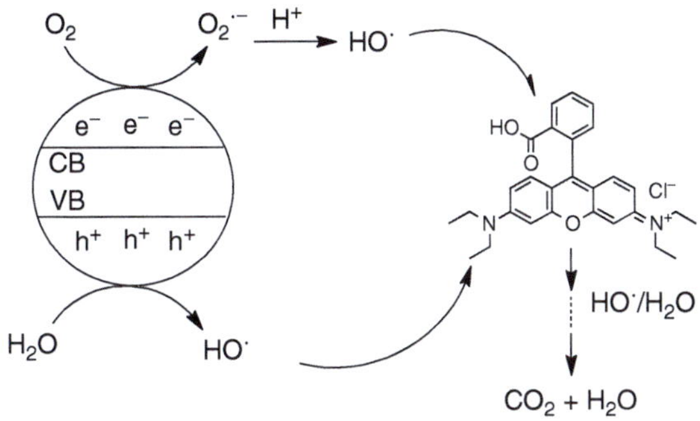

Scheme 4.8 Proposed mechanism for the photodegradation of organic dyes using MOFs as catalysts [67].

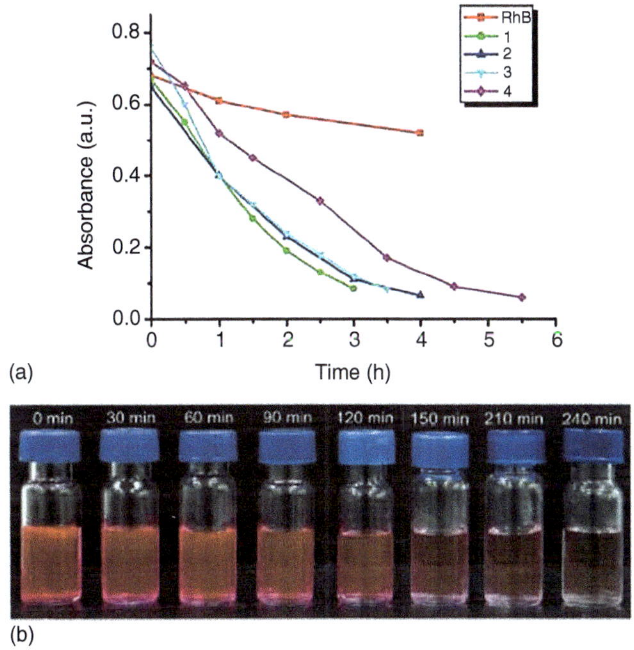

Figure 4.10 (a) Curves of absorbance of the RhB solutions degraded by 1–4 under UV light irradiation, and (b) image showing the catalytic photodegradation under UV light irradiation for 0, 30, 60, 90, 120, 150, 210, and 240 minutes using 3 as a catalyst. Source: Lü et.al. [67]/Royal Society of Chemistry.

acetone with 2-CBA or 4-NBA was catalyzed by UiO-66-S_{30}Pr. With 4-NBA as the aldehyde, the catalyst performance was very good (80% ee and 73% conversion), though both ee and conversion were lower using 2-CBA (60% ee and 59% conversion). Although the catalytic activity of UiO-66-S_{30}Pr was slightly lower than that of neat L-proline, it showed excellent performance as a heterogeneous catalyst since it can be recycled for three-run reactions without significant decreases in stereoselectivity (ee) and reaction conversion (for 2-CBA) and only slight decreases

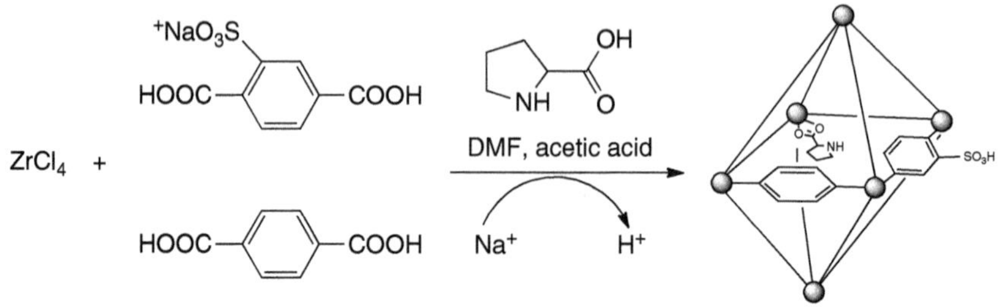

Scheme 4.9 Proline immobilization by directly solvothermal method. Source: Jaluddin et al. [51]/with permission of Elsevier.

for 4-NBA, confirming that the use of an OSN membrane is a promising separation method for the novel zirconium-based sulfonated MOF-proline catalyst [51].

Chen and his coworkers have reported a new Cu(II)-based mixed-MOF [Cu_2(1,2-BDC)$_2$(Fbtx)$_2$] $3H_2O\}_n$ (denoted as Cu-FMOF, 1,2-BDC = 1,2-benzenedicarboxylate, Fbtx = 1,4-bis(1,2,4-triazole-1-ylmethyl)-2,3,5,6-tetrafluorobenenze) that could be used as a highly efficient heterogeneous catalyst for the TEMPO-assisted oxidation of alcohols to aldehydes. Furthermore, the Cu-FMOF catalyst could be easily recovered and reused several times without a significant loss of catalytic activity or structural deterioration [68].

The incorporation of sulfonic acid functional groups into MOFs has a significant impact on gas sorption properties and potential catalytic and proton conduction applications [69, 70].

Within this context, Zhou and coworkers reported a series of sulfonic acid-functionalized mixed-linker MOFs having the DUT-4 topology by using different ratios of 2,6-naphthalenedicarboxylic acid (H_2-NDC) and 4,8-disulfonaphthalene-2,6-dicarboxylic acid (H_2-NDC-2SO_3H) in one-pot reactions. Due to the strong acidic character of the sulfonic acid groups, the DUT-4-SO_3H(30) (30 corresponds to the theoretical percentage of H_2-NDC-2SO_3H used during the synthesis) has been evaluated as a catalyst for the ring opening of styrene oxide under mild conditions, reaching full conversion (99%) after five hours of reaction (Figure 4.11). Moreover, the DUT-4-SO_3H(30) catalyst could be recovered and reused for three additional cycles without significant loss in activity or stability [71]. Comparison of the catalytic performance of DUT-4-SO_3H(30) with other reported MOF catalysts used for the ring opening of styrene oxide has been reported in Table 4.5 [71].

Photoelectrochemical alcohol oxidation by mixed-linker MOFs has been investigated by Morris and coworkers, recently. A mixed-linker MOF comprising a photosentisizer [Ru(dcbpy)(bpy)$_2$]$^{2+}$ (bpy = 2,2′-bipyridine, dcbpy = 5,5′-dicarboxy-2,2′-bipyridine) and catalyst [Ru(tpy)(dcbpy)Cl]$^+$ (tpy = 2,2′:6′,2″-terpyridine) were incorporated into the UiO-67 framework and grown as thin films on a TiO_2-coated, fluorine-doped tin oxide (FTO) electrode (RuB-RuTB-UiO-67-TiO_2/FTO). When used as an electrode for the photoelectrochemical oxidation of benzyl alcohol, the mixed-linker MOF film showed a Faradaic efficiency of 34%, corresponding to a threefold increase in efficiency relative to a RuB-UiO-67-TiO_2/FTO control [72].

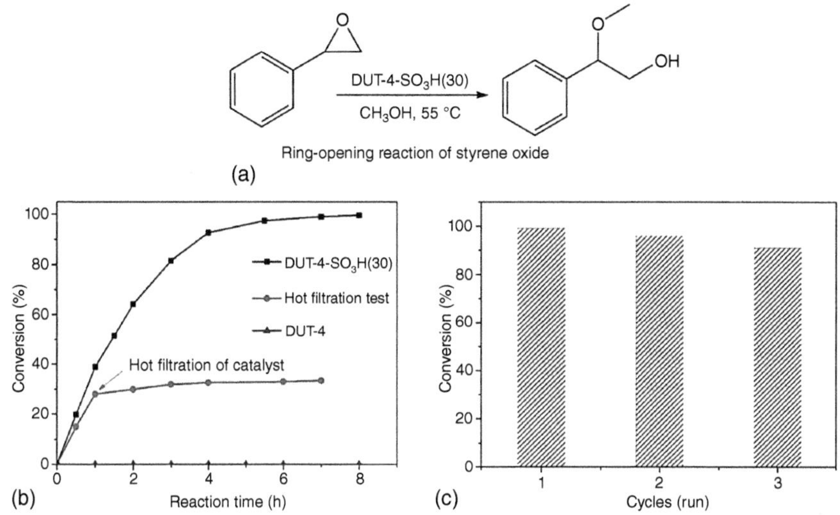

Figure 4.11 (a) Styrene oxide ring-opening reaction with methanol, catalyzed by MOFs; (b) time-conversion plot for the ring-opening reaction of styrene oxide using DUT-4 and DUT-4-SO3H(30) as catalysts and the performed hot filtration test; and (c) reusability of the catalyst in three successive runs for the ring opening of styrene oxide with methanol. Source: Wang et al. [71]/Royal Society of Chemistry.

Table 4.5 Comparison of the catalytic performance of DUT-4-SO_3H(30) with other reported MOF catalysts used for the ring opening of styrene oxide. CH_3OH, 55 °C.

Entry	Catalyst	T (min)	T (°C)	Conv. (%)
1	Cu-MOF	3 h	RT	99
2	Fe(BTC)	60	40	>99
3	HKUST-1	2.5 h	40	90
4	HKUST-1/ATP	20	50	98.9
5	MIL-101-SO_3H	30	RT	>99
6	MIL-101(HPW)	20	40	99 (dropped to 62 in the 3rd run)
7	HP-CuBTC-0.25	4 h	40	35.6
8	Zr-UiO-66	24 h	55	40
9	Zr-MOF-808	24 h	55	100
10	UiO-66-X (X = H, NH_2, NO_2, Br, Cl)	24 h	50	96
11	DUT-4-SO_3H(30)	6 h	55	>99

4.6 Conclusion

Since the modifications on the organic struts of MOFs with specific catalytic moieties provide an accessible and efficient method to introduce new catalytic capabilities into MOFs, rational design of MOFs as heterogeneous catalysts has been flourishing. It can be foreseen that in the near future, examples of cascade reactions with mixed-linker MOFs will be much more abundant, and it is envisioned that in the long term, the design of a MOF based on the knowledge of the required catalytic sites will precede the synthesis of these materials and probably more than two metals, two linkers, or contributions of both metals and linkers will be synthesized to develop new generations of advanced MOF catalysts for catalyzing cascade reactions.

References

1 Long, J.R. and Yaghi, O.M. (2009). *Chem. Soc. Rev.* 38: 1213–1214.
2 Furukawa, H., Cordova, K.E., O'Keeffe, M., and Yaghi, O.M. (2013). *Science* 341: 1230444.
3 Deng, H., Doonan, C.J., Furukawa, H. et al. (2010). *Science* 327: 846–850.
4 Koh, K., Wong-Foy, A.G., and Matzger, A.J. (2008). *Angew. Chem. Int. Ed.* 47: 677–680.
5 Koh, K., Wong-Foy, A.G., and Matzger, A.J. (2009). *J. Am. Chem. Soc.* 131: 4184–4185.
6 Koh, K., Wong-Foy, A.G., and Matzger, A.J. (2010). *J. Am. Chem. Soc.* 132: 15005–15010.
7 Hirai, K., Furukawa, S., Kondo, M. et al. (2011). *Angew. Chem. Int. Ed.* 123: 8207–8211.
8 Burrows, A.D. (2011). *CrystEngComm* 13: 3623–3642.
9 Bunck, D.N. and Dichtel, W.R. (2013). *Chem. Eur. J.* 19: 818–827.
10 Burrows, A.D., Frost, C.G., Mahon, M.F., and Richardson, C. (2008). *Angew. Chem. Int. Ed.* 47: 8482–8486.
11 Chun, H., Dybtsev, D.N., Kim, H., and Kim, K. (2005). *Chem. Eur. J.* 11: 3521–3529.
12 Fukushima, T., Horike, S., Inubushi, Y. et al. (2010). *Angew. Chem. Int. Ed.* 49: 4820–4824.
13 Burrows, A.D., Fisher, L.C., Richardson, C., and Rigby, S.P. (2011). *Chem. Commun.* 47: 3380–3382.
14 Marx, S., Kleist, W., Huang, J. et al. (2010). *Dalton Trans.* 39: 3795–3798.
15 Lescouet, T., Kockrick, E., Bergeret, G. et al. (2012). *J. Mater. Chem.* 22: 10287–10293.
16 Taylor-Pashow, K.M., Della Rocca, J., Xie, Z. et al. (2009). *J. Am. Chem. Soc.* 131: 14261–14263.
17 Kleist, W., Jutz, F., Maciejewski, M., and Baiker, A. (2009). *Eur. J. Inorg. Chem.* 2009: 3552–3561.
18 Kleist, W., Maciejewski, M., and Baiker, A. (2010). *Thermochim. Acta* 499: 71–78.

19 Farha, O.K., Wilmer, C.E., Eryazici, I. et al. (2012). *J. Am. Chem. Soc.* 134: 9860–9863.
20 Liao, Y., Yang, S.K., Koh, K. et al. (2012). *Nano Lett.* 12: 3080–3085.
21 Qin, J.-S., Yuan, S., Wang, Q. et al. (2017). *J. Mater. Chem. A* 5: 4280–4291.
22 Seo, J., Matsuda, R., Sakamoto, H. et al. (2009). *J. Am. Chem. Soc.* 131: 12792–12800.
23 Guillerm, V., Kim, D., Eubank, J.F. et al. (2014). *Chem. Soc. Rev.* 43: 6141–6172.
24 Yan, Y., Telepeni, I., Yang, S. et al. (2010). *J. Am. Chem. Soc.* 132: 4092–4094.
25 Zhang, Y.-B., Zhou, H.-L., Lin, R.-B. et al. (2012). *Nat. Commun.* 3: 1–9.
26 Jiang, G., Wu, T., Zheng, S.-T. et al. (2011). *Cryst. Growth Des.* 11: 3713–3716.
27 Taddei, M., Costantino, F., Ienco, A. et al. (2013). *Chem. Commun.* 49: 1315–1317.
28 Chen, Q., Chang, Z., Song, W.C. et al. (2013). *Angew. Chem.* 125: 11764–11767.
29 Park, H.J., Cheon, Y.E., and Suh, M.P. (2010). *Chem. Eur. J.* 16: 11662–11669.
30 Zheng, S.-T., Zhao, X., Lau, S. et al. (2013). *J. Am. Chem. Soc.* 135: 10270–10273.
31 Zhao, X., Bu, X., Nguyen, E.T. et al. (2016). *J. Am. Chem. Soc.* 138: 15102–15105.
32 Wang, H., Xu, J., Zhang, D.S. et al. (2015). *Angew. Chem. Int. Ed.* 54: 5966–5970.
33 Ranocchiari, M. and Bokhoven, J.A.V. (2011). *Phys. Chem. Chem. Phys.* 13: 6388–6396.
34 Bhattacharjee, S., Lee, Y.-R., Puthiaraj, P. et al. (2015). *Catal. Surv. Asia* 19: 203–222.
35 Dhakshinamoorthy, A., Asiri, A.M., and Garcia, H. (2016). *Catal. Sci. Technol.* 6: 5238–5261.
36 Roy, A.S., Mondal, J., Banerjee, B. et al. (2014). *Appl. Catal., A* 469: 320–327.
37 Diaz, U., Boronat, M., and Corma, A. (2012). *Proc. R. Soc. A Math. Phys. Eng. Sci.* 468: 1927–1954.
38 Huang, Y., Gao, S., Liu, T. et al. (2012). *ChemPlusChem* 77: 106–112.
39 Kozachuk, O., Luz, I., Llabres i Xamena, F.X. et al. (2014). *Angew. Chem. Int. Ed.* 53: 7058–7062.
40 Marx, S., Kleist, W., and Baiker, A. (2011). *J. Catal.* 281: 76–87.
41 Xu, X., van Bokhoven, J.A., and Ranocchiari, M. (2014). *ChemCatChem* 6: 1887–1891.
42 Sun, D., Fu, Y., Liu, W. et al. (2013). *Chem. Eur. J.* 19: 14279–14285.
43 Goh, T.W., Xiao, C., Maligal-Ganesh, R.V. et al. (2015). *Chem. Eng. Sci.* 124: 45–51.
44 Servalli, M., Ranocchiari, M., and Van Bokhoven, J.A. (2012). *Chem. Commun.* 48: 1904–1906.
45 Gascon, J., Aktay, U., Hernandez-Alonso, M.D. et al. (2009). *J. Catal.* 261: 75–87.
46 Llabrés i Xamena, F.X., García Cirujano, F., and Corma Canós, A. (2012). *Microporous Mesoporous Mater.* 157: 112–117.
47 Wang, J.-L., Wang, C., and Lin, W. (2012). *ACS Catal.* 2: 2630–2640.
48 Wang, S. and Wang, X. (2015). *Small* 11: 3097–3112.
49 Dhakshinamoorthy, A., Asiri, A.M., and Garcia, H. (2016). *Angew. Chem. Int. Ed.* 55: 5414–5445.

50 Lee, Y., Kim, S., Kang, J.K., and Cohen, S.M. (2015). *Chem. Commun.* 51: 5735–5738.

51 Jaluddin, S.N., Kassim, K., and Ibrahim, W.N.W. (2015). *Proc. Chem.* 16: 8–14.

52 Sun, R., Liu, B., Li, B.-G., and Jie, S. (2016). *ChemCatChem* 8: 3261–3271.

53 Li, X., Van Zeeland, R., Maligal-Ganesh, R.V. et al. (2016). *ACS Catal.* 6: 6324–6328.

54 Huang, Z., Yuan, S., Zhang, T. et al. (2020). *Appl. Catal., B* 272: 118976.

55 Rasero-Almansa, A.M., Corma, A., Iglesias, M., and Sánchez, F. (2014). *ChemCatChem* 6: 3426–3433.

56 Siu, P.W., Brown, Z.J., Farha, O.K. et al. (2013). *Chem. Commun.* 49: 10920–10922.

57 Haldar, R., Reddy, S.K., Suresh, V.M. et al. (2014). *Chem. Eur. J.* 20: 4347–4356.

58 Lili, L., Xin, Z., Shumin, R. et al. (2014). *RSC Adv.* 4: 13093–13107.

59 Le, H.T., Tran, T.V., Phan, N.T., and Truong, T. (2015). *Catal. Sci. Technol.* 5: 851–859.

60 Masoomi, M.Y., Bagheri, M., and Morsali, A. (2015). *Inorg. Chem.* 54: 11269–11275.

61 Xuan, W., Ye, C., Zhang, M. et al. (2013). *Chem. Sci.* 4: 3154–3159.

62 Bhunia, A., Dey, S., Moreno, J.M. et al. (2016). *Chem. Commun.* 52: 1401–1404.

63 Cui, G.-H., He, C.-H., Jiao, C.-H. et al. (2012). *CrystEngComm* 14: 4210–4216.

64 Qin, L., Zheng, J., Xiao, S.-L. et al. (2013). *Inorg. Chem. Commun.* 34: 71–74.

65 Wang, X.X., Yu, B., Van Hecke, K., and Cui, G.H. (2014). *RSC Adv.* 4: 61281–61289.

66 Wang, X.-L., Liu, D.-N., Luan, J. et al. (2015). *Inorg. Chim. Acta* 426: 39–44.

67 Lü, C.-N., Chen, M.-M., Zhang, W.-H. et al. (2015). *CrystEngComm* 17: 1935–1943.

68 Chen, S.-C., Lu, S.-N., Tian, F. et al. (2017). *Catal. Commun.* 95: 6–11.

69 Akiyama, G., Matsuda, R., Sato, H. et al. (2011). *Adv. Mater.* 23: 3294–3297.

70 Phang, W.J., Jo, H., Lee, W.R. et al. (2015). *Angew. Chem. Int. Ed.* 127: 5231–5235.

71 Wang, G.-B., Leus, K., Hendrickx, K. et al. (2017). *Dalton Trans.* 46: 14356–14364.

72 Lin, S., Cairnie, D.R., Davis, D. et al. (2021). *Faraday Discuss.* 225: 371–383.

5

Acid-Catalyzed Diastereoselective Reactions Inside MOF Pores

Herme G. Baldoví, Sergio Navalón, and Francesc X. Llabrés i Xamena

Universitat Politècnica de València, Consejo Superior de Investigaciones Científicas, Instituto de Tecnología Química CSIC-UPV, Departamento de Química, Av. de los Naranjos s/n, Valencia 46022, Spain

5.1 Introduction

Enzymes are the paradigm of highly selective catalysts and represent the model to follow when designing a metal–organic framework (MOF) for catalytic applications. The first stage in enzymatic catalysis consists of the formation of an enzyme–substrate complex when the substrate is adsorbed on the active site located inside well-defined enzymatic cavities. The specific substrate recognition by the enzyme relies on reversible conformational changes and secondary (weak) interactions in which both the active center and the residues of specific binding amino acids cooperate to enable the formation of the desired transition state, as shown in Figure 5.1. In this way, the high specificity of enzymes depends on the multiple and very precise weak interactions involving the substrate and these binding amino acids. These weak, noncovalent interactions can be electrostatic, hydrogen bonding, or van der Waals forces involved via the hydrophobic effect.

The development of novel bioinspired catalytic systems has been addressed by designing homogeneous molecular complexes with sophisticated (chiral) ligands resembling the enzyme active sites. However, these systems can hardly reproduce long-range effects beyond the first or second coordination sphere, and the long-range effects are typically found in enzymes. One strategy to overcome these limitations has been the use of porous, crystalline materials as heterogeneous catalysts, in which well-defined actives sites can be combined with confinement effects arising from cavity encapsulation.

In this sense, the high tunability of MOFs compositions, functional groups, pore architectures, and chemical modifications by post-synthesis treatments offers altogether unprecedented possibilities for the design and engineering of artificial enzymes. Thus, MOFs are excellent playgrounds for developing novel bioinspired catalytic systems in which chemical reactions take place inside well-defined pore

Catalysis in Confined Frameworks: Synthesis, Characterization, and Applications, First Edition.
Edited by Hermenegildo Garcia and Amarajothi Dhakshinamoorthy.

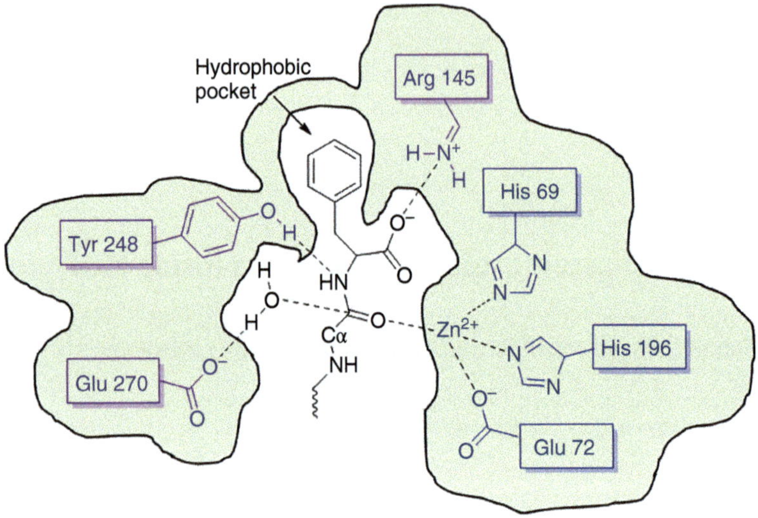

Figure 5.1 Active site and binding amino acids involved in the enzyme–substrate complex of carboxypeptidase A. The active site is formed by a Zn^{2+} ion coordinated to His69, Glu72, and His196. Binding amino acids Arg145, Tyr248, and Glu270 also participate in the formation of the enzyme–substrate complex. The nonpolar environment increases the polarizing power of Zn^{2+}, which facilitates the hydrolysis of the peptide bond. Glu270 assists in the deprotonation of a H_2O molecule, and the resulting –OH group attacks the Zn^{2+}-activated C=O bond of the peptide, while Tyr248 donates a proton to the –NH group of the peptide bond to be hydrolyzed. Source: Reproduced from Rechac et al. [1] with permission from the Royal Society of Chemistry.

systems of strictly regular dimensions, shapes, and chemical environments. This can be achieved by placing specific active centers at fixed positions within the MOF structure but also by modifying the spatial environment surrounding these active sites by tuning secondary, noncovalent interactions between these centers and the reaction substrates, by analogy with enzyme catalysis. In doing this, it is possible to improve the adsorption of reaction substrates within the MOF cavities, to determine their orientation with respect to the active sites, and to stabilize specific transition states, thus driving the reaction toward the desired product among various possible side reactions.

Noncovalent interactions in MOFs can sometimes modify the pore architecture of the MOF itself, in a process reminiscent of the *induced fit model* of enzyme catalysis proposed by Koshland [2]. According to this model, the initial interaction between the enzyme and the substrate is relatively weak, but once the substrate binds to the active site of the enzyme, both the substrate and the enzyme undergo slight conformational changes that strengthen the binding and maximize the fit to improve the catalysis. A similar process can also take place in MOF during a catalytic reaction, thanks to the flexible nature of these hybrid materials. In fact, it is well known that noncovalent interactions with certain substrates adsorbed inside the pore cavities can modify the pore architecture of the MOF in a reversible way, which is the so-called *dynamic porosity* or *breathing effect* [3, 4]. These adsorption-induced interactions allow switching between open and closed states of the MOF pores, which ultimately controls the molecular traffic in and out of the material surface. Note that such a dynamic behavior could endow the material with highly desired shape-selective properties triggered by external stimuli. For instance, the well-known metal terephthalates MIL-53 feature one-dimensional channels that are closed in the presence of water (MIL-53lt), but open wide when the solid is evacuated (MIL-53ht) [5]. A more complex situation is found in the ZnCar compound, containing Zn^{2+} ions and dipeptide carnosine ligands (β-alanyl-L-histidine) [6]. This 3D compound shows an interesting structural adaptability of the pore network in the presence of different guest molecules (DMF, H_2O, or methanol), thanks to the torsional flexibility of the main His–β-Ala chains, while the rigidity of the framework is maintained due to the Zn–imidazole chains. Note that such adaptive pore properties of MOFs are facilitated by their flexible frameworks, which are seldom found in other porous crystalline materials, such as zeolites, mesoporous silicates, or metal oxides.

Although dynamic/flexible confinement effects inside the pore cavities have not been exploited so far in catalysis, in this chapter we will show throughout selected examples taken from the literature, how MOFs can be specifically selected or designed *ad hoc* to impart selectivity to a catalytic reaction throughout confinement effects. It is necessary to clarify first that confinement effects include all the observed local properties introduced by the MOF that go beyond the nature of the active site itself. We thus include here effects related to the pore structure (pore size/shape and eventual dynamic behavior) as well as the presence of additional functional groups in the proximity of the active sites that can influence the reaction outcome. This can be achieved either by steric effects or by secondary noncovalent interactions with the functional groups of the organic linkers forming the pore walls. Since this is a very broad term, and it would encompass a large number of examples, we have limited ourselves to comment on those cases in which confinement effects in MOFs have been exploited to bestow diastereoselectivity to a catalytic process. Thus, we will generally refer to the use of MOFs as catalysts in reactions where a prochiral group of a substrate is selectively converted into one of the two possible pairs of diastereomers. Note that we will not include here the use of MOFs with homochiral pore systems or with chiral functional groups that are extensively used in enantioselective processes, which is actually the subject of another Chapter 6 of this book.

Sometimes it is not easy to understand the exact origin of the observed (diastereo)selectivity in the examples mentioned below, since in many cases it is the product of several factors acting together, or because the exact mechanism is not completely elucidated or it is simply overlooked in the original paper. For this reason, the examples discussed in this chapter have been organized according to the chemical transformation taking place during the reaction catalyzed by the MOF, rather than by the exact factor controlling the selectivity of the process.

5.2 Diastereoselective Reactions Catalyzed by MOFs

5.2.1 Meerwein–Ponndorf–Verley Reduction of Carbonyl Compounds

The Meerwein–Ponndorf–Verley reaction (MPV) [7–9] is an important resource in preparative organic chemistry, since it allows reducing a carbonyl compound (an aldehyde or a ketone) to the corresponding (primary or secondary) alcohol through a catalytic transfer hydrogenation reaction, using sacrificial alcohol as a source of hydride ions (Scheme 5.1). Moreover, this reduction is *chemoselective*, which means that the carbonyl group can be reduced in the presence of other easily reducible functional groups, such as unsaturated C—C bonds, esters, and nitro groups, which are not altered during typical MPV reaction conditions. This is a clear advantage of MPV with respect to other chemical reducing agents, such as $NaBH_4$, which are much less selective.

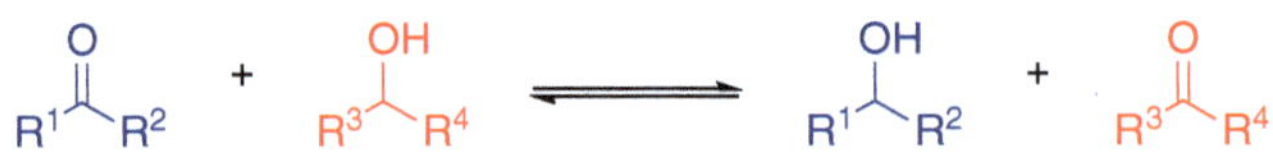

Scheme 5.1 General scheme of the MPV reaction.

First reports on MPV reaction used aluminum alkoxides as catalysts, but large (even stoichiometric) quantities of the alkoxides were usually required to obtain good yields. Thus, development of new efficient catalysts for MPV reactions is an area of active research. Those have included metal oxides (γ-Al_2O_3 [10], La_2O_3 [11], MgO [12], and hydrous ZrO_2 [13]); zeolites and mesoporous silicates with Al [14], Ti [15], Sn [16], Zr [17], Nb, or Ta [18] centers; and recently also MOFs.

Given the prochirality of the carbonyl group, the MPV reaction can also be diastereoselective, even without the need of using chiral reducing agents or catalysts, due to chelation or steric reaction control. In this line, substituted cyclohexanones have been considered as substrates to evaluate the diastereoselective properties of various MPV heterogeneous catalysts. In those cases, the diastereoselectivity observed was usually attributed to steric effects during the formation of the transition states required for both pairs of diastereomers. Thus, various beta zeolites containing Al, Ti, Sn, or Zr [15–17, 19, 20] have been used for reducing various *ortho-*, *meta-*, and *para*-substituted cyclohexanones, yielding the diastereomer with the hydroxyl group in axial position; i.e. the less thermodynamically stable alcohol. These results were interpreted as being due to the restricted available space inside the zeolite pores, which drove the reaction preferentially through the less bulky transition state. Accordingly, when metal-containing mesoporous silicates were used as catalysts, the opposite diastereoselectivity was observed, i.e. with much wider pores and without space restrictions, the thermodynamically more stable alcohol was the main reaction product [17, 21–25].

MOF-808 is a zirconium compound formed by hexameric $[Zr_6O_4(OH)_4]$ octahedral oxoclusters featuring a tridimensional network of adamantane-type cavities of 18.4 Å with apertures of 14 Å. Each cluster is connected by six trimesate linkers to other oxoclusters and six formate molecules bridging two Zr^{4+} ions within the same cluster. These capping formate molecules can be removed by simple solvent washing, leaving one coordination vacancy on each Zr^{4+} ion and allowing them to act as Lewis acid sites. Indeed, the activity of MOF-808 as a Lewis acid catalyst has been reported for a number of reactions, including hydrolysis of dipeptides [26] and chemical warfare agent simulants [27, 28], conversion of glucose into 5-hydromethylfurfural [29], transesterification of dimethyl carbonate [29], as well as MPV reactions [30–34], in some cases with very good results and high diastereoselectivities, as we will illustrate below.

Thus, MPV reduction of 3-methylcyclohexanone over MOF-808 with either isopropanol or 2-butanol as hydride sources produced 3-methylcyclohexanol quantitatively (>99%) after six hours, with activities comparable to other state-of-the-art catalysts for this reaction, such as Zr- and Ti-zeolites and mesoporous silicates. The reaction was selective for the thermodynamically more stable *cis*-diastereomer, as in the case of wide mesoporous silicate Zr-SBA, and opposite to what was observed for narrow pore beta zeolites (see Table 5.1). Reduction of 2-methylcyclohexanone is much more demanding than 3-methylcyclohexanone, since the methyl group in *ortho* position to the ketone introduces a severe steric hindrance. For this reason, the catalytic activity observed for Zr- and Ti-zeolites and mesoporous Zr-SBA was considerably low (see Table 5.1). The reaction was still governed by a transition state selectivity, leading to the *cis* alcohol in narrow pore Zr- and Ti-beta zeolites and *trans* alcohol in wide pore Zr-SBA. Conversely, MOF-808 showed a much higher catalytic performance than any catalyst previously described. Interestingly, the alcohol used as hydride source in the reaction governed the diastereoselectivity: iPrOH produced the *trans*-alcohol, while 2-BuOH produced the *cis* isomer. Given the excellent results obtained for the reduction of sterically hindered 2-methylcyclohexanone over MOF-808, we also explored the introduction of a still bulkier substituent in *ortho* position, a phenyl group. Despite the large size of the phenyl group, MOF-808 afforded excellent conversion and diastereoselectivity to the *cis*-alcohol.

The excellent chemo- and diastereoselectivities obtained for MPV reduction using MOF-808 as a catalyst have also been exploited to prepare steroid derivatives of interest for the pharmaceutical industry [30, 31]. This was possible due to the relatively large pore system of MOF-808, which allows the diffusion of bulky substrates such as steroids to the internal surface where the catalytic sites are located. In particular, we demonstrated that 17-ketosteroids are reduced in one single step to the corresponding 17-hydroxysteroids. Interestingly, we observed that reduction of estrone over MOF-808 and using 2-BuOH as reducing alcohol produced 17α-estradiol diastereomer with excellent selectivity (see Scheme 5.2). This is a very interesting result since the production of 17α-hydroxysteroids is, in general, a challenging process. Most chemical-reducing agents, such as $NaBH_4$, $Zn(BH_4)_2$, or $LiAlH_4$, produce the 17β isomer due to the steric hindrance of the 18-methyl group that blocks hydride attack from the upper face. Therefore, the preparation of 17α-hydroxysteroids usually requires several reaction steps to invert the configuration from 17β to 17α, and the final yields are very low. Conversely, the use of a MPV reduction over MOF-808 allows producing the 17α isomer selectively

Table 5.1 Summary of the catalytic results of MPV reduction of various substituted cyclohexanones over MOF-808 and other catalysts.

Catalyst	Conditions	Substrate	*cis*-Alcohol / *trans*-Alcohol	Reference
		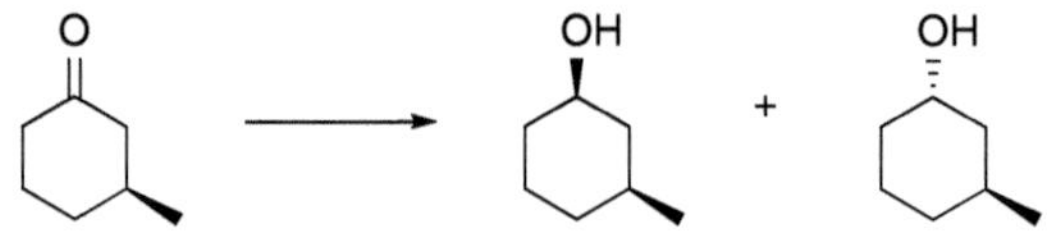		
		Conversion, time	**Diastereoselectivity**	
MOF-808	iPrOH, 80 °C	99%, 6 h	82 (*cis*)	[32]
	2-BuOH, 80 °C	99%, 6 h	68 (*cis*)	[32]
Zr-BEA	iPrOH, reflux	54.4%, 0.5 h	71 (*trans*)	[17]
Ti-BEA	iPrOH, reflux	25.8%, 6 h	70 (*trans*)	[15]
Zr-SBA	iPrOH, reflux	94.1%, 6 h	75 (*cis*)	[17]
		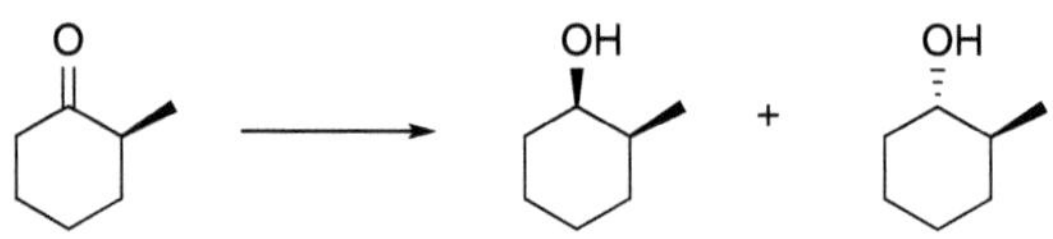		
		Conversion, time	**Diastereoselectivity**	
MOF-808	iPrOH, 80 °C	96%, 6 h	53 (*trans*)	[32]
	2-BuOH, 80 °C	99%, 24 h	61 (*cis*)	[32]
Zr-BEA	iPrOH, reflux	6.1%, 0.5 h	55 (*cis*)	[17]
Ti-BEA	iPrOH, reflux	8.8%, 6 h	60 (*cis*)	[15]
Zr-SBA	iPrOH, reflux	3.6%, 6 h	58 (*trans*)	[17]
		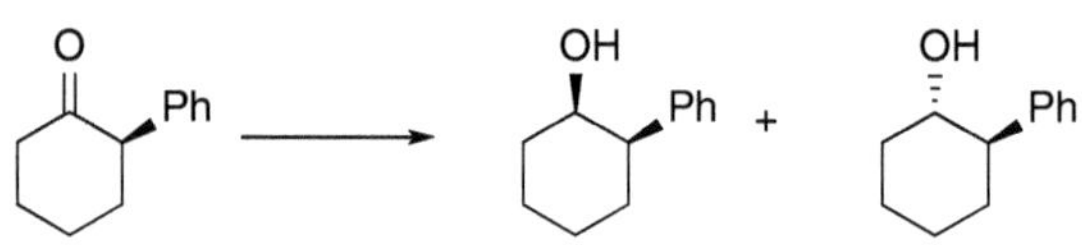		
		Conversion, time	**Diastereoselectivity**	
MOF-808	iPrOH, 80 °C	96%, 24 h	90 (*cis*)	[32]
	2-BuOH, 80 °C	97%, 24 h	94 (*cis*)	[32]
Zr-BEA	2-BuOH, 120 °C	12%, 24 h	72 (*cis*)	[32]
$NaBH_4$	EtOH, r.t.	100%, 2 h	62 (*trans*)	[32]

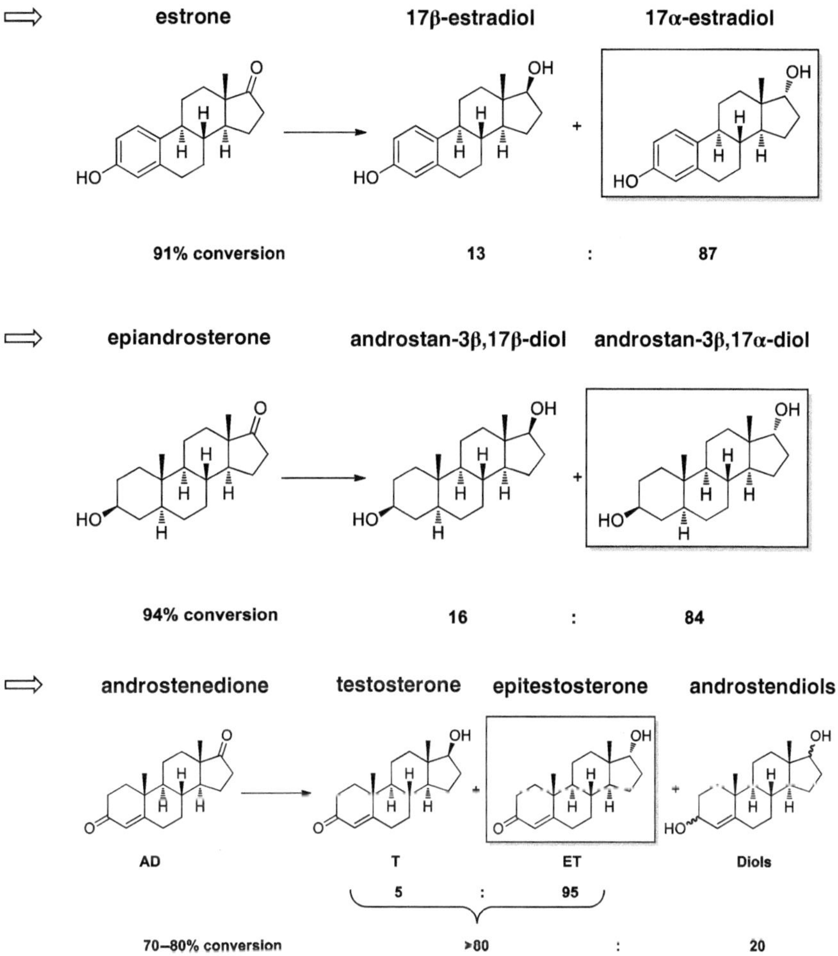

Scheme 5.2 Summary of the MPV reduction of various 17-ketosteroids over MOF-808.

in on single reaction step, which represents an important breakthrough in the targeted preparation of these compounds.

A detailed kinetic study of estrone reduction allowed us to conclude that the observed diastereoselectivity was due to the different accommodation inside the cavities of MOF-808 of the transition states leading to α and β isomers [31]. A strong contribution from the entropic term pointed out that confinement effects and steric hindrance are responsible for the preferred orientation and stabilization of the transition state leading to the 17α diastereomer. Preliminary molecular mechanic calculations indicated that the transition state leading to 17α was more tightly locked in place inside MOF-808 cavities and stabilized by up to four hydrogen bonds with the oxygen atoms of the Zr_6 oxoclusters, with d(O⋯H) comprised between 2.4 and 2.6 Å. Meanwhile, in the case of the 17β isomer, there is only one short hydrogen bond [35].

These studies were extended to the preparation of other hydroxysteroid derivatives. Thus, MPV reduction of epiandrosterone over MOF-808 produced again the corresponding 17α isomer with excellent selectively (Scheme 5.2), which demonstrated the general applicability of MOF-808 for the production of the elusive 17α-alcohols from ketosteroids. Furthermore, the study was completed with the reduction of androstenedione, which contains two keto groups in positions 3 and 17 and a sensible C=C bond in position 4. Therefore, this substrate is appropriate to address other relevant aspects of the reaction, such as the chemo- and regioselectivity. Our results (Scheme 5.2) evidenced again the great potential of MOF-808, since:

(i) The reaction is chemoselective (the C=C bond remains unaltered during the reaction);
(ii) The reaction is regioselective (the keto group in position 17 is selectively reduced over that in position 3);
(iii) The reaction is diastereoselective (the 17-keto group is reduced to the 17α isomer (epitestosterone) with over 95% selectivity; and
(iv) By properly adjusting the reaction conditions, it is possible to minimize the eventual production of androstenediols, achieving up to 70–80% conversion of androstenedione with selectivities above 80% to monohydroxylated compounds, as a mixture of testosterone + epitestosterone, of which over 95% is epitestosterone.

5.2.2 Aldol Addition Reactions

The Henry or nitroaldol reaction is a classical tool in organic chemistry toward the formation of new C—C bonds. Commonly, this reaction takes place between a nitroalkane having α-hydrogens and a carbonyl compound (aldehyde or ketone) in the presence of a base to form a β-nitro alcohol. In the area of heterogeneous catalysis also including MOFs, one of the general methods to induce diastereo- or enantioselectivity in this reaction includes the use of supported chiral organocatalysts such as L-proline or chiral metal catalysts. These strategies are, however, out of the scope of this chapter that is devoted to the study of acid-catalyzed diastereoselective reactions inside the cavities of a MOF in the absence of chiral moieties. In this context, a series of studies have reported the use of MOFs as nanoreactors having catalytically active metal nodes together in some cases with organic ligands that can also participate during the reaction pathway [36–38]. Importantly, the occurrence of the reaction in the confined pore network determines to some extent the observed diastereoselectivity of the process.

In one of these examples, a 2D Zn-based MOF with formula $[Zn_2(2\text{-acetamidoterephthalate})_2(4,4'\text{-bipyridine})_2(H_2O)(DMF)]_n$ was found to be an active catalyst for the nitroaldol reaction between a series of aromatic and aliphatic aldehydes with nitroethane [38]. The Zn-MOF was prepared by hydrothermal reaction of 2-acetamidoterephthalic acid with $Zn(NO_3)_2 6H_2O$ in the presence of 4,4′-bipyridine, resulting in a 2D MOF structure with two identical interpenetrated networks. Initially, this MOF was employed as heterogeneous catalyst for the aldol reaction of benzaldehyde and nitroethane in the presence of MeOH as solvent at 70 °C

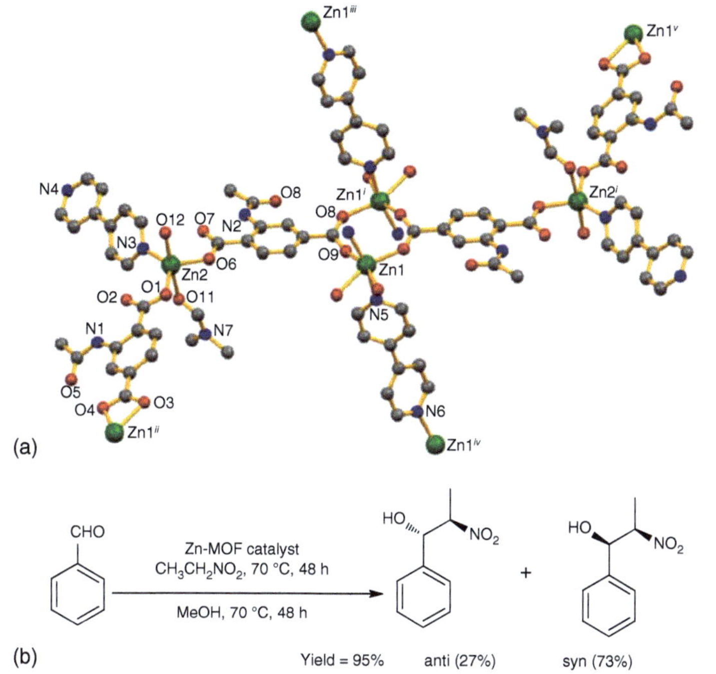

Figure 5.2 (a) Coordination scheme in the Zn-MOF with partial atom labeling. Hydrogen atoms are omitted for clarity. Symmetry operations to generate equivalent atoms: (i) $1-x$, $1-y$, $-z$; (ii) $1-x$, $1-y$, $1-z$; (iii) $1-x$, $2-y$, $-z$; and (iv) x, $-1+y$, z. (b) Nitroaldol (Henry) reaction. Source: Reproduced from Karmakar et al. [38] with permission from the Royal Society of Chemistry.

for 48 hours. Control experiments using $Zn(NO_3)_2$ or the MOF organic ligand as catalyst did not result in the observation of catalytic activity. In contrast, the presence of the MOF as catalyst (3 mol%) resulted in the formation of a reaction crude with a mixture of two products with predominant formation of the *syn* form of β-nitroalkanol diastereoisomer with respect to the *anti*-one (Figure 5.2) as revealed by ^{1}H-NMR. The use of *para*-substituted benzaldehydes with electron-withdrawing (NO_2 or Cl) or electron-donor groups (CH_3O– or CH_3–) resulted in a yield increase (98–93%) or decrease (39–62%), respectively, compared to unsubstituted benzaldehyde (78%). These experiments highlight the importance of the electrophilicity of the benzaldehydes with electron-withdrawing functional groups to achieve high yields. For the aldol reaction between *p*-nitrobenzaldehyde and nitroethane, the MOF catalyst was reused two times without observing significant decrease in catalytic activity and preserving its structural integrity based on powder XRD and FTIR measurements.

Regarding the reaction mechanism using the Zn-MOF (Figure 5.3), it was proposed that the Zn^{2+} metal centers present in the MOF activate both the aldehyde and the nitro compound. Subsequently, C—C bond formation occurs through nucleophilic addition, leading to the formation of the corresponding β-nitroalkanol. It is likely that the basic character of the amide group in the MOF organic ligand could favor the proton transfer steps during the reaction mechanism. Regardless of these

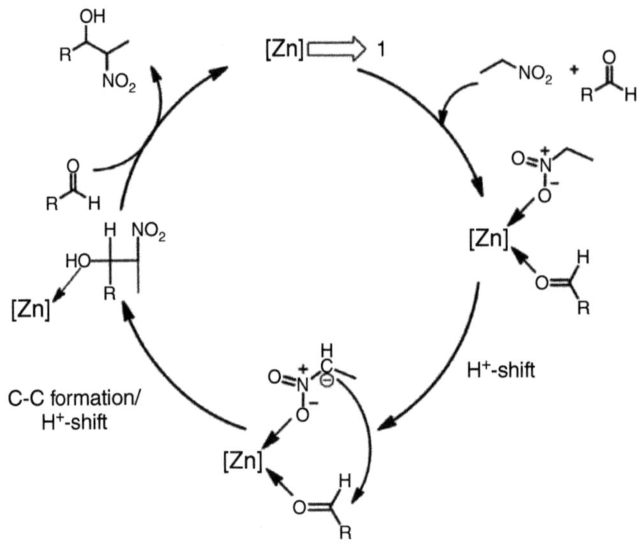

Figure 5.3 Proposed reaction pathway for the formation of the β-nitro-alkanol in the Henry reaction catalyzed by the Zn-MOF. Source: Reproduced from Karmakar et al. [38] with permission from the Royal Society of Chemistry.

comments, further studies are required to understand the origin of the preferential formation of the *syn* diastereomer.

In a related study, an amide-functionalized 2D Zn-based MOF was found to be an active and stable heterogeneous catalyst for the diastereoselective nitroaldol (Henry) reaction in aqueous medium [37]. Specifically, the 2D amide functionalized MOF featured a bridged amide pyridyl benzoate ligand, L, with the formula $[Zn_2L_2(1,4\text{-}BDC)]_n \cdot 2n(DMF)$ (BDC = benzenedicarboxylate) and containing a threefold interpenetrated framework (Figure 5.4). The 2D-Zn MOF acts as a heterogeneous catalyst for the Henry reaction between aldehydes and nitroethane in water with predominant formation of the *syn* β-nitroalkanol diastereomers (*syn*:*anti* ratios of 65 : 35 to 78 : 22). A control experiment using $Zn(NO_3)_2 6H_2O$ as homogeneous catalyst showed lower yields (18%; 82 : 18 *syn*:*anti*) with respect to the heterogeneous catalyst (93%: 74 : 26). Best results were obtained with H_2O as solvent (93% yield; 74 : 26), with respect to MeOH (75% yield; 75 : 25), THF (25% yield; 74 : 26), or CH_3CN (10% yield; 73 : 27), highlighting the importance of using polar protic solvents to favor the catalytic reaction. Furthermore, when H_2O was used as solvent, the catalyst retained most of its initial activity and stability for five consecutive uses, as evidenced by FTIR spectroscopy and XRD measurements. Overall, the good performance of the MOF was attributed to the Lewis acidity of the Zn^{2+} centers, the basic character of the organic ligand with the amide bridge, as well as to the presence of water involved in the protonation/deprotonation steps of the reaction mechanism.

Interestingly, another study has addressed the importance of having an adequate MOF porosity to favor the diffusion of reagents and products in such a way that this

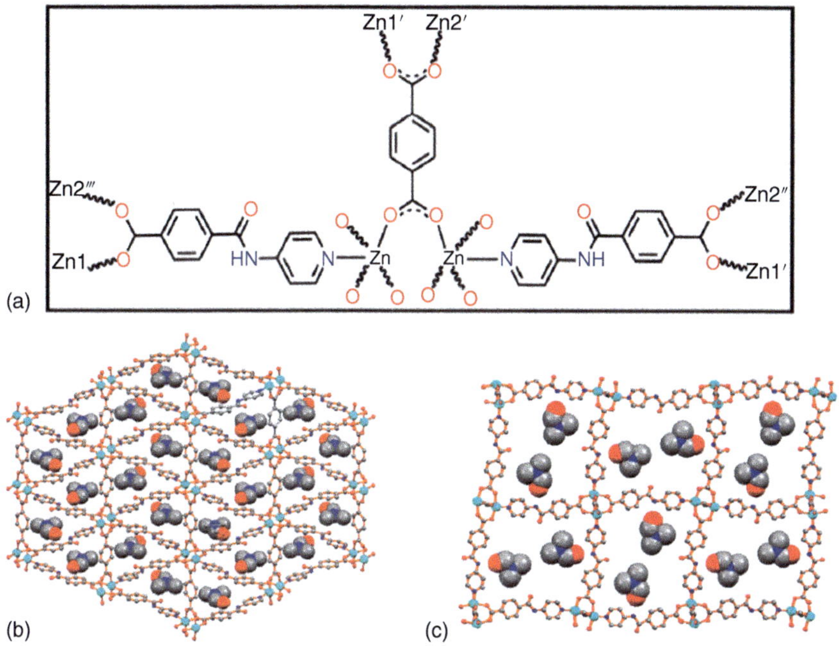

Figure 5.4 (a) Schematic representation of the asymmetric unit of 2D Zn-MOF (excluding DMF molecules). (b) Illustration of the 3D network of 2D Zn-MOF along the crystallographic *a*-axis (b) and *b*-axis (c) (DMF molecules are represented as spacefill model). Hydrogen atoms are omitted for clarity. Source: Reproduced from Paul et al. [37] with permission from the Royal Society of Chemistry.

factor determines the observed catalytic activity for the Henry reaction. For this purpose, a Cu(II)-based MOF termed HKUST-1 (Hong Kong University of Science and Technology) was evaluated as heterogeneous catalyst for the reaction between nitromethane and salicylaldehyde or the bulkier 9-anthracenecarbaldehyde [36]. This MOF is built up by Cu(II) nodes and 1,3,5-benzenecarboxylate linkers and exhibits good porosity ($1300\,m^2\,g^{-1}$, $0.49\,cm^3\,g^{-1}$). In addition, this MOF has open metal sites in the node structure that can act as Lewis catalytic centers to perform various organic transformations. In fact, this MOF exhibited catalytic activity for the reaction between salicylaldehyde and nitromethane (23% conversion after 300 hours at 100 °C), leading to the formation of *trans*-nitrovinylphenol as the only reaction product. This observation agrees with the sorption capacity of the reagents in the porous structure of the MOF as revealed by experimental sorption measurements. In addition, molecular mechanic calculations also revealed negligible activation barrier during the diffusion of either salicylaldehyde or *trans*-nitrovinylphenol through the MOF pores. Attempts to perform the catalytic aldol reaction with nitromethane and the bulky 9-anthracenecarbaldehyde instead of salicylaldehyde under similar reaction conditions showed very low reactivity (<4% conversion), due to the inaccessibility of the bulky aldehyde to the pores as revealed by adsorption experiments.

5.2.3 Diels–Alder Reaction

Diels–Alder reaction is another important synthetic method of C–C formation in organic chemistry. Commonly, this process involves the reaction between a conjugated diene and an alkene frequently termed as dienophile to form unsaturated six-member rings. The more reactive dienophiles for this reaction have an electron-poor character due to the presence of electron-withdrawing functional groups. In the area of homogeneous and heterogeneous catalysis, both Lewis and Brønsted acid sites have been reported as active centers for this reaction. In this context, some MOFs have also found application as heterogeneous catalysts for this purpose [39–45] and, interestingly, the obtained results are associated in some cases due to the occurrence of the reaction in their confined cages. In 2012, the first report appeared on the use of a MOF termed MIL-100(Fe) as heterogeneous catalysts for the Diels–Alder reaction [46]. MIL-100(Fe) is a tridimensional Fe(III) trimesate material of ideal formula $[Fe_3(\mu\text{-}O)(X)(H_2O)_2(btc)_2\ nH_2O]_n$ constituted by mesoporous cavities (2.5 and 2.9 nm) accessible through pentagonal (5.5 Å) or hexagonal (8.6 Å) microporous windows. Thermal activation of MIL-100(Fe) under vacuum results in the formation of Fe(III) Lewis acid centers together with weaker Fe(III)-OH Brønsted acids, both active centers responsible to catalyze the Diels–Alder reaction. In the first step, MIL-100(Fe) was employed as an active and reusable heterogeneous catalyst for the Diels–Alder cyclization of 1,3-cyclohexadiene and dimethyl fumarate using dichlorobenzene as solvent. Additional experiments using simultaneously dimethyl fumarate and the bulkier dibutyl fumarate as dienophiles resulted in the formation of dimethyl bicyclo[2.2.2]oct-5-ene-2,3-dicarboxylate and dibutyl bicyclo[2.2.2]oct-5-ene-2,3-dicarboxylate. The obtained product ratios of these compounds using MIL-100(Fe) as heterogeneous catalyst or $Fe(NO_3)_2$ as homogeneous catalysts were 2.1 and 1.6, respectively. As expected, the reaction using dimethyl fumarate is favored under both homogeneous and heterogeneous conditions. The higher preference formation in the case of using MIL-100(Fe) with respect to $Fe(NO_3)_2$ was associated with the important role of catalytic confinement inside the porous solid, which offers steric hindrance during the reaction. Additional experiments performed using 1,3-cyclohexadiene with methyl acrylate resulted in a slightly higher preferential formation of the *endo* product when using MIL-100(Fe) (*endo*:*exo* = 88 : 12) with respect to the use of the homogenous Fe(III) catalyst (*endo*:*exo* = 85 : 15), suggesting again that the reaction occurs inside the pores of the MOF or at a sterically congested active site.

Interestingly, the same study reported the possibility of introducing post-synthetically new Brønsted and Lewis acid centers on the MIL-100(Fe) structure by a simple acid treatment using HCl [46]. The resulting solids exhibited higher catalytic activity than the parent MOF for a series of Diels–Alder reactions of 1,3-cyclohexadiene with various oxygenated dienophiles, such as dimethyl fumarate, methyl acrylate, and methacrolein. The enhanced catalytic activity of acid-treated MIL-100(Fe) was associated with the formation of new Lewis and Brønsted acid sites, as revealed by *in situ* FTIR spectroscopic study of CO adsorption (Scheme 5.3).

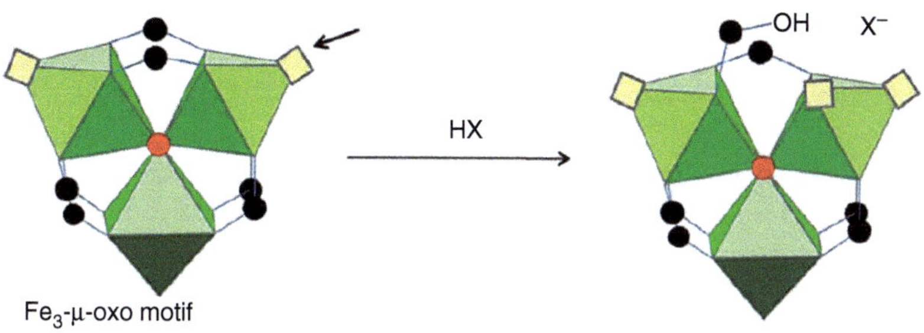

Scheme 5.3 Proposed acid activation of the Fe_3-μ-oxo cluster by a protonic acid (HX). A new coordinatively unsaturated site is opened in one iron octahedron (yellow square). Color code: black is the carboxylate C-atom; red is the central oxo anion.
Source: Vermoortele et al. [46] with permission from the Royal Society of Chemistry.

In some studies [39, 45], MOF-based catalysts for the Diels–Alder reaction have been specifically designed to favor the encapsulation of the reactants in the confined space of the framework and boost the catalytic activity. In one of these studies, an electron-rich MOF based on an anthracene-based rigid tetracarboxylate linkers coordinated to copper paddlewheel dimers was prepared and used as catalyst for the Diels–Alder reactions of 9-hydroxyanthracene and *N*-substituted maleimides as dienophiles [39]. The MOF exhibits a 3D structure with cross-linking hexagonal channels along the *a*, *b*, and *c* axis (Figure 5.5) with good porosity (BET 1058 m^2 g^{-1} and 0.793 cm^3 g^{-1}) and thermal stability (<320 °C).

High to good yields of *N*-phenyl-based maleimides (75–86%) as well as *N*-(1-naphthalene)maleimide (77%) and *N*-(1-pyrenyl)maleimide (65%) were obtained

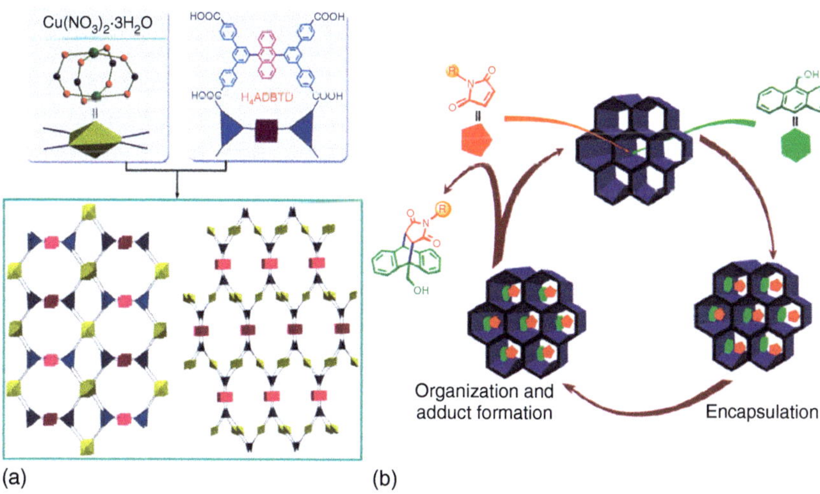

Figure 5.5 (a) Schematic representation of the formation of a 3D Cu-MOF and the hexagonal channels in the structure along the *b* and *c* axis (bottom). Color codes: Cu green, O red, and C blue. (b) Proposed mechanism for the catalytic Diels–Alder reaction. Source: Reproduced from Gole et al. [39] with permission from the Royal Society of Chemistry.

even working under mild reaction conditions (40–50 °C in ethanol a solvent). The good yields achieved for the polyaromatic *N*-maleimides, frequently exhibiting relative low reactivity, were attributed to the occurrence of strong π–π interaction of the confined reactants with the organic ligand of the MOF inside the channels. The use of *N*-cyclohexylmaleimide resulted, however, in lower reactivity (60% yield) compared with phenyl analogous. The reasons of these observations were associated with the relatively low interaction of the aliphatic cyclohexyl group with the aromatic organic ligand of the MOF, although other possible electronic effects can also influence this observation and deserved further investigation. Nevertheless, the catalytic results presented in this work pointed out to a significant enhancement of the catalytic performance of the MOF due to confinement effects and secondary π–π interactions between the substrate and the MOF walls. The observed catalytic activity when using $Cu(OAc)_2$ as homogeneous catalysts having the paddlewheel backbone as the MOF was employed to conclude the important role of the copper metal centers in the MOF as Lewis acid sites of the reaction.

MOFs have also been employed as heterogeneous catalysts for the inverse electron demand Diels–Alder reaction. In this cycloaddition reaction, an electron-rich dienophile reacts with an electron-poor diene leading to the formation of a six-membered ring. Following this strategy, the MOFs UiO-66(Zr) and UiO-66(Zr)-NH_2 were employed as catalysts for the preparation of a pyrano[3,2-*c*]quinoline through an inverse electron-demand aza-Diels–Alder [4 + 2] cycloaddition (Povarov reaction) of an aryl imine, formed *in situ* from aniline and benzaldehyde, and 3,4-dihydro-2*H*-pyran (DHP) [41]. Frequently, a pair of *cis*/*trans* diastereomers are obtained, depending on the reaction conditions (Scheme 5.4).

Interestingly, the use of UiO-66 catalysts leads to the formation of the *trans* isomer in diastereomeric excesses of 90–95%. The authors proposed a reaction mechanism using UiO-66 catalysts in which the observed diastereoselectivity would be explained by the preferential *exo* approach of DHP to the aryl amine coordinated to the Zr^{4+} sites of the MOF. The *endo* approach would be less favorable due to the steric hindrance with the MOF walls. In this sense, the MOF would be acting as a "macroligand" for the Zr^{4+} ions, thereby controlling the orientation of the reaction intermediates and determining the diastereoselectivity of the process.

5.2.4 Isomerization Reactions

(−)-Menthol is a fragrance of high interest for the pharmaceutical and flavoring industries. The catalytic enantioselective cyclization of (+)-citronellal toward (−)-isopulegol followed by catalytic hydrogenation is an important industrial process. During the catalytic cyclization of (+)-citronellal to isopulegol, four isopulegol stereoisomers can be formed (Scheme 5.5) and, therefore, a high stereoselectivity toward the production of (−)-isopulegol is required. For this purpose, several reports have studied the use of inorganic solids having Lewis and/or Brønsted acid sites to perform the cyclization of (+)-citronellal to (−)-isopulegol and/or the presence of active sites such as Pd NPs to promote the hydrogenation of (−)-isopulegol to (−)-menthol.

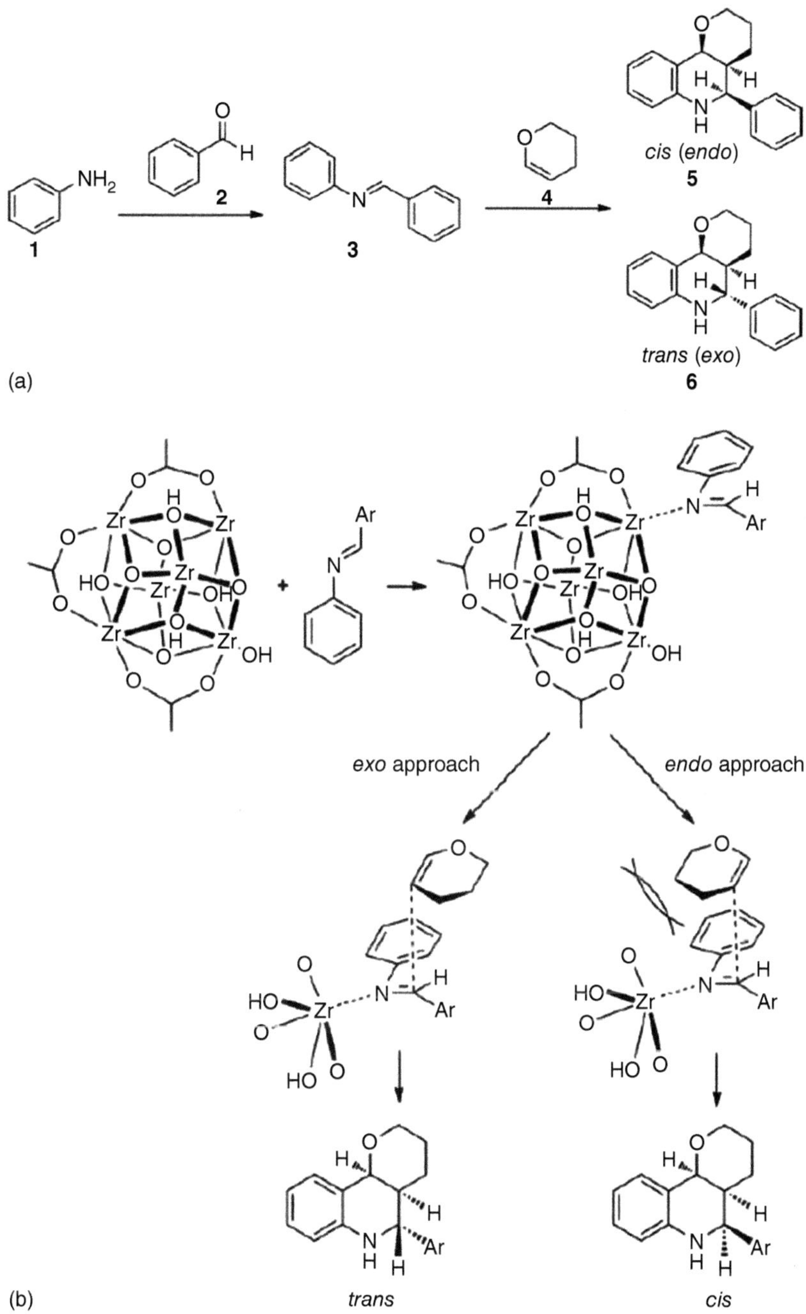

Scheme 5.4 (a) Povarov synthesis of pyranoquinolines; (b) reaction mechanism proposed to explain the observed diastereoselectivity [41]. Source: Reproduced with permission from John Wiley and Sons.

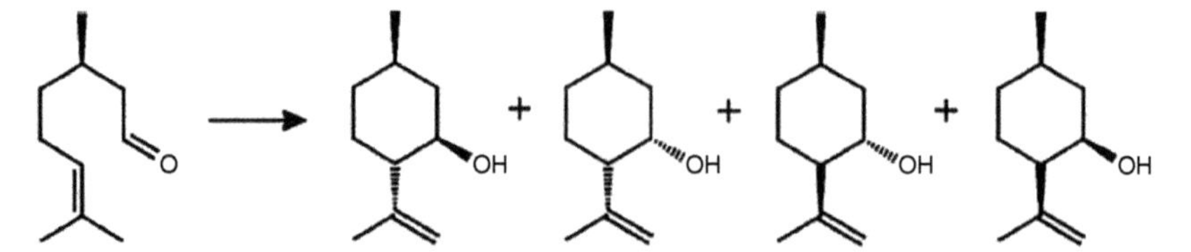

Scheme 5.5 Isomerization of citronellal to isopulegol [47]. Source: Reproduced with permission from John Wiley and Sons.

In the area of MOFs, some studies have reported the possibility of tuning the acidity of their metal acid sites to boost the catalytic activity of the carbonyl-ene reaction [47–52]. In a seminal study by de Vos and coworkers, it was found that the functionalization of pristine UiO-66(Zr) with electron-withdrawing (NO_2, Cl, Br, and F) or electron-donor (CH_3, OCH_3, and NH_2) functional groups is an appropriate strategy to tune the Lewis acidity of the metal nodes and determines their catalytic activity for the cyclization of (+)-citronellal to (−)-isopulegol [47]. Before the catalysis, the solid materials were activated at 220 °C under vacuum, resulting in the dehydration of the $Zr_6O_4(OH)_4$ cluster into Zr_6O_6. Characterization of the as-prepared UiO-66 solids revealed that they are not ideal crystals and among the expected 12 organic ligands surrounding the metal cluster systematically three linkers are missing in all of them. Interestingly, the most active catalysts for the cyclization of (+)-citronellal to (−)-isopulegol are those having electron-withdrawing functional groups in the terephthalate organic ligand, specially UiO-66(Zr)-NO_2. Remarkable is also the observation that the diastereoselectivity toward (−)-isopulegol is also favored by the presence of electron-withdrawing functional groups, especially the NO_2 one, although no explanation for this important finding was given in the original report. In order to rationalize the observed catalytic activity and the electronic effects of the linkers due to the presence of electron-donor or electron-withdrawing functional groups, the meta-Hammett constant of the ligand substituent was considered. As it can be seen in Figure 5.6, a good relationship was observed by plotting the log (k_x/k_H) and the σ_m values. k is the first-order rate constant determined from initial kinetics at low conversion of citronellal for the series of UiO-66 solids under study. These data allowed us to conclude that the Lewis acid sites are stronger in those materials having electron-withdrawing groups close to the Zr sites, especially in the presence of the nitro functional group. Additionally, molecular modeling calculations confirmed that the presence of the nitro group in the UiO-66(Zr) solid lowers both the adsorption and activation-free energy of the reaction, favoring the cyclization of citronellal to isopulegol. This study exemplifies that the concept of electronic modulation in active sites reported in homogeneous catalysis can be transferred to a large extent to the field of MOFs.

As in the field of inorganic heterogeneous catalysts, a step forward in the use of MOFs for (−)-menthol production has been the development of multifunctional heterogeneous catalysts for the one-pot tandem isomerization/hydrogenation of citronellal to menthol. For this purpose, the heterogeneous MOF catalyst has

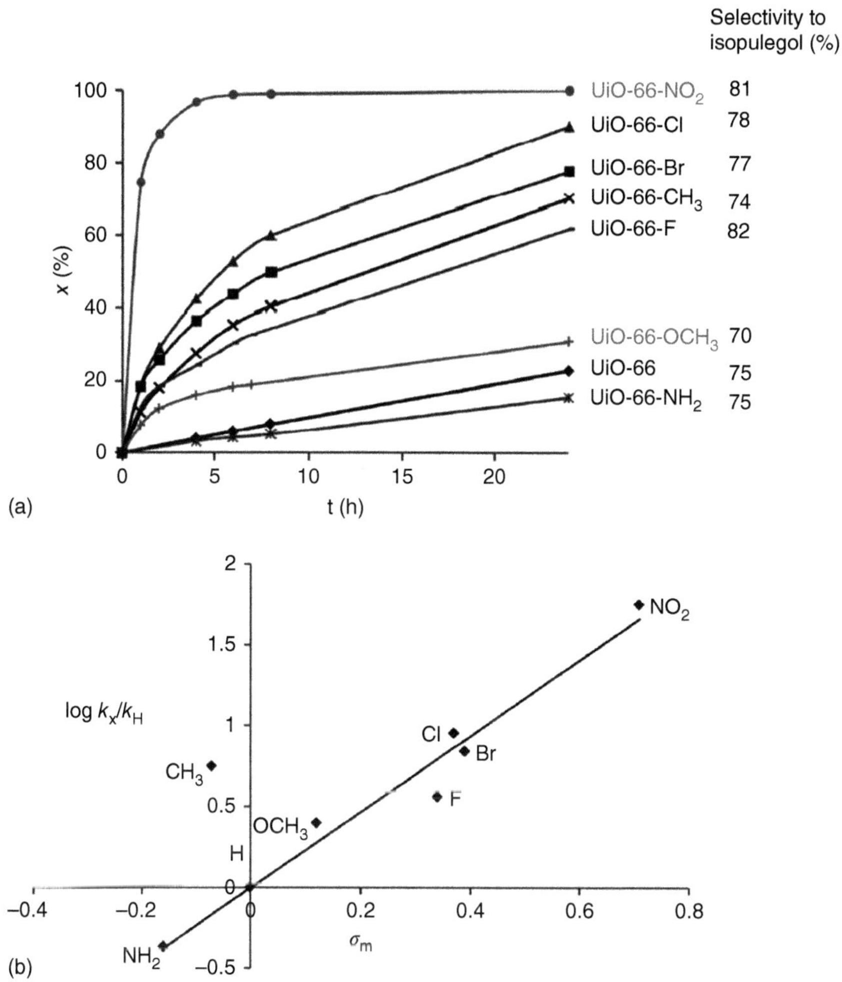

Figure 5.6 (a) Figure 5.1. Conversion of citronellal over UiO-66-X (75 mg) vs. time (5 ml of toluene, 373 K, citronellal:Zr = 10). (b) Figure 5.2. Hammett plot for cyclization of citronellal with different UiO-66-X catalysts [47]. Source: Reproduced with permission from John Wiley and Sons.

both acid and metal sites. Specifically, the crystalline and porous solid termed as MIL-101(Cr) supported Pd NPs was found to be active as a bifunctional solid catalyst for this one-pot tandem reaction [49]. MIL-101(Cr) was prepared by hydrothermal synthesis mixing terephthalic acid and a Cr^{3+} precursor in H_2O in the presence of HF. The previously washed MIL-101(Cr) was further modified by Pd NPs prepared by impregnation of $Pd(NO_3)_2 2H_2O$ followed by thermal reduction under H_2 flow. Based on previous precedents, coordinatively unsaturated Cr^{3+} centers present in the MIL-101(Cr) solid can act as Lewis sites for the selective cyclization of citronellal to isopulegol, while the presence of Pd NPs is responsible for the hydrogenation of isopulegol to menthol. In this study, racemic citronellal was employed as starting material toward the formation of the desired (−)-menthol.

The racemic citronellal mixture cyclization can lead to the formation of desired isopulegol accompanied by the other three stereoisomers (Scheme 5.5). Then, the catalytic hydrogenation results in the formation of four menthol stereoisomers, the desired one is derived from isopulegol. With these comments in mind, it can be understood the importance of the diastereoselectivity ratio obtained toward the desired isopulegol and its corresponding menthol stereoisomer.

The results obtained confirmed that Pd@MIL-101(Cr) is an active catalyst for the full conversion of citronellal in 18 hours with both high selectivity to menthols (86%) and good diastereoselectivity (81%) (−)-menthol product.

5.2.5 Cyclopropanation

Cyclopropane is found as a basic structural unit in a wide range of naturally occurring compounds. This small cycloalkane is not only present in natural products, but it is also an important synthetic target for many applications, such as the synthesis of antibiotics derived from quinolone (ciprofloxacin and sparfloxacin) [53], in biomimetic household insecticides (natural compounds derived from pyrethroids) [54], and in fragrance chemistry (cyclopropanation of dienol groups) [55]. The special features of cyclopropanes are related to the specific reactivity of the three-membered ring and its ability to induce conformational constraints in acyclic alkanes. That is why cyclopropanes have been used as versatile synthetic intermediates in the synthesis of more functionalized cycloalkanes and acyclic compounds [53].

Cyclopropanation reactions (CPRs) correspond to a group of chemical reactions that result in the formation of cyclopropane rings. Although the reaction mechanism is still under debate, it is generally accepted the catalytic transfer of carbene fragments or strong nucleophile to olefins through a cheletropic mechanism that involves the formation of metal complexes comprising the metal, carbene, and double bond of alkenes (Eqs. (1) and (2) in Scheme 5.6). Among the different reactions

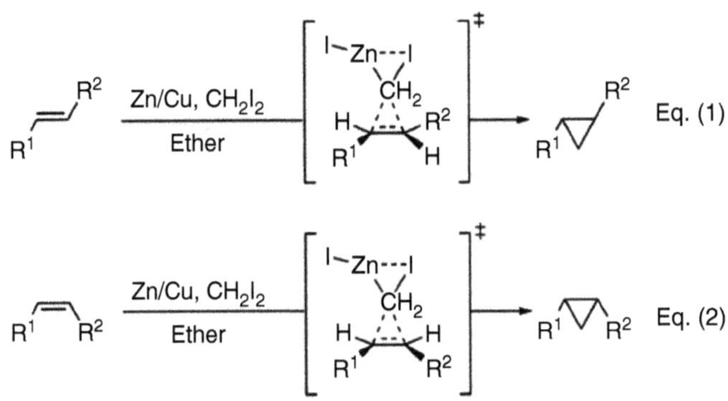

Scheme 5.6 Synthesis of cyclopropanes. Simmons and Smith were the first to appreciate that this reagent ($IZnCH_2I$) could be used for the stereospecific conversion of alkenes to cyclopropanes (Eqs. (1) and (2)). Source: Reproduced Lebel et al. [53] with permission from American Chemical Society.

that can induce the formation of cyclopropanes, it must be highlighted "Simmons–Smith reaction," [53, 56] transition metal-catalyzed decomposition of diazo compounds, "Buchner ring expansion," [53] "Johnson–Corey–Chaykovsky reaction" also known as "Michael initiated ring closure" that is used for electron-poor olefins [53, 57, 58], the "Kulinkovich reaction" [59], and "Bingel reaction" for fullerenes/graphene-like materials and intramolecular cyclization [60].

Diastereoselectivity and stereoselectivity of CPRs have been explored in heterogenous catalysis, especially the formation of cyclopropane by decomposition of diazo compounds. In these works, CPRs have been catalyzed by transition metals from groups 8–11 elements, being copper one of the most used. In particular, metallic nodes of MOFs have been utilized to achieve high yields in this reaction [61]. Also, high productions of diastereoselective products of the CPR have been produced in mesoporous materials by depositing the metal catalyst or by adding specific chiral or not chiral complexes to MOFs [62, 63]. In this context, many studies have centered their research to achieve higher diastereoselectivity by tuning the coordination of active open metallic sites through ligand exchange [64]. However, these strategies are out of the scope of this chapter. In this chapter, we will focus on works in which authors have studied the catalytic stereoselectivity in the confined spaces of MOFs. In published studies, researchers have followed several strategies to enhance diastereoselective in CPR that can be summarized in (i) tailoring the chemical environment of the metal nodes, (ii) modifying pores or cavities size and topology, and (iii) defect engineering to induce distortions in frameworks of pristine MOFs.

As example of the catalytic behavior due to chemical environment, we [65] were among the first groups that studied CPR utilizing $Cu_3[(BTC)_2]$ (HKUST-1). We found out that the good activity of this catalyst was due to the availability of large amounts of metallic accessible sites, the reversible coordination of organic linkers with copper, and the influence of electrostatic cavity, since Cu^{2+} cations of the MOF act as Lewis acids centers. Compared to other copper-loaded mesoporous materials, $Cu_3[(BTC)_2]$ MOF exhibited similar rates but higher diastereoselectivities to *trans*-cyclopropanes (Table 5.2). Also, we found that utilizing bigger molecules as reactants $Cu_3[(BTC)_2]$ induces 100% selectivity toward *trans* products. We justified this behavior due to steric hindrance of internal double bonds or aromatic units within the pores.

Other research groups have investigated this reaction dealing with pore size of MOFs [66]. Herein, researchers have proposed the synthesis of a chiral metal–organic framework (CMOF) constructed by redox-active ruthenium(II)/salen-based ligand and $[Zn_4(\mu_4\text{-}O)O_2CR_6]$ as secondary building units (SBUs). In this way, they synthesized two MOFs, a twofold interpenetrated material named CMOF1 and another non-interpenetrated CMOF2, resulting in dramatic changes in cavity diameters (Figure 5.7). Whereas non-interpenetrated CMOF2 had larger pores diameter of 1.7 nm and open channel dimension of $1.4 \times 1.0\,nm^2$, and interpenetrated CMOF1 exhibited narrow cavities of 0.8 nm in diameter connected to each other with $0.4 \times 0.3\,nm^2$ windows.

Table 5.2 Catalytic results for the cyclopropanation reaction involving styrene and alkyl diazoacetates.

Catalyst[a)]	Diazo compound	Conversion (%)[b)]	*t* (h)	Selectivity (%)	d.r. (%)[c)]
$[Cu_3(BTC)_2]$	EDA	98	3	98	71
Cu–H	EDA	74	2	49	59
Cu–USY	EDA	32	20	100	55
Cu–MCM-41	EDA	60	21	100	59
$[Cu_3(BTC)_2]$	*t*BuDA	25	24	100	68
Cu–H	*t*BuDA	74	2	48	61
Cu–USY	*t*BuDA	23	19	100	62
Cu-MCM-41	*t*BuDA	57	19	100	59

a) Catalyst loading: 5 mol%.
b) Selectivity toward cyclopropanes; the remaining diazo compound was converted into coupling products.
c) Diastereomeric selectivity: (*trans-cis*)/(*trans+cis*).
Source: Reproduced with permission from John Wiley and Sons [65].

From the catalytic point of view, authors described that Ru(II)-salen complex in MOFs was oxidized to Ru(III)-salen and inactivated shortly during reaction, suggesting that Ru(II) was the redox active catalytic species for this reaction. In this way, in order to increase the catalytic yields in CMOFs, researchers added $NaBH(OMe)_2$ as reducing agent to promote, during catalysis, the reduction of Ru(III) to Ru(II). In the case of CMOF2, this addition of reducing agent increased conversion rate from 7% to 54%. Also, CMOFs exhibited significant differences between them in terms of catalytic yields, diastereomeric ratios (d.r.), and enantiomeric excess values. On one hand, reduced CMOF1 had poor catalytic efficiencies (1% yield) compared with reduced CMOF2 (7%). On the other hand, CMOF1 and CMOF2 showed substantial d.r. values toward *trans* products: 1.2 and 4.2, respectively. Diastereoselectivity is even higher when CMOF2 was utilized in the presence of reducing agent, achieving values of d.r. of 7. The authors justify the catalytic efficiency differences between MOFs due to framework interpenetration in CMOF1, which significantly reduces channel openings and prevents the diffusion of reagents to the active sites in the interior of the mesoporous material.

A recent study [67] elegantly described how topology of cavities and electrostatic interactions inside the pores are able to control diastereoselectivity in

Figure 5.7 (a) ORTEP drawing of the $[Ru^{III}(L\text{-}H_2)(py)_2]Cl$ ligand as determined from single-crystal X-ray diffraction. The thermal ellipsoids are shown at 50% probability. (b) Synthesis and single-crystal to single-crystal reduction/oxidation of CMOF1 and CMOF2. The photographs illustrate the typical colors and morphologies of CMOF2 (green) and reduced CMOF2 (red). (c) Space-filling model of CMOF1 as viewed perpendicular to the (1 0 2) plane. (d) Space-filling model of CMOF2 as viewed perpendicular to the (1 0 2) plane. Source: Reproduced with permission from Falkowski et al. [66]/John Wiley & Sons.

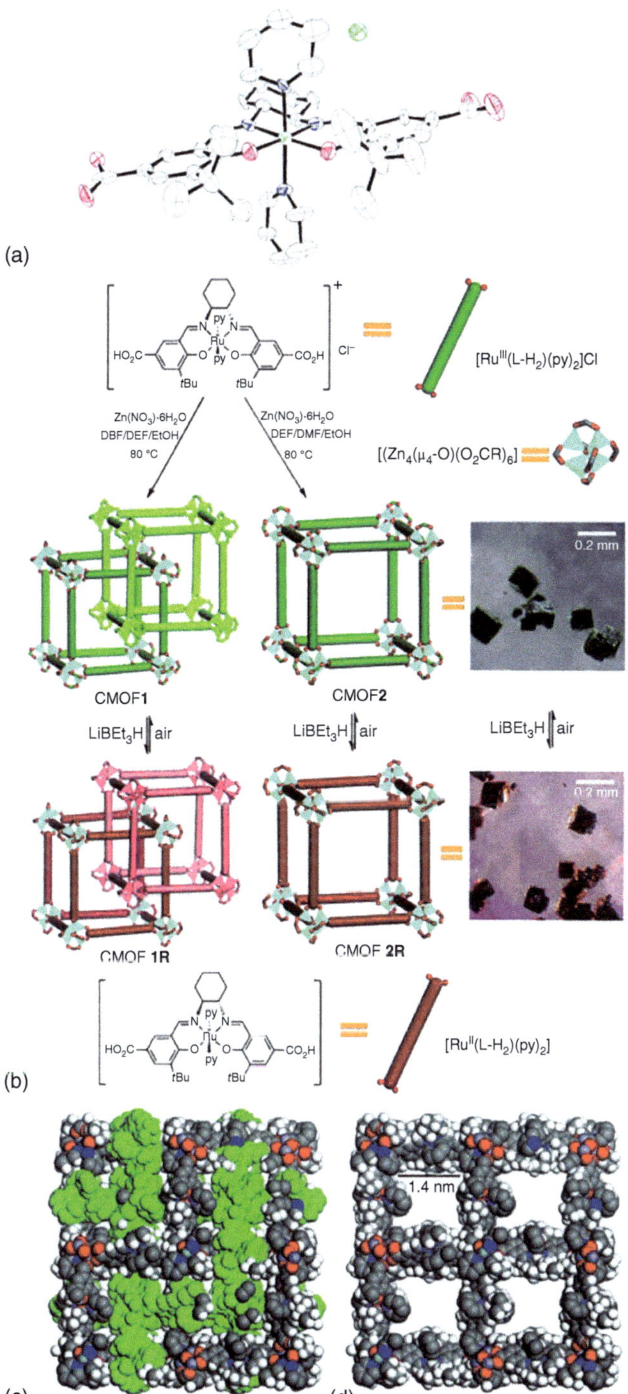
(a)
py
N
Ru
O
py
HO2C
CO2H
tBu
Cl−
[RuIII(L-H2)(py)2]Cl
Zn(NO3)·6H2O
DBF/DEF/EtOH
80 °C
Zn(NO3)·6H2O
DEF/DMF/EtOH
80 °C
[(Zn4(μ4-O)(O2CR)6]
0.2 mm
CMOF1
CMOF2
LiBEt3H
air
CMOF 1R
CMOF 2R
[RuII(L-H2)(py)2]
(b)
1.4 nm
(c)
(d)

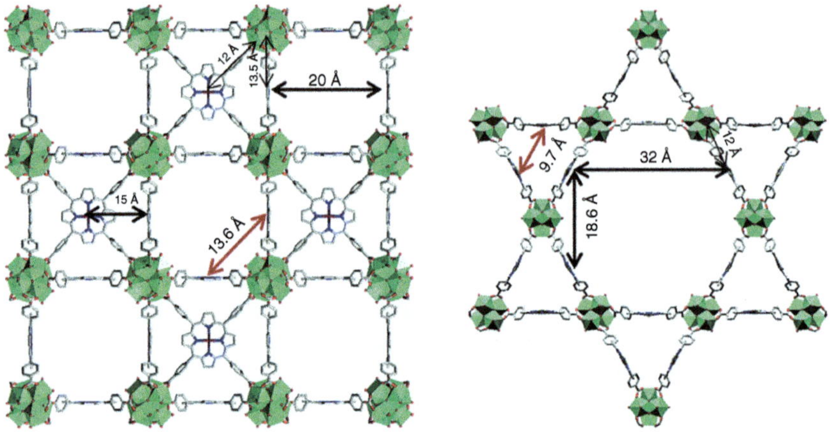

Figure 5.8 Schematic representation of PCN-224lJRh (left) exhibiting squared micropores and PCN-222lJRh revealing its small trigonal micropores and large hexagonal mesopores (right). Black and bold arrows indicate Rh–Rh distances, while black arrows represent Zr–Rh distances, respectively. Red arrows highlight the most feasible (shortest possible) Rh–Rh distances, which would favor the assisted binding of a coordinating substrate molecule. Source: Reproduced from Epp et al. [67] with permission from the Royal Society of Chemistry.

porphyrin-based MOFs. In this work, the authors synthesized two Zr-MOF utilizing a tetradentate Rh-porphyrin complex termed PCN-222(Rh) and PCN-224(Rh) with different micropores topologies. PCN-224(Rh) exhibits square pores and distances between Rh atoms of 13.8, 15, and 20 Å, whereas PCN-222(Rh) revealed a small trigonal micropore and hexagonal mesopores positioning Rh active sites at 9.4, 18.6, and 32 Å (Figure 5.8).

Researchers tested PCN MOFs against isolated Ru-porphyrin resulting in higher conversion yields for the MOF due to confinement effects. This fact was more notorious when more voluminous olefins were employed as substrates, producing an increase in catalytic yields (Table 5.3).

As free porphyrin metallic complex lacks chirality, diastereoselectivity was poor in most cases. However, when olefins with polar functional groups able to form hydrogen bonds with PCN frameworks were employed, such as 4-hydroxystyrene or 4-aminostyrene, the d.r. sharply increased to 21 and 23, respectively. Presumably, functional groups that could form hydrogen bonding can anchor to Rh or Zr atoms present in PCN frameworks. Such coordination mode is likely to align substrates in a certain way. Sterically oriented Rh-carbene plus confinement effects result in a symmetry breaking toward one diastereomer over the other, compared to free Rh-porphyrin, which lacks any pore confinement. Additionally, in the case of 4-aminostyrene, diastereoselectivity in PCN-222(Rh) was doubled in comparison to PCN-224(Rh). Authors justified this fact with the dimensions of the substrate 8.1 Å plus the bond length between double bond and Rh of 2 Å summing a length of 10 Å that is the same distance between Rh–Rh in the trigonal micropores of PCN-222(Rh). In this way, PCN-222(Rh) allowed closer substrate-redox center contacts resulting in major efficiencies and diastereoselectivity.

Table 5.3 Catalytic results for the cyclopropanation of olefins and ethyl-diazoacetate (EDA) using MP-MOFs as catalyst.

Olefin	Catalyst[a]	Yield[c] (%)	d.r.[d]
Styrene	Rh-TCCPOMe	20	1 : 1
4-Methoxystyrene	Rh-TCCPOMe	22	1 : 1
4-Hydroxystyrene	Rh-TCCPOMe	5	1 : 1
4-Aminostyrene (4-AS)	Rh-TCCPOMe	28	1 : 1
3-Aminostyrene (3-AS)	Rh-TCCPOMe	30	2 : 1
2-Aminostyrene (2-AS)	Rh-TCCPOMe	10	1 : 1
N,*N*-Dimethyl-4-AS	Rh-TCCPOMe	33	1 : 1
4-Methylstyrene	PCN-224(Rh)	58	1 : 1
4-Methoxystyrene	PCN-224(Rh)	52	1 : 1
4-Nitrostyrene	PCN-224(Rh)	68	1 : 1
4-Chlorostyrene	PCN-224(Rh)	52	1 : 1
Styrene	PCN-224(Rh)	71	1 : 1
Styrene[b]	PCN-224(Rh)	29	1 : 1
4-Hydroxystyrene	PCN-224(Rh)	4	21 : 1
4-Aminostyrene (4-AS)	PCN-224(Rh)	35	23 : 1
3-Aminostyrene (3-AS)	PCN-224(Rh)	36	5 : 1
2-Aminostyrene (2-AS)	PCN-224(Rh)	12	14 : 1
N,*N*-Dimethyl-4-AS	PCN-224(Rh)	31	2 : 1
1-Octene	PCN-224(Rh)	22	1 : 1
Cyclooctene	PCN-224(Rh)	49	1 : 1
3,4-Dihydro-2*H*-pyran	PCN-224(Rh)	20	1 : 1
4-Aminostyrene (4-AS)	PCN-222(Rh)	29	42 : 1
3-Aminostyrene (3-AS)	PCN-222(Rh)	23	10 : 1
2-Aminostyrene (2-AS)	PCN-222(Rh)	12	15 : 1
N,*N*-Dimethyl-4-AS	PCN-222(Rh)	16	1 : 1

All reactions were carried out at room temperature and stirring (rpm = 500) operated under air after 6 h reaction time. Yields are obtained applying an excess of substrate: 135 μl (1.1 mmol) EDA diluted in 3 ml CH_2Cl_2 was added via a motorized syringe pump (rate = 0.5 ml h^{-1}) to a suspension of catalyst in 2 ml substrate.

a) Catalyst loading: 0.0033 mmol, 0.4 mol% Rh.
b) Manual addition of 135 μl EDA (1.1 mmol) in 2 ml CH_2Cl_2 to a solution of 75 μl styrene in 2 ml CH_2Cl_2 within 20 min.
c) GC-yield; based on EDA as the limiting reagent expect entry 13, whereas styrene is the limiting reagent.
d) Diastereomeric ratio (*trans*: *cis*) of the products.

Source: Reproduced from Epp et al. [67] with permission from the Royal Society of Chemistry.

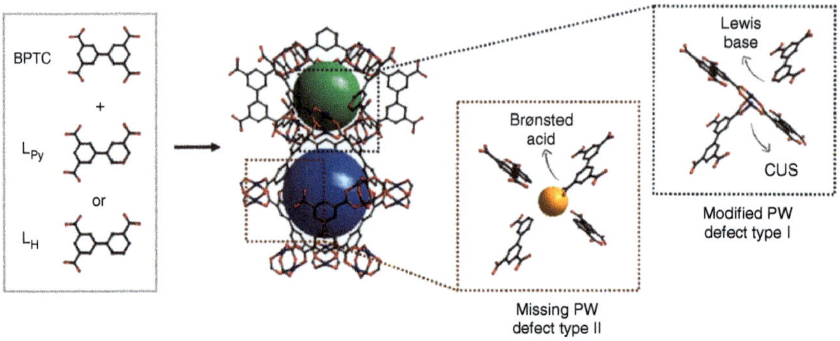

Figure 5.9 Schematic illustration of defect creation in NOTT-100 by mixed-linker approach and conceptual structure of the two types of paddlewheel-related defects. Color codes of the atoms: Cu, blue; C, gray; O, red; and N, brown. CUS: coordinatively unsaturated site. Source: Reprinted with permission American Chemical Society [72].

Defects engineering is a common procedure to enhance catalytic efficiencies broadly utilized in crystalline mesoporous materials inducing modifications over cavities and porous channels [68–70]. Particularly in MOFs, this method has been utilized to tune properties of pristine materials and achieve major catalytic efficiencies and even diastereoselectivity [71, 72]. Several researchers have developed multiple methodologies to produce defects or distortions in the MOF molecular framework but all of them could be encompassed in two categories: synthetic or post-synthetic.

In this way, there is a brilliant study [72] in which the authors promote defect engineering in MOFs by employing truncated linkers during synthesis of NOTT-100 (Figure 5.9). Starting with biphenyl-3,3′,5,5′-tetracarboxylates (BPCT) linkers used in the preparation of pristine NOTT-100, researchers used biphenyl-3,3′,5-tricarboxilates (L_H) lacking a ligating group, or 5-(5-carboxypyridin-3-yl) isophthalates (L_{py}) bearing a weaker coordinator linker, to synthesize defected-engineered MOFs (DE-NOTT-MOFs), utilizing the mixed-linker approach method. Implementation of truncated linkers during synthesis originates two groups of new defects in original NOTT-100 molecular network: (i) secondary active sites as Brønsted acid due to protonated carboxylate groups and Lewis basic sites present in nitrogen in pyridine rings. (ii) New extra coordinatively unsaturated metal sites due to the missing carboxylate group. Incorporation of L_H and L_{py} results in DE-NOTT-100 with superior surface area and appearance of bigger micropores of 19.5 and 10.9 Å, respectively, in comparison to pristine NOTT-100 (pore size around 6.8 Å). Catalytic efficiencies differ among NOTT-100 and DE-NOTT-100, moving from the initial 26% conversion in NOTT-100 up to almost 100% conversion in DE-NOTT-100, although diastereoselectivity was the same favoring *trans*-products. The authors atrtibute these results to the presence of additional open coordination positions, bigger pores facilitating diffusion of reactant along confined spaces, and additional Lewis and Bønsted active sites.

An interesting example of post-synthetic modification published by Heinz et al. [71] described the production of framework defects over precious group metal paddlewheel-based HKUST-1 through thermal defect engineering (TDE). In this chapter, researchers used ruthenium, rhodium, or combination of both to form the characteristic metal atoms dimer that forms HKUST-1, adding during synthesis additional axial ligands (Cl^- and AcO^-) that coordinate to metal cations. Pristine Ru MOF typically possessed mixed +2 and +3 valence and an additional axial ligand for charge compensation, being either acetate or chloride. During TDE, axial acetates are readily removed providing access to univalent $Ru^{II,II}$ sites with two open metal sites per dimer. In contrast, the axial chloride ligands remain strongly bounded. Moreover, information derived from vibrational spectroscopy suggests partial decarboxylation of BTC linkers forming isophthalate and removal of axial and equatorial connectors that did not affect crystallinity. Although materials kept their high porosity, TDE induced slight shifts toward larger pores size due to partial carboxylates linker removal from pore wall. Pristine Ru-KHUST-1, Ru,Rh-KHUST-1, and Rh-KHUST-1 were used as catalysts for the reaction between styrene and ethyldiazoacetate, leading to the formation of homo coupling product, cyclopropanes or linear coupling products. Authors noticed that after TDE treatments of defective materials, the chemoselectivity of the reaction changed toward the formation of cyclopropanes. This fact is evident in the case of Ru-KHUST-1 that went from 68% to 85% yield. However, in comparison with "perfect" materials such as Cu-KHUST-1 that favors diastereoselectivity (68%) to *trans*-isomers in cyclopropanation, the increase of defects in precious metal-based KHUST-1 led to an overall decrease in diastereoselectivity. The reason for this drop in selectivity in defective materials was attributed to a loss of the direct effect of steric "wall" formed by the metal and its four equatorial oxygen ligands due to defect production (Figure 5.10). Additionally, the authors discarded confinement effects accounting for diastereoselectivity, since both homogeneous $Ru_2(OAc)_4$ and Cu-HKUST-1 produce almost equally high diastereoselectivity.

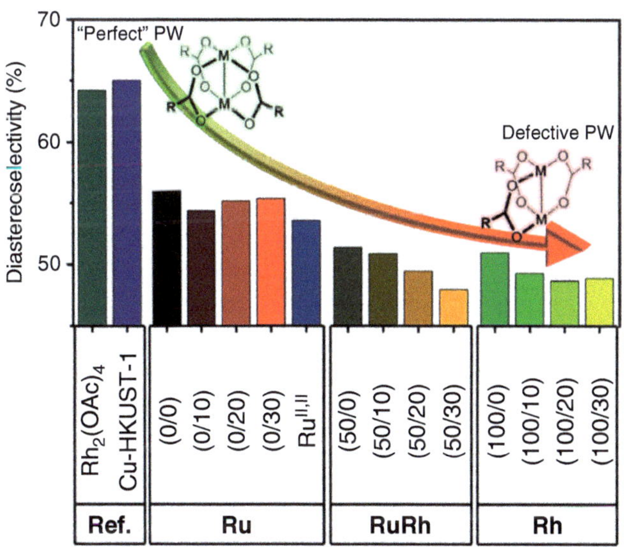

Figure 5.10 Diastereoselectivities of paddlewheel (PW) systems studied herein after four hours. The progressive decrease in diastereoselectivity can be correlated with the assumed structural changes with increasing defectiveness of the PWs as indicated schematically. The trend is visible in the order Cu ≫ Ru > RuRh > Rh as well as upon increasing TDE. Source: Reproduced from Heinz et al. [71] with permission from the Royal Society of Chemistry.

5.3 Conclusions and Outlook

In this chapter, we have illustrated through selected examples the potential of achiral MOFs as heterogeneous catalysts for diastereoselective transformations, in which a prochiral functional group of the reaction substrate is selectively transformed into one pair of diastereomers. This is, in general, achieved through the imposition of confinement effects by the MOF structure, which encompasses size/shape pore restrictions, steric effects, as well as the formation of secondary noncovalent interactions between the substrates and the MOF components in the proximity of the active sites. Altogether, these confinement effects can drive the catalytic reaction through the preferential formation of a specific transition state to the desired diastereomer pair.

To date, most of the existing MOF-based catalytic processes have focused on the design of the active site itself, while the design of the pore cavity in which these centers are located has received much less attention. To us, it is evident that a deeper understanding and control of the basic principles introduced by confinement effects will provide valuable tools for developing novel MOF-engineered materials for (diastereo)selective applications. The examples mentioned throughout this chapter are a clear demonstration of the possibilities, and we can expect more reports in this direction toward MOF-based artificial enzymes.

Acknowledgments

Grant PID2020-112590GB-C21 funded by MCIN/AEI/10.13039/501100011033. S.N. thanks the support of grant PID2021-123856OB-I00 funded by MCIN/AEI/10.13039/501100011033 and by ERDF A way of making Europe.

References

1 Rechac, V.L., Peris, E., and Llabrés i Xamena, F.X. (2019). Cavity effects in Metal Organic Frameworks. *Noncovalent Interactions in Catalysis* (ed. K.M. Mahmudov, M.N. Kopylovich, M.F.C.G. da Silva, and A.J.L. Pombeiro), 440–454. Cambridge: RSC.
2 Koshland, D.E. (1995). *Angew. Chem. Int. Ed.* 33: 2375–2378.
3 Kitagawa, S. and Kondo, M. (1998). *Bull. Chem. Soc. Jpn.* 71: 1739–1753.
4 Murdock, C.R., Hughes, B.C., Lu, Z., and Jenkins, D.M. (2014). *Coord. Chem. Rev.* 258, 259: 119–136.
5 Serre, C., Millange, F., Thouvenot, C. et al. (2002). *J. Am. Chem. Soc.* 124: 13519–13526.
6 Katsoulidis, A.P., Park, K.S., Antypov, D. et al. (2014). *Angew. Chem. Int. Ed.* 53: 193–198.
7 Ponndorf, W. (1926). *Angew. Chem.* 39: 138.
8 Verley, A. (1925). *Bull. Soc. Chim. Fr.* 37: 537.

9 Meerwein, H. and Schmidt, R. (1925). *Justus Liebigs Ann. Chem.* 444: 221–238.

10 Posner, G.H., Runquist, A.W., and Chapdelaine, M.J. (1977). *J. Org. Chem.* 42: 1202–1208.

11 Gargano, M., D'orazio, V., Ravasio, N., and Rossi, M. (1990). *J. Mol. Catal.* 58: L5–L8.

12 Kaspar, J., Trovarelli, A., Lenarda, M., and Graziani, M. (1989). *Tetrahedron Lett.* 30: 2705–2706.

13 Kuno, H., Takahashi, K., Shibagaki, M. et al. (1990). *Bull. Chem. Soc. Jpn.* 63: 1943–1946. https://doi.org/10.1246/bcsj.63.1943.

14 Shabtai, J., Lazar, R., and Biron, E. (1984). *J. Mol. Catal.* 27: 35–43.

15 Van Der Waal, J.C., Tan, K., and Van Bekkum, H. (1996). *Catal. Lett.* 41: 63–67.

16 Corma, A., Domine, M.E., Nemeth, L., and Valencia, S. (2002). *J. Am. Chem. Soc.* 124: 3194–3195.

17 Zhu, Y., Chuah, G., and Jaenicke, S. (2004). *J. Catal.* 227: 1–10.

18 Corma, A., Llabrés i Xamena, F.X., Prestipino, C. et al. (2009). *J. Phys. Chem. C* 113: 11306–11315.

19 Creyghton, E.J., Ganeshie, S.D., Downing, R.S., and Van Bekkum, H. (1995). *J. Chem. Soc. Chem. Commun.* 18: 1859–1860.

20 Van Der Waal, J.C., Creyghton, E.J., Kunkeler, P.J. et al. (1997). *Top. Catal.* 4: 261–268.

21 Anwander, R., Palm, C., Gerstberger, G. et al. (1998). *Chem. Commun.* 1811–1812.

22 De Bruyn, M., De Vos, D.E., and Jacobs, P.A. (2002). *Adv. Synth. Catal.* 344: 1120–1125.

23 Karatas, B.U., Oksal, B.S., and Karatas, E. (2017). *J. Incl. Phenom. Macrocycl. Chem.* 87: 85–94.

24 Morey, M.S., Stucky, G.D., Schwarz, S., and Fröba, M. (1999). *J. Phys. Chem. B* 103: 2037–2041.

25 Zhu, Y., Jaenicke, S., and Chuah, G.K. (2003). *J. Catal.* 218: 396–404.

26 Ly, H.G.T., Fu, G., Kondinski, A. et al. (2018). *J. Am. Chem. Soc.* 140: 6325–6335.

27 Kalinovskyy, Y., Cooper, N.J., Main, M.J. et al. (2017). *Dalton Trans.* 46: 15704–15709.

28 Moon, S.Y., Liu, Y., Hupp, J.T., and Farha, O.K. (2015). *Angew. Chem. Int. Ed.* 54: 6795–6799.

29 Gong, J., Katz, M.J., and Kerton, F.M. (2018). *RSC Adv.* 8: 31618–31627.

30 Mautschke, H.-H., Drache, F., Senkovska, I. et al. (2018). *Cat. Sci. Technol.* 8: 3610–3616.

31 Mautschke, H.H. and Llabrés i Xamena, F.X. (2021). *Chem. Eur. J.* 27: 10766–10775.

32 Mautschke, H.H. and Llabrés i Xamena, F.X. (2022). *Molecules* 27: 6315.

33 Plessers, E., Fu, G., Tan, C.Y.X. et al. (2016). *Catalysts* 6: 104.

34 Valekar, A.H., Cho, K.-H., Chitale, S.K. et al. (2016). *Green Chem.* 18: 4542–4552.

35 Mautschke, H.H. (2019). Zirconium metal organic frameworks as heterogeneous catalysts for Meerwein-Ponndorf-Verley Reactions. PhD Dissertation, Universitat Politècnica de València. https://doi.org/10.4995/Thesis/10251/130203.

36 Sotnik, S.A., Gavrilenko, K.S., Lytvynenko, A.S., and Kolotilov, S.V. (2015). *Inorg. Chim. Acta* 426: 119–125.

37 Paul, A., Karmakar, A., Guedes da Silva, M.F.C., and Pombeiro, A.J.L. (2015). *RSC Adv.* 5: 87400–87410.

38 Karmakar, A., Guedes Da Silva, M.F.C., and Pombeiro, A.J.L. (2014). *Dalton Trans.* 43: 7795–7810.

39 Gole, B., Bar, A.K., Mallick, A. et al. (2013). *Chem. Commun.* 49: 7439–7441.

40 Hatano, M. and Ishihara, K. (2012). *Chem. Commun.* 48: 4273–4283.

41 Rechac, V.L., Cirujano, F.G., Corma, A., and Llabrés i Xamena, F.X. (2016). *Eur. J. Inorg. Chem.* 2016, https://doi.org/10.1002/ejic.201600372.

42 Tanaka, K., Nagase, S., Anami, T. et al. (2016). *RSC Adv.* 6: 111436–111439.

43 Tanaka, K., Yanamoto, D., Yoshimura, K. et al. (2015). *CrystEngComm* 17: 1291–1295.

44 Xia, Q., Liu, Y., Li, Z. et al. (2016). *Chem. Commun.* 52: 13167–13170.

45 Zhang, W., Dynes, J.J.J., Hu, Y. et al. (2019). *Nat. Commun.* 10: 1913.

46 Vermoortele, F., Ameloot, R., Alaerts, L. et al. (2012). *J. Mater. Chem.* 22: 10313–10321.

47 Vermoortele, F., Vandichel, M., Van de Voorde, B. et al. (2012). *Angew. Chem. Int. Ed.* 51: 4887–4890.

48 Alaerts, L., Seguin, E., Poelman, H. et al. (2006). *Chem. Eur. J.* 12: 7353–7363.

49 Cirujano, F.G., Llabrés Xamena, F.X., and Corma, A. (2012). *Dalton Trans.* 41: 4249–4254.

50 Ermer, M., Mehler, J., Rosenberger, B. et al. (2021). *ChemistryOpen* 10: 233–242.

51 Rasero-Almansa, A.M., Iglesias, M., and Sánchez, F. (2016). *RSC Adv.* 6: 106790–106797.

52 Vandichel, M., Vermoortele, F., Cottenie, S. et al. (2013). *J. Catal.* 305: 118–129.

53 Lebel, H., Marcoux, J.-F., Molinaro, C., and Charette, A.B. (2003). *Chem. Rev.* 103: 977–1050.

54 Arlt, D., Jautelat, M., and Lantzsch, R. (1981). *Angew. Chem. Int. Ed.* 20: 703–722.

55 Schröder, F. (2014). *Chem. Biodivers.* 11: 1734–1751.

56 Pellissier, H. (2008). *Tetrahedron* 64: 7041–7095.

57 Das, I., Pal, T.K., and Pathak, T. (2007). *J. Org. Chem.* 72: 9181–9189.

58 García Ruano, J.L., Alonso de Diego, S.A., Martín, M.R. et al. (2004). *Org. Lett.* 6: 4945–4948.

59 Kulinkovich, O.G. and de Meijere, A. (2000). *Chem. Rev.* 100: 2789–2834.

60 Bingel, C. (1993). *Chem. Ber.* 126: 1957–1959.

61 Shi, F.-N., Silva, A.R., and Rocha, J. (2011). *J. Solid State Chem.* 184: 2196–2203.

62 Wang, H., Sun, W., and Xia, C. (2003). *J. Mol. Catal. A Chem.* 206: 199–203.

63 Verde-Sesto, E., Maya, E.M., Lozano, Á.E. et al. (2012). *J. Mater. Chem.* 22: 24637–24643.

64 Levchenko, V.A., Siah, H.-S.M., Øien-Ødegaard, S. et al. (2020). *Mol. Catal.* 492: 111009.

65 Corma, A., Iglesias, M., Llabrés i Xamena, F.X., and Sánchez, F. (2010). *Chem. Eur. J.* 16: 9789–9795.

66 Falkowski, J.M., Wang, C., Liu, S., and Lin, W. (2011). *Angew. Chem. Int. Ed.* 50: 8674–8678.

67 Epp, K., Bueken, B., Hofmann, B.J. et al. (2019). *Cat. Sci. Technol.* 9: 6452–6459.

68 García-García, P. and Corma, A. (2018). *Isr. J. Chem.* 58: 1062–1074.

69 Kokuryo, S., Miyake, K., Uchida, Y. et al. (2022). *Mater. Today Sustain.* 17: 100098.

70 Wang, J., Liu, L., Chen, C. et al. (2020). *J. Mater. Chem. A* 8: 4464–4472.

71 Heinz, W.R., Junk, R., Agirrezabal-Telleria, I. et al. (2020). *Cat. Sci. Technol.* 10: 8077–8085.

72 Fan, Z., Wang, J., Wang, W. et al. (2020). *ACS Appl. Mater. Interfaces* 12: 37993–38002.

6

Chiral MOFs for Asymmetric Catalysis

Kayhaneh Berijani and Ali Morsali

Tarbiat Modares University, Department of Chemistry, Faculty of Sciences, Jalal AleAhmad, Nasr, Tehran, Iran

6.1 Chiral Metal–Organic Frameworks (CMOFs)

Chirality is an inherent and fundamental property in some compounds that living organisms are the most important. Louis Pasteur, a qualified and famous scientist (the nineteenth century) was the first to describe chemical handedness, or "chirality." Generally, a simple optically active chiral compound has at least a tetrahedral atom (for example carbon as a chiral center) bonded to four different groups (Figure 6.1a). These kinds of materials have nonsuperimposable mirror images. It must be pointed out that most biological compounds are chiral, such as sugars, starch, cellulose, amino acids (the building blocks of proteins), DNA, and enzymes (Figure 6.1b). But there is a question of why chirality is important in compounds [1]. In summary, the importance of chiral molecules in biological systems or the demand for enantiomerically pure compounds that are important candidates for drugs cannot be overstated. For example, in the past, thalidomide as a sedative was a widely used drug for pregnant women. It is interesting that each of the enantiomers of thalidomide showed different features (R-enantiomer with positive effect [sedative]; S-enantiomer with negative effect [teratogenic]) [2]. As a consequence, the judicious choice of the synthesis method to produce enantiomerically pure materials is very important. Until now, intense research has been conducted in the chiral materials field, with a great scope from simple compounds to complex materials, porous and nonporous, with high enantioselectivity and efficiency. Because of the important role of porous materials (such as zeolite, carbon, and mesoporous silica) in daily activities, from producing materials for energy technologies to various fields like catalysis, separation, and biomedicine, their chiral counterparts have been also investigated [3–7]. A class of inorganic–organic hybrid materials that are porous, crystalline, and used in a wide range of applications are metal–organic frameworks (MOFs).

These coordination polymers with organic and inorganic components can be fabricated with unique properties for different applications such as catalysis,

Catalysis in Confined Frameworks: Synthesis, Characterization, and Applications, First Edition.
Edited by Hermenegildo Garcia and Amarajothi Dhakshinamoorthy.

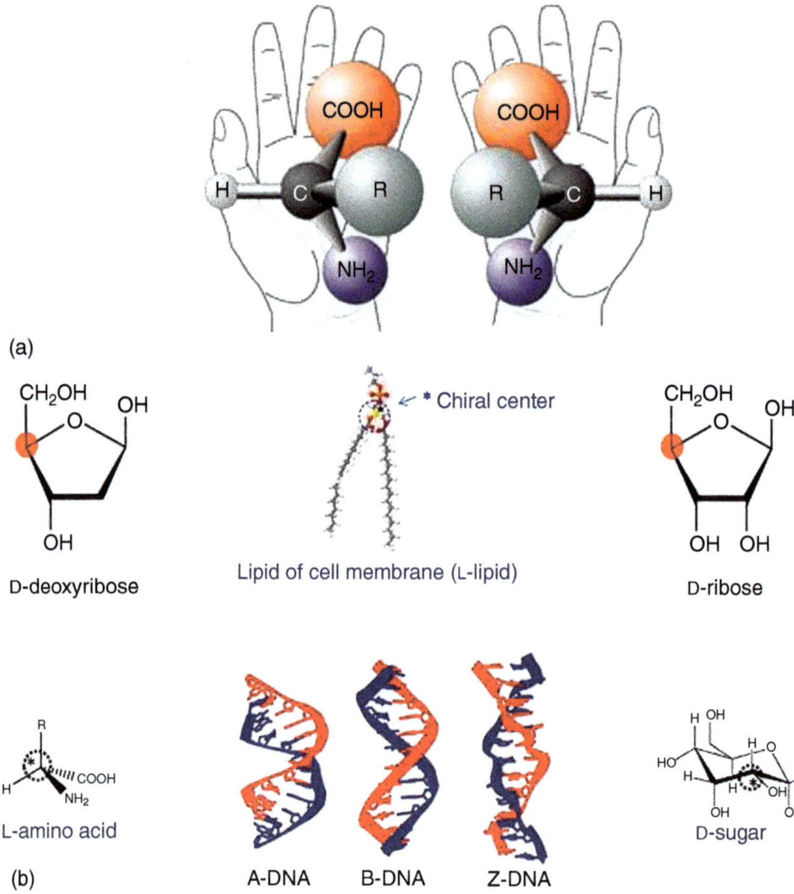

Figure 6.1 (a) Two enantiomers of an amino acid with a chiral center (https://en.wikipedia.org/wiki/Chirality). (b) Examples of chiral molecules.

sensing, and gas storage. Their features originate from the kind and size of the components used. For instance, the unprecedented porosity of MOFs can be tuned to produce pores with various dimensions. Or MOFs depending on their structural properties can accept a broad range of functional groups that convert them into effective candidates for different uses. While in other porous materials like zeolites, the structural diversity is not observed like in MOFs. Like MOFs, chiral MOFs have also attracted considerable attention in comparison to traditional chiral porous solids due to their properties such as high crystallinity, large surface areas, porosity, high ability in chiralization, composition tunability, and diverse functionality. Chiral metal–organic frameworks (CMOFs) with tunable characteristics have been converted into versatile chiral porous candidates for practical asymmetric applications, especially catalysis, resolution, adsorption, biomedicine, nonlinear optics, etc. (Figure 6.2) [8]. Indeed, chirality is one of the features that MOFs can have, not only does it limit the performance of MOFs but also increases their ability

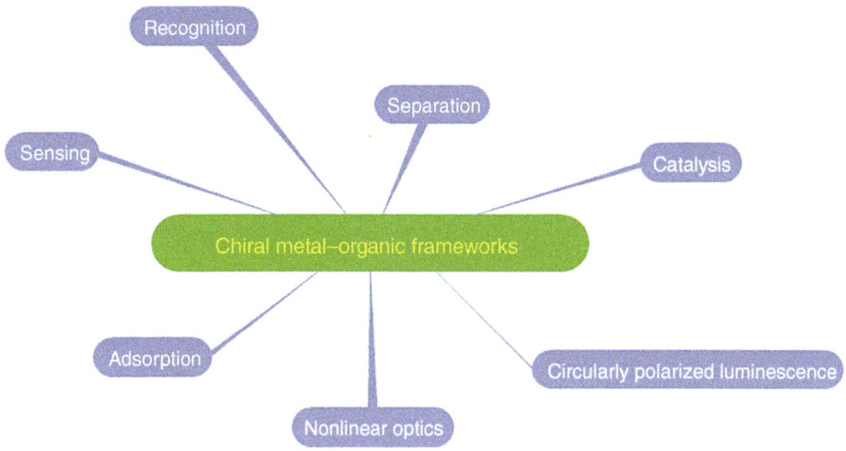

Figure 6.2 Some practical asymmetric applications of CMOFs.

in asymmetric applications. From 1999 (Aoyama et al.) to date, various chiral MOFs have been extensively designed and prepared using chiral and achiral building blocks through different synthesis methods (spontaneous resolution, direct, and indirect methods) [9].

For example, in 1999, Aoyama and coworkers reported the first chiral helical MOF using achiral components [10]. This chiral coordination polymer was produced via homochiral crystallization, and its chirality control was achieved through seeding. Despite the production of chirality in the structure, one possible mechanism takes place, namely racemization. This phenomenon is one problem in the synthesis of chiral MOFs without any chiral additive. Two other synthesis methods are direct and indirect ways that make major contributions to preparing CMOFs. In these methods, chiral components and/or chiral additives are used. The use of the chiral compounds in the chiralization of the framework is an effective method than the inherent chiralization. Ideally, the suitable choice of chiral species helps the facile construction of CMOFs. These methods allow the flexible and reasonable design of chiral MOFs, as both the architecture and chemical functionality of frameworks can be controlled. For example, in 2000, Kim and coworkers reported the preparation of a homochiral metal–organic porous material using enantiopure metal–organic clusters as secondary building blocks. They claimed their synthesis strategy can be used for the chiralization of a wide range of porous organic materials in chiral-related fields [11]. For example, chiral co-ligands have a key role in introducing chirality into frameworks. Rosseinsky and Kim with their co-workers used 1,2-propanediol and L-lactic acid as chiral co-ligands, respectively [12, 13]. Or the cavities of chiral MOFs are also used in asymmetric processes like enantiomeric separation by selective recognition/sorption [14]. Totally, the successful syntheses of CMOFs have continued to date. Thousands of chiral MOFs as tunable platforms have been reported with different designs, such as post-chiral modified MOFs [15], homochiral MOF thin films [16], cyclodextrin-based metal–organic frameworks

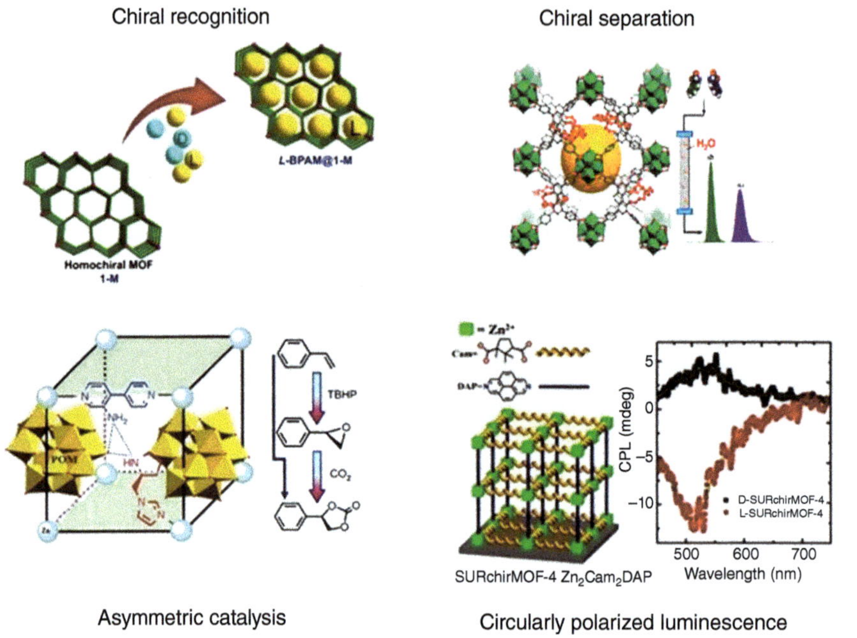

Figure 6.3 Four different structures of chiral MOFs with asymmetric applications (Chiral recognition: this article is open access. It is freely available, and permission is not required to access this data source [19]; Asymmetric catalysis: this article is open access. It is freely available, and permission is not required to access this data source [20]; Circularly polarized luminescence, Source: Reprinted with permission from Chen et al. [21].; Chiral separation, Source: Reprinted with permission from Jiang et al. [22].)

(CD-MOFs) [17], and chiral guest-encapsulated MOFs [18]. Although the realm of chiral MOFs for extensive asymmetric uses is still under development, there are still quite a lot of challenges (Figure 6.3).

In this chapter, we discuss the performance of chiral MOF-based catalysts, in terms of design and chemical nature. Successful use of CMOFs in enantioselective heterogeneous catalysis is one of the most important applicable issues for the research community that is active in the production of pure chiral compounds. We hope this chapter will be useful for active researchers in the asymmetric catalysis field.

6.2 Synthesis Methods of CMOFs with Achiral and Chiral Building Blocks

In addition to the appropriate selection of organic linkers and metal nodes to produce chiral MOFs, the precise choice of their synthesis strategies is also necessary. Generally, there are three methods to introduce chirality into MOFs: (i) spontaneous resolution, where achiral starting materials are utilized and framework chiralization happens through crystallization. Crystal chirality depends on the space group of the crystal (the crystallization in a chiral space group). Although achiral components can produce chiral MOFs, chiral, optically pure building blocks are preferred. (ii) The direct method is one of them, where chiral organic linkers are used to induce absolute chirality in the final frameworks. In this case, an enantiopure organic linker can be employed along with a metal source as a metal node. However, rigidity and the length of the chiral organic linker should be considered. But the vast majority of the reported chiral MOFs are prepared via (iii) the indirect method. Post-synthetic modification (PSM) relates to the indirect method that has a high contribution in producing chirality in the achiral framework structures. This method can help to solve the lack of active sites in MOFs. For example, considerable investigations have been conducted on the post-chiral modification of the framework such as five kinds of chiral MIL-101(Al)–NH_2 that were produced using five different chiral compounds through PSM. Herein, the organic linker was achiral that was chiralized through its functional groups, namely NH_2. As a matter of fact, the PSM does not only happen in the organic linkers but can also happen in metal nodes. This method is a simple and straight pathway in comparison to the direct method, which has uncontrollable conditions and diverse complexities [15, 23]. Given that the synthesis of chiral MOFs has great importance, their synthesis methods are explained, in the following, in separate Subsections 6.2.1–6.2.3 along with the examples related.

6.2.1 Spontaneous Resolution

In the past years, there has been much interest in helical architectures such as coordination polymers and polynuclear complexes. When the components have no intrinsic chirality, the most common case is produced, namely a racemic mixture or racemate (a mixture of equal quantities of two enantiomers). Indeed, the spontaneous resolution to produce crystalline CMOFs with achiral components is also a hard and complex strategy. In this process, crystal packing or space groups can induce chirality into the framework. The crystallization of the framework in a chiral space group leads to structural deformation in the units. For example, Cao and coworkers reported the synthesis of a 3D framework with chiral tetrahedral secondary building units (SBUs), in 2009 [24]. These units were generated by different connections between their components (herein carboxylates as bridging groups and zinc ions as metallic sites) (Figure 6.4).

The MOF synthesized (**1**: $[Zn_2(X)(CH_3CH_2OH)]\cdot 3H_2O$; H4X: tetrakis[4-(carboxyphenyl) oxamethyl] methane acid) has ferroelectric and nonlinear optical features. A usual nondestructive analytical technique, namely, single-crystal XRD showed the point group Cs (for details, please see Supporting Information of the ref. 24). It is interesting that in the asymmetric units of the framework, two crystallographic-independent zinc ions are seen. In case 1, the ZnO_6 octahedron is formed using zinc ion and the carboxylate groups that are considered as the chelate and bridging ligands. In case 2, the EtOH molecule and carboxylate have an important role in the production of the five-coordinated zinc in a distorted trigonal bipyramid. In this synthesis, the kind of component such as an organic linker is an effective

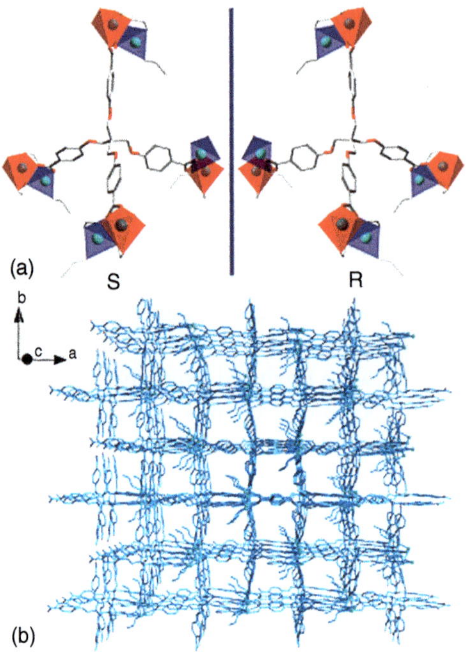

Figure 6.4 (a) Two kinds of chiral secondary building units. (b) Microporous crystalline metal-organic framework (perspective view): O: red; C: gray; Zn: azury. For the sake of clarity, Lattice H_2O molecules have been omitted. Source: Reprinted with permission from Guo et al. [24].

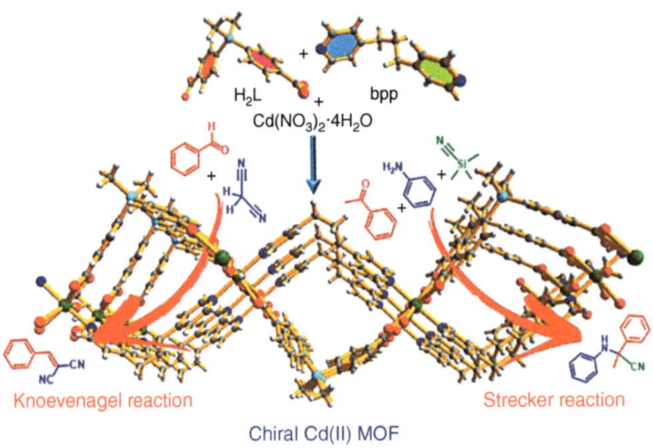

Figure 6.5 A Chiral Cd(II)-MOF with achiral precursors, as a heterogeneous catalyst for Knoevenagel condensation and Strecker reactions. Source: Reprinted with permission from Gupta and Mandal [26].

factor. For instance, Bharadwaj and coworkers reported chiral MOFs using Cd as a metal source and bis[4-(3,5-dicarboxyphenyl)-1*H*-3,5-dimethylpyrazolyl]methane as an achiral organic ligand. They claimed that the rotation of this V-shaped ligand is restricted via the coordination of Cd ions, which leads to chirality in the structure. As a result, the direction of the metal ion coordination with the organic linker can affect the chirality creation. Their investigations showed there are open metal sites (OMSs) in the framework that themselves can be catalytic active sites for the synthesis of α-aminonitriles [25]. Or, in 2019, Mandal and Gupta reported the design and synthesis of a CMOF with achiral precursors (a V-shaped dicarboxylic acid, a flexible bidentate linker, and cadmium centers as metal nodes) (Figure 6.5) [26]. This 2D CMOF was constructed through solvothermal synthesis. To characterize, two important techniques were also used crystallography and gassorption. The linked 1D helical chains to the metal centers led to a chiral 2D framework formation. In MOFs, the high density of OMSs has been widely used as catalytic active sites. Herein, taking into account gas-sorption measurements, the metal centers (OMSs) can perform as Lewis acid sites (due to removing DMF and water molecules). To confirm its catalytic activity, the Knoevenagel condensation reaction and Strecker reaction were investigated. The stability of this CMOF during the reaction, facile separation, recycling, and the obtained catalytic results were acceptable findings for this heterogeneous catalyst.

It is worth noting that a pair of doubly interpenetrated Ni-MOFs ($\Lambda - 1$ and $\Delta - 1$) synthesized by Huang and coworkers are assigned to the spontaneous resolution method [27]. They showed the catalytic activity in the Knoevenagel condensation with good results at room temperature. Owing to the intrinsic properties of chiral MOFs synthesized by the spontaneous resolution method, the relationship between structure and targeted applications is revealed. Asymmetric practical applications that are important such as sensing, asymmetric catalysis, methane purification, photocatalytic activity, ferroelectric activity, drug delivery, and enantioseparation. Although there have been several research articles related to these kinds of CMOFs and several reviews that summarize the synthesis methods of CMOFs, to address the issues such as twisting of the organic linker, V-shaped ligand, "DNA-like" chiral double helix structure of MOF, symmetry-breaking, and the helicity of structure (like helical SBU, chain, channels, and arrangement), the further discussion is needed [9, 25, 28–35]. It is quite clear that each synthesis method has advantages and disadvantages. But investigations show that two other categories of synthesis of CMOFs, namely, the direct and indirect methods are mostly used; they will be discussed as follows.

6.2.2 Direct Synthesis

One of the synthesis strategies of chiral MOFs is the direct method. In this route, the enantiopure ligands play an important role in the creation of CMOFs. This chiralization can be considered a straightforward and effective synthesis. Therefore, it is of great importance to extend the design, preparation, and use of chiral ligands. Until now, several kinds of chiral ligands have been designed and synthesized to construct CMOFs that some of them are shown in Figure 6.6.

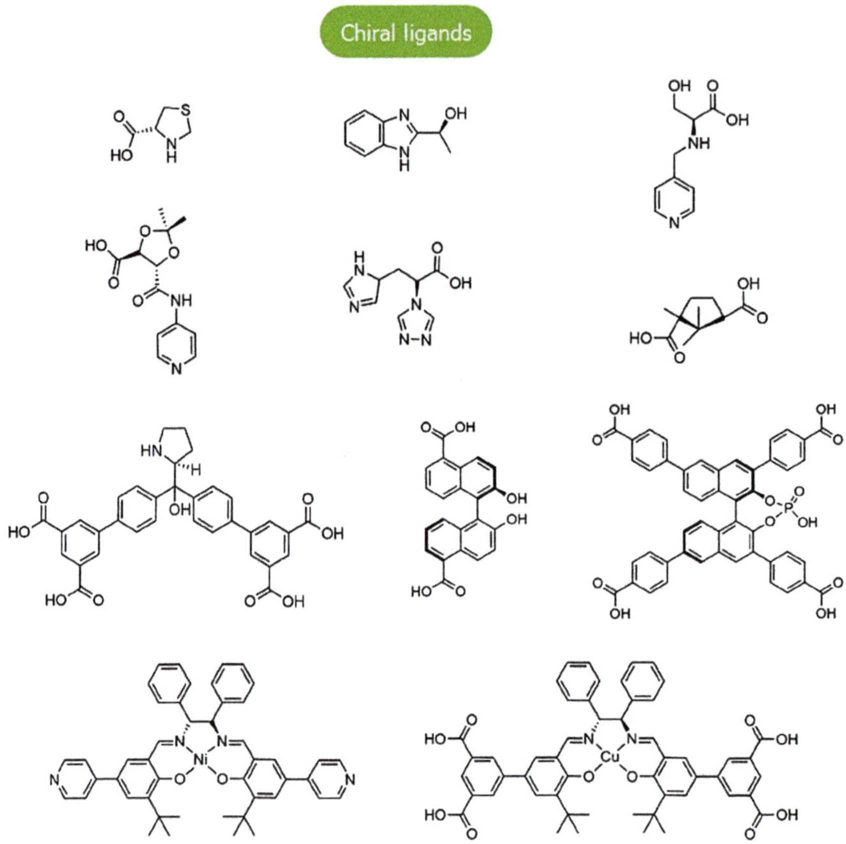

Figure 6.6 Some chiral ligands to synthesize CMOFs.

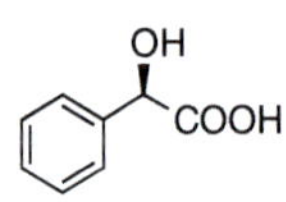

R-mandelic acid

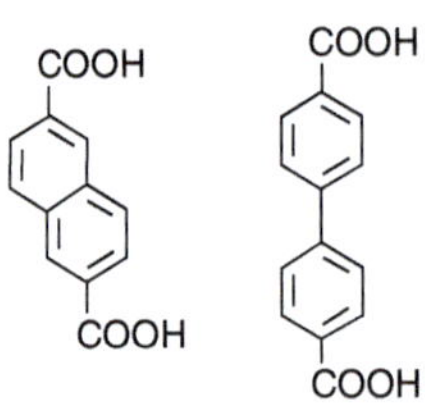

Figure 6.7 Two kinds of dicarboxylic acids with different lengths.

Optical active organic linkers can be used alone in the structure as bridging ligands that link metallic parts, or they can be employed in mixed-ligand form, namely, chiral and achiral organic linkers, simultaneously. For example, Xiong and coworkers synthesized a chiral 2D network using inherent chiral linkers (enantiopure lactate) [36]. They also reported several homochiral 2D layered networks by chiral alkaloid quinine and its derivatives [37, 38]. Enantiopure tartaric acid derivatives are other chiral organic linkers that Kim and coworkers used for preparing chiral 2D frameworks with catalytic activity in the transesterification reaction [11]. As mentioned, the mixed-ligand method can also be used to construct chiral MOFs. This strategy can produce new opportunities to create three-dimensional chiral frameworks. For instance, chiral mandelic acid as a medicinally important chiral molecule was utilized in this kind of synthesis along with two rigid dicarboxylic acids to generate chiral porous frameworks.

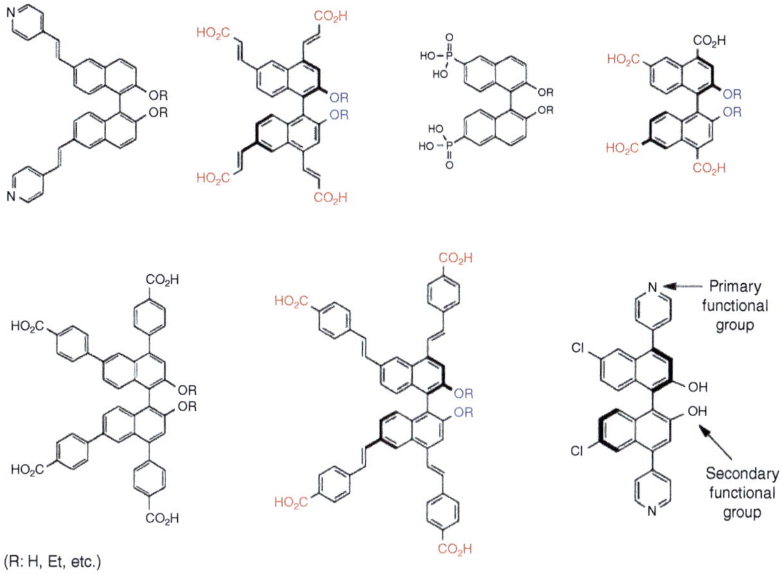

Figure 6.8 Examples of BINOL-derived chiral ligands with different functions.

It must be pointed out that the above-mentioned CMOFs were different in the viewpoint of their pore. Because the length of the used dicarboxylic acids was different (Figure 6.7) [39].

Examples of BINOL-derived chiral ligands with different functions like pyridyl, carboxyl, and phosphonic acid have been shown in Figure 6.8, and some of them were also used in the Lin group [40]. Such ligands are of great importance in preparing CMOFs with tunable porosity that can be employed in different asymmetric applications, especially enantioselective catalysis. For example, Lin et al. reported the synthesis of a chiral, porous, noninterpenetrating MOF using chiral bipyridine bridging ligands.

The accessible channels and chiral dihydroxy groups in the structure converted it into an effective asymmetric catalyst for the production of chiral alcohols with high enantiomeric excesses (up to 93%) [41–43]. DUT-129 with its sod topology and high density of stereogenic centers, is one of the chiral MOFs for which, in its synthesis, chiral-organic linker (H_2bodc) has been employed. According to the calculations, the resulting CMOF with a pore diameter of 5.2 Å showed high selectivity in enantiomeric discrimination in different applications such as adsorption and separation [44]. In spite of chiral MOFs, chiral covalent–organic frameworks and chiral hydrogen-bonded frameworks can be also synthesized via the direct route for different applications [45–47]. In direct creating CMOFs, salen ligands, unary, and multiple metallosalen ligands are other chiral ligands that should be mentioned [48–52]. Although many examples have been reported in this field, there are undesirable effects in this method. One of them is the chemical stability of the framework. But another synthesis route, namely the indirect method, solves these undesirable effects, which have been explained, as follows.

6.2.3 Indirect Synthesis

Another synthesis method that has had great success in producing chiral MOFs, specially CMOF catalysts, is indirect chiralization. The maintenance of structural and chemical stability as well as chiral functional group diversity are the advantages of this method. PSM, or herein post-chiral modification that is in this classification, is utilized to graft chiral compounds on the metal node or organic linker. This strategy also includes the incorporation of chiral compounds in the pores of achiral MOF. The significant progress of this strategy started with its introduction in 1990 and has been formally continued with different materials specially MOFs until now by different researchers such as Hoskins, Robson, Cohen, and Wang [53, 54]. This chemical modification happens after the synthesis of MOF's framework. Through this method, different chiral MOFs can be prepared with a variety of functional moieties for a broad range of practical applications like catalysis, sensing, separation, etc. without effectively reducing the structural uniformity. For example, PSM of an achiral MOF with a kind of organic linker that has a functional group such as amine, can generate a position for reaction with various groups like carbonyl. This modification can also introduce new performances and features into a framework for practical applications. For instance, until now, the importance of the existence of chiral stationary phases in chromatography is to a great extent determined. Therefore, the development of their design with new properties is crucial. To further understand, we can imply five chiral MOFs synthesized using an achiral backbone of MIL-101–NH_2 that were reported by Yan et al. They synthesized these chiral MOFs with PSM and then used them as chiral capillary columns for asymmetric gas chromatography. The selected achiral framework was chiralized through grafting various chiral molecules ((S)-2-Phenylpropionic acid, (R)-1,2-epoxyethylbenzene, (+)-diacetyl-L-tartaric anhydride, L-proline, and (1S)-(+)-10-camphorsulfonyl chloride) that had significant roles in chiral separation [23]. Given the diversity and rapid growth of the obtained CMOFs through PSM in recent years, there are many investigations that are mainly focused on their performances, especially catalytic activity. To realize the relationship between these kinds of CMOFs and asymmetric catalysis in a logical way, herein, the related examples have been summarized. MIL-101(Cr)-tart is one of these types of CMOFs that has been reported (Scheme 6.1) [15].

They prepared this CMOF as an asymmetric heterogeneous catalyst with two kinds of catalytic sites, simultaneously. After hydrothermal synthesis of achiral MIL-101(Cr) based on the reported method using the mixture of H_2BDC (164 mg), $Cr(NO_3)_3{\cdot}9H_2O$, HF, and D.I.H_2O, the concentrated silver tartrate solution was used. Indeed, MIL-101(Cr)-tart was prepared by the replacement process. In this work, they proposed an anion exchange hypothesis through the post-synthetic exchange: the elimination of the fluorides of an achiral backbone using chiral tartrate anions. After careful structural analysis, it is found that this CMOF can be selected as a candidate for chiral catalytic reactions. The coordinatively unsaturated metal sites and chiral functional groups could show catalytic activity. Due to the presence of the OMSs (Cr) as Lewis acid sites, asymmetric epoxidation of olefins

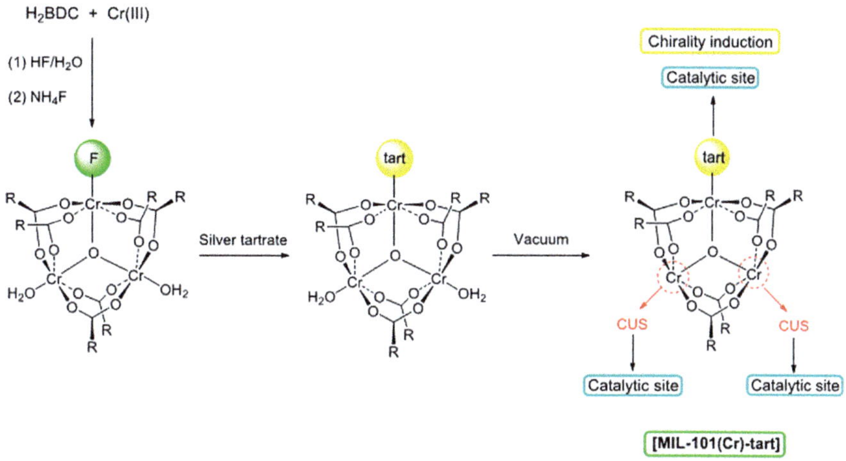

Scheme 6.1 Synthetic route of chiral heterogeneous catalyst (MIL-101(Cr)-tart). Source: Reprinted with permission from Berijani and Morsali [15].

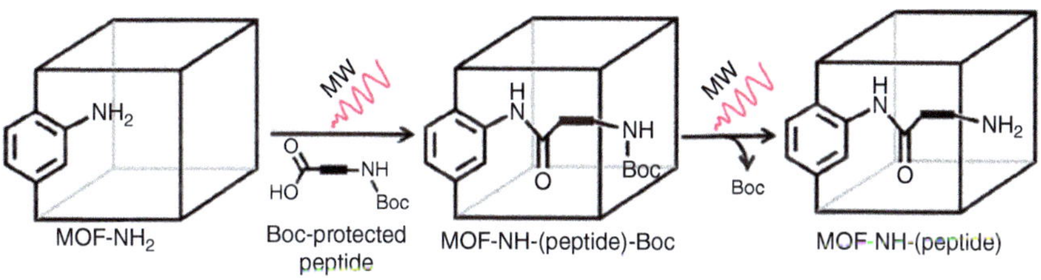

Figure 6.9 Post-covalent modification of achiral MOF-NH_2 cavities using a chiral oligopeptide. Source: Reprinted with permission from Bonnefoy et al. [55].

was investigated. In addition to asymmetric epoxidation, asymmetric methanolysis was also studied because the tartrate groups showed Brønsted acidity. Asymmetric catalysis of olefins epoxidation using O_2 and isobutyraldehyde (IBA) under the optimized conditions gave good results (69–100% conv and 75–100% ee). Also, in the enantioselective methanolysis of styrene oxide, 100% conversion and 90% enantiomeric excess were obtained. A similar approach (post-synthetic method) was also taken to produce biofunctionalized CMOFs (the post-covalent modification of achiral MOF cavities using chiral amino acids and various oligopeptides) (Figure 6.9) [55].

In 2016, Yaghi and coworkers investigated seven post-synthetic reactions within the pores of multivariate MOF and its catalytic activity in the selectively cleaving peptide bonds [56]. Zn-PYI1 and Zn-PYI2 are also two photoactive chiral MOFs that were chiralized using chiral organocatalysts (L- or D-proline derivatives) through PSM, and they were utilized for asymmetric α-alkylation of aldehydes [57]. It is worth mentioning that such a method is an effective and usual route that is also used for other materials that do not have rich surface chemical functionality, like zeolites, porous ceramics, and nanoporous carbon. Until now, many examples have been reported with rich chemical diversity in this field that some of them with catalytic activity will be also presented, as follows.

6.3 Chiral MOF Catalysts

6.3.1 Brief History of CMOF-Based Catalysts

As a unique subset of important applications, heterogeneous catalysis is one of the earliest suggested applications for materials. Among The kinds of materials, the nature of MOFs enables them to produce suitable platforms for the further development of heterogeneous catalysts with high quality. For example, in 1994, for the shape-selective cyanosilylation of aldehydes, a Lewis acid catalyst from the MOFs family was reported. The quest continued in this field by different researchers for asymmetric heterogeneous catalysis. Until 2000, the first chiral MOF catalyst was reported by Kim et al. They synthesized a homochiral metal–organic porous material (POST-1) for the enantioselective transesterification reaction. The pyridinic moieties in the framework's channels were catalytic active sites [11]. One year later, in 2001, Lin and coworkers reported a series of chiral lanthanide-MOFs (lanthanide = La, Ce, Pr, Nd, Sm, Gd, and Tb), among which Sm-MOF showed catalytic activity in asymmetric cyanosilylation of aldehydes with moderate conversions and small enantioselectivity [58]. In 2003, the same research group presented another work with a new design, namely chiral porous zirconium phosphonates containing Ru–BINAP–DPEN moieties (DPEN: 1,2-diphenylethylenediamine; BINAP: 2,2′-bis-(diphenylphosphino)-1,1′-binaphthyl) [42]. The synthesis approach used was based on the molecular building block (MBB) method. The performance of the obtained structures was investigated in asymmetric hydrogenation of aromatic ketones with high enantiomeric excesses (up to 99%). In another work, in 2005, they designed and synthesized CMOFs using metal ions and chiral-organic linkers (BINOL) that had chiral dihydroxy groups. Assembly of the components led to the production of 3D CMOFs with high activity in the enantioselective asymmetric catalysis, to produce chiral secondary alcohols (enantiomeric excess = up to 93%) [43]. In these kinds of frameworks, due to the presence of dihydroxy groups on the organic linker, the various catalytically active sites can be immobilized through post-modification. Therefore, the kind of organic backbone used is very important. Among them, biphenol [59], diene [60], SPINOL [61], and metallosalen [62] can be mentioned. These pioneering works motivated many researchers to create chiral MOF catalysts with new designs, various chiral functional groups, and rich structural topologies, such as CMOF catalysts synthesized through chirality induction by Duan and coworkers (2010) [63], BINAP-based CMOF catalysts prepared by Lin and coworkers (2015) [64], the synthesis of MTV-MOFs with catalytic performance for asymmetric sequential catalysis by Cui and coworkers (2017) [52], etc. In the following, we will introduce them using chiral MOF-catalyzed asymmetric catalytic reactions, based on the classification of simple and complex asymmetric reactions.

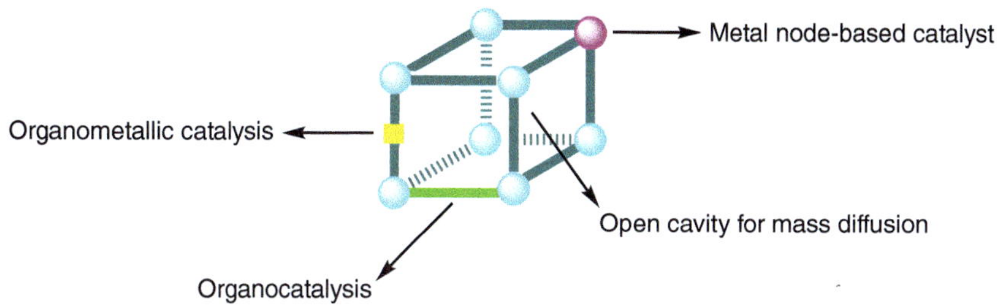

Figure 6.10 The suggested classification of catalytic sites in chiral MOF-based catalysts.

6.3.2 Designing CMOF Catalysts

Chiral MOFs, as a class of chiral porous organic–inorganic hybrids with crystalline structures, have significantly enriched the domain of asymmetric catalysis. In general, for designing effective chiral MOF-based catalysts, the presence of accessible catalytic active sites and chiral centers in the framework structures or pore environment is essential, and their performance can be adjusted by elaborately choosing components. Briefly, their classification can be divided based on their nature, namely metal node- and organometallic-based catalysts, organocatalysis, and encapsulated catalytic active species in the pores of MOFs, whose schematic illustration has been shown in Figure 6.10.

Numerous metallic centers have been reported as Lewis acids and used as catalysts in a wide range of catalytic reactions. These kinds of catalytic centers are in MOFs' structures, too. One kind of these centers is the OMS or coordinatively unsaturated site (CUS) or open coordination site (OCS) in MOFs. Their production is due to vacant sites on the metal (ions or clusters) that are not occupied during the preparation of framework by organic linkers (such as Ce-MDIP-(1 and 2): cyanosilylation of aldehydes [63]; UTSA-32a: ring-opening of meso-epoxides [65]). Generally, guest solvent molecules can occupy these unsaturated positions, which can be easily removed through the activation processes [66]. In another case, organometals can be used to create CMOF-based catalysts via their incorporation into the organic linkers of frameworks that, for example, metallosalen, BINOL, and diene have a significant role in the synthesis of these kinds of chiral catalysts [67–69] (such as KUMOF-1: carbonyl–ene reaction and hetero Diels–Alder reaction [59]; ZSF-1–4: cycloaddition of CO_2 and epoxides and C–H azidation [62]). BINAP is also one of the kinds of privileged chiral ligands that have been widely used in the synthesis of chiral catalysts [70]. For the first time, this chiral organic linker was used to synthesize a chiral MOF catalyst by Lin and coworkers [71]. To realize organocatalysts as another active site in asymmetric catalysis, Duan and coworkers developed these kinds of chiral catalysts based on the synthesis of chiral MOFs, using pyrrolidine (chiral organocatalyst) and polyoxometalate (oxidation catalyst) [20]. However, it should be mentioned that proline is one of the most important organocatalysts that is of great importance to extend, and until now, the various chiral MOF catalysts have been synthesized using proline and its derivatives (L- or D-pyrrolidin-2-ylimidazole) [57, 72]. In this class, BINOL and Biphenol can also be acted. For example, Tanaka et al. reported a CMOF using a ditopic carboxylate linker (BINOL) and Cu-based SBU [73]. This structure showed high performance in the asymmetric ring-opening (ARO) reactions of epoxide with amine, which is the most straightforward route to prepare β-amino alcohols. In the last case, the cavities and pores of MOFs can be a suitable space for the catalytically active species. In this regard, Morsali et al. reported the synthesis of chiral MOFs through the encapsulation of chiral nanoparticles (*in situ* ultrasonication synthesis). It is evident that the precise selection of functionalities in the construction of chiral catalysts has an effective role. In this work too, the right selection of functionalities is observed by producing the synergic effect between L-proline-nanoparticles and the azine groups of the backbone. The enhanced basicity converted this chiral MOF into a chiral basic catalyst [18]. The encapsulation strategy is widely employed to prepare MOFs with different applicability domains. For instance, recently, Cheng et al. synthesized a bifunctional catalyst Pd-SCNP@DUT-67-Pro (SCNPs: single-chain polymer nanoparticles). Herein, Pd-SCNP and chiral proline were introduced into the channels and on the metal node part of DUT-67 [74]. In another work, the same group presented CP@Pd@MOF-808 (CP: chiral polymer) in which the CP was threaded within the host pore of the framework as a heterogeneous support [75]. But, among them, enzymes should not be forgotten. As is known, they are efficient biocatalysts that are used due to their nature. Although they are widely employed, their biological activity can be changed by external factors like pH, the existence of special chemical compounds, etc. These changes can have positive and/or negative effects on their performances. To resolve this limitation, MOFs were characterized as useful supports to protect them [76, 77]. As a matter of fact, MOFs with large pores have attracted remarkable attention due to their potential applications in different fields, especially catalysis, but efforts in the field of their new designs are still going on. Throughout the progress of chiral catalysts, different chiral MOFs have been constructed, and we mentioned some of them. In the following, further examples of asymmetric catalytic reactions that are catalyzed by CMOFs will be reported.

6.4 Examples of Enantioselective Catalysis Using CMOF-Based Catalysts

6.4.1 Type I: Chiral MOFs in Simple Asymmetric Reactions

Catalytic asymmetric **Oxidation** processes are very important in chemical reactions, especially organic syntheses. Without a doubt, enantioselective epoxidation reactions as atom transfer reactions are one of the most valuable asymmetric processes in this class. Chiral epoxides as reactive intermediates can be used in different areas, such as producing fragrances, pharmaceuticals, food additives, etc. [78] Therefore, there is a great interest in producing new chiral heterogeneous catalysts for the enantioselective oxidation of olefins. The first catalytic system based on MOF for the enantioselective epoxidation of olefins was reported by Hupp and coworkers in 2006 [79]. Their synthesized framework was a microporous pillared MOF that was synthesized by chiral (salen) Mn struts, dinuclear zinc paddlewheel SBUs, and biphenyldicarboxylate (bpdc) as the second ligand. Their selected substrates were chromene derivatives that their oxidation was performed using 2-(tertbutylsulfonyl)iodosylbenzene, an oxidizing agent. The considerable yields and high selectivities were the obtained results in this work, (71% and 82%, respectively). The considerable decrease in selectivity and leaching of the initial manganese were not observed in the recycling framework. Also, the resulting turnover numbers were another advantage of this chiral MOF catalyst. C-NU-1000-Mo is one of the chiral oxidation catalysts that was reported for enantioselective epoxidation of various prochiral olefins [80]. The materials used in its construction included NU-1000, L-tartaric acid, $MoO_2(acac)_2$. At first, NU-1000 was chiralized through solvent-assisted ligand incorporation (SALI) by tartaric acid. Then, the catalytic active centers were immobilized on chiral support produced. After its structural characterization, it was employed as a chiral catalyst to access chiral epoxides.

In this catalytic system, TBHP was a strong oxidant, and dichloroethane was selected as a solvent. The obtained results of the oxidation of the various olefins showed good conversions, enantioselectivities, and high ee (80–100%). From the viewpoint of reusability and stability, the considered CMOF demonstrated high stability because, after five cycles, a significant decrease was not observed in the enantioselectivity and ee of the styrene oxidation as a model substrate. In another work, the same group reported a chiral MIL with Lewis and Brønsted acidic sites for enantioselective epoxidation and methanolysis. This CMOF was constructed from BDC organic linker, Cr metal node, and tartrate anion in 2019 [15]. After the preparation of MIL-101(Cr), reacting tartrate anions with Cr–F centers happened to eliminate F anions on Cr. After the chiralization of achiral MIL and its characterization, it was used as a chiral catalyst. In asymmetric epoxidation of the various olefins, according to the optimized conditions mentioned in the article and by varying the oxidants, IBA/O_2 was selected because considerable enantiomeric excesses were obtained (up to 100%) (Table 6.1). The ideal asymmetric induction was due to the number of tartrate anions. The results demonstrate that the enantioselectivity parameter is very sensitive to the pore/surface surrounding the catalyst [81]. Surely, the rate of substrate diffusion can also affect the rate of epoxidation.

The MOFs were found to be active epoxidation catalysts with a broad scope of alkene substrates. A chiral porphyrinic MOF was reported in this class [82]. Herein, PCN-224 was selected as an achiral framework that its organic linker was an achiral polytopic linker (TCPP: tetrakis(4-carboxyphenyl)porphyrin). Its chiralization happened through the post-synthetic exchange. The exchange process was performed by chiral organic anions. This CMOF demonstrated considerable porosity, high surface area, and stability. Also, it could catalyze two asymmetric catalytic reactions, one of which was the epoxidation reaction. In addition to its structural stability in the catalytic reactions performed, high epoxide selectivity and enantiomeric excess were the other important parameters. For example, these results for the styrene as a model substrate were 89% and 96%, respectively. These parameters can be related

Table 6.1 Enantioselective epoxidation of several olefins.[a)]

Entry	Substrate	Conversion (%)/ Time (h)	Epoxide selectivity (%)	Ee (%) (Conf.)[b)]
1	α-methyl styrene	83/8	68	85 (S)
2	1-Phenyl-1-cyclohexene	79/8	100	78 (R, R)
3	1-Octene	77/8	100	100 (R or S)
4	1-Decene	69/8	100	91 (R)
5	trans-stilbene	100/30 min	100	75 (R, R)

a) Reaction conditions: catalyst 50.0 mg (0.0013 mmol Cr/g catalyst), ^{i}PrCHO/substrate 3 mmol, CH_3CN 5 ml, molecular oxygen 1 atm.
b) Determined by GC on a chiral SGE-CYDEX-B capillary column. Absolute configuration of the epoxides was determined by comparison with the GC data of R-(+)-limonene.

to the framework pore size, abundant chiral functional groups, close vicinity of the pro-S or -R face of the olefin to the catalytic sites, and the degree of non-covalent interactions. The catalytic findings showed the S enantiomer is the major product, and a racemic mixture was not observed. Briefly, the used oxidative catalytic system included Mn^{III}/O_2/aldehyde and its probable mechanism has been previously presented [83]. Acyl and acylperoxy radicals, peroxy acid as well as Mn^{V}-oxo are the active species in this epoxidation mechanism. Although, until now, the various chiral MOF oxidation catalysts have been reported, owing to the importance of this catalytic reaction, trying to create new catalysts with different designs is continuing.

Asymmetric **Hydrogenation** is an important method for the synthesis of chiral molecules that has remained a challenge. Different catalysts have been designed in this field, the most significant of which are chiral MOFs. For example, recently, Ma and coworkers reported the synthesis of a chiral MOF with the name CFLP@MOF (CFLP: chiral frustrated Lewis pair) [84]. For the first time, they discussed the concept of chiral frustrated Lewis pairs from a homogeneous system to heterogeneous catalysis. The high activity of this chiral catalyst, the obtained enantioselectivity, as well as the recyclability are important properties of this new chiral platform. The combination strategy can be an effective method to produce active chiral catalysts in asymmetric transfer hydrogenation, namely metal and ligand design. This route was reported by Liu and coworkers in 2021 [85]. Their synthesized structures were four isostructural chiral MOFs with Ca, Sr, and Zn as metal nodes and the functionalized phosphono-carboxylate ligands of 1,1′-biphenolas enantiopure organic linkers. The employed functional groups of organic linkers were 2,4,6-trimethyl- and 2,4,6-trifluoro-phenyl groups. The presence of the channels as well as metal phosphonates converts the resulting structures into active chiral catalysts in the asymmetric hydrogenation of imines. The hydrogen source used was Hantzsch ester. The obtained results showed that the Ca-based MOF catalyst with the functional group of 2,4,6-trimethyl on the organic linker, has high performance (enantioselectivity: up to 97%). Apart from the importance of three parameters in chiral catalysis, namely enantioselectivity, high activity, and enantiomeric

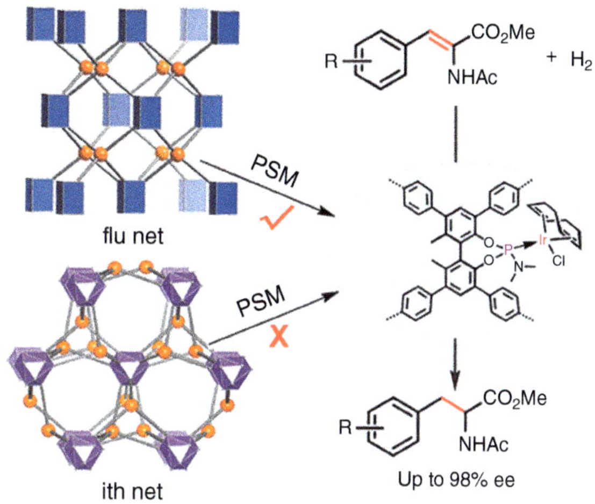

Figure 6.11 Robust chiral Zr-Based MOF catalysts for enantioselective hydrogenation. Source: Reprinted with permission from [86].

excess, the structural stability of the catalyst is also important. Therefore, the creation of chiral robust and porous solids such as chiral Zr-MOFs, with different designs, can be a priority in the catalytic platforms. In 2020, Cui et al. reported the preparation of chiral Zr(IV)-based MOFs with different topologies (**flu** or **ith**) to support Ir complexes for asymmetric hydrogenation of α-dehydroamino acid esters (Figure 6.11) [86]. The clusters of five chiral frameworks obtained were Zr_6, Zr_9, or Zr_{12} clusters. The organic linkers used were also the tetracarboxylates that were derived from 1,1′-biphenol. These chiral tetratopic linkers were prepared through the modification of biphenol using benzoate and naphthoate groups. The chiral monophosphorus ligands had suitable positions to incorporate metallic catalytic active centers ($[Ir(1,5\text{-cyclooctadiene})Cl]_2$). The hydrogen source was H_2 (5 bar). Totally, the reported isolated yields and enantiomeric exccsses showed high activity of the used chiral catalysts. For instance, (Z)-methyl-α-acetamidocinnamate was hydrogenated to generate (S)-*N*-acetylphenyl- alaninc methyl ester with a yield of 96% and enantiomeric excess of 98%. Or the catalytic hydrogenation of (Z)-ethyl-α-acetamidocinnamate carried out with high yield and enantiomeric excess (95% and 97%, respectively). Although most of the obtained results werc good, low or moderate enantioselectivity and enantiomeric excess were also observed. For example, when the aliphatic substrate was employed, high yield (97%) was produced but the enantiomeric excess was 40%. Its reason is related to the steric hindrance of the substrate that does not allow the useful chirality transfer. It must be pointed out that among these CMOFs, the frameworks with the **ith** network did not show high catalytic activity owing to the steric crowding. Asymmetric hydrogenation of functionalized olefins is one of the reactions that is used to generate chiral molecules. Rh catalysts are the most widely employed catalysts in this field, but there are also challenges because each site may plausibly feature one or two phosphorus ligands. Recently, Cui and coworkers reported a sequential PSM to incorporate single-site Rh species into a framework [87]. The selected framework was from the zirconium MOF family with chiral spinol-based ligands. These Rh-monophosphorus materials obtained were used in the asymmetric

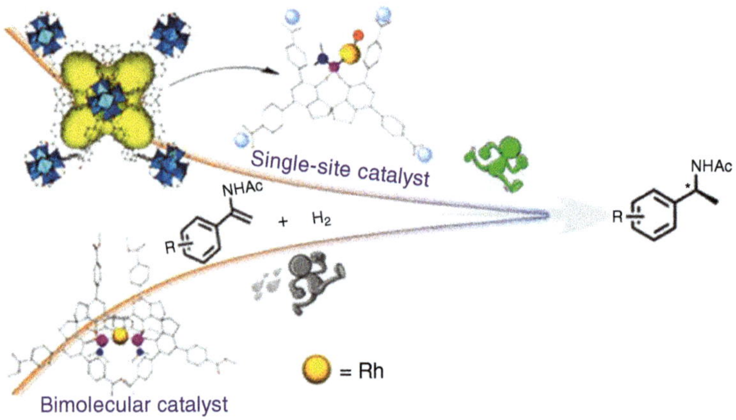

Figure 6.12 Leveraging chiral Zr-MOFs to investigate Rh species in asymmetric hydrogenation of the functionalized olefins. Source: Reprinted with permission from Jiang et al. [87].

hydrogenation of enamides and α-dehydroamino acid esters. The excellent yields and enantioselectivities were the considerable results (up to 99.9% enantiomeric excess) (Figure 6.12).

Also, they made a comparison between the Rh-monophosphorus catalyst and the homogeneous Rh-biphosphorus. Based on the findings, they claimed that the high activity of the Rh-monophosphorus species can be related to the single-site Rh-monophosphorus species and the confined framework cavities. In addition, this specific construction has been recognized as an excellent candidate for the synthesis of a first-in-class drug ((R)-cinacalcet hydrochloride) with 99.1% enantiomeric excess.

Cyanation and Cyanosilylation are important chemical reactions. Catalytic asymmetric cyanation reactions of aldehydes and ketones are useful processes to produce chiral cyanohydrins. The obtained products are versatile compounds that are used to prepare chemical materials, specially natural products and chemical pharmaceuticals. Chiral MOF catalysts are one of the kinds of chiral heterogeneous catalysts that due to their performance can be employed in various areas. One of the fields is catalytic asymmetric cyanation. For example, Cui and coworkers reported two chiral MOFs using dipyridyl-functionalized Ti(salan) linker and Cd metal node [88]. After the synthesis through solvothermal reactions, they were characterized by different methods. Due to the abundant Ti(salan) catalytic active centers, these chiral porous frameworks were utilized in the asymmetric catalysis of aldehydes (cyanation). At first, *p*-methoxy benzaldehyde was selected as a model substrate to determine initial optimal conditions. In the presence of the structure 1 ($[Cd_2(O_2CCH_3)_4(TiL)_2O(OMe)_2]\cdot 2H_2O$ [5 mol%]), cyanohydrin silyl ether was obtained with good conversion and enantioselectivity (82% and 90%, respectively). It is important that the reaction could not take place without the chiral MOF catalyst. Benzaldehyde, aromatic aldehydes, and their derivatives were other substrates that were investigated (Table 6.2). Briefly, their two chiral

Table 6.2 Asymmetric cyanation of aldehydes by 1, 2, and $TiL(OBu)_2$.[a)]

Entry	Catalyst	R	Conv. (%)[b)]	Ee (%)[c)]
1	(R)-1	Ph	78	79
2	(R)-2	Ph	77	77
3	(R)-$TiL(OBu)_2$	Ph	99	5
4	(R)-1	*p*-MeC_6H_4	89	75
5	(R)-2	*p*-MeC_6H_4	78	68
6	(R)-1	*m*-$MeOC_6H_4$	75	60
7	(R)-2	*m*-$MeOC_6H_4$	74	59
8	(R)-1	*p*-$MeOC_6H_4$	82	90
9	(S)-1	*p*-$MeOC_6H_4$	81	91
10	(R)-2	*p*-$MeOC_6H_4$	80	77
11	(R)-$TiL(OBu)_2$	*p*-$MeOC_6H_4$	96(99)[d)]	9(3)[d)]
12	(R)-1	*p*-BrC_6H_4	85	36
13	(R)-2	*p*-BrC_6H_4	73	42
14	(R)-$TiL(OBu)_2$	*p*-BrC_6H_4	99	3
15	(R)-1	1-naphthyl	78	34
16	(R)-2	1-naphthyl	77	76
17	(R)-1	2-naphthyl	81	59
18	(R)-$TiL(OBu)_2$	2-naphthyl	99	5
19	(R)-1	2-thiophyl	77	56
20	(R)-2	2-thiophyl	81	82
21	(R)-1	3-$G_0C_6H_4$	<3	n.d.
22	(R)-$TiL(OBu)_2$	3-$G_0C_6H_4$	95	n.d.

G_0:

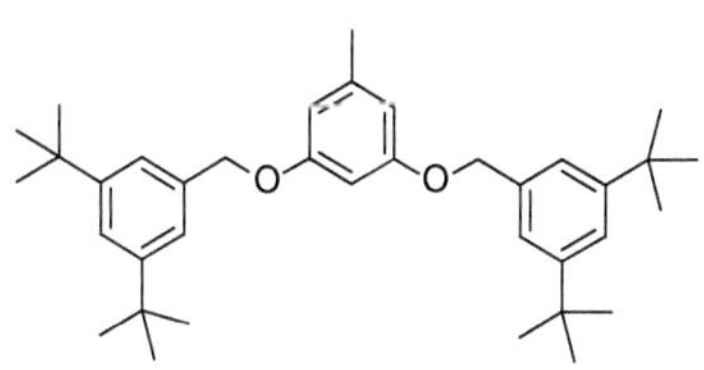

Note: n.d., not determined.

a) Solvents: CH_2Br_2 for **1** and Ti**L**$(OBu)_2$; CH_2Cl_2 for **2**; Catalyst and Ph_3PO loading based on aldehyde.
b) Calculated by ^1H NMR.
c) Determined by HPLC.
d) The results in brackets were obtained by using CH_2Cl_2 as solvent.

MOFs with μ-oxo-bis[Ti(salan)] dimers showed higher performance in the enantioselective cyanation of aldehydes than homogeneous analogs (up to 91% ee). Cyanosilylation of carbonyl compounds is a significant Lewis acid-catalyzed C—C bond formation process that is rarely investigated by chiral MOFs. For instance, the first porphyrin-salen-based chiral MOF was synthesized by Jiang and coworkers through the incorporation of the metalloporphyrin and metallosalen struts into the framework of MOF [89]. This chiral MOF demonstrated high quality in the asymmetric cyanosilylation of aldehydes. Herein, *para*-methoxy benzaldehyde was selected as a model substrate. At first, 95% conversion and very low enantioselectivity (almost 10% enantiomeric excess) were afforded with 1 mol% catalyst loading (RT/24 h). But when Lewis base PPh_3O is added to the reaction, both conversion and enantioselectivity, especially the enantioselectivity efficiently improved (100% and 25%, respectively). As is known, temperature is one of the main factors in the enantiomeric excess value. The authors reported that at a higher temperature, racemization happens. Therefore, the reaction was studied at low temperatures (to −20 °C). A significant increase in the enantiomeric excess was obtained (up to 98%), but a decreased conversion was observed.

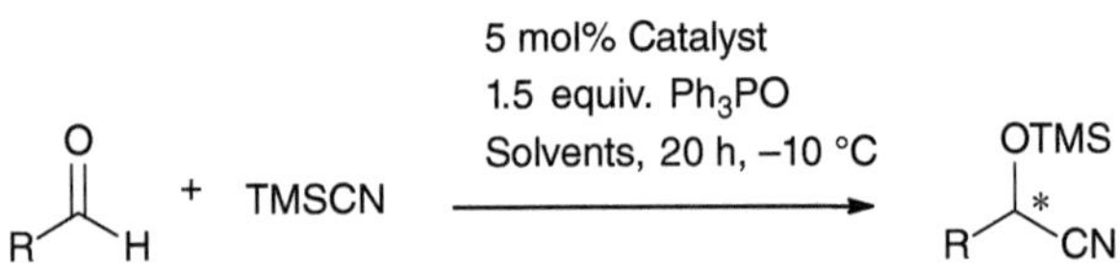

For better results, the optimal reaction conditions were 1 mol% catalyst, 1 mmol substrate, 1.2 equiv. TMSCN, 1.0 equiv. PPh_3O, 2 ml CH_2Cl_2, −20 °C and 48 hours. The catalytic performance of the considered chiral MOF catalyst was investigated for the asymmetric cyanosilylation of a wide range of the various aldehydes (conversion and ee, up to 96% and 98%, respectively). The acceptable results relate to the synergistic function between Lewis acid activation and chiral induction. In another work, Dong and coworkers reported a homochiral MOF (BINAPDA-Zr-MOF) using a different chiral organic linker from the dibenzoic acid family [90]. The space group of this chiral framework was *F*23. Also, it had the same topological structure as UiO-66. Asymmetric aldehyde cyanosilylation was the catalytic reaction selected. They investigated the optimization of the model reaction with benzaldehyde. As well as they achieved remarkable findings and compared their results with the reported catalysts. Time, temperature, the kind of solvent, reusing at least five times without loss of the catalytic activity, and enantioselectivity (five hours, room temp, MeCN, recycle five times, yield 99%, ee 98%) were its advantages in comparison to others [63, 89, 91].

For the first time, the **Strecker** reaction as a multi-component process was introduced by Adolph Strecker in 1850. Since the discovery of the first amino acid (alanine) through this synthesis method, the Strecker reaction has been selected as one of the most attractive synthesis strategies for chemists. This consideration is due to the direct and economical access to various α-amino acids. Owing to the ever-increasing request for the produced products of this synthesis, research efforts have been made in the asymmetric Strecker reactions field [92]. With respect to the importance of chiral MOF catalysts, asymmetric Strecker reactions have been investigated using CMOFs, too. Although the reports are not considerable. Bharadwaj and coworkers synthesized a chiral Cd-MOF without any chiral source using an achiral ligand (V-shaped ligand L) via spontaneous resolution [25]. This chiral MOF was a thermally stable framework with the chemical formula $\{[Cd_2(L)(H_2O)\text{-}(N,N\text{-dimethylformamide})]\cdot 3DMF\cdot 2H_2O\}n$. Considerable properties were observed such as microporous nature, 1D open channels, active OMSs, 3,6-c binodal net, and 3,6-conn topology that is rare. The high performance of this MOF for the synthesis of α-aminonitriles was ensured by the OMSs. Herein, the asymmetric Strecker reaction was investigated in suitable conditions, solvent-free and at room temperature, with excellent conversion yields. Their investigation of the three-component Strecker process started using three reactants, namely benzaldehyde, trimethylsilyl cyanide (TMSCN), and aniline. TLC analysis and ^1H NMR spectra were used to monitor and confirm the development of the reaction and production of the desired product (α-amino nitrile), respectively. The various ketones with different electron-withdrawing and donating groups attached were also investigated with good conversion yields (Table 6.3). Therefore, it is concluded that these groups exert a strong electronic effect on the catalytic results. Although the catalytic findings were good, according to their report no enantiomeric excess in the products was observed. In this case, the authors suggested that the bulky phenyl groups in the substrates can produce limitations in inducing asymmetric catalysis.

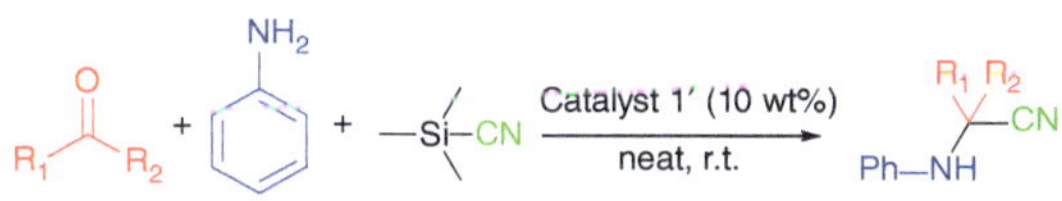

In general, the presence of active catalytic centers in the framework structures can be finely adjusted by selecting desirable components. One of the chemical reactions that can be catalyzed by chiral MOFs is asymmetric **Friedel–Crafts**, developed by Charles Friedel and James Crafts in 1877 to attach substituents to an aromatic ring [93]. Brønsted and Lewis's acidity of chiral MOFs can be useful for catalyzing these kinds of reactions. For instance, Lin et al. reported the synthesis of a pair of chiral MOFs (CMOF-1 and 2) that were prepared using SBU based on Cu and chiral ligands derived from 1,1′-binaphthyl-2,2′-phosphoric acid [94]. Both chiral frameworks were employed for the reaction between indole and *N*-sulfonyl aldimines, which is one kind of the Friedel–Crafts process (Scheme 6.2). Different yields and enantiomeric excesses were obtained. They demonstrated that CMOF-1 as a chiral catalyst produced the major enantiomers with the opposite chirality in comparison to the corresponding homogeneous catalyst. It is worth mentioning that in this work modest enantiomeric excesses were observed. Asymmetric Friedel–Crafts alkylation of indole with nitroalkene has also been investigated using different chiral catalysts, such as chiral metal–bisoxazoline complex [95]. For example, in another research work, Tanaka et al. presented the first asymmetric Friedel–Crafts alkylation between *N,N*-dialkyl anilines and

Table 6.3 The obtained results for the Strecker reaction catalyzed by 1′.[a)]

Entry	Ketone	Yield (%)
1	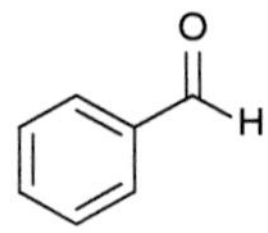	99
2	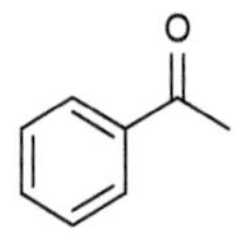	98
3	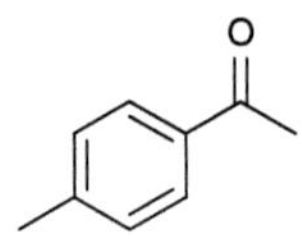	94
4	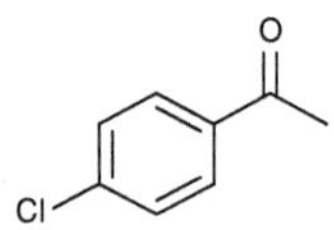	90
5	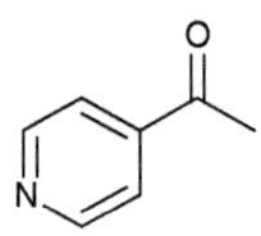	88
6	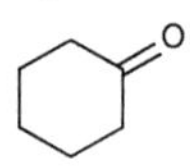	90
7	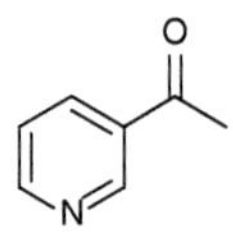	86
8	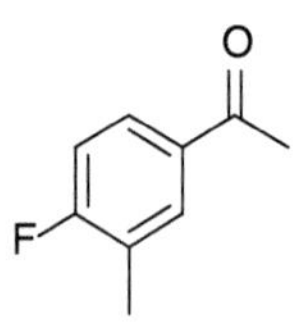	95
9	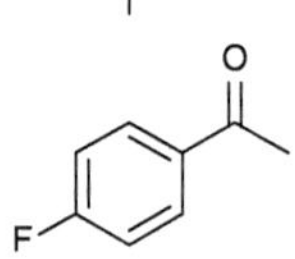	97
10	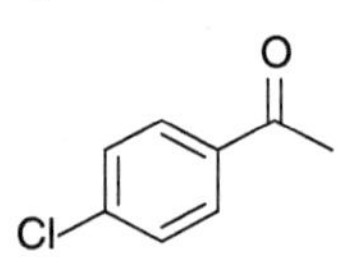	86

a) Reaction conditions: ketone (0.1 mmol), amine (0.1 mmol), TMSCN (0.12 mmol), and catalyst **1′** (10 wt %) at 25 °C for 4 h.

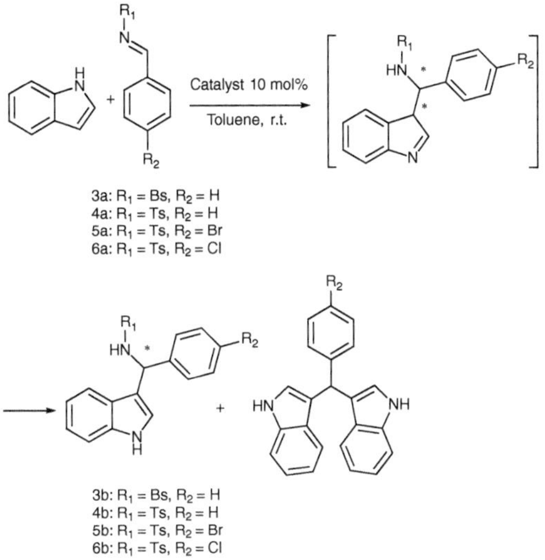

Scheme 6.2 Asymmetric Friedel–Crafts between indole and *N*-sulfonyl aldimines.

trans-β-nitrostyrene by a chiral MOF [96]. This heterogeneous recyclable catalyst was synthesized using one kind of metal node, Cu, and one kind of organic linker, namely (R)-2,2′-dihydroxy-1,1′-binaphthyl-6,6′-dicarboxylic acid. Excellent yields and enantioselectivities were obtained (~96% and 98% ee, respectively).

CO_2 conversion and Azidation are other significant reactions. Considering the rapid increase in the concentration of CO_2, the conversion of CO_2 to suitable mate rials is one of the chemical processes that is a priority. These conversion processes can be performed with achiral and chiral materials. In this field, MOFs owing to their tunability, are considered promising materials [97, 98]. As is known, the engineering of chiral MOFs can also convert them into models for practical applications, specially for asymmetric heterogeneous catalysis. One of the important reasons to select MOFs is the existence of random defects, which can be used for different goals. Modulators and missing components can generate them, which leads to vacancies and OMSs. These accessible centers like metal centers can act as active Lewis acids in catalytic reactions such as CO_2 chemical fixation. The enantioselective cycloaddition reaction of CO_2 to an epoxide is usually carried out under rough catalytic conditions, such as high pressure and temperature, with low yield. So, the good design of catalysts is an important factor in catalytic systems. For example, PCN-224-Mn(tart) is one of the chiral catalysts that was able to proceed with CO_2 fixation to the epoxide asymmetrically [82]. In this structure, two kinds of metal centers (Zr and Mn) and tartrate species could act as Lewis and Brønsted acids, respectively. Briefly, the selected reaction conditions to catalyze included a temperature of $60 \pm 5\,°C$, a pressure of CO_2 1 bar, and Bu_4NBr as the cocatalyst. The various olefin epoxides were investigated (Table 6.4). It is interesting that, apart from metallic active catalytic centers, tartrate anions could also act as active catalytic centers through the synergistic effect with the other acidic sites. Indeed, the findings of enantiomeric excess showed that the chirality induction happened successfully based on the interaction between

Table 6.4 CO_2 cycloaddition to the various olefin epoxides.[a)]

Entry	Substrate	Entry	Substrate
1	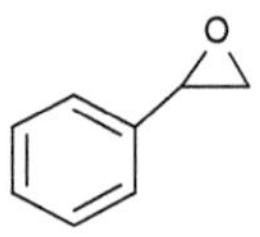 Conv.: 96%, Cyclic carbonate selectivity: 100%, Ee.: 94% (S)	4	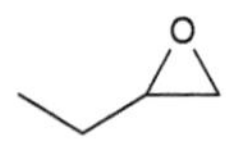 Conv.: 91%, Cyclic carbonate selectivity: 100%, Ee.: 97% (R)
2	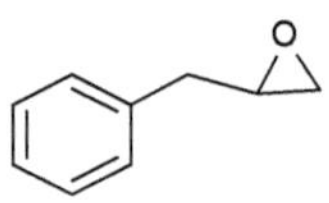 Conv.: 87%, Cyclic carbonate selectivity: 100%, Ee.: 90% (S)	5	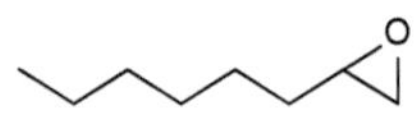 Conv.: 78%, Cyclic carbonate selectivity: 100%, Ee.: 96% (R)
3	 Conv.: 99%, Cyclic carbonate selectivity: 100%, Ee.: 98% (S)		

a) Reaction conditions: catalyst 0.03 g, substrate 2 mmol, co-catalyst 2.5 mmol, CO_2 1 bar, temperature $60 \pm 5\,°C$, time 15 h. Conversions (Conv.), cyclic carbonate selectivity and enantiomeric excess (Ee.) were determined with GC/FID (chiral column) by using chlorobenzene as a usual internal standard. All catalytic experiments were repeated two times.

the active acidic centers (Zr and Mn) and auxiliary OH groups of tartrate with the epoxide (O atom).

In another work, Jiang and coworkers reported the synthesis of four robust and stable CMOFs for CO_2 conversion (ZSF-1–4) [62]. In these frameworks, there are clusters based on Zr—O bonds and privileged chiral metallosalen ligands. One of the frameworks (ZSF-1) has salen-Ni moieties. Its stability in aqueous solutions with a wide pH range was an important feature that caused the direct fixation of CO_2. To determine the optimal catalytic conditions, the styrene oxide model substrate was used. Briefly, in addition to temperature (100 °C), time (20 hours), and CO_2 (1 atm), tetrabutylammonium bromide was also utilized as responsible to open the ring of the epoxides. A yield of 94% was a good result for this framework in producing cyclic carbonate. On the basis of the observation, the obtained result was considerable at a low catalyst loading. Also, they investigated replacing ZSF-1 with the ligand Cy-salen-Ni species. The obtained yield by this structure was 61%. Although the catalyst loading was higher. The structural features of ZSF-1 and ZSF-2 converted them into strong Zr-MOFs to produce cyclic carbonates. On the other hand, ZSF-3 and ZSF-4 were employed for C–H azidation process with high enantiomeric excesses (up to 94%). The importance of azides relates to their applications in biological chemistry and material science for specialty pharmaceutical companies. Their applications are due to their versatile synthetic functions that can be transformed into nitrogen-containing groups like amides, imines, nitrile, amines, and triazoles.

Therefore, herein also the direct and simple synthetic routes are a priority for the C–H activation in producing azides groups in a single step. Until now, various catalysts have been reported, for example, a palladium-catalyzed regioselective azidation of allylic C—H bonds with NaN_3 as the azide source under O_2 was reported by the same group [99]. But ZSF-3 and ZSF-4 catalyzed the asymmetric benzylic C–H azidation using NaN_3 and PhIO with moderate yields (51–64%). In all cases, these four heterogeneous catalysts were better in comparison to the corresponding homogeneous catalysts.

The asymmetric **Nitroaldol reaction** is another chemical reaction that can be catalyzed by chiral MOFs. Indeed, the catalytic asymmetric aldol processes using chiral MOFs as chiral heterogeneous catalysts have been ubiquitous in the last decade. Different designs of CMOFs have been reported that proline has a key role in them. But, given the rapid growth of CMOF catalysts for this kind of asymmetric catalysis in recent years, there are several reasons. One of the reasons is producing nitroalkanols, which have great pharmaceutical importance. So, there are different synthesis methods to construct CMOF catalysts for producing these chiral products. For example, Kim and coworkers reported new chiral MOFs based on the MIL-101 framework using auxiliary chiral ligands (L_{30} and L_{31} from the family of proline) through post-synthetic chiral modification in 2009 [100]. Herein, the characterization methods showed the incorporation of the chiral units into the achiral framework. The obtained structures demonstrated their performance in asymmetric aldol reactions with considerable yields and enantiomeric excesses (up to 90% and 80%, respectively) (Figure 6.13). It should be mentioned that homogeneously catalyzed reactions with L_{30} and L_{31} were poor from the viewpoint of enantioselectivities.

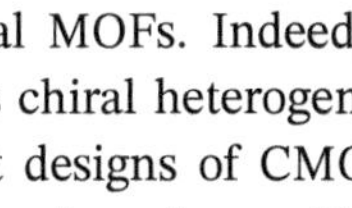

L_{30}

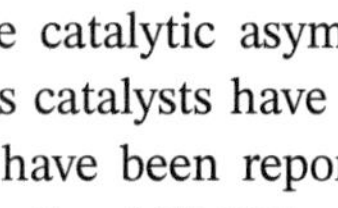

L_{31}

In another research work, the asymmetric aldol reaction of aromatic aldehydes with cyclohexanone was investigated by Duan and coworkers in 2010 [63]. Their chiral Cd-MOF was synthesized using (S)-*N*-tert-butoxy-carbonyl-2-(imidazole)-1-pyrrolidine as a chiral agent along with the other reactants. The resulting chiral

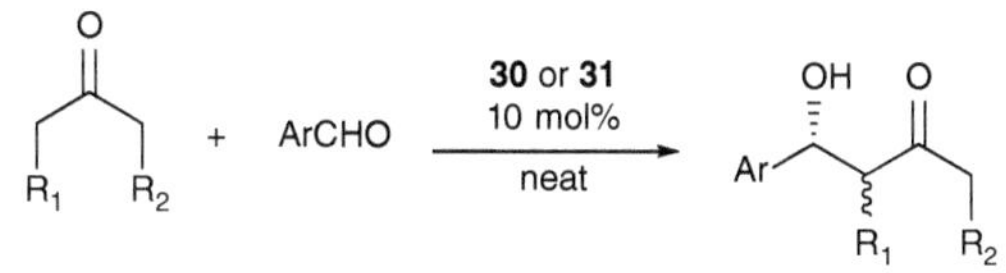

Substrates	MOF: yield/ee (%) 30	31
R_1 = H, R_2 = H, Ar = 4–NO_2Ph	66–69	59–63
R_1 = H, R_2 = H, Ar = 4–Py	91–76	87–58
R_1 = H, R_2 = H, Ar = 4–ClPh	74–70	69–52
R_1, R_2 = –$(CH_2)_3$–, Ar = 4–NO_2Ph	81–66	
R_1 = H, R_2 = Me, Ar = 4–NO_2P	49–81	

Figure 6.13 The catalysis of asymmetric aldol reaction with different substrates by chiral MOF catalysts.

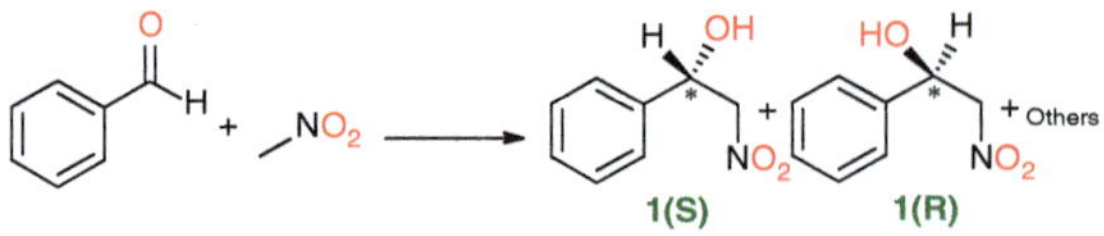

Figure 6.14 Asymmetric Henry (nitroaldol) condensation.

catalyst showed moderate enantioselectivities with enantiomeric excesses up to 61%. Kaskel and coworkers reported the synthesis of chiral trifunctional linkers through the introduction of chiral oxazolidinones into 4,4′,4″-benzene-1,3,5-triyltribenzoate backbone. The resulting chiral structures with Lewis acid sites could be utilized for the asymmetric Mukaiyama aldol reaction. These three-dimensional structures, owing to their chiral functionalization, demonstrated significant levels of enantioselectivity, and enantiomeric excesses of up to 40% were obtained [101]. Post-functionalized MOFs by peptides were the first examples that were reported by Canivet et al. and were used for the chiral aldol reaction. Al-MIL-101-NH_2, In-MIL-68-NH_2, and Zr-UiO-66-NH_2 were their selected frameworks [55]. Although in that time, their design was good, enantiomeric excesses in this work were up to 35%. In another research, Ren and coworkers reported the performance of the synthesized chiral Cu(salen) MOF for the asymmetric Henry (nitroaldol) reaction (up to 98% enantiomeric excess) [102]. Finally, we want to explain one of our works that is about preparing a chiral MOF via the encapsulation of Fe_3O_4@L-proline NPs by TMU-5 ((Fe_3O_4@L-proline)@TMU-5) that happened through an *in-situ* ultrasonication synthesis [18]. The Henry (nitroaldol) condensation is the reaction that was examined with this CMOF (Figure 6.14). The catalytic activity of all parts of this CMOF was investigated: The conversion of the reaction with achiral TMU-5 was high, and the main product was produced with good selectivity (75%) without any enantiomeric excess. In comparison to TMU-5, LPF NPs demonstrated lower catalytic activity, but due to the presence of L-proline on the magnetic NPs surface, enantioselectivity was observed. About the main catalyst (0.1-LPF@5: 0.1 g initial amount of LPF NPs), excellent results were obtained with 98% enantiomeric excess. After the results obtained by GC as an analytical technique with the chiral column, the main chiral product was also confirmed by circular dichroism (CD) analysis. But the factors such as atmospheric pressure and H_2O as solvent are two important parameters in this chiral catalytic system. Finally, the improved Lewis basicity in this system is owing to the existence of the synergic cooperation between azine functions of the TMU-5 framework and L-proline amine groups. In the next Section 6.4.2, we will describe some reports about complex asymmetric reactions along with related examples.

6.4.2 Type II: Chiral MOFs in Complex Asymmetric Reactions

Until now, we have reported some asymmetric catalytic reactions that were relatively simple. But this part is related to asymmetric multicatalysis systems that have multiactive sites with the performance for sequential/tandem transformations [103].

We begin this section with **Sequential alcohol oxidation/asymmetric cyanation of aldehyde** catalysis. For the first time, Cui and coworkers reported the synthesis of a chiral MOF using organic radicals and chiral ligands based on metallosalen [104]. It is interesting that the radical part can help to catalyze the oxidation of alcohols, and the salens have a key role in the asymmetric cyanation of aldehyde. Their model substrate was *p*-bromobenzyl alcohol. Briefly, in the considered catalytic system, TEMPO, tert-butyl nitrite, and ethylene dichloride were used under atmospheric O_2 at temperature of 80 °C. At different times, various conversions were produced, from 46% to 99%. After the oxidation reaction is complete, for continuing the reaction to asymmetric cyanation form, at 0 °C, TMSCN with the formula $(CH_3)_3SiCN$, triphenylphosphine oxide (Ph_3PO or PPh_3O) are needed. Therefore, the temperature was cooled to 0 °C and then the above-mentioned compounds were added. The reaction proceeded under a nitrogen atmosphere. The considered product was obtained with 80% enantiomeric excess (conversion 98%). They also investigated this kind of catalytic reaction with benzyl alcohols that had the electron-withdrawing and -donating groups on the benzene ring. The conversions and enantioselectivities obtained were considerable. But, the authors studied that the pores or the surface of the framework can be or NOT used as places for the sequential reaction. Their investigation shows that the reaction does not occur within pores owing to the large diameter of the bulky substrate. Therefore, the surface of the framework is a suitable position for catalysis (Table 6.5).

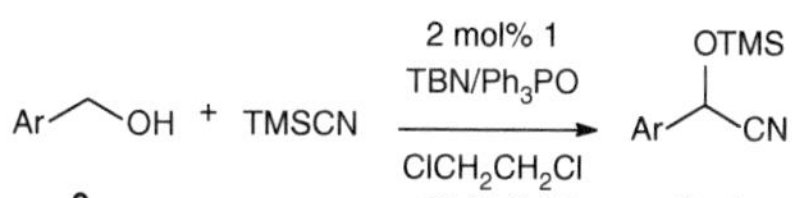

2 3a–j

Table 6.5 Sequential oxidation/asymmetric cyanation reactions of alcohols.[a)]

Entry	Ar	3/conv (%)[b)]	ee (%)[c)]
1	*p*-Br-Ph	a/98 (78)	80 (65)
2	*p*-C1-Ph	b/91	75
3	Ph	c/94 (85)	77 (72)
4	*p*-Me-Ph	d/93	77
5	*p*-OMe-Ph	e/96 (82)	83 (52)
6	*m*-OMe-Ph	f/91	81
7	*o*-OMe-Ph	g/94 (84)	87 (23)
8	1-naphthyl	h/75	66
9	2-naphthyl	i/78	77
10	*m*-G_0-Ph	j/<5 (45)	n.d. (n.d.)

Note: n.d., not determined.

a) For reaction details, see the Supporting Information; the data in parentheses are results catalyzed by a mixture of Me_3L^1 and VOL^2 (the same amount of catalyst loadings as the MOF system).

b) Determined by ^1H NMR.

c) Determined by HPLC.

Sequential alkene epoxidation/epoxide ring-opening in asymmetric form is an interesting reaction that has been examined by different chiral catalysts. Cui and coworkers reported the preparation of an isostructural family of chiral, porous, and interpenetrating MTV-MOFs using metallosalen active sites, in 2017 that had this ability [52]. At first, they investigated the binary MTV-MOFs (R)-1^{CuM} namely 1^{CuMn} and 1^{CuFe}. These chiral structures could catalyze the epoxidation of alkenes with high enantiomeric excesses (up to 90%). Oxidation of sulfides (1^{CuFe}) and cyanosilylation of aldehydes (1^{CuV}) were the other catalytic reactions that were studied in this work. Aminolysis of trans-stilbene oxide with anilines was another important chemical process that was investigated. Among the obtained isostructures, 1^{CuCr} could be used in this catalytic reaction as an effective catalyst. Although most amino alcohols were produced with excellent enantiomeric excesses (up to 99%), aminolysis of 2,2-dimethyl benzopyran oxides did not show good enantiomeric excess. The hydrolysis of epoxides was also evaluated, and the results were considerable. But another significant study of the authors was combining three different catalytic metal centers, namely (R)-1^{CuMnM_3} (M_3 = Cr or Co), and their investigation of sequential reactions. Both chiral structures could promote the epoxidation of 2,2-dimethyl-2*H*-chromene and its derivatives as well as the ring opening of epoxide with aniline and water. For example, for the considered sequential transformation, (R)-1^{CuMnM_3} (M3 :Cr, (0.5 mol%)), CH_2Cl_2, temperature 0 °C, 2-(tertbutylsulfonyl)iodosylbenzene (oxidant) and aniline (nucleophile) were used. At 15 minutes intervals, adding the oxidant to the reaction is important because it has a positive effect on the conversion of the epoxide. Then, aniline was added to the catalytic reaction, after the process of epoxidation. Finally, only one pair of enantiomers was obtained among all four possible pairs, and the considered amino alcohol was produced with high conversion and enantiomeric excess (85% and 94%, respectively). To further understand their findings, the study of this article is necessary. But briefly, the produced ternary MTV-MOF were asymmetric catalysts for a broad scope of asymmetric sequential processes such as the olefin epoxidation/epoxide ring-opening. The twofold interpenetration of the frameworks led to a dense arrangement of M(salen) units. In this case, with the platforms of MTV-MOFs, the catalytic activity could be improved in comparison to the other structures (for example, the binary MOFs). Therefore, these multifunctional frameworks, with different designs, can be utilized in different applications, specially in various catalytic reactions. Sequential is another study in this field (Figure 6.15). In 2022, Cheng et al. reported the synthesis of a chiral composite, CP@Pd@MOF-808, with the catalytic ability in this kind of sequential reaction [75]. Briefly, CP, Pd nanoparticles, and MOF-808 were its components. The polymer with chiral sites was within the pores of the framework of MOF-808. This method led to the creation of a chiral heterogeneous catalyst that showed high activity than homogeneous catalysts. To obtain these results, first, they investigated the Suzuki coupling reaction. Their model reaction was the reaction between bromobenzene and 4-formylphenylboronic acid. To get the optimized reaction conditions, all the catalytic parameters were investigated such as solvent, temperature, time, and the amount of catalyst. The mixed solvent system served as the best solvent (81%

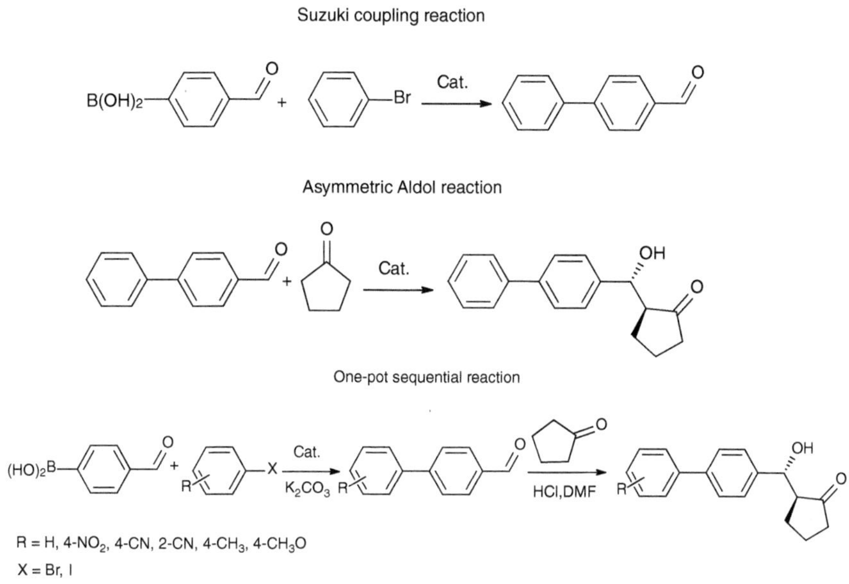

Figure 6.15 Suzuki coupling (top), asymmetric Aldol (middle), and one-pot Sequential reactions (bottom).

yield). For the other parameters, the suitable base in the reaction was K_2CO_3. With 20 mg catalyst, the best yield was produced (10 mg: 62%; 30 mg: 80%; 40 mg: 55%; 20 mg/80 °C: 81; 20 mg/100 °C: 82). According to the results obtained in different temperatures and times, 80 °C and eight hours were better than the others. Also, they investigated different parts of the final catalyst in different cases, such as MOF-808, Pd NPs, Pd@MOF-808, and CP@Pd@MOF-808. CP@Pd@MOF-808 demonstrated excellent ability in comparison to the others in this kind of catalytic reaction. Another reaction examined was the Aldol reaction in which CP within MOF had played an important role. In the Aldol reaction, first, the reaction between biphenyl-4-carbaldehyde and cyclopentanone was studied as a model reaction. Herein, the optimum conditions were also investigated. The useful reaction medium was the mixed-solvent system (dimethyl formamide/water: 2/1) with 43% yield and 90% ee(anti) than the product. Based on the obtained findings, the reaction time of three days, catalyst dosage of 20 mg, and temperature of 10 °C were selected as the optimal conditions. Finally, this chiral solid demonstrated considerable catalytic results in one-pot Suzuki coupling/asymmetric Aldol reactions: coupling yield 89%, the Aldol yield 45%, and enantiomeric excess(anti) 94%. As a result, this work can provide new opportunities for the design of CP-mediated catalysts with different properties.

Ultimately, **Tandem oxidation/asymmetric aldol reactions** are one of the complex reactions that will be discussed as the last example. In 2023, Cheng et al. presented a precise design to produce a chiral multifunctional catalyst for a multi-step tandem reaction. Their design was based on using NPs and chiral MOFs [105]. They synthesized two metal alloy NPs@chiral MOFs as chiral bifunctional heterogeneous catalysts. The used parent MOF was NH_2-UiO-66 and the chiral agent was proline, which was added into the framework through the coordination of Zr-nodes. Also, tiny AuPd alloy NPs with different ratios were used and encapsulated. Yields up to 93% in the oxidation reaction and enantioselectivities up to 91% in tandem selective oxidation of aromatic alcohol/asymmetric Aldol reactions were excellent results of this work. The significant factor in these results relates to the synergistic effect of Au and Pd because their investigations confirmed that the catalytic reaction occurred inside the pores of the catalysts. Apart from the results obtained, recyclability at least three times without activity loss or structural damage was another advantage of these new chiral MOF-based catalysts.

6.5 Conclusion

Although remarkable progress has been made in asymmetric catalysis, designing enantioselective catalysts and their synthesis remain important challenges. Therefore, the development of various kinds of chiral homogeneous and heterogeneous catalysts largely relies on their design and performance. Among chiral catalysts, chiral MOFs are suitable candidates for asymmetric catalytic transformations due to their intriguing properties such as regular crystalline structure, porosity, high surface area, catalytic sites with high accessibility, and structural tunability.

In this chapter, we have reported the synthesis strategies of CMOFs (spontaneous resolution, direct and indirect methods) and a brief history of CMOF-based catalysts. Then, the kinds of catalytic active sites that can be found in chiral MOF-based catalysts have been introduced (such as metal node- and organometallic-based catalysts, organocatalysis, and encapsulated catalytic active species in the pores of MOFs). Finally, examples of enantioselective catalysis using CMOF-based catalysts have been reported, along with the classification of simple and complex asymmetric catalytic reactions. The examples illustrate the importance and advancements of CMOF catalysts. Findings show that CMOF catalysts as chiral effective catalysts can be utilized in larger-scale asymmetric transformations because the obtained optically pure products are so essential in different fields, especially in the drug industry. It is anticipated that this chapter can help active researchers further understand the relationship between the structure of chiral MOF-based catalysts and chirality induction with high enantioselectivity in asymmetric catalysis using CMOFs.

References

1 Davankov, V. (2006). Chirality as an inherent general property of matter. *Chirality* 18 (7): 459–461.
2 Kim, J.H. and Scialli, A.R. (2011). Thalidomide: the tragedy of birth defects and the effective treatment of disease. *Toxicol. Sci.* 122 (1): 1–6.

3 Morris, R.E. and Bu, X. (2010). Induction of chiral porous solids containing only achiral building blocks. *Nat. Chem.* 2 (5): 353–361.

4 Li, C. (2004). Chiral synthesis on catalysts immobilized in microporous and mesoporous materials. *Catal. Rev.* 46 (3, 4): 419–492.

5 Lu, T., Yan, W., and Xu, R. (2019). Chiral zeolite beta: structure, synthesis, and application. *Inorg. Chem. Front.* 6 (8): 1938–1951.

6 Fuchs, I., Fechler, N., Antonietti, M., and Mastai, Y. (2016). Enantioselective nanoporous carbon based on chiral ionic liquids. *Angew. Chem. Int. Ed.* 55 (1): 408–412.

7 Lacasta, S., Sebastián, V., Casado, C. et al. (2011). Chiral imprinting with amino acids of ordered mesoporous silica exhibiting enantioselectivity after calcination. *Chem. Mater.* 23 (5): 1280–1287.

8 Sharifzadeh, Z., Berijani, K., and Morsali, A. (2021). Chiral metal–organic frameworks based on asymmetric synthetic strategies and applications. *Coord. Chem. Rev.* 445: 214083.

9 Berijani, K., Chang, L.M., and Gu, Z.G. (2023). Chiral templated synthesis of homochiral metal-organic frameworks. *Coord. Chem. Rev.* 474: 214852.

10 Ezuhara, T., Endo, K., and Aoyama, Y. (1999). Helical coordination polymers from achiral components in crystals. Homochiral crystallization, homochiral helix winding in the solid state, and chirality control by seeding. *J. Am. Chem. Soc.* 121 (14): 3279–3283.

11 Seo, J.S., Whang, D., Lee, H. et al. (2000). A homochiral metal–organic porous material for enantioselective separation and catalysis. *Nature* 404 (6781): 982–986.

12 Kepert, C.J., Prior, T.J., and Rosseinsky, M.J. (2000). A versatile family of interconvertible microporous chiral molecular frameworks: the first example of ligand control of network chirality. *J. Am. Chem. Soc.* 122 (21): 5158–5168.

13 Dybtsev, D.N., Nuzhdin, A.L., Chun, H. et al. (2006). A homochiral metal–organic material with permanent porosity, enantioselective sorption properties, and catalytic activity. *Angew. Chem. Int. Ed.* 118 (6): 930–934.

14 Xie, S.M. and Yuan, L.M. (2017). Recent progress of chiral stationary phases for separation of enantiomers in gas chromatography. *J. Sep. Sci.* 40 (1): 124–137.

15 Berijani, K. and Morsali, A. (2019). Dual activity of durable chiral hydroxyl-rich MOF for asymmetric catalytic reactions. *J. Catal.* 378: 28–35.

16 Gu, Z.G., Fu, W.Q., Wu, X., and Zhang, J. (2016). Liquid-phase epitaxial growth of a homochiral MOF thin film on poly (L-DOPA) functionalized substrate for improved enantiomer separation. *Chem. Commun.* 52 (4): 772–775.

17 Chen, Y., Su, J., Dong, W. et al. (2022). Cyclodextrin-based metal–organic framework nanoparticles as superior carriers for curcumin: study of encapsulation mechanism, solubility, release kinetics, and antioxidative stability. *Food Chem.* 383: 132605.

18 Razavi, S.A.A., Berijani, K., and Morsali, A. (2020). Hybrid nanomaterials for asymmetric purposes: green enantioselective C–C bond formation by chiralization and multi-functionalization approaches. *Catal. Sci. Technol.* 10 (24): 8240–8253.

19 Xia, Z., Jing, X., He, C. et al. (2017). Coordinative alignment of chiral molecules to control over the chirality transfer in spontaneous resolution and asymmetric catalysis. *Sci. Rep.* 7 (1): 1–11.

20 Han, Q., Qi, B., Ren, W. et al. (2015). Polyoxometalate-based homochiral metal-organic frameworks for tandem asymmetric transformation of cyclic carbonates from olefins. *Nat. Commun.* 6 (1): 1–8.

21 Chen, S.M., Chang, L.M., Yang, X.K. et al. (2019). Liquid-phase epitaxial growth of azapyrene-based chiral metal–organic framework thin films for circularly polarized luminescence. *ACS Appl. Mater. Interfaces* 11 (34): 31421–31426.

22 Jiang, H., Yang, K., Zhao, X. et al. (2020). Highly stable Zr(IV)-based metal–organic frameworks for chiral separation in reversed-phase liquid chromatography. *J. Am. Chem. Soc.* 143 (1): 390–398.

23 Kou, W.T., Yang, C.X., and Yan, X.P. (2018). Post-synthetic modification of metal–organic frameworks for chiral gas chromatography. *J. Mater. Chem. A* 6 (37): 17861–17866.

24 Guo, Z., Cao, R., Wang, X. et al. (2009). A multifunctional 3D ferroelectric and NLO-active porous metal– organic framework. *J. Am. Chem. Soc.* 131 (20): 6894–6895.

25 Verma, A., Tomar, K., and Bharadwaj, P.K. (2017). Chiral cadmium(II) metal–organic framework from an achiral ligand by spontaneous resolution: an efficient heterogeneous catalyst for the strecker reaction of ketones. *Inorg. Chem.* 56 (22): 13629–13633.

26 Gupta, V. and Mandal, S.K. (2019). Design and construction of a chiral Cd(II)-MOF from achiral precursors: synthesis, crystal structure and catalytic activity toward C–C and C–N bond forming reactions. *Inorg. Chem.* 58 (5): 3219–3226.

27 Tao, Y., Yang, C., Fang, H. et al. (2019). Spontaneous resolution and structure transformation of niii metal–organic frameworks from an achiral precursor. *Cryst. Growth Des.* 19 (6): 3358–3364.

28 Ma, M., Jiahuan, C., Liu, H. et al. (2022). A review on chiral metal-organic frameworks: synthesis and asymmetric applications. *Nanoscale.* 14: 13405–13427.

29 Gong, W., Chen, Z., Dong, J. et al. (2022). Chiral metal–organic frameworks. *Chem. Rev.* 122 (9): 9078–9144.

30 Martell, J.D., Zasada, L.B., Forse, A.C. et al. (2017). Enantioselective recognition of ammonium carbamates in a chiral metal–organic framework. *J. Am. Chem. Soc.* 139 (44): 16000–16012.

31 Meng, L., Niu, Z., Liang, C. et al. (2018). Integration of open metal sites and Lewis basic sites for construction of a Cu MOF with a rare chiral Oh-type cage for high performance in methane purification. *Chem. Eur. J.* 24 (50): 13181–13187.

32 Sun, Y.Q., Zhong, J.C., Ding, L., and Chen, Y.P. (2015). Syntheses, structures, photoluminescence and photocatalysis of chiral 3D Cd(II) frameworks from achiral mixed flexible ligands by spontaneous resolution. *Dalton Trans.* 44 (26): 11852–11859.

33 Xu, G.C., Ma, X.M., Zhang, L. et al. (2010). Disorder– order ferroelectric transition in the metal formate framework of [NH_4][Zn (HCOO)3]. *J. Am. Chem. Soc.* 132 (28): 9588–9590.

34 Sun, C.Y., Qin, C., Wang, C.G. et al. (2011). Chiral nanoporous metal-organic frameworks with high porosity as materials for drug delivery. *Adv. Mater.* 23 (47): 5629–5632.

35 Bradshaw, D., Prior, T.J., Cussen, E.J. et al. (2004). Permanent microporosity and enantioselective sorption in a chiral open framework. *J. Am. Chem. Soc.* 126 (19): 6106–6114.

36 Chen, Z.F., Zhang, J., Xiong, R.G., and You, X.Z. (2000). A novel two-dimensional chiral coordination polymer: bis (S-(-)-lactate) zinc (II). *Inorg. Chem. Commun.* 3 (9): 493–496.

37 Xie, Y.R., Wang, X.S., Zhao, H. et al. (2003). Unprecedented homochiral olefin–copper(I) 2D coordination polymer grid based on chiral ammonium salts as building blocks. *Organometallics* 22 (22): 4396–4398.

38 Qu, Z.R., Chen, Z.F., Zhang, J. et al. (2003). The first highly stable homochiral olefin–copper(I) 2D coordination polymer grid based on quinine as a building block. *Organometallics* 22 (14): 2814–2816.

39 Perez Barrio, J., Rebilly, J.N., Carter, B. et al. (2008). Control of porosity geometry in amino acid derived nanoporous materials. *Chem. Eur. J.* 14 (15): 4521–4532.

40 Ma, L., Abney, C., and Lin, W. (2009). Enantioselective catalysis with homochiral metal–organic frameworks. *Chem. Soc. Rev.* 38 (5): 1248–1256.

41 Ngo, H.L. and Lin, W. (2002). Chiral crown ether pillared lamellar lanthanide phosphonates. *J. Am. Chem. Soc.* 124 (48): 14298–14299.

42 Hu, A., Ngo, H.L., and Lin, W. (2003). Chiral porous hybrid solids for practical heterogeneous asymmetric hydrogenation of aromatic ketones. *J. Am. Chem. Soc.* 125 (38): 11490–11491.

43 Wu, C.D., Hu, A., Zhang, L., and Lin, W. (2005). A homochiral porous metal organic framework for highly enantioselective heterogeneous asymmetric catalysis. *J. Am. Chem. Soc.* 127 (25): 8940–8941.

44 Kutzscher, C., Janssen-Müller, D., Notzon, A. et al. (2017). Synthesis of the homochiral metal-organic framework DUT-129 based on a chiral dicarboxylate linker with 6 stereocenters. *CrystEngComm* 19 (18): 2494–2499.

45 Xu, H.S., Ding, S.Y., An, W.K. et al. (2016). Constructing crystalline covalent organic frameworks from chiral building blocks. *J. Am. Chem. Soc.* 138 (36): 11489–11492.

46 Zhang, H., Lou, L.L., Yu, K., and Liu, S. (2021). Advances in chiral metal–organic and covalent organic frameworks for asymmetric catalysis. *Small* 17 (22): 2005686.

47 Yang, J., Wang, J., Hou, B. et al. (2020). Porous hydrogen-bonded organic frameworks (HOFs): from design to potential applications. *Chem. Eng. J.* 399: 125873.

48 Xuan, W., Zhang, M., Liu, Y. et al. (2012). A chiral quadruple-stranded helicate cage for enantioselective recognition and separation. *J. Am. Chem. Soc.* 134 (16): 6904–6907.

49 Liu, Y., Li, Z., Yuan, G. et al. (2016). Chiral Cu (salen)-based metal–organic framework for heterogeneously catalyzed aziridination and amination of olefins. *Inorg. Chem.* 55 (24): 12500–12503.

50 Zhu, C., Yuan, G., Chen, X. et al. (2012). Chiral nanoporous metal–metallosalen frameworks for hydrolytic kinetic resolution of epoxides. *J. Am. Chem. Soc.* 134 (19): 8058–8061.

51 Song, F., Wang, C., Falkowski, J.M. et al. (2010). Isoreticular chiral metal–organic frameworks for asymmetric alkene epoxidation: tuning catalytic activity by controlling framework catenation and varying open channel sizes. *J. Am. Chem. Soc.* 132 (43): 15390–15398.

52 Xia, Q., Li, Z., Tan, C. et al. (2017). Multivariate metal–organic frameworks as multifunctional heterogeneous asymmetric catalysts for sequential reactions. *J. Am. Chem. Soc.* 139 (24): 8259–8266.

53 Hoskins, B.F. and Robson, R. (1990). Design and construction of a new class of scaffolding-like materials comprising infinite polymeric frameworks of 3D-linked molecular rods. A reappraisal of the zinc cyanide and cadmium cyanide structures and the synthesis and structure of the diamond-related frameworks $[N(CH_3)_4][CuIZnII(CN)_4]$ and CuI[4,4′,4″,4‴-tetracyanotetraphenylmethane] BF4. $xC_6H_5NO_2$. *J. Am. Chem. Soc.* 112 (4): 1546–1554.

54 Wang, Z. and Cohen, S.M. (2007). Postsynthetic covalent modification of a neutral metal– organic framework. *J. Am. Chem. Soc.* 129 (41): 12368–12369.

55 Bonnefoy, J., Legrand, A., Quadrelli, E.A. et al. (2015). Enantiopure peptide-functionalized metal–organic frameworks. *J. Am. Chem. Soc.* 137 (29): 9409–9416.

56 Fracaroli, A.M., Siman, P., Nagib, D.A. et al. (2016). Seven post-synthetic covalent reactions in tandem leading to enzyme-like complexity within metal–organic framework crystals. *J. Am. Chem. Soc.* 138 (27): 8352–8355.

57 Wu, P., He, C., Wang, J. et al. (2012). Photoactive chiral metal–organic frameworks for light-driven asymmetric α-alkylation of aldehydes. *J. Am. Chem. Soc.* 134 (36): 14991–14999.

58 Evans, O.R., Ngo, H.L., and Lin, W. (2001). Chiral porous solids based on lamellar lanthanide phosphonates. *J. Am. Chem. Soc.* 123 (42): 10395–10396.

59 Jeong, K.S., Go, Y.B., Shin, S.M. et al. (2011). Asymmetric catalytic reactions by NbO-type chiral metal–organic frameworks. *Chem. Sci.* 2 (5): 877–882.

60 Sawano, T., Ji, P., McIsaac, A.R. et al. (2015). The first chiral diene-based metal–organic frameworks for highly enantioselective carbon–carbon bond formation reactions. *Chem. Sci.* 6 (12): 7163–7168.

61 Gong, W., Chen, X., Jiang, H. et al. (2019). Highly stable Zr(IV)-based metal–organic frameworks with chiral phosphoric acids for catalytic asymmetric tandem reactions. *J. Am. Chem. Soc.* 141 (18): 7498–7508.

62 Li, J., Ren, Y., Yue, C. et al. (2018). Highly stable chiral zirconium–metallosalen frameworks for CO_2 conversion and asymmetric C–H azidation. *ACS Appl. Mater. Interfaces* 10 (42): 36047–36057.

63 Dang, D., Wu, P., He, C. et al. (2010). Homochiral metal– organic frameworks for heterogeneous asymmetric catalysis. *J. Am. Chem. Soc.* 132 (41): 14321–14323.

64 Sawano, T., Thacker, N.C., Lin, Z. et al. (2015). Robust, chiral, and porous BINAP-based metal–organic frameworks for highly enantioselective cyclization reactions. *J. Am. Chem. Soc.* 137 (38): 12241–12248.

65 Regati, S., He, Y., Thimmaiah, M. et al. (2013). Enantioselective ring-opening of meso-epoxides by aromatic amines catalyzed by a homochiral metal–organic framework. *Chem. Commun.* 49 (84): 9836–9838.

66 Kökçam-Demir, Ü., Goldman, A., Esrafili, L. et al. (2020). Coordinatively unsaturated metal sites (open metal sites) in metal–organic frameworks: design and applications. *Chem. Soc. Rev.* 49 (9): 2751–2798.

67 Katsuki, T. (2002). Chiral metallosalen complexes: Structures and catalyst tuning for asymmetric epoxidation and cyclopropanation. *Adv. Synth. Catal.* 344 (2): 131–147.

68 Chen, Y., Yekta, S., and Yudin, A.K. (2003). Modified BINOL ligands in asymmetric catalysis. *Chem. Rev.* 103 (8): 3155–3212.

69 Hayashi, T., Ueyama, K., Tokunaga, N., and Yoshida, K. (2003). A chiral chelating diene as a new type of chiral ligand for transition metal catalysts: Its preparation and use for the rhodium-catalyzed asymmetric 1,4-addition. *J. Am. Chem. Soc.* 125 (38): 11508–11509.

70 Noyori, R. and Takaya, H. (1990). BINAP: an efficient chiral element for asymmetric catalysis. *Acc. Chem. Res.* 23 (10): 345–350.

71 Falkowski, J.M., Sawano, T., Zhang, T. et al. (2014). Privileged phosphine-based metal–organic frameworks for broad-scope asymmetric catalysis. *J. Am. Chem. Soc.* 136 (14): 5213–5216.

72 Han, Q., He, C., Zhao, M. et al. (2013). Engineering chiral polyoxometalate hybrid metal–organic frameworks for asymmetric dihydroxylation of olefins. *J. Am. Chem. Soc.* 135 (28): 10186–10189.

73 Tanaka, K., Oda, S., and Shiro, M. (2008). A novel chiral porous metal–organic framework: asymmetric ring opening reaction of epoxide with amine in the chiral open space. *Chem. Commun.* 7: 820–822.

74 Cheng, L., Wu, R.J., Li, Y.M. et al. (2022). Single-chain polymer nanoparticles-encapsulated chiral bifunctional metal-organic frameworks for asymmetric sequential reactions. *Inorg. Chem. Commun.* 141: 109577.

75 Li, Y.M., Cao, L., Ren, H. et al. (2023). Chiral polymer-mediated Pd@ MOF-808 for efficient sequential asymmetric reaction. *Catal. Lett.* 153 (4): 1193–1204.

76 Silva, A.R., Alexandre, J.Y., Souza, J.E. et al. (2022). The chemistry and applications of metal–organic frameworks (MOFs) as industrial enzyme immobilization systems. *Molecules* 27 (14): 4529.

77 Zheng, L., Wang, F., Jiang, C. et al. (2022). Recent progress in the construction and applications of metal-organic frameworks and covalent-organic frameworks-based nanozymes. *Coord. Chem. Rev.* 471: 214760.

78 Mookherjee, B.D. and Wilson, R.A. (1996). *Encyclopedia of Chemical Technology Kirk-Orthmer*, 4e, vol. 17, 603. New York: Wiley.

79 Cho, S.H., Ma, B., Nguyen, S.T. et al. (2006). A metal–organic framework material that functions as an enantioselective catalyst for olefin epoxidation. *Chem. Commun.* 24: 2563–2565.

80 Berijani, K., Morsali, A., and Hupp, J.T. (2019). An effective strategy for creating asymmetric MOFs for chirality induction: a chiral Zr-based MOF for enantioselective epoxidation. *Catal. Sci. Technol.* 9 (13): 3388–3397.

81 Berijani, K. and Hosseini-Monfared, H. (2018). Collaborative effect of Mn-porphyrin and mesoporous SBA-15 in the enantioselective epoxidation of olefins with oxygen. *Inorg. Chim. Acta* 471: 113–120.

82 Berijani, K. and Morsali, A. (2020). Construction of an asymmetric porphyrinic zirconium metal–organic framework through ionic postchiral modification. *Inorg. Chem.* 60 (1): 206–218.

83 Berijani, K. and Hosseini-Monfared, H. (2017). Aerobic enantioselective epoxidation of olefins mediated by an easy-to-prepare recyclable manganese-porphyrin. *Mol. Catal.* 433: 136–144.

84 Zhang, Y., Chen, S., Al-Enizi, A.M. et al. (2023). Chiral frustrated Lewis pair@ metal-organic framework as a new platform for heterogeneous asymmetric hydrogenation. *Angew. Chem. Int. Ed.* 135 (2): e202213399.

85 Chen, X., Qiao, Z., Hou, B. et al. (2021). Chiral metal-organic frameworks with tunable catalytic selectivity in asymmetric transfer hydrogenation reactions. *Nano Res.* 14 (2): 466–472.

86 Jiang, H., Zhang, W., Kang, X. et al. (2020). Topology-based functionalization of robust chiral Zr-based metal–organic frameworks for catalytic enantioselective hydrogenation. *J. Am. Chem. Soc.* 142 (21): 9642–9652.

87 Gong, W., Chen, X., Zhang, W. et al. (2022). Leveraging chiral Zr(IV)-based metal–organic frameworks to elucidate catalytically active Rh species in asymmetric hydrogenation reactions. *J. Am. Chem. Soc.* 144 (7): 3117–3126.

88 Du, X., Li, Z., Liu, Y. et al. (2015). Chiral porous metal–organic frameworks containing μ-oxo-bis [Ti (salan)] units for asymmetric cyanation of aldehydes. *Dalton Trans.* 44 (29): 12999–13002.

89 Li, J., Ren, Y., Qi, C., and Jiang, H. (2017). The first porphyrin–salen based chiral metal–organic framework for asymmetric cyanosilylation of aldehydes. *Chem. Commun.* 53 (58): 8223–8226.

90 Jin, F.Z., Zhao, C.C., Ma, H.C. et al. (2019). Homochiral BINAPDA-Zr-MOF for heterogeneous asymmetric cyanosilylation of aldehydes. *Inorg. Chem.* 58 (14): 9253–9259.

91 Zhu, C., Xia, Q., Chen, X. et al. (2016). Chiral metal–organic framework as a platform for cooperative catalysis in asymmetric cyanosilylation of aldehydes. *ACS Catal.* 6 (11): 7590–7596.

92 Wang, J., Liu, X., and Feng, X. (2011). Asymmetric strecker reactions. *Chem. Rev.* 111 (11): 6947–6983.

93 Friedel, C. and Crafts, J.M. (1877). Sur une methode generale nouvelle de synthbse d'hydrocarbures, d'acetones, etc. *C. R. Acad. Sci.* 84: 1392.

94 Zheng, M., Liu, Y., Wang, C. et al. (2012). Cavity-induced enantioselectivity reversal in a chiral metal–organic framework Brønsted acid catalyst. *Chem. Sci.* 3 (8): 2623–2627.

95 Jia, Y.X., Zhu, S.F., Yang, Y., and Zhou, Q.L. (2006). Asymmetric Friedel–Crafts alkylations of indoles with nitroalkenes catalyzed by Zn(ii)– bisoxazoline complexes. *J. Org. Chem.* 71 (1): 75–80.

96 Tanaka, K., Sakuragi, K., Ozaki, H., and Takada, Y. (2018). Highly enantioselective Friedel–Crafts alkylation of N,N-dialkylanilines with trans-β-nitrostyrene catalyzed by a homochiral metal–organic framework. *Chem. Commun.* 54 (49): 6328–6331.

97 Bezuidenhout, C.X., Smith, V.J., Bhatt, P.M. et al. (2015). Extreme carbon dioxide sorption hysteresis in open-channel rigid metal–organic frameworks. *Angew. Chem. Int . Ed.* 127 (7): 2107–2111.

98 Liang, L., Liu, C., Jiang, F. et al. (2017). Carbon dioxide capture and conversion by an acid-base resistant metal-organic framework. *Nat. Commun.* 8 (1): 1–10.

99 Chen, H., Yang, W., Wu, W., and Jiang, H. (2014). Palladium-catalyzed regioselective azidation of allylic C–H bonds under atmospheric pressure of dioxygen. *Org. Biomol. Chem.* 12 (21): 3340–3343.

100 Banerjee, M., Das, S., Yoon, M. et al. (2009). Postsynthetic modification switches an achiral framework to catalytically active homochiral metal–organic porous materials. *J. Am. Chem. Soc.* 131 (22): 7524–7525.

101 Gedrich, K., Heitbaum, M., Notzon, A. et al. (2011). A family of chiral metal–organic frameworks. *Chem. Eur. J.* 17 (7): 2099–2106.

102 Fan, Y., Ren, Y., Li, J. et al. (2018). Enhanced activity and enantioselectivity of henry reaction by the postsynthetic reduction modification for a chiral Cu (salen)-based metal–organic framework. *Inorg. Chem.* 57 (19): 11986–11994.

103 Huang, Y.B., Liang, J., Wang, X.S., and Cao, R. (2017). Multifunctional metal–organic framework catalysts: synergistic catalysis and tandem reactions. *Chem. Soc. Rev.* 46 (1): 126–157.

104 Li, Z., Liu, Y., Kang, X., and Cui, Y. (2018). Chiral metal–organic framework decorated with TEMPO radicals for sequential oxidation/asymmetric cyanation catalysis. *Inorg. Chem.* 57 (16): 9786–9789.

105 Cheng, L., Guo, Q., Zhao, K. et al. (2023). AuPd alloys and chiral proline dual-functionalized NH_2-UiO-66 catalysts for tandem oxidation/asymmetric aldol reactions. *Catal. Lett.* 153 (4): 1024–1035.

7

MOF-Supported Metal Nanoparticles for Catalytic Applications

Danyu Guo, Liyu Chen, and Yingwei Li

South China University of Technology, School of Chemistry and Chemical Engineering, Wushan load, Tianhe District, Guangzhou 510640, China

7.1 Introduction

Metal nanoparticles (NPs) are an important category of nanomaterials with interesting properties for a wide range of applications [1–5]. However, metal NPs possess high surface free energy and thus exhibit low thermodynamic stability, leading to easy aggregation in practical applications. The immobilization of metal NPs on porous supports can mitigate the challenge of metal aggregation [5–8]. Metal–organic frameworks (MOFs) are porous materials that connect metal ions or clusters with multidentate organic ligands [9]. Using MOFs as a stabilizing host material can provide a limiting space that restricts particle growth and impedes agglomeration. Loading metal NPs into porous matrices of MOFs can obtain properties that are unattainable with individual materials. Further tuning the size and shape of the MNPs in the pores of the framework can enhance the catalytic activity [10].

A variety of MNP@MOF composites have been developed and applied in catalytic reactions. In MNP@MOF composites, metal NPs are generally used as the active centers for catalytic conversions. MOFs play the role of stabilizing the MNPs in the pores and adjusting the electronic environment activity of MNPs to enhance the catalytic activity [11]. Therefore, MNP@MOF composites show synergistic effect and exhibit enhanced activity than their single counterparts or physical mixtures.

In this chapter, we provide an overview of MNP@MOF composites for various catalytic applications. The synergistic effect between MOF and metal NPs in MNP@MOF composites is briefly reviewed and discussed. Next, the applications of MNP@MOF in electrocatalysis, photocatalysis, and thermocatalysis are presented. In addition, the current problems and future challenges in this field are proposed. We hope that this chapter will provide valuable information for readers to gain a deeper understanding of the design of MNP@MOF for catalytic applications.

Catalysis in Confined Frameworks: Synthesis, Characterization, and Applications, First Edition.
Edited by Hermenegildo Garcia and Amarajothi Dhakshinamoorthy.

7.2 Synergistic Catalysis by MNP@MOF Composites

7.2.1 The Inorganic Nodes of MOFs Cooperating with Metal NPs

The inorganic nodes of MOFs can cooperate with metal NPs for catalysis. Some MOFs possess abundant open metal sites (i.e. Lewis acid sites) after removal of coordinative solvent molecules. The Lewis acid sites of MOF can work synergistically with metal NPs to show interesting functions [12]. An et al. [13] proposed that the metal-oxo clusters in MOFs can stabilize Cu NPs through the confinement effect by the pores and the strong metal-support interactions, which acted as efficient catalysts for carbon dioxide (CO_2) hydrogenation to methanol. Tsumori et al. [14] fabricated a "quasi-MOF" through a controlled deligandation process to greatly expose inorganic nodes to guest metal NPs, resulting in a strong interaction and synergistic effect between the immobilized metal NPs and the inorganic nodes. The resulting Au/MIL-101(573) composites showed dramatically enhanced catalytic activity for CO oxidation at temperatures even as low as 193 K, maintaining 100% CO conversion up to 4000 minutes.

7.2.2 The Organic Linkers of MOFs Cooperating with Metal NPs

Metal NPs can also cooperate with organic linkers of MOFs. The catalytic activity and selectivity of heterogeneous catalysts depend largely on the chemical properties of the catalyst surface. The interactions of functional ligands with metal NPs can modulate the electronic properties at the atomic level, resulting in improvement in the catalytic activity and selectivity of MNP@MOF composites. For example, Yaghi and coworkers [15] demonstrated that different chemical environments of the catalysts led to different pathways in the isomerization reaction of methylcyclopentane. The modification of organic ligands with different functional groups (e.g. $-SO_3H$ and $-NH_3^+$) can make the MOFs strongly or weakly acidic. Pt@UiO-66 with single functional group and Pt@UiO-66 with bifunctional groups both showed significantly improved performance compared with the unmodified Pt@UiO-66. Pt@UiO-66 showed different catalytic selectivities when changing the functional groups.

The synergistic interaction of different functional groups with metal NPs can affect the electron density of metal NPs and thus the adsorption strength of reactants, resulting in different selectivity. Huang's group demonstrated that Pd NPs were stabilized in an atomically tunable chemical environment by UiO-66-type structures to give Pd/MOFs (denoted as Pd@UiO-66-X, X = H, NH_2, and OMe) for the aerobic reaction between benzaldehyde and ethylene glycol [16]. For Pd@UiO-66-NH_2, the selectivity to benzaldehyde ethylene acetal reached 94%, while Pd@UiO-66 and Pd@UiO-66-OMe exhibited high selectivity to the ester 2-hydroxyethyl benzoate (90% and 97%, respectively). By changing the functional groups on the organic linkers in UiO-66-X from $-NH_2$ to –H/–OMe, the selectivity of the products can be completely changed from acetal to ester.

7.2.3 The Nanostructures of MOFs Cooperating with Metal NPs

Based on the highly flexible and tunable pore structures of MOFs, MOFs can show a molecular sieve-like effect for selective transformation. The synthesis of MOFs with suitable pore size can selectively allow substrates smaller than the MOF pores to diffuse into the pores to reach metal NPs (active sites), thus achieving selective catalysis [17, 18]. Lu et al. [19] synthesized a series of core–shell MNP@ZIF-8 nanostructures. Combining the catalytic properties of Pt NPs and the molecular sieving ability of ZIF-8, the hydrogenation performance of nanocomposites was investigated for *n*-hexene and *cis*-cyclooctene. The reaction results showed that the conversion of *n*-hexene (13.3%) was higher than *cis*-cyclooctene (1.7%), mainly due to the molecular sieving effect of ZIF-8. *cis*-cyclooctene with a size larger than the pore size of ZIF-8 could not reach the active site in the pores, thus showing lower activity. As another example, MNP@MOF composites can be used as catalysts for selective hydrogenation of cinnamaldehyde. Pt@UiO-66-NH_2 with Pt NPs encapsulated in the pores of UiO-66-NH_2 exhibited both high conversion and selectivity of cinnamaldehyde to cinnamyl alcohol. However, Pt/UiO-66-NH_2 with Pt NPs loaded onto the external surface of UiO-66-NH_2 show much lower selectivity to cinnamyl alcohol. Due to the steric hindrance of UiO-66-NH_2, the C=C bond in the middle of cinnamaldehyde was hardly accessible to the surface of Pt compared to the C=O bond at the end of the molecule [20].

7.3 Electrocatalysis Applications

7.3.1 Hydrogen Evolution Reaction

One of the solutions to the growing environmental problem is exploring clean and sustainable energy sources. Hydrogen (H_2) is being intensively studied as a new energy carrier due to its high energy density and renewable property [21–25]. Electrochemical water splitting is an environmentally friendly way to produce H_2. In order to improve the energy conversion efficiency of water splitting, efficient electrocatalysts are needed to facilitate H_2 evolution reaction (HER) [26, 27]. Porous MOFs loaded with noble metal NPs have been studied as catalysts for HER [28–30].

The morphology of MOFs plays an important role in the catalysis. In this regard, 2D MOF nanosheets (NSs) are more advantageous than bulk MOFs as carriers due to their good accessibility of active sites and fast mass/charge transfer [31]. Rui et al. [32] obtained Ni-MOF@Pt NSs by *in situ* deposition of Pt NPs onto 2D MOF NSs, which were directly used for HER. Ni-MOF@Pt NSs showed a low overpotential of 43 mV at 10 mA cm^{-2}, while Ni-MOF-Pt (the physical mixture of Pt NPs and 2D Ni-MOF NSs) showed an overpotential of 83 mV under the same conditions. Rather than growing Pt NPs randomly on the surface of MOF NSs, Wang et al. [33] preferentially deposited Pt NPs on the edges of Ni MOF NSs to obtain Pt/MOF-O (Figure 7.1a–d), by using ethylene glycol and diethylene glycol as weak reducing agents. For comparison, Pt NPs were randomly deposited on the Ni MOF NSs to obtain Pt/MOF-C. The overpotentials of Pt/MOF-O and Pt/MOF-C in alkaline

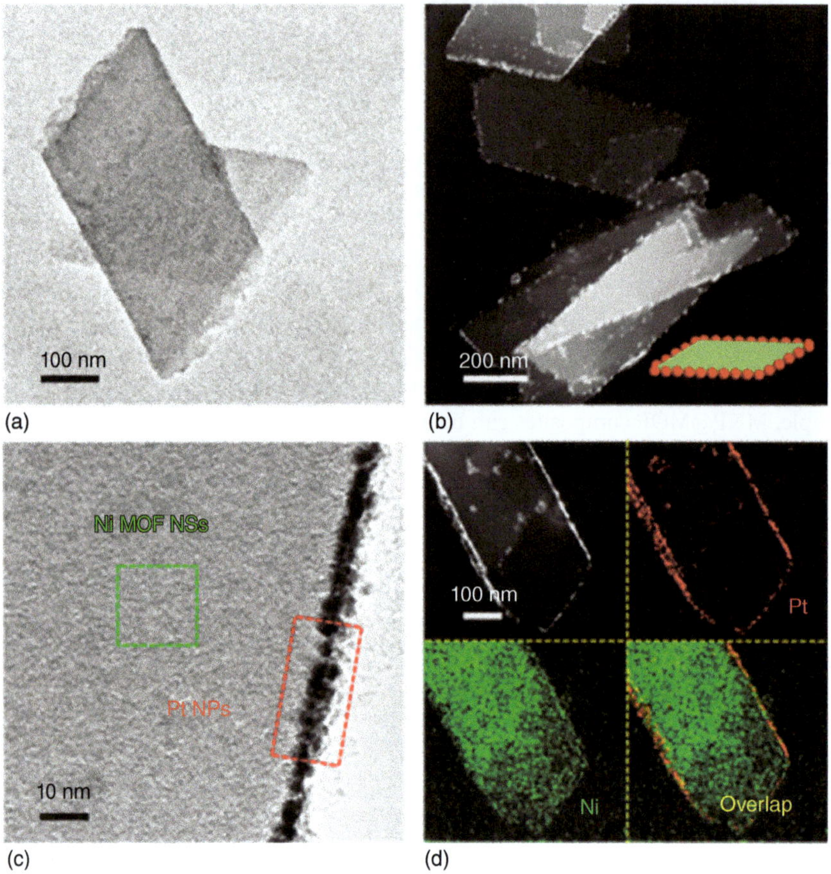

Figure 7.1 Morphology and structure characterizations of Pt/MOF-O: (a) TEM image of the Ni MOF NSs; (b) HAADF-STEM image; (c) HRTEM images; (d) EDS-elemental mappings. Source: Reproduced with permission from Wang et al. [33], © 2021/American Chemical Society.

solution at the current density of $10\,mA\,cm^{-2}$ were 66 and 81 mV, respectively. Detailed characterizations revealed that the significant increase in HER activity may be due to the fact that the O atom can optimize the electronic structure of Pt by modulation of the coordination environment.

The modification of the MOF ligand in MNP@MOF composites can further increase the activity. For example, the incorporation of metal phthalocyanines (MPcs) into porous MOFs can bring new functionalities in the composites [34]. Monama et al. [35] first prepared CuPc/MOF composites by employing a simple impregnation method, followed by chemical electroplating of Pd to synthesize Pd@CuPc/MOF. The Tafel slope of $mV\,dec^{-1}$ and could be stable in $0.300\,mol\,l^{-1}$ H_2SO_4 at high exchange current density ($8.9\,A\,m^{-2}$), indicating that the catalyst exhibited good HER activity and stability. The presence of CuPc enhanced the electron transfer and increased the electrochemically accessible surface area in the composite, which facilitated the charge transfer at the catalyst/electrolyte interface and ultimately led to the enhancement of HER catalytic activity.

The above examples have shown great promise of MOF-supported noble metals for HER. However, the use of MOF-supported transition metals for HER remains to be explored. Transition metals have shown good catalytic activity for HER [36]. Therefore, the preparation of transition metal-based MNP@MOF composites for HER is highly expected.

7.3.2 Oxygen Evolution Reaction

The oxygen evolution reaction (OER) is a key electrochemical reaction in water splitting, fuel cells, and rechargeable metal–air cells. OER is a complex four-electron transfer process with multiple adsorption intermediates, resulting in slow reaction kinetics [37–41]. MOF-based materials have attracted much attention as electrocatalysts for promoting OER. However, the poor electrical conductivity and unsatisfactory activity of pristine MOFs have greatly limited their electrocatalytic applications [39, 42–45]. To improve the electrocatalytic efficiency, MOFs can be integrated with metal NPs to form MNP@MOF composites [46].

The synergistic interaction between Pt and metal nodes in the composite improves the OER catalytic performance by embedding Pt NPs in ultrathin 2D Ni-MOF NSs (NiNSMOFs) [47]. The Pt@NiNSMOFs loaded with 3.2 wt% Pt showed superior OER performance with an overpotential 55 mV at a current density of 10 mA cm^{-2}, much lower than that of commercial RuO_2.

MOFs and Pt quantum dots (QDs) can form core–shell structures. The catalyst showed good catalytic performance even with small amount of Pt [48]. The obtained Pt QDs@Fe-MOF/NF required an overpotential of only 144 mV to reach a current density of 10 mA cm^{-2}, while 346 and 172 mV were required for NF and Fe-MOF/NF, respectively. Pt QDs@Fe-MOF/NF showed long-term stability at constant current density of 20 mA cm^{-2} over 100 hours, which may be due to the stabilization effect offered by the Fe ions of the MOFs.

MOF-supported transition metal catalysts have also been developed for OER. Ma et al. [49] synthesized bimetallic MOF-encapsulated Cu–Ni NPs (Ni–Cu@Cu–Ni-MOF) by *in situ* etching using Cu–Ni spherical nanostructures as precursors. The current density of Ni–Cu@Cu–Ni-MOF was increased by 1.48, 1.60, and 71.18 times at an overpotential of 624 mV compared with the Cu–Ni seeds, Ni-MOF-74 and Cu-MOF-74, respectively. After 30 000 seconds of stability testing, Ni–Cu@Cu–Ni-MOF maintained 87% of the current density of 6.6 mA cm^{-2}, almost double that of the Cu–Ni seeds (3.4 mA cm^{-2}). The encapsulated bimetallic NPs promoted the electrical conductivity and stability of the MOFs and provide depletable metal active sites, thus enhancing the OER activity.

The development of MNP@MOF electrocatalysts for OER is still in progress. In order to realize the practical application of these electrocatalysts, further reduction of overpotential and improvement of stability are needed. Based on the consideration of the high price and scarcity of precious metals, effective MNP@MOF catalysts based on transition metals for OER are expected in the future.

7.3.3 Oxygen Reduction Reaction

In order to solve the problems of energy shortage and environmental pollution caused by traditional fossil fuels, proton exchange membrane fuel cells with high specific power density, high energy conversion efficiency, and environmentally friendly have attracted increasing attention [50–54]. The oxygen reduction reaction (ORR) occurs mainly at the cathode of a fuel cell. The reaction mechanism of ORR is relatively complex and can be divided into a four-electron pathway and a two-electron pathway [55–58]. The four-electron reaction pathway is the conversion of O_2 to water (acidic) and hydroxide (basic) with high output voltage and energy conversion efficiency, but the overpotential of the reaction is also relatively high [31]. Pt/C is the most effective commercial electrocatalyst for ORR. [59] MOFs can be used as carriers for Pt NPs to develop new Pt-based electrocatalysts.

2D Ni-based MOFs can increase electron and mass transfer and enhance the exposure of unsaturated coordination metal nodes on the surface. Pt NPs were embedded on a 2D Ni-based MOF NSs for the synthesis of Pt@NiNSMOFs [47]. Pt@NiNSMOFs showed good OER catalytic performance due to the synergistic interaction between the MOF and metal nodes. The ORR cyclic voltammetry curve results showed that the Pt@NiNSMOFs samples with different Pt loadings exhibited significantly better performance than the MOF NSs without Pt decoration. The catalytic activity was enhanced as the Pt loading increased from 0.7% to 3.2% and the onset potential shifted from 0.902 to 1.024 V vs. the reversible hydrogen electrode (RHE). The Pt@ NiNSMOFs (3.2 wt% of Pt) showed a positive half-wave potential shift of 110 mV compared to commercial Pt@C.

To date, various efficient MNP@MOF catalysts have been developed for ORR. However, most of these catalysts require the use of Pt, which would increase the cost of catalysts [59]. In addition, durability is also important factor for the large-scale application of MNP@MOF catalysts. The development of MNP@MOF catalysts with good electron conductivity and high stability is highly important for their industrial applications [60]. Moreover, advanced analytical techniques are also required to detailed explore the catalysis mechanism of MNP@MOF to guild the design of efficient catalysts.

7.3.4 CO_2 Reduction Reaction

The overuse of fossil fuels generates large amounts of CO_2. Electrochemical CO_2 reduction reaction (CO_2RR), which uses renewable electrical energy resources to drive the conversion of CO_2 into useful chemicals and fuels, can realize carbon recycling under mild conditions [61–66]. However, the large kinetic barriers and strong competitive HER hinder the efficiency of CO_2RR. Efficient electrocatalysts are required to improve the catalytic efficiency of the reaction [67, 68]. The high surface-area-to-volume ratio of metal NPs provides a large number of active sites. However, small metal NPs with high surface energy might agglomerate during the catalytic process. The high porosity as well as the tunable chemical composition and structure make MOFs a promising carrier for the stabilization metal NPs [69–72].

Different products (e.g. CO, HCOOH, HCHO, CH_3OH, CH_4, and C_2H_4) can be obtained depending on the active center of the catalyst.

7.3.4.1 CO

Electrochemical CO_2 reduction to CO can deliver high current density at mild overpotential with a high Faraday efficiency (FE). CO can be used as a chemical raw material to produce fuel, and the CO_2 produced after combustion can be recycled, thus achieving a carbon loop. Therefore, the selective conversion of CO_2 to CO is a promising path [73–76]. Guntern et al. [77] synthesized Ag@Al-PMOF composites and investigated their CO_2RR catalytic performance. The intimate contact at the interface upon removal of the native ligand using acetone allows electron transfer from the Al-PMOF to the Ag nanoclusters (NCs). Electron-rich active Ag NCs could facilitate the activation of CO_2 molecules by enhancing electron movement from the active sites to the antibonding orbitals of CO_2, leading to an increased selectivity for CO_2RR relative to HER. Over the entire potential range, Ag@AlPMOF exhibited increased selectivity for CO. At −1.1 V vs. RHE, the highest measured FE of CO for Ag@Al-PMOF was 55.8%, with a 2.2-fold enhancement compared to bare Ag NCs (Figure 7.2a).

7.3.4.2 HCOOH

Formic acid is a renewable energy carrier that can be easily stored and transported and is in wide demand in various fields [80, 81]. Kung et al. [78] used solvothermal deposition method to prepare Cu NPs encapsulated in hexagonal channels of NU-1000 and used it for CO_2 reduction to obtain formates. The FE of formates reached 31% at an applied potential of −0.82 V vs. RHE (Figure 7.2b). After the reaction at −0.82 V vs. RHE for 30 minutes, the structure of the catalyst was characterized, and it was found that the crystallinity and morphology remained the same before and after the reaction, indicating that the catalyst has excellent stability.

7.3.4.3 C_2H_4

Ethylene (C_2H_4) is an important component of plastics and other chemicals with high market demand. However, the effective C–C coupling during the hydrogenation of CO_2 is difficult to occur. Selective CO_2RR catalysts are required [82, 83]. Xie et al. [79] synthesized AuNN@PCN-222(Cu) by impregnating Au nanoneedles (AuNNs) into zirconium-based PCN-222 MOF for CO_2RR. AuNN@PCN-222(Cu) mainly produced C_2H_4, reaching the highest FE of 52.5% at −1.2 V vs. RHE. In contrast, the FE of C_2H_4 for AuNP@PCN-222(Cu) with Au NPs deposited on the surface and PCN-222(Cu) over the entire potential range is below 25%. The density functional theory (DFT) calculations reveal that the Cu–N interaction was weakened due to both AuNNs insertion and intermediate adsorption, making the N position in the metalloporphyrin to have catalytic activity (Figure 7.2c). Meanwhile, the AuNNs further increase the concentration of CO inside the MOF tunnel through the spatial restriction effect. The increase in CO concentration raises the probability of C–C coupling when the active sites are sufficiently close.

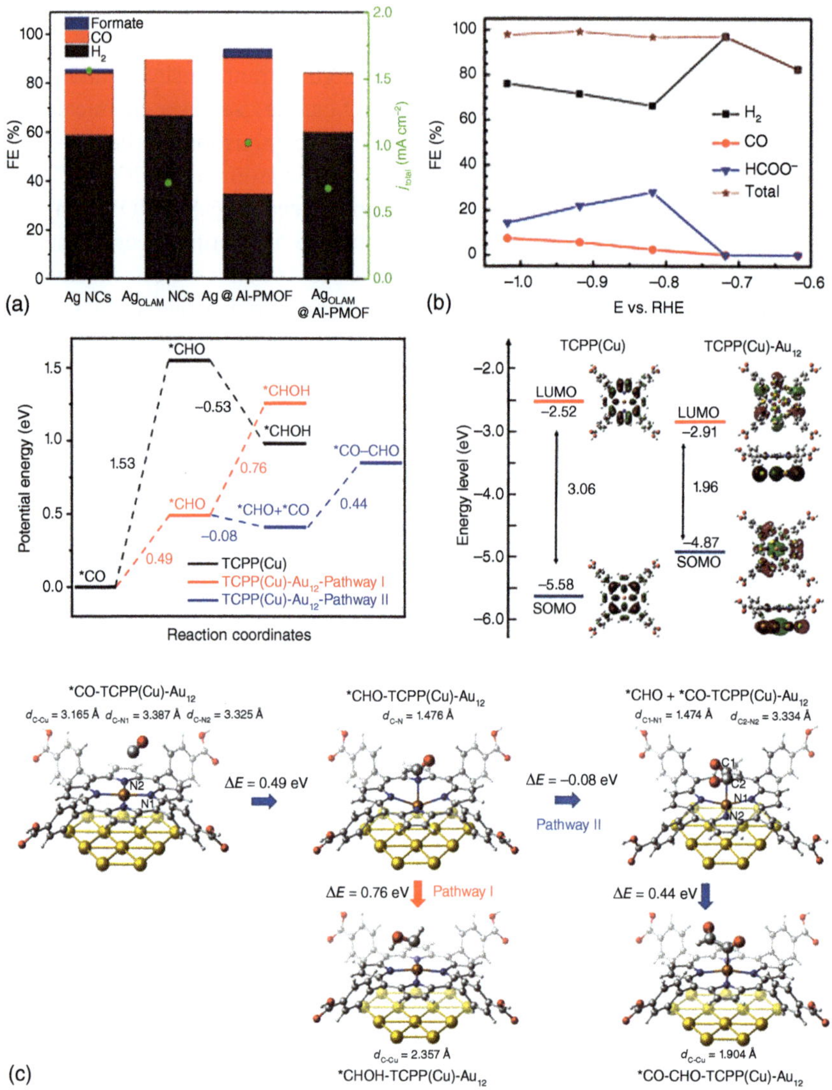

Figure 7.2 (a) FEs and total current densities for Ag NC and Ag@Al-PMOF catalysts, with and without OLAM ligands, measured at @1.1 V vs. RHE. Source: Reproduced from Guntern et al. [77] with permission from Wiley, copyright 2019. (b) Faradaic efficiencies (FE) of generated products based on 30 minutes of electrolysis of electrode-supported Cu-SIM NU-1000 thin films at each of several applied potentials in CO_2-purged electrolyte. Source: Reproduced from Kung et al. [78] with permission from American Chemical Society, copyright 2017. (c) DFT calculations of the configuration and energetics of key intermediates involved in the C1 and C2 paths. Source: Reproduced from Xie et al. [79] with permission from Springer Nature, copyright 2022.

Although considerable progress has been made in CO_2RR, there are still challenges and opportunities for the development of MNP@MOF catalysts in this field. After the introduction of different MNPs into the MOF material, the charge redistribution at the interface can lead to a change in the electronic structure of the active site, achieving an increase in catalytic activity and selectivity. However, the "synergistic effect" between the MOFs and the post-added guest is still an ambiguous concept without sufficient experimental data to make a clear explanation. In general, the preparation of efficient and stable MNP@MOF catalysts and the elucidation of their catalytic mechanisms and active sites through *in situ* characterization and theoretical calculations are of vital importance.

7.3.5 Nitrogen Reduction Reaction

Ammonia is widely used in the synthesis of pharmaceuticals and fertilizers and is also a high energy density carrier of hydrogen energy [84–86]. Currently, ammonia is mainly produced using the traditional Haber–Bosch process for global supply, which requires harsh operating conditions resulting in high energy consumption and CO_2 emissions [87–89]. Nitrogen reduction reaction (NRR) has become a focus of research because it can be proceeded under mild conditions, reducing energy consumption and emissions [90–92]. The progress of electrochemical NRR has been inhibited by the strong competition from HER. MOFs and their composites are very promising NRR electrocatalysts with advantages such as good regulation and N_2 enrichment ability in environmental electrocatalytic NRR [93–96]. The introduction of hydrophobicity to block the entry of water into the catalytic system is an effective strategy for HER inhibition [97].

Among them, ZIFs with superhydrophobic properties can be used to coat Ag for efficient electrocatalytic NRR [98]. The obtained Ag–Au@ZIF electrode showed excellent selectivity for NRR (~90%) with an FE of about 10%. In the case of Pt/Au@ZIF, ZIF not only acted as a hydrophobic layer to inhibit HER but also reduced the d-band of the Pt/Au to weaken the H adsorption on Pt/Au. The abundant electron deficient sites on Pt made the N_2 molecules enriched at the catalyst-ZIF interface to improve the catalyst activity [99] (Figure 7.3a). The change in d-band electronic structure with the addition of ZIF resulted in a 44-fold increase in FE of Pt/Au@ZIF NRR relative to the catalyst without the addition of ZIF. Pt/Au@ZIF possessed good catalytic performance at a wide range of reduction potentials with FE >40% and ammonia yields >130 μg mg_{cat}^{-1} h^{-1}. Experiments and DFT calculations also demonstrated the presence of N_2 adsorption sites through the creation of electron-deficient regions on the Pt/Au-ZIF interacting Pt catalysts.

In addition to the direct use of hydrophobic MOFs, hydrophobic treatment of MOFs is another method. He et al. [100] synthesized a MOF crystal with disulfide trimer units for the encapsulation of well-dispersed ultrafine Au NPs for NRR. The surface modification of the obtained Au@MOF using silicone resulted in a final HT Au@MOF composite with hydrophobic properties (Figure 7.3b). The affinity between the S–S groups in MOF and Au NPs allows the Au NPs with a small size to be efficiently loaded in the MOF matrix. Moreover, the organosilicon layer can

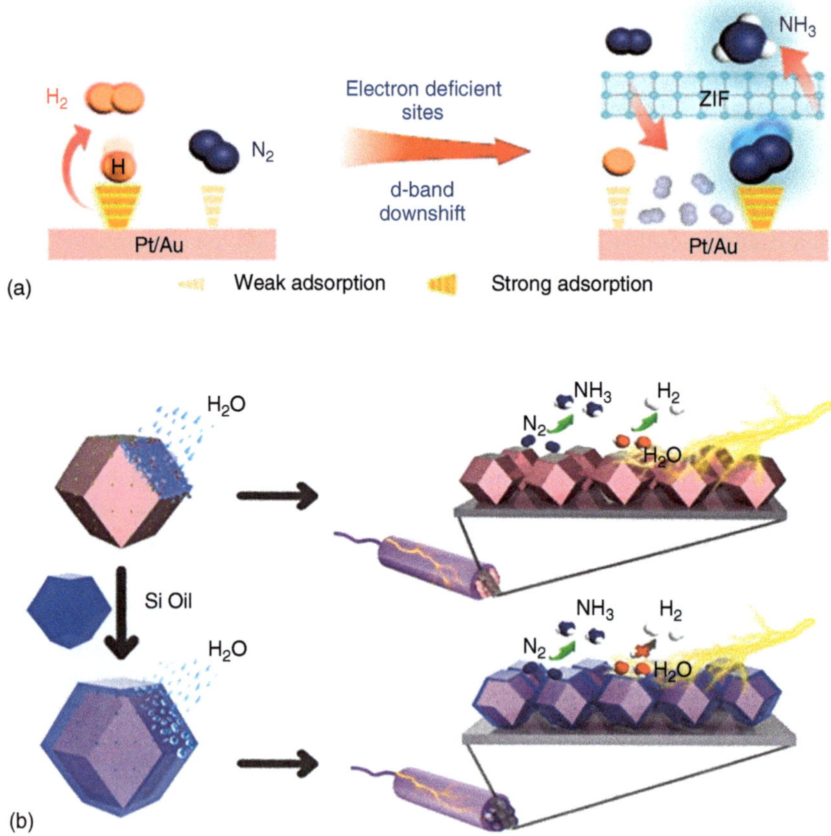

Figure 7.3 (a) The functions of ZIF: lower the d-band of Pt/Au to weaken metal–H interactions while facilitating metal–N_2 interaction. Source: Reproduced from Sim et al. [99] with permission from Wiley, copyright 2020. (b) Illustration of the electrocatalytic NRR by HT Au@MOF composite. Source: Reproduced from He et al. [100] with permission from Elsevier Inc, copyright 2020.

suppress the HER. The highest ammonia yield was 49.5 $\mu g\, mg_{cat}^{-1}\, h^{-1}$ at −0.3 V vs. RHE and the FE was 60.9%, which was 3.5-fold and 5.9-fold of that of Au@MOF, respectively.

As discussed above, MNP@MOF catalysts show high NRR activity while effectively suppressing HER. In order to achieve wide application, low-cost and simple synthesis methods are of vital importance. MNP@MOF materials using transition metals are highly desired for NRR.

7.3.6 Oxidation of Small Molecules

The electrochemical oxidation of small molecules can be classified as electrochemical synthesis reactions (e.g. alcohol oxidation, aldehyde oxidation, and carbohydrate oxidation) and sacrificial agent oxidation reactions (e.g. urea oxidation and hydrazine oxidation) [101]. Noble metal catalysts and specialized facilities are usually required to prepare these oxidation products with high energy consumption and high costs [102, 103].

Hydrazine is a high hydrogen storage fuel. Hydrazine oxidation remains highly dependent on the utilization of precious metal-based catalysts. The high cost of precious metals and their tendency to aggregate hinder their application in hydrazine oxidation [104–108]. Loading metal NPs into the pores of MOFs can restrict the migration and aggregation of metal NPs, thus improving the catalytic performance. Han et al. [109] synthesized Au/NH_2-MIL-125(Ti) nanocomposites with good electrocatalytic activity. In the presence of hydrazine, Au/NH_2-MIL-125(Ti)/GCE showed a large anodic wave at an onset potential of −0.034 V (vs. Ag/AgCl), much lower than that of the pristine MOF (about 0.28 V). The sensor constructed using this material has a wide linear range of 10 nM to 100 mM and a low detection limit of 0.5 nM.

So far, there have been very few reports on the use of MNP@MOF for electrocatalytic small molecule oxidation. The development of transition metal-based catalysts with low cost and good catalytic performance is desired. Meanwhile, understanding and elucidating the mechanism of molecular oxidation reactions is of vital importance. The understanding of the structure–activity relationship can help to precisely control the reaction process, adjust the selectivity of target products, and design target catalysts.

7.4 Photocatalytic Applications

7.4.1 Photocatalytic Hydrogen Production

Photocatalytic hydrogen production, especially in the visible spectrum region, provides a sustainable development for environmental and energy problems [110–114]. The modification of organic linkers of MOFs makes them photoresponsive to UV and even visible light. In addition, metal NPs may act as electron acceptors to facilitate the separation of photogenerated electrons and holes in semiconductors. Metal NPs stabilized by MOFs usually behave as excellent electron acceptors and active centers, accepting electrons from excited MOFs. Different methods have been developed for the synthesis of MNP@MOF, showing high photocatalytic activity for hydrogen production.

The design of photocatalysts with high activity under visible light irradiation is attractive [115–117]. MOFs are regarded as semiconductors, which can transform to a charge-separated state with a positive hole in the valence band and an electron in the conduction band after photon absorption [118–122]. Pristine MOFs can be used for photocatalytic hydrogen production. For example, both UiO-66 and UiO-66-NH_2 can be used as stable photocatalysts, but the H_2 production rates are quite low [123].

The incorporation of Pt NPs into MOFs can greatly enhance the water-splitting activity [124–127]. Xiao et al. [128] deliberately encapsulated ultrafine Pt NPs inside MOFs to obtain Pt@UiO-66-NH_2. The Pt@UiO-66-NH_2 exhibited a drastically enhanced activity (257.38 μmol g^{-1} h^{-1}), about 150 and 5 times higher than that of UiO-66-NH_2 and Pt/UiO-66-NH_2, respectively. Moreover, Pt@UiO-66-NH_2 has good stability, showing no significant change in hydrogen production rate during four catalytic runs of 10 hours. The high catalytic performance may be originated from the easy access of Pt NPs to protons and easy electron transfer between the MOF and the internal Pt NPs inside the highly porous MOF.

Due to the high surface energy, Pt NPs tends to aggregate during the catalytic process, which leads to the decay of performance. Huang et al. [129] encapsulated highly dispersed Pt clusters with ultra-small sizes in MIL-125-NH-CH_2OH by an *in situ* reduction method (Figure 7.4a). The synergistic effect between the Pt clusters and MOFs promotes the transfer and separation of photoinduced charge. Pt/MIL-125-NH-CH_2OH catalyst exhibited a H_2 production yield of 4496.4 μmol g^{-1} h^{-1}, much higher than that of the pristine MIL-125-NH-CH_2OH (144 μmol g^{-1} h^{-1}).

Ni is a promising alternative to Pt as a hydrogen precipitation active catalyst. Zhen et al. [132] synthesized tiny nickel particles (~9 nm) with a high dispersion (41.8%) in MOF-5. The rate of hydrogen evolution reached 30.22 mmol h^{-1} g^{-1} over Ni@MOF-5 irradiated under visible light irradiation ($\lambda \geq 420$ nm) in two hours, with activity similar to Pt@MOF-5. Moreover, the activity of Ni@MOF-5 can retain 99% at the second cycle under the same conditions, revealing the good stability of Ni@MOF-5 during the photocatalytic H_2 evolution.

Charge transfer/separation efficiency is the key to achieve high photocatalytic activity and is important for improving the performance of photocatalytic reactions [133–137]. Introducing a new guest to form a hybrid or heterostructure can accelerate H_2 generation by promoting electron transfer between photosensitizer and cocatalyst, such as MOFs-containing POMs and metal NPs (PNPMOF) (Figure 7.4b) [130]. In PNPMOF, POMs accelerate catalytic H_2 evolution by promoting electron transfer between NH_2-MIL-53 and catalytic Pt NPs.

When the electron mediator ferrocene carboxylic acid (Fc) as a guest is directly decorated on the outer surface of Pt@UiO-66-NH_2 (Pt@UiO-66-NH_2/Fc), the rate of hydrogen production was decreased. In this case, the transfer of electrons mainly from MOF to Fc rather than Pt (Figure 7.4c) [131]. When Fc was introduced into the pores of UiO-66-NH_2, the Fe portion in Fc was oxidized to give Pt-Fc@UiO-66-NH_2, which had a higher photocatalytic H_2 production rate of 514.8 mmol g^{-1} h^{-1} than that of Pt@UiO-66-NH_2/Fc (102.6 mmol g^{-1} h^{-1}).

Heterojunctions in photocatalysts also play an important role in improving photocatalytic activity. In a previous study, p–n heterojunctions between porous Co_3O_4 and $BiVO_4$ in ZIF-67 MOFs-derived materials significantly promoted charge

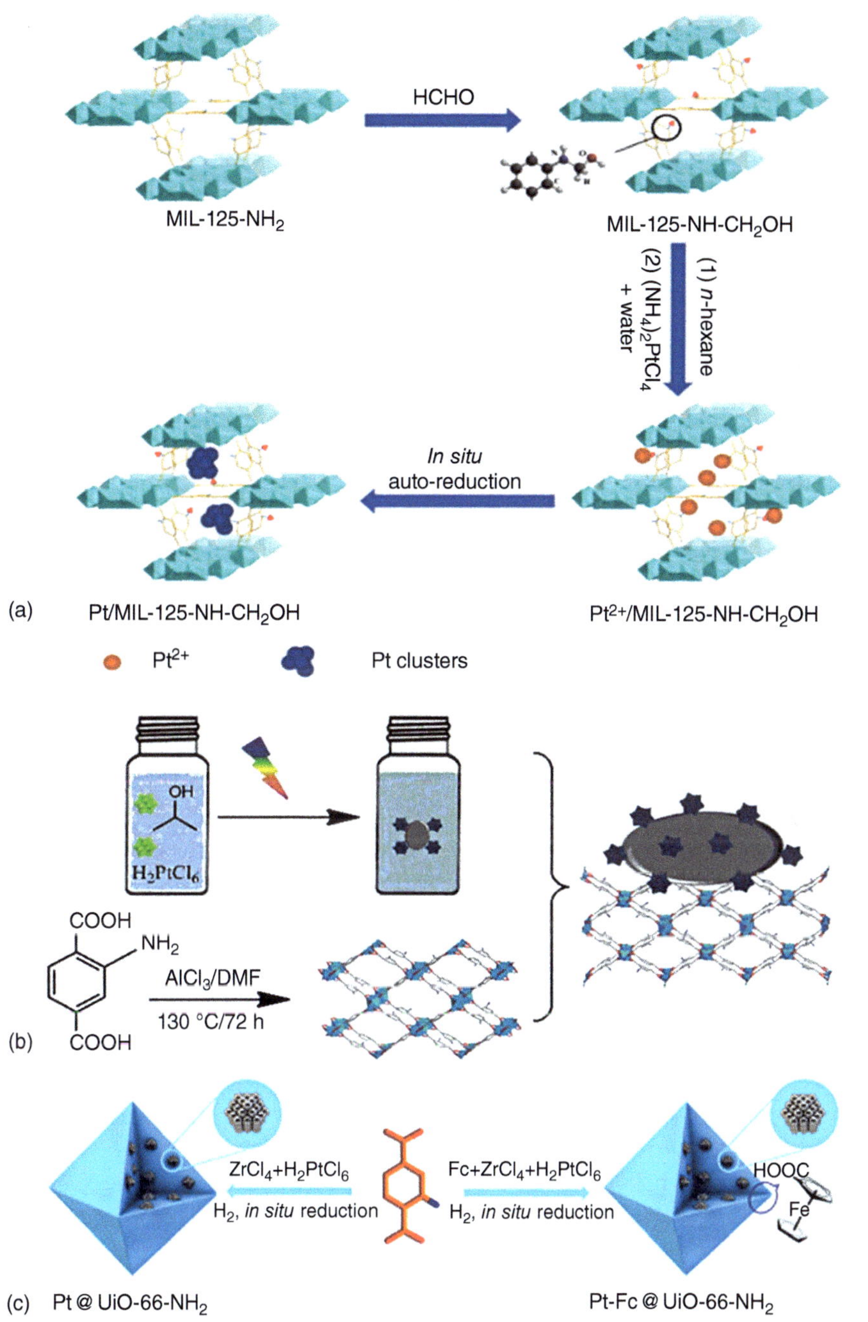

Figure 7.4 (a) Schematic representation of highly dispersed Pt clusters inside the MIL-125-NH-CH_2OH. Source: Reproduced from Huang et al. [129] with permission from Springer Nature, copyright 2021. (b) Route to PNPMOF (symbols not to scale). Green polyhedra: $H_3PW_{12}O_{40}$; dark blue polyhedra: reduced POM units, $[PW_{12}O_{40}]^{4-}$; gray ellipses: Pt NPs; light blue octahedra: $AlO_4(OH)_2$; gray sticks: carbon connector units; dark blue sticks: $-NH_2$ side chains; red sticks: oxygen. Source: Reproduced from Guo et al. [130] with permission from Royal Society of Chemistry, copyright 2016. (c) Illustration of the synthetic strategies for Pt@UiO66-NH_2, and Pt-Fc@UiO-66-NH_2. Source: Reproduced from Xu et al. [131] with permission from Wiley, copyright 2021.

transfer and reduced electron-hole recombination [138]. After this, a highly reactive Au@CdS/MIL-101 heterostructure was successfully prepared with a high H_2 yield of 25 000 μmol h^{-1} g^{-1} under visible light irradiation due to the enhanced electron separation and transfer [135]. Zhang et al. [139] designed a Pt@NH_2-UiO-66@MnO_x heterostructured photocatalyst with Pt and MnO_x as co-catalysts, which can produce hydrogen at a rate of 1340.6 μmol g^{-1} h^{-1}. In contrast, the hydrogen production rate of Pt @NH_2-UiO-66 was 299.4 μmol g^{-1} h^{-1}. NH_2-UiO-66 spatially separates Pt and MnO_x. The heterojunction between Pt NPs and NH_2-UiO-66 inhibits the overall recombination of photogenerated electrons and holes through electron extraction. The MnO_x NPs promote the transfer of photogenerated holes from the interior to the surface of the MOF crystal and inhibit electron–hole recombination.

MOFs can be directly used as photocatalysts, but their performance was not satisfactory. The incorporation of metals into MOFs can greatly improve the photocatalytic activity of MOFs. Functionalized MOFs can act as excellent supports to prevent metal NPs aggregation. Charge transfer/separation efficiency is the key to achieve high photocatalytic activity. To improve the charge transfer/separation efficiency, the tuning of the interface between metal and MOFs by incorporation of guest species is effective.

7.4.2 Photocatalytic CO_2 Reduction

Photocatalytic CO_2 reduction is one of the ideal approaches to solve the energy crisis caused by fossil fuels and to reduce excessive CO_2 emissions [140, 141]. MOF catalysts are widely used for the photoreduction of CO_2 due to their excellent CO_2 capture capability, unique electronic energy band structure, and customizable light absorption capability [142–146]. However, the efficiency and selectivity of pure MOF materials are still limited. To further improve the photocatalytic performance, metal NPs can be introduced to MOFs for synergistic catalysis [147–149]. CO_2 can be reduced to simple C1 products such as CO, CH_3OH, and $HCOO^-$ by MNP@MOF photocatalyst.

7.4.2.1 CO_2 Photoreduction to CO

Incorporating metal NPs, e.g. Pt, Ag, Au, and Cu can enhance the catalytic performance of MOFs. Metal NPs can not only provide catalytic active sites but also promote light absorption [150, 151]. Deng et al. [152] introduced Ag NPs into Co-MOF-74 with a hollow structure for photocatalytic CO_2 reduction. The catalytic activity of AgNPs@MOF-74 was 3.8 times higher than that of MOF-74, due to the localized surface plasmon resonance (LSPR) effect of Ag NPs.

Tuning the structure of the MOFs to achieve effective charge transfer within the hybrid catalyst is an effective way to improve the catalytic performance [153, 154]. Guo et al. [155] synthesized a nanoscale MIL-101(Cr)-Ag with tunable size to achieve good photocatalytic performance. The sample exhibited high CO_2 photocatalytic reduction activity with remarkable production rates of CO 808.2 μmol g^{-1} h^{-1} and CH_4 427.5 μmol g^{-1} h^{-1}, while the MIL-101 gave the CO, CH_4, and H_2 yields of 29.6, 17.2, and 2.1 μmol g^{-1} h^{-1}, respectively (Figure 7.5a). The introduction of Ag led to

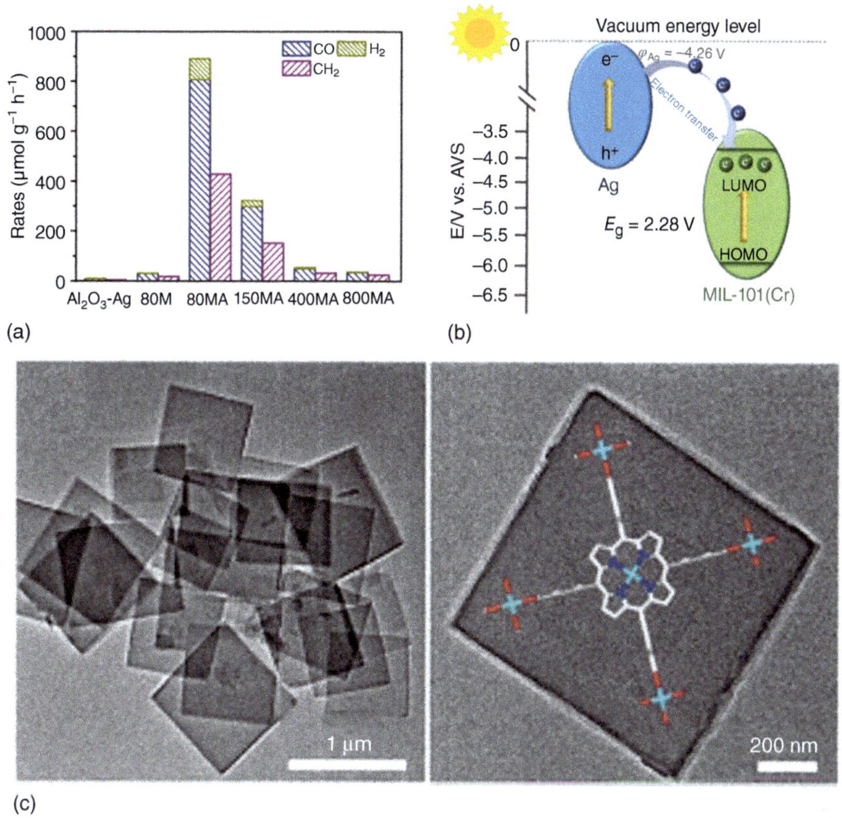

Figure 7.5 (a) Comparison of photocatalytic CO_2 reduction performances for varied catalysts; (b) Schematic illustration of the electron transfer process in MIL-101(Cr)-Ag hybrids. Source: Reproduced from Guo et al. [155] with permission from American Chemical Society, copyright 2019. (c) TEM images with different magnifications of PPF-3_1. Source: Reproduced with permission from Chen et al. [156], © 2019/Royal Society of Chemistry.

the appearance of defect states in MIL-101(Cr)-Ag, inhibiting charge recombination, promoting charge separation, and thus improving the photocatalytic activity for CO_2 reduction (Figure 7.5b).

The use of semiconductor-like NH_2-UiO-68 as a carrier with amino-functionalized linkers and permanent porosity allows Pt NPs to be stably embedded inside MOFs with good charge-transfer ability [157]. Pt@NH_2-UiO-68 enhanced the photocatalytic activity (400.2 mmol g^{-1}) compared to Pt deposited externally as Pt/NH_2-UiO-68 (121.2 mmol g^{-1}). Effective charge transfer from NH_2-UiO-68 to the excited Pt NPs can accordingly inhibit the recombination of photogenerated charge carriers, leading to the enhancement of photocatalytic activity.

7.4.2.2 CO_2 Photoreduction to CH_3OH

CO_2 is produced in an environmentally friendly way by one-step reaction or two-step reaction to obtain methanol. The main challenges of this process are to enhance charge separation and efficient visible light absorption [158–163]. To address these issues, Becerra et al. [164] deposited Au NPs on stable ZIF-67 to improve semiconductor properties through LSPR effect of Au NPs, leading to improved photocurrent and enhanced charge separation. Au_{20}@ZIF-67 obtained a maximum generation rate of 2.5 mmol g^{-1} h^{-1} of methanol and 0.5 mmol g^{-1} h^{-1} of ethanol. However, no products were formed using pristine ZIF-67.

Mohammad et al. [165] synthesized Cu and Au-modified ZIF-8 nanocomposites (Au/ZIF-8 and Cu/ZIF-8), which exhibited narrower band gap energy compared to pristine ZIF-8 and could absorb a wider range of visible light for photocatalytic reactions. Both Au/ZIF-8 and Cu/ZIF-8 photocatalysts showed exceptional yields of 2650 and 2240 μmol g^{-1} when 0.1 M Na_2SO_3 solution is used, respectively, much higher than that of pristine ZIF-8 (1300 μmol g^{-1}) at the same medium.

7.4.2.3 CO_2 Photoreduction to $HCOO^-$/HCOOH

Enhanced charge separation and efficient visible light absorption can also facilitate the reduction of CO_2 to formate or formic acid [166]. NH_2-MIL-125(Ti) can photocatalyze hydrogen production in the presence of sacrificial agents, while the photocatalytic reduction of CO_2 to formate can be obtained after loading Pt or Au on NH_2-MIL-125(Ti) [167]. The introduction of Pt and Au noble metals to form Schottky junctions promotes the separation of photogenerated carriers and suppresses photogenerated electron–hole complexation, ultimately improving photocatalytic performance. Compared with noble metals, Cu is an abundant and cheap metal and has been used for the reduction of CO_2 [168–170]. Dai et al. [171] introduced ultra-small Cu NCs into the aminated UiO-66-NH_2. The photoreduction experiments were carried out using triethanolamine as a sacrificial agent, and the evolution rate of formic acid of Cu NCs@MOF-801 was about 94 μmol h^{-1} g^{-1}. In contrast, the aminated Cu NCs@UiO-66-NH_2 showed enhanced reactivity toward formic acid products with an evolutionary rate of 128 μmol h^{-1} g^{-1}, probably because the electron-accepting amino groups exhibited strong interactions with formic acid and/or formate to promote the formation of these species.

Slow mass transfer of CO_2 and products in the micropores of MOFs would inhibit the photocatalytic CO_2 reduction. To overcome this dilemma, ultrathin porphyrin paddlewheel framework (PPF) NSs can be used as carriers to load plasma Au NPs for Au/PPF-3 hybrids (Figure 7.5c) [156]. The NS structure improved charge transfer and mass transport during photocatalytic CO_2 reduction, while PPF-3 contained porphyrins as organic chromophore can further enhance its light absorption capacity. Using Au/PPF-3 hybrids as photocatalysts, the average production rate of formic acid was about 42.7 mmol g^{-1} h^{-1} in four hours. However, with the increase of reaction time, the average yield of formic acid partially decreased to about 34 mmol g^{-1} h^{-1} within 24 hours.

Enhanced charge separation and efficient visible light absorption are the main approaches to enhance photocatalysis. Many catalysts have been shown to have good catalytic performance through both pathways. The stability of the catalyst still needs to be improved. Besides, CO_2 reduction catalytic reactions still require the use of sacrificial agents. Stable catalysts that do not require sacrificial agents will be a hot spot for future research.

7.4.3 Photocatalytic Organic Reactions

Under light irradiation, MOFs can be photoexcited to generate electron holes, providing a platform for photocatalytic organic reactions [172–175]. MNP@MOF composites improve the catalytic effect through the synergistic effect generated by strong interactions between different components while retaining the advantages of each component [5, 176]. In recent years, MNP@MOF composites have been used to photocatalyze various organic reactions.

7.4.3.1 Photocatalytic Hydrogenation Reactions

The production of aniline derivatives by selective hydrogenation of nitroaromatics is an important class of reactions in modern industry [177–180]. However, due to the simultaneous presence of nitro and other unsaturated groups in most cases, it is difficult to achieve a high selectivity of nitro hydrogenation by conventional catalysts [181–184].

In yolk-shell PdAg@ZIF-8 complexes, Pd showed good hydrogen and nitroaromatic activation ability, and the introduced Ag had strong visible light absorption [185]. Due to the LSPR properties of Ag to achieve effective photothermal conversion, the selectivity of the composite for the hydrogenation of nitrostyrene to aminostyrene under visible light irradiation was 97.5%. DFT calculations showed that the highly electronegative PdAg nanocages facilitated the adsorption of $-NO_2$ groups, leading to an improved catalyst selectivity.

7.4.3.2 Photocatalytic Oxidation Reactions

MOFs have also been applied to photocatalytic organic oxidation reactions. In an earlier work, Wu and coworkers [186] synthesized three M/MIL-125(Ti) (M = Au, Pd, Pt) composites with highly dispersed metal NPs. Compared with the original MIL 125(Ti), the formed M/MIL-125(Ti) composites showed enhanced activity in the photo-oxidation of benzyl alcohol to benzaldehyde, and the selectivity could reach more than 99%. The enhanced photoactivity is due to the formation of a Schottky junction between MNPs and MIL-125(Ti), which prolongs the spatial separation and lifetime of the photogenerated carriers.

In recent years, copper has been found to display surface plasmon resonance upon visible light irradiation, extending light absorption from the UV to the visible region [187, 188]. UiO-66, a similar semiconductor with a wide band gap, was chosen as a carrier to load Cu NPs to form Cu@UiO-66. Light absorption was extended to the visible region while the redox capability of MOF was maintained. A Schottky junction was also formed which offered strong photoexcitation electron transfer ability while suppressing photoinduced electron and hole complexation (Figure 7.6a) [189]. The conversion of benzyl alcohol over UiO-66 and 0.1%Cu@UiO-66 was 30.8% and 53.3%, respectively. The catalytic performance of 0.1% Cu@UiO-66 was improved compared to the UiO-66 precursor.

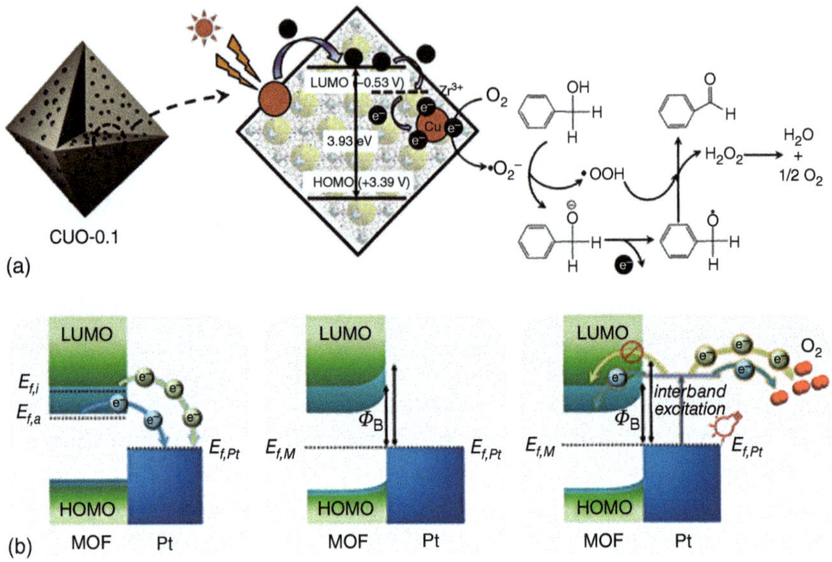

Figure 7.6 (a) Migration of electrons in CUO-0.1 according to energy levels and proposed mechanism for the selective oxidation of benzyl alcohol. Source: Reproduced from Xiao et al. [189] with permission from Elsevier Inc, copyright 2019. (b) Schematic illustration of band diagrams for the Pt/MOF and the migration of electrons originated from interband excitation of Pt NPs under light irradiation. Source: Reproduced from Sun et al. [190] with permission from Wiley, copyright 2022.

7.4.3.3 Photocatalytic Coupling Reaction

Suzuki-coupling reaction is the construction of a C=C bond between aryl halides and boronic acid reaction. Pd catalysts have shown high activity in Suzuki-coupling reactions. Using NH_2-UiO-66 as a carrier, well-dispersed Pd NPs could be obtained with a high number of active sites [191]. The obtained Pd@NH_2-UiO-66(Zr) showed high catalytic activity under visible light irradiation, with 90.4% conversion of iodobenzene in 30 minutes. When the reaction was expanded eight times, more than 99% of iodobenzene was converted to biphenyl using Pd@NH_2-UiO-66(Zr) with a high turnover number of 2514, indicating that the expansion of the reaction did not sacrifice its excellent catalytic performance. To further promote the transfer of photogenerated electrons from the excited state of NH_2-UiO-66 (Zr) to the metal Pd NCs, Sun et al. further introduced Cu into Pd@NH_2-UiO-66 (Zr) [192]. When extending the reaction time to four hours and increasing the amount of TEA (0.45 mmol), the reaction using CuPd@NH_2-UiO-66 (Zr) achieved almost complete conversion (99%) with a selectivity of more than 99%. Cu metal as an electron mediator promoted the transfer of photoexcited electrons from NH_2-UiO-66 (Zr) to Pd to form electron-rich Pd thus showing higher catalytic activity. It is suggested that the interaction between metals and MOFs can modulate the electronic properties of the metal.

Photocatalytic coupling of benzylamines is significant in pharmacy and fine chemical manufacturing. Schottky junctions can be formed between Pt and ZIF-8 to promote photogenerated carrier separation and improved photocatalytic performance

[190]. The aniline conversion of Pt/ZIF-8 was improved to 99.7%, much higher than that of Pt/UiO-66 and Pt/MIL-125. The electrons excited between the Pt bands cannot be injected backward into the MOF due to the high energy gap of ZIF-8, resulting in an increase in the electron density of Pt, which leads to an increase in the catalytic performance of Pt/ZIF-8 (Figure 7.6b).

Direct alkylation of amines using alcohols as alkylating agents is an efficient way to synthesize the important chemical *N*-alkyl amines [193–195]. Wang [196] encapsulated Pd NPs in MIL-100(Fe) and applied them to photoinduced alkylation of alcohols and amines. After 24 hours of visible light irradiation, the conversion of Pd@MIL-100(Fe) aniline was 88%, and the selectivity to *N*-benzylaniline was 76%. MIL-100 can be excited as a semiconductor to generate electrons to transfer into the Pd NPs, forming electron-rich Pd species to improve their catalytic performance.

There are two main pathways for MNP@MOF photocatalytic organic reactions, one is to enhance the photothermal effect to drive organic reactions and the other is to promote photogenerated carrier generation to drive organic reactions. MNP@MOF is increasingly used for photocatalytic organic reactions with high activity and selectivity due to the tunable MOF carriers and the properties of metal NPs. The combination of different MOFs and metal NPs will bring more possibilities to develop composites for photocatalytic organic reactions.

7.4.4 Photocatalytic Degradation of Organic Pollutants

With the development of industry, more and more toxic organic pollutants are released into the environment, such as organic dyes, heavy metals, pharmaceuticals, and personal care products (PPCPs) [197–201]. Compared with traditional degradation technology, photocatalytic technology can convert waste organics into nontoxic and nonhazardous substances and reduce secondary pollution [202]. MOFs with semiconductor properties were used as carriers for metal NPs to obtain MNP@MOF complexes that can be used for photocatalytic degradation of organic pollutants to harmless components under mild conditions.

7.4.4.1 Degradation of Pollutants in Wastewater

The presence of various dyes and heavy metals in industrial wastewater that are difficult to degrade has caused great concern as it adversely affects human health and the ecological environment [203–206]. Methyl orange (MO) is one of the commercial dyes with aromatic structure, which is difficult to be completely removed by conventional wastewater treatment processes [207]. For this purpose, Liang et al. [208] prepared a series of M@MIL-100(Fe) (M = Au, Pd, Pt) nanocomposites using the photodeposition technique at room temperature, and the obtained materials could exhibit significantly enhanced visible light photoactivity for the oxidation of MO. The Pt@MIL-100(Fe) nanocomposites showed a high photocatalytic activity in the degradation of MO aqueous solutions. MO is almost 100% degradable after 40 minutes of visible light irradiation. MIL-100(Fe) is photoactive, which enhances visible light absorption, and the noble metal particles act as electron reservoirs,

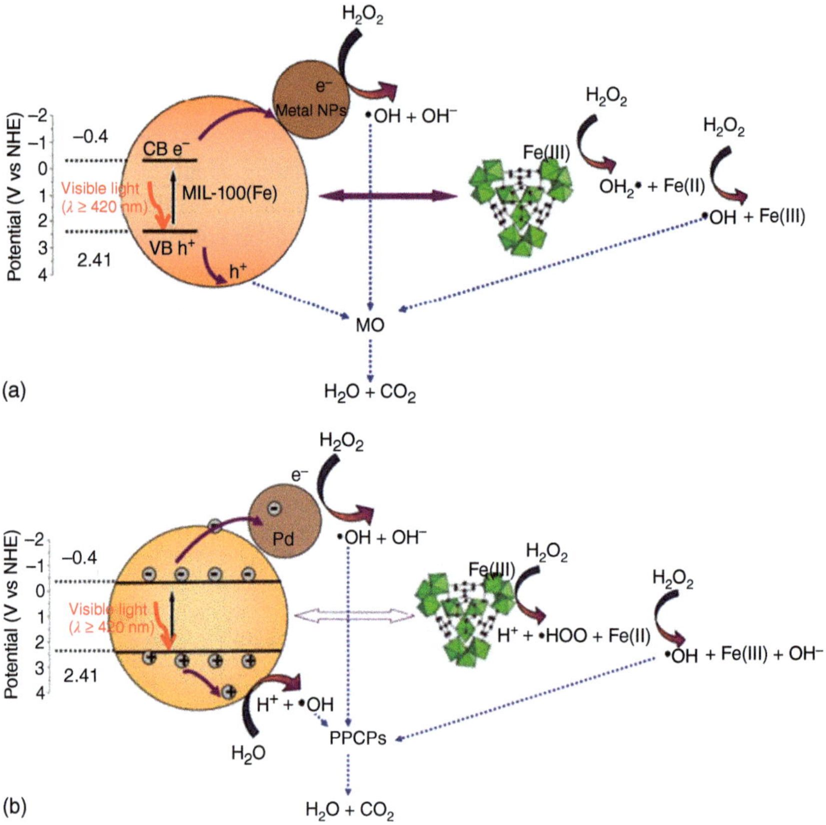

Figure 7.7 (a) Schematic diagram of the proposed mechanism for MO photocatalytic degradation over the M@MIL-100(Fe) (M = Au, Pd, and Pt) under visible-light irradiation ($\lambda \geq 420$ nm). Source: Reproduced from Liang et al. [208] with permission from Springer Nature, copyright 2015. (b) Proposed mechanism for the photocatalytic degradation of PPCPs over Pd@MIL-100(Fe) under visible light irradiation. Source: Reproduced from Liang et al. [209] with permission from Elsevier Inc, copyright 2019.

thus extending the lifetime of photogenerated carriers and allowing Pt@MIL-100 to exhibit excellent performance in photocatalytic degradation of MO (Figure 7.7a).

Cr(VI) is often discharged with MO in industrial wastewater, which makes the removal of Cr(VI) even more costly [210]. Shen et al. [211] immobilized highly dispersed Pd NPs on UiO-66(NH_2) and used the obtained Pd@UiO-66(NH_2) nanocomposites for Cr(VI) reduction under mild conditions. More importantly, Pd@UiO-66(NH_2) could retain its catalytic performance or even enhance it in the presence of organic pollutants such as MO, methylene blue (MB), and Cr(VI) at the same time. The reduction rate was 70% when Cr(VI) was the only substrate, whereas the reduction of Cr(VI) was significantly enhanced when organic dyes were added to the reaction system. In the binary systems of Cr(VI)/MO and Cr(VI)/MB, the reduction rates of Cr(VI) were 79% and 100%, respectively, which were significantly higher than those of the single system. Pd@UiO-66(NH_2) nanocomposites upon visible light irradiation can be used for photocatalytic degradation of multiple pollutants. It is probably due to enhanced light absorption intensity, extended carrier lifetime, and increased BET surface area upon loading of metal NPs.

Like MO and Cr(VI), PPCPs are a widespread chemical contaminant [212–214]. MIL-100(Fe) with chemical stability and water stability was used as a carrier for Pd NPs. The synthesized Pd@MIL-100(Fe) exhibited excellent photocatalytic degradation of three PPCPs [209]. The degradation efficiency of theophylline was about 65%, 95%, and 100% at pH 6, 4, and 2, respectively, after 120 minutes of visible light irradiation, which increased with decreasing pH value. The synergistic effect of metal NPs and MOF in Pd@MIL-100(Fe) enhanced the light absorption intensity and separated carriers more efficiently, leading to higher photoactivity (Figure 7.7b).

Up to now, many MNP@MOFs have been used to degrade various other organic pollutants in wastewater. For example, the unique MOF structure of Ag/UiO-66-NH_2 improves the electron transfer process to promote the photocatalytic reduction of carcinogenic 4-nitrophenol [215], and MIL-125(Ti) modified by Ag NPs exhibits good visible light absorption to improve the degradation of Rhodamine B [216].

7.4.4.2 Degradation of Gas-Phase Organic Compounds

In addition to organic dyes in water, gas-phase organic compounds can also be degraded by photocatalysis. Highly dispersed PtPd alloy nanocrystals were encapsulated in ZIF-8, and the resulting PtPd@ZIF-8 catalysts exhibited good photocatalytic activity for the conversion of adsorbed ethylene to CO_2 and water at room temperature [217]. More than 83.2% ethylene conversion was achieved using very low amounts of catalyst, and there was no significant decrease in conversion after 10 catalytic cycles, indicating the good stability.

This section reviews the research progress of MNP@MOF in photocatalytic reactions in recent years. MNP@MOF catalysts are receiving increasing attention due to their unique structure and properties. Currently, MNP@MOF degradation of organic pollutants in wastewater is well established and gradually expanding in application, while its use for degradation of organic compounds in the gas phase is still a great challenge.

7.5 Thermocatalytic Applications

7.5.1 Oxidation Reactions

MOFs embedded with noble metals can be used as catalysts for gas- and liquid-phase oxidation reactions [70, 218].

7.5.1.1 Gas-Phase Oxidation Reactions

Au NPs loaded on metal oxides, zeolites, mesoporous materials, and activated carbon have been extensively studied as CO catalysts [219–222]. MOFs can also be employed as support for the loading of Au NPs for CO catalysts. For example, Jiang et al. [223] reported the deposition of Au NPs on a zeolite-type MOF by a simple solid grinding method, representing the first example of an active catalyst for CO oxidation by using a microporous MOF as a host for noble metal NPs. The CO oxidation activity is improved with increasing Au loadings, the temperatures for 50% conversion were approximately 225, 200, 185, and 170 °C, respectively, for 0.5, 1.0, 2.0, and 5.0 wt% Au@ZIF-8 samples. There was no serious sintering after three to four times of repeated catalysis. Since Au@ZIF-8 was prepared using only a simple solid grinding method, the dispersion of Au NPs was not high, potentially leading to low Au loadings. Wu et al. [224] used a two-solvent approach to incorporate metal NPs into the pores of MOFs without agglomeration of the NPs on the outer surface of the host framework. The small Au NPs obtained using this method are highly dispersed on the pores of the MOF host and have high catalytic activity for the oxidation of CO, which was better than that reported by Au@ZIF-8 [223].

7.5.1.2 Liquid-Phase Oxidation Reactions

7.5.1.2.1 The Oxidation of Alcohols

In addition to applications in catalysis in the gas phase, MNP@MOF composites can also be used in catalytic reactions in the liquid phase, for example, the oxidation of alcohols. Nanoscale MOFs have well-defined properties and uniform size and morphology and can be well dispersed in aqueous media or other solvents to show good catalytic properties [225, 226]. Luan et al. [227] reported a one-time absorption/reduction synthesis of gold-deposited MOFs at the nanoscale level, where the MOFs were functionalized with amino acids to stabilize the metal NPs and obviate the use of PVP. The catalytic performance of Au@UiO-66(NH_2) in the aerobic oxidation of various alcohols was evaluated under optimized reaction conditions. Benzyl alcohol was transformed to the desired benzaldehyde in 94% yield. This Au@UiO-66(NH_2)-catalyzed aerobic oxidation approach was compatible with electron-rich aromatic alcohols, such as 4-methoxybenzyl alcohol, with an isolated yield of 94% to the desired 4-methoxy benzaldehyde. In addition, other substrates such as 4-methoxybenzyl alcohol, aromatic alcohol, 4-fluorobenzyl alcohol, cinnamyl, and 1-phenylethanol could also afford good yields.

Noble metals have special electronic structures and physicochemical properties [228–230]. Since precious metals are expensive and scarce, combining transition metals with precious metals in alloys or core–shell structures to form bimetallic NPs is an effective strategy. This strategy not only reduces the consumption of precious metals but also largely improves the catalytic performance due to the synergistic effect between the two metals [231, 232]. The integration of Pt with Cu into bimetallic PtCu alloy NPs exhibited enhanced performance than Pt in the oxidation of benzyl alcohol [233]. Noteworthy, this catalytic system could be extended to various substrates, different substituted benzyl alcohols and heteroaryl alcohols were completely oxidized to the corresponding aldehydes within five hours. In addition to aromatic alcohols, MNP@MOF composites can be applied for the catalysis of bifunctional linear molecules, e.g. diols. The special pore size and pore environment of MOFs can provide site-selectivity for diol oxidation by protecting secondary functional groups through simple physical spatial constraints to form the desired products (Figure 7.8) [234]. The use of MNP@MOF materials as site-selective catalysts may be extended to other organic transformation and drug preparation.

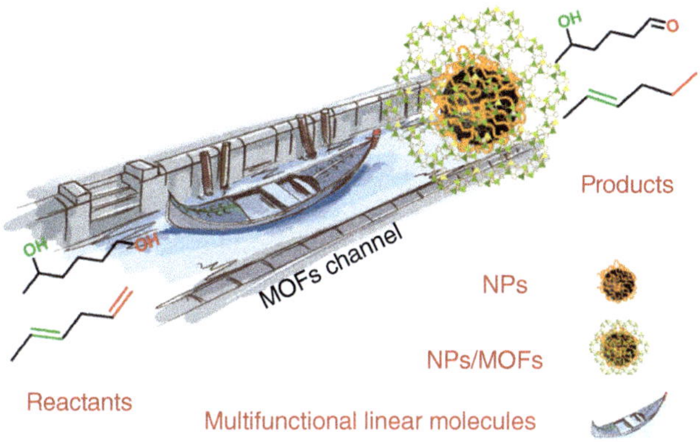

Figure 7.8 Schematic representation of the site-selective catalysis of multifunctional linear molecule. Source: Reproduced from Zhang et al. [234] with permission from Wiley, copyright 2018.

7.5.1.2.2 The Oxidation of Furfural

Furfural, one of the most promising biomass-derived platform chemicals, also has great potential to replace various fossil monomers [235–237]. Tong and coworkers [238] designed and prepared several MOF-based Pt catalysts such as Pt@MOF-5, Pt@UiO-66, and Pt@UiO-66-NH_2 for the oxidation of furfural with ethanol to prepare furan-2-acrolein. The reaction results indicated that Pt@MOF-5(Zn) achieved an 84% furfural conversion and 75.7% yield to furan-2-acrolein without any considerable activity loss even after five cycles. The high catalytic efficiency of this Pt@MOF-5(Zn) catalyst was attributed to the uniform distribution of Pt NPs within the MOF-5(Zn) channels. Nevertheless, there is a lack of theoretical and data support for the stability of MOF-5, and more data are needed to explain the reason for its stability.

So far, various reported metal@MOF composites exhibited better performance than metal NPs embedded in conventional materials in selective oxidation reactions. However, the stability of the catalyst cannot be guaranteed during the catalytic process, which may be due to metal aggregation and framework collapse of MOFs. In order to obtain high thermocatalytic selective oxidation performance under relatively harsh conditions, a stable MOF needs to be selected as a carrier. The synergistic relationship between MOF and metal NPs needs to be clarified to guild the design of MNP@MOF catalysts.

7.5.2 Hydrogenation Reactions

7.5.2.1 Hydrogenation of C=C and C≡C Groups

Hydrogenation reactions are critical for both science and industry [10, 239, 240]. The use of MNP@MOF as a catalyst for olefin hydrogenation has been studied for many years [231, 241, 242]. Zhang et al. [243] successfully incorporated Pt NPs inside various MOFs, such as UiO-66, NH_3-UiO-66, and NH_3-MIL-53. Uniformly dispersed metal NPs are completely encapsulated in the MOF matrix without adhering to the outer surface. The pore structure of MOFs can achieve selective hydrogenation by limiting the diffusion of molecules. The Pt NP@MOF materials exhibited outstanding olefin hydrogenation activity. For the hydrogenation of hexene, the Pt/UiO-66 composites were active after 24 hours of reaction with 100% conversion, indicating that the UiO-66 framework had no significant diffusion limitation for hexene. The Pt/UiO-66 composites had no significant catalytic activity for the larger dimension of tetraphenylene (6.7 Å) because the pore size of UiO-66 (6 Å) limited its diffusion, also indicating that the absence of Pt NPs on the outer surface of the composites (Figure 7.9a). In addition, the pore structures of MOFs can limit the random rotation of molecules, achieving selective hydrogenation. Li et al. [176] used a general mechanochemically assisted encapsulation method to synthesize Pd@ZIF-8. The small pores of ZIF-8 could restrict the random rotation of the molecule to partially hydrogenate hex-1,4-dienes to hex-2-enes with a selectivity of 50%. When Pd/ZnO was employed as a catalyst, only the complete hydrogenation product (hexane) was obtained.

However, the pores in 3D MOF crystals would also limit the diffusion of substrate molecules and hinder their access to the internal active sites. Compared with 3D MOF, 2D MOF NSs with larger surface area and easier access to the active sites are promising carrier materials for metal NPs [246]. Deng et al. [247] loaded Pd NPs onto Zr-Fc MOF NSs via *in situ* reduction of K_2PdCl_4 by 1, 1'-ferrocene dicarboxylic acid (Fc') ligands. The obtained Pd/Zr-Fc MOF NSs with a high number of active sites could completely transform 0.6 mmol of styrene at 30 °C in three minutes. For comparison, the bulk Pd@UiO-66 yielded only a small amount of product even after 10 minutes of reaction.

Compared to noble metals, Cu, a transition metal with high earth abundance, has gained much attention [244, 248, 249]. CuNPs@NU-1000 were used in the synthesis of ethylene from semi-hydrogenated acetylene to solve the problem of acetylene contamination [250]. Under conditions simulating the industrial removal of acetylene from ethylene, acetylene was completely consumed and a relatively high ethylene purity (>99.5%) was eventually achieved.

7.5.2.2 The Reduction of $-NO_2$ Group

The reduction of nitroaromatics is also an important hydrogenation reaction [179, 251–254]. How to improve the adsorption and dissociation of H_2 on the catalyst is the key to improve the catalytic activity of hydrogenation reaction. By combining the oxidative etching of Pt–Ni alloy with the *in situ* precipitation of MOF, a new frame@MOF, Pt–Ni frame@MOF was successfully synthesized [255]. In Pt–Ni frame@MOF, MOFs could maximize the exposure of Pt–Ni bimetallic nano-framework and increase H_2 concentration to accelerate the hydrogenation reactions. Pt–Ni frame@MOF exhibited a catalytic activity with 3.3 and 2.4 times higher than that of bare Pt–Ni nano-frame and Pt–Ni frame on MOF catalysts. Compared to 3D bulk MOFs, the recently emerged 2D MOF NSs offer a larger outer surface and highly exposed active sites, facilitating contact between reactants and catalytic active sites [256]. Liu et al. [257] reported a rapid preparation of 2D UiO-67 NSs under microwave-assisted conditions and formed Au/UiO-67 NS by *in situ* reduction of $HAuCl_4$ in the pores of UiO-67 NS (Figure 7.9b). In the presence

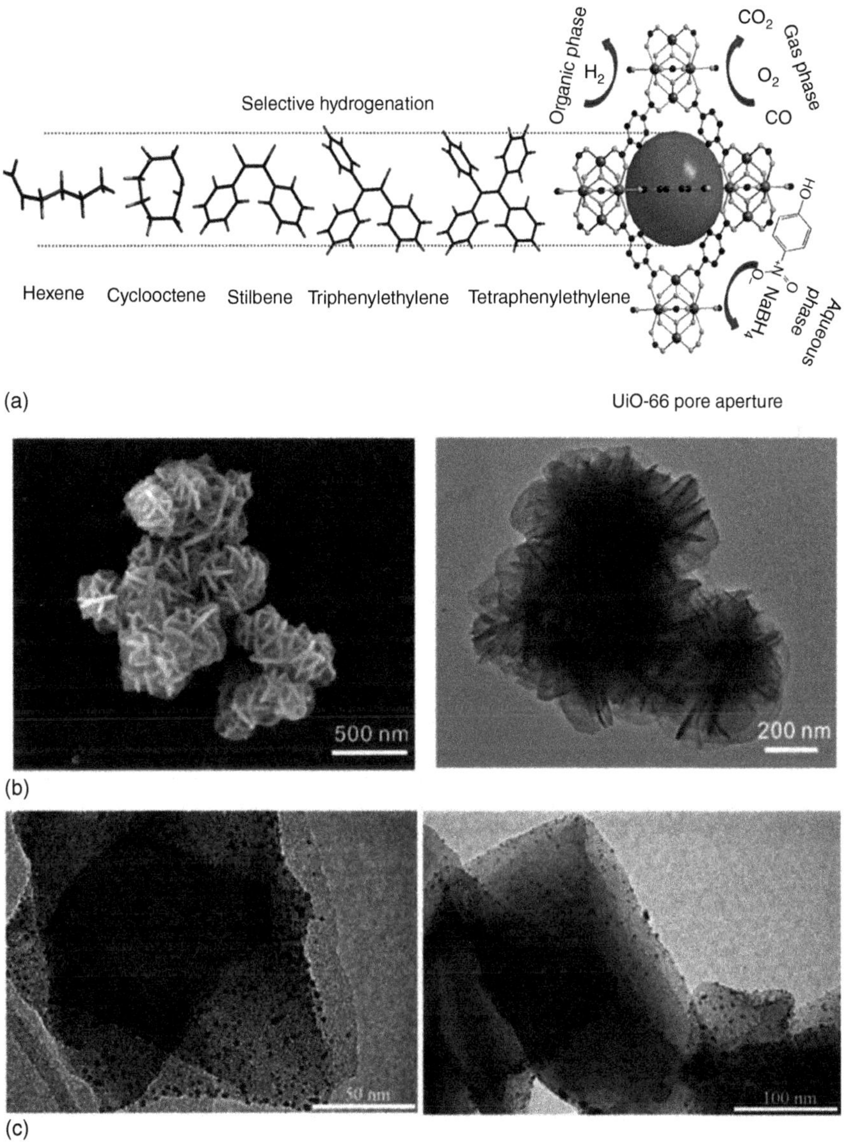

Figure 7.9 (a) The catalytic reactions of Pt/UiO-66 composites. Source: Reproduced from Zhang et al. [243] with permission from Wiley, copyright 2014. (b) SEM image and TEM images of UiO-67 NS. Source: Reproduced with permission from Natesakhawat et al. [244], © 2022/American Chemical Society. (c) TEM images of Ru-MIL-101 at different magnifications. Source: Reproduced with permission from Fang et al. [245], © 2015/Royal Society of Chemistry.

of sodium borohydride (sodium borohydride) as a reducing agent, 4-nitrophenol (high concentration up to 0.01 M) could be completely converted to 4-aminophenol within 100seconds. The high activity of Au/UiO-67 NS should be attributed to the synergistic effect of 2D MOFs with reduced diffusion distance and highly exposed Au NPs.

7.5.2.3 The Reduction of C=O Groups

Furfural is considered one of the most important biomass-derived platform compounds. In recent years, the hydrogenation of 5-hydroxymethylfurfural (HMF) by MNP@MOFs catalysts has been extensively studied [258, 259]. Li and coworkers [245] designed a catalyst loaded with Ru NPs on an acidic MOF MIL-101. Complete conversion of furfural with a selectivity higher than 96% was achieved within 2.5 hours at 160 °C and 4.0 MPa H_2 pressure. MIL-101 served as an ideal support for homogeneous dispersion of Ru NPs in the system and provided Lewis acid sites for the required dehydration step of 4-hydroxy-2-cyclopentenone. In addition, Ru NPs also showed good performance in the hydrogenation of furfural to furfuryl alcohol when loaded on UiO-66 [260]. Ru/UiO-66 showed good catalytic activity, with a conversion of furfural by 94.9%. Li et al. [261] reported the preparation of Pd NPs loaded on layered porous H-UiO-66 for the hydrogenation of furfuryl alcohols (Figure 7.9c). A complete furfural conversion and >99% furfuryl alcohol yield were obtained over Pd/H-UiO-66 at 60 °C and 0.5 MPa H_2 in the aqueous medium within three hours of reaction.

MNP@MOF has been designed for other hydrogenation reactions, such as hydrogeneration of levulinic acid to γ-valerolactone over Ru/2D Zr-MOF NS hybrid nanomaterials [262], hollow cubic structure of Pd/MOF140-AA catalyzed phenol hydrogenation [263], and selective hydrogenation of benzoic acid over isoreticular UiO-66-X with various linker groups and altered M-oxo clusters [264].

As discussed above, various MNP@MOF catalysts have been synthesized for many hydrogenation reactions. The design MOF supports with high accessibility of embedded metal NPs is of vital importance to achieve high catalytic activity. For further development, the deeper understanding of the reaction mechanism and structure–activity relationship is significantly important.

7.5.3 Coupling Reactions

7.5.3.1 Suzuki–Miyaura Coupling Reactions

Suzuki coupling is the cross-coupling of aryl or alkenyl boronic acids or boronic esters with chlorinated-, brominated-, or iodo-substituted aromatics or olefins. Palladium-based catalysts have shown excellent ability in Suzuki–Miyaura coupling reactions for the construction of diaryl units [265–268]. MOFs can serve as good carriers for MNPs to catalyze coupling reactions. Li and coworkers [269] reported the water-mediated coupling reactions of aryl chlorides over a heterogeneous palladium catalyst deposited on MIL-101, which represented the first example of a MOF-supported metal NPs for the coupling reactions. In the coupling of 4-chloroanisole with phenylboronic acid, Pd/MIL-101 exhibited good yield of cross-coupled product (82%) within six hours, while only 35% yield was obtained using Pd/C. The functionalization of the linkers of MOFs can modify the electronic environment of the embedded metal NPs for enhanced performance [270]. Huang et al [271] demonstrated the enhanced stability of metal NPs loaded on MIL-53(Al)-NH_2 compared with MIL-53(Al) as a carrier. The obtained Pd/MIL-53(Al)-NH_2 catalysts were easily recovered and reused in the coupling reaction of bromobenzene and phenylboronic acid for five cycles without significant loss of activity.

In addition to functionalized organic linkers, surface modification of MOFs can improve the catalytic properties of the composites. Mishra introduced polydopamine (PDA) to modify the surface of MOF based on amine-based modification [272], and complete conversion was observed within 1.5 hours at 60 °C (Figure 7.10a). The PDA layer provided strong binding sites for metal ion precursors for the stabilization

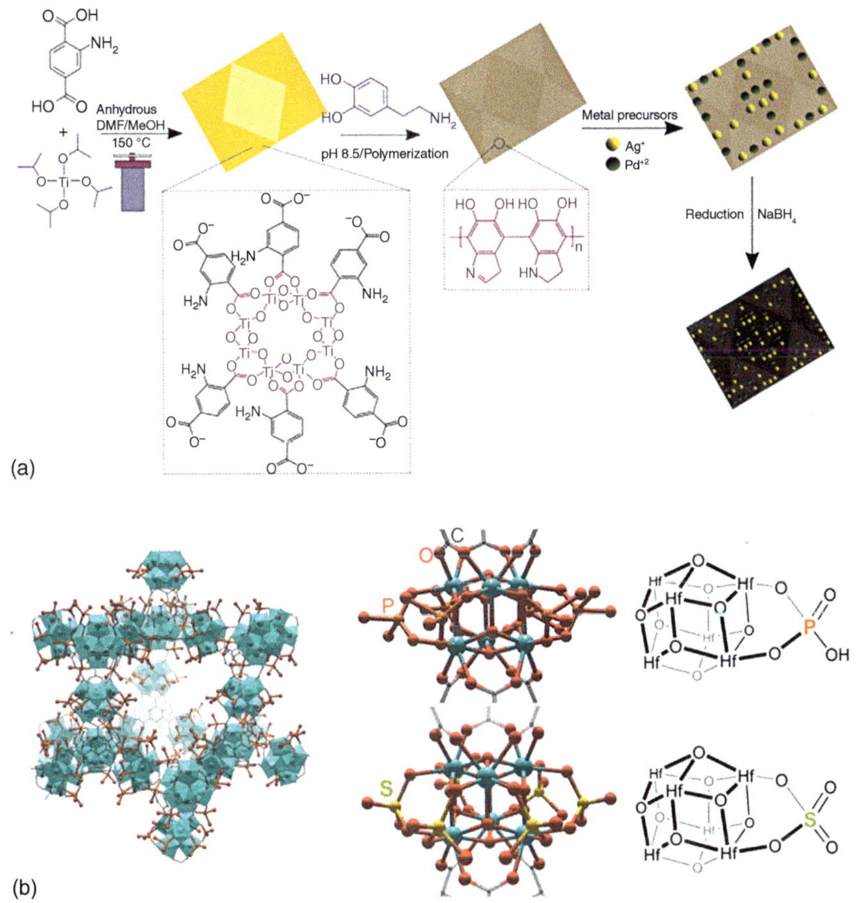

Figure 7.10 (a) Schematic representation for the steps involved in the synthesis of AgPd@MIL-125-NH_2-PDA. Source: Reproduced from Mishra et al. [272] with permission from Elsevier Inc, copyright 2022. (b) Crystal structures of Hf-MOF-808 after phosphation and sulfation. Spheres are gray (carbon), red (oxygen), cyan (hafnium), yellow (sulfur), and orange (phosphorus). Hydrogens were omitted for clarity. Source: Reproduced from Otake et al. [273] with permission from American Chemical Society, copyright 2019.

of metal NPs. Furthermore, the abundant catechol and amino groups on PDA can enrich the substrates near the active sites to further enhance the catalytic activity.

7.5.3.2 Heck Coupling Reactions

The Heck reaction is a coupling reaction between an aryl or vinyl halogenate and a reactive olefin. The direct use of Pd catalysts in heck reactions usually leads to the aggregation of metal, which results in reduced recyclability [274]. MOFs can serve as good support for the stabilization of Pd catalysts for the Heck reaction. Gole et al. [275] reported the functionalization of ligands to construct MOFs with the ethyl fraction pointing toward the pores. Pd ions could strongly interact with ethynyl through the unique π-donor and π-acceptor effects, thus promoting significant loading of Pd NPs into MOFs with large amounts without altering the monodispersity and particle size. In the presence of a very small amount of catalyst, the conversion of Pd@MOF catalyst could reach 90–99%. The Pd@MOF-3 samples were heated in a programmable oven at 150 °C for three hours under vacuum. The Pd particle size did not change significantly, indicating the thermal stability of the material.

The acidic oxides modified on the MOFs can also interact with the metal and serve to stabilize the metal sites. Otake et al. [273] incorporated a palladium in phosphated and sulfated MOFs, hafnium-based MOF-808 (Hf-MOF-808-PO_4 and Hf-MOF-808-SO_4) (Figure 7.10b). The stable Pd(II) sites in Hf-MOF-808-PO_4 and Hf-MOF-808-SO_4 exhibited better catalytic activity in the oxidation of Heck reaction. Turnover frequencies (TOFs) of Pd@Hf-MOF-808-SO_4 and Pd@Hf-MOF-808-PO_4 were 0.06 and 0.93 h^{-1} under differential conditions (10% conversion), respectively. DFT calculations showed that the reduced mobility of Pd(0) atoms compared to the pristine MOF does not coalesce into unreactive NPs, thus increasing the catalyst lifetime.

7.5.3.3 Glaser Coupling Reactions

The Glaser reaction is the coupling of two alkynes into a double alkyne in the presence of a base and a copper salt. Chen et al. [276] proposed a general synthesis strategy to immobilize ultrafine alloyed NPs of transition and noble metals within an MOF. As an example, for the homocoupling reaction of phenylacetylene, the optimized Cu–Pd@MIL-101 exhibited an extremely high activity with 98% yield of 1,4-diphenylbuta-1,3-diyne, which was much more active than its monometallic counterparts and other alloyed catalysts (all less than 20%).

7.5.3.4 Knoevenagel Condensation Reaction

Li et al. [277] prepared a bifunctional Pd(0)@UiO-68-AP catalyst for benzyl alcohol oxidation–Knoevenagel condensation. The Pd NPs were crystalline and highly dispersed with an average particle size of ≤2 nm. The Zr-Lewis acid site and the secondary amino group of UiO-MOF provided Knoevenagel condensation activity, and the final 2-benzylidenemalononitrile was obtained from benzyl alcohol. Under the condition of stirring for another eight hours at room temperature, 2-benzylidenemalononitrile was catalyzed to obtain more than 99% with both conversion and selectivity.

There is also an oxidative condensation reaction in the presence of molecular oxygen. Ning et al. [278] developed a novel catalyst system for oxidative condensation of furfural with ethanol in the presence of molecular oxygen. The Au@UiO-66 is beneficial to the occurrence of oxidative esterification and generation of ethyl-2-furoate. The oxidative condensation process was dominated by Au@UiO-66-COOH as the catalyst, in which the selectivity of furan-2-acrolein reached 84.1%.

7.5.3.5 Three-Component Coupling Reaction

Au catalysts have shown great potential in catalyzing three-component (A^3) coupling reactions of aldehydes, alkynes, and amines [279–281]. The structural customizability and functional functionality of MIL-101-NH_2 allow the binding of $Au^{\delta+}$ in oxidation state by PSM method [282]. The TOF value of the A^3 coupling reaction by MIL-101-SA-$Au^{\delta+}$ was 11-fold higher than that of MIL-101-SA-Au^0. This result clearly indicates that $Au^{\delta+}$ rather than Au^0 is a more efficient active site.

When catalysts are applied to catalytic coupling reactions, additional groups can be grafted onto the MOFs host of metal clusters or modify the MOFs surface to improve the activity and selectivity of the encapsulated MNPs by various methods. The grafted functional groups strongly interact with the metal NPs to reduce the aggregation and precipitation, resulting in better activity and stability of the catalyst. On this basis, the surface of MOFs can be further modified to improve their ability to adsorb reactants and further enhance the catalytic activity. Nevertheless, the high stability and catalytic activity of MNP@MOFs for coupling reactions are yet to be explored.

7.5.4 CO_2 Cycloaddition Reactions

CO_2 is responsible for the Earth's greenhouse effect. In addition to CO_2 reduction and collection, the conversion of CO_2 into high-value products or fuels offers a promising pathway for recovering carbon resources [283–287]. In recent years, many MNP@MOFs with superior performance as catalysts for CO_2 cycloaddition reactions have been developed.

Tang et al. [288] prepared an Au/Zn-MOF nanocage catalyst by manipulating the structural evolution of MOF and metal NPs in a cation exchange reaction by the Kirkendall effect. The Au/Zn-MOF nanocages were used to catalyze the cycloaddition reaction of CO_2 and propylene oxide at mild temperature (70 °C) and pressure conditions (CO_2 pressure = 30 atm) with good catalytic performance in 95–99% yields. After six consecutive cycles, the Au/Zn-MOF nanocages were still highly active and the reaction yields remained to be above 90%, indicating good stability. After replacing the substrates with other epoxides of larger molecular sizes, the catalyst all exhibited high catalytic activity, indicating the broad feasibility of the catalytic system.

When MNP@MOF catalysts are used for CO_2 epoxidation and CO_2 to cyclic carbonates, most of them require specific conditions such as high CO_2 concentration, high temperature, and co-catalysts. The co-catalyst, such as a halogenated salt (tetrabutylammonium bromide) [289–291], usually would leach from the catalysts after catalyst separation, leading to a reduction in catalytic activity and durability. Therefore, the exploration of MNP@MOF catalysts that can perform CO_2 epoxidation under mild and co-catalyst-free conditions requires further research.

7.5.5 Tandem Reactions

Multistep catalytic reactions typically require two or more catalysts to achieve, resulting in time-consuming reactions and tedious intermediate purification/separation processes. Multistep reactions can be catalyzed more efficiently if different functions are integrated into a single catalyst. Both MOFs and metal NPs in MNP@MOF composites can be used as active sites to synergistically catalyze tandem reactions [5, 292–294].

Zhao and coworkers [295] used a core–shell Pd@IRMOF-3 nanocatalyst for the tandem reaction to produce 2-(4-aminobenzylidene)-malononitrile. The Pd NP cores were uniformly loaded on the amino-functionalized IRMOF-3 shell. The alkaline IRMOF-3 shell achieved Knoevenagel condensation of the reactants with malononitrile, followed by preferential selective hydrogenation of the intermediate $-NO_2$ groups by the Pd NPs to obtain the products. The reaction reached 99.9% conversion and 99.9% selectivity after six hours, reflecting good stability. As another example, Pd metal loaded on UiO-66-NH_2 was employed as a catalyst for the tandem reaction. The oxidation of benzyl alcohol to benzaldehyde was catalyzed by Pd NPs in Pd@UiO-66-NH_2, followed by the acetalization of benzaldehyde and ethylene glycol catalyzed by Lewis acid sites on UiO-66-NH_2 [296]. In addition, the interaction of the amino-containing UiO-66-NH_2 with Pd NCs prevents metal aggregation with better stability.

Bimetallic PdAg@MOF with three active sites (Lewis acid-Pd-Ag) was successfully used as a multifunctional catalyst for the synthesis of secondary arylamines in a one-pot multistep tandem reaction [297]. In PdAg@MIL-101, MOF provided Lewis acid sites, Pd possessed hydrogenation activity, while Ag greatly improves the selectivity for the desired product (Figure 7.11a). Comparing the selectivity of Pd@MIL-101, the selectivity of Pd_1Ag_1@MIL-101 increased from 56% to 90%.

The introduction of C_{60} with high hydrogen capacity into MNP@MOFs may result in catalysts with good catalytic performance for hydrogenation [299, 300]. Zheng et al. [301] synthesized a C_{60}Pd@UiO-67 catalyst using MOFs, NPs, and C_{60} as catalytic sites. It was applied to the tandem reaction for the synthesis of secondary arylamines with 100% conversion and the highest selectivity for secondary arylamines (76%). The selectivity of Pd@UiO-67 was 60%, while the C_{60}@UiO-67 without Pd was inactive under similar conditions, with conversions below 10%. Pd NPs have intrinsic hydrogenation activity, and their catalytic activity (both conversion and selectivity) will be significantly increased when combined with C_{60}, which is further enhanced by the Lewis acid site provided by UiO-67.

All of the above materials contain noble metals, and transition metals have unique electronic structures and low cost that have great potential for applications in catalysis. Trimetallic catalysts, especially Cu@Co@Ni/MIL-101 with a triple core–shell structure, may be promising candidates to replace noble metals because they have more active sites on their surfaces (Figure 7.11b) [298]. The average size of

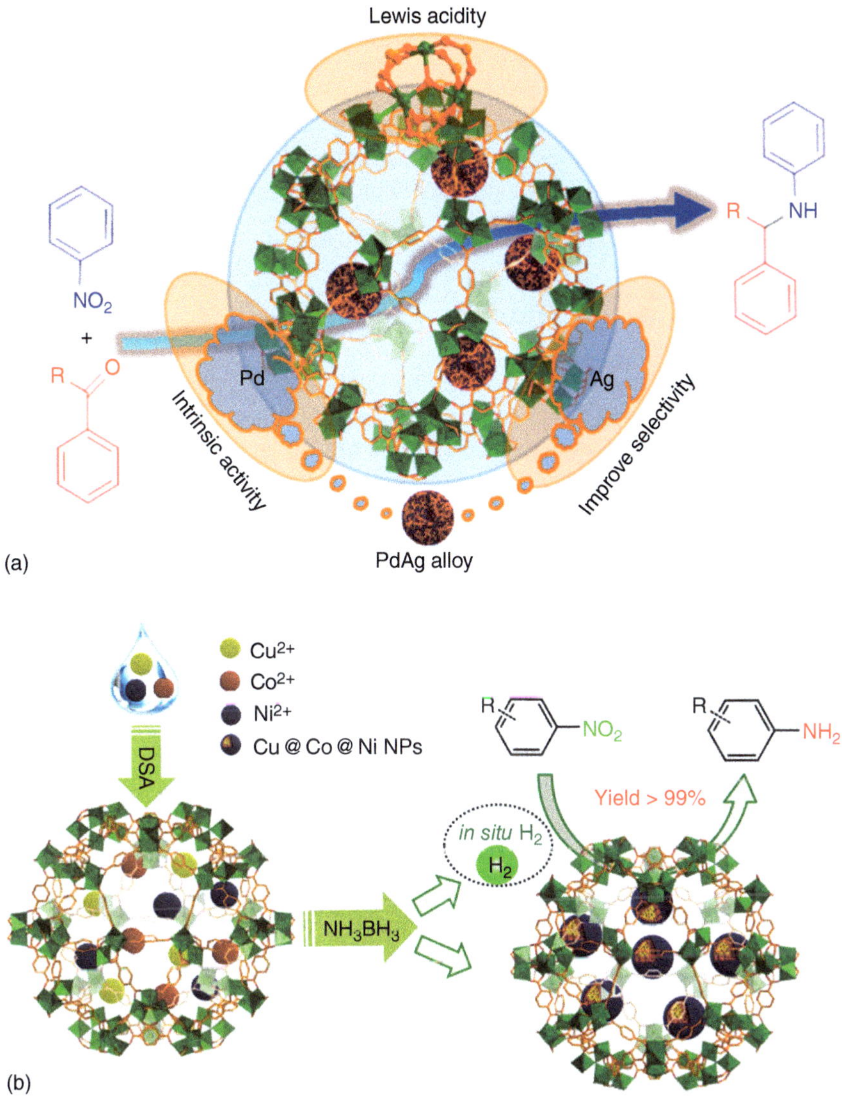

Figure 7.11 (a) Schematic illustration showing the multistep reaction over PdAg@MIL-101 involving Lewis acid and Pd/Ag Sites. Source: Reproduced from Chen et al. [297] with permission from American Chemical Society, copyright 2015. (b) Schematic illustration showing the *in situ* preparation of Cu@Co@Ni/MIL-101 for cascade catalysis involving dehydrogenation of NH_3BH_3 and subsequent hydrogenation of nitroarenes. Source: Reproduced from Sun et al. [298] with permission from American Chemical Society, copyright 2019.

Cu@Co@Ni NPs located within the pores of MIL-101 was 3.3 nm, and the tiny NPs had higher catalytic activity. Various nitro compounds with different substituents could be completely reduced to the corresponding anilines within five minutes with >99% selectivity.

The integration of different metal nodes, functional organic linkers, and metal NPs into a single nanostructure is able to provide more functionalities for multistep catalytic reactions. The key to achieve excellent catalytic efficiency should depend on the strong synergy between them. In future research, if the strong synergy between different components can be elucidated, it will provide a valid theoretical guidance for new tandem reaction catalysts.

7.6 Conclusions and Outlooks

MOFs have attracted attention as an emerging catalysts due to their unique properties. The acid/base sites on MOFs cannot satisfy a wide range of catalytic reactions. It is very promising to immobilize metal NPs on MOFs. The regulation of the size and distribution of metal NPs in MOFs can greatly affect the catalytic performance. The strong interaction between MOFs and metal NPs can show synergistic effect to greatly promote the reactions. In this chapter, the applications of MNP@MOF in electrocatalysis, photocatalysis, and thermocatalysis are systematically reviewed and discussed.

MNP@MOF catalysts have shown superior performance than their single counterparts in electrocatalysis, photocatalysis, and thermocatalysis applications. The development of MNP@MOF in catalytic applications is still underway, and many important issues are remained to be addressed.

(1) The loading of metal NPs, especially noble metal NPs, on MOFs exhibits good catalytic effects. However, the scarcity of precious metals limits their large-scale applications. So far, many transition metals have shown great potential as catalysts. From an industrial point of view, the use of transition metal NPs instead of noble metal NPs should be developed for the preparation of MNP@MOF.
(2) The synergistic interaction of MOFs with MNPs has not been thoroughly clarified to elucidate the structure–performance relationship. If metal NPs are precisely encapsulated into MOFs with a well-defined interface, their interaction may be well characterized by advanced techniques. In-depth studies to characterize these interactions can improve the fundamental understanding of synergistic interactions through *in-situ* and operando methods. The combination of the experiment results and theoretical calculations can give important guidelines for the design of advanced MNP@MOF composites.
(3) Up to date, only ZIF, UiO, and MIL series are commonly used. The development of other MOF-based catalysts is highly expected. Since the catalytic activity of MNPs is greatly influenced by the carriers, the selection of suitable carriers or modification of the carriers will improve the catalytic performance of MNPs.
(4) Currently known MNP@MOF composites exhibit excellent properties, but the low yield, and time-consuming preparation procedures only allow them to be

applied on a laboratory scale. A facile and rapid method for large-scale preparation of MNP@MOF is needed in the future.

In summary, MNP@MOF shows great potential in the field of catalysis due to the synergistic interaction between its components. However, the progress is still at the laboratory research stage. Researchers need further studies to find MNP@MOFs with low cost, high preparation efficiency, and excellent performance for practical applications.

References

1 Haruta, M., Yamada, N., Kobayashi, T., and Iijima, S. (1989). Gold catalysts prepared by coprecipitation for low-temperature oxidation of hydrogen and of carbon monoxide. *J. Catal.* 115 (2): 301–309.
2 Liu, X., Iocozzia, J., Wang, Y. et al. (2017). Noble metal–metal oxide nanohybrids with tailored nanostructures for efficient solar energy conversion, photocatalysis and environmental remediation. *Energy Environ. Sci.* 10 (2): 402–434.
3 Prieto, G., Zecevic, J., Friedrich, H. et al. (2013). Towards stable catalysts by controlling collective properties of supported metal nanoparticles. *Nat. Mater.* 12 (1): 34–39.
4 Seh, Z.W., Kibsgaard, J., Dickens, C.F. et al. (2017). Combining theory and experiment in electrocatalysis: Insights into materials design. *Science* 355: eaad4998.
5 Yang, Q., Xu, Q., and Jiang, H.L. (2017). Metal–organic frameworks meet metal nanoparticles: synergistic effect for enhanced catalysis. *Chem. Soc. Rev.* 46 (15): 4774–4808.
6 Kim, C.R., Uemura, T., and Kitagawa, S. (2016). Inorganic nanoparticles in porous coordination polymers. *Chem. Soc. Rev.* 45 (14): 3828–3845.
7 Falcaro, P., Ricco, R., Yazdi, A. et al. (2016). Application of metal and metal oxide nanoparticles@MOFs. *Coordin. Chem. Rev.* 307: 237–254.
8 Chen, L. and Xu, Q. (2019). Metal-organic framework composites for catalysis. *Matter* 1 (1): 57–89.
9 Chen, L., Luque, R., and Li, Y. (2017). Controllable design of tunable nanostructures inside metal-organic frameworks. *Chem. Soc. Rev.* 46 (15): 4614–4630.
10 Dhakshinamoorthy, A. and Garcia, H. (2012). Catalysis by metal nanoparticles embedded on metal-organic frameworks. *Chem. Soc. Rev.* 41 (15): 5262–5284.
11 Bavykina, A., Kolobov, N., Khan, I.S. et al. (2020). Metal-organic frameworks in heterogeneous catalysis: recent progress, new trends, and future perspectives. *Chem. Rev.* 120 (16): 8468–8535.
12 Zhu, N.N., Liu, X.H., Li, T. et al. (2017). Composite system of Ag nanoparticles and metal-organic frameworks for the capture and conversion of carbon dioxide under mild conditions. *Inorg. Chem.* 56 (6): 3414–3420.

13 An, B., Zhang, J., Cheng, K. et al. (2017). Confinement of ultrasmall Cu/ZnO_x nanoparticles in metal-organic frameworks for selective methanol synthesis from catalytic hydrogenation of CO_2. *J. Am. Chem. Soc.* 139 (10): 3834–3840.

14 Tsumori, N., Chen, L., Wang, Q. et al. (2018). Quasi-MOF: exposing inorganic nodes to guest metal nanoparticles for drastically enhanced catalytic activity. *Chem* 4 (4): 845–856.

15 Choi, K.M., Na, K., Somorjai, G.A., and Yaghi, O.M. (2015). Chemical environment control and enhanced catalytic performance of platinum nanoparticles embedded in nanocrystalline metal-organic frameworks. *J. Am. Chem. Soc.* 137 (24): 7810–7816.

16 Li, X., Goh, T.W., Li, L. et al. (2016). Controlling catalytic properties of pd nanoclusters through their chemical environment at the atomic level using isoreticular metal–organic frameworks. *ACS Catal.* 6 (6): 3461–3468.

17 Wang, P., Zhao, J., Li, X. et al. (2013). Assembly of ZIF nanostructures around free Pt nanoparticles: efficient size-selective catalysts for hydrogenation of alkenes under mild conditions. *Chem. Commun.* 49 (32): 3330–3332.

18 Yang, Q., Xu, Q., Yu, S.H., and Jiang, H.L. (2016). Pd nanocubes@ZIF-8: integration of plasmon-driven photothermal conversion with a metal-organic framework for efficient and selective catalysis. *Angew. Chem. Int. Ed.* 55 (11): 3685–3689.

19 Lu, G., Li, S., Guo, Z. et al. (2012). Imparting functionality to a metal-organic framework material by controlled nanoparticle encapsulation. *Nat. Chem.* 4 (4): 310–316.

20 Guo, Z., Xiao, C., Maligal-Ganesh, R.V. et al. (2014). Pt nanoclusters confined within metal–organic framework cavities for chemoselective cinnamaldehyde hydrogenation. *ACS Catal.* 4 (5): 1340–1348.

21 Zhou, W., Lu, J., Zhou, K. et al. (2016). $CoSe_2$ nanoparticles embedded defective carbon nanotubes derived from MOFs as efficient electrocatalyst for hydrogen evolution reaction. *Nano Energy* 28: 143–150.

22 Zhai, X., Yu, Q., Liu, G. et al. (2021). Hierarchical microsphere MOF arrays with ultralow Ir doping for efficient hydrogen evolution coupled with hydrazine oxidation in seawater. *J. Mater. Chem. A* 9 (48): 27424–27433.

23 Senthil Raja, D., Lin, H.W., and Lu, S.Y. (2019). Synergistically well-mixed MOFs grown on nickel foam as highly efficient durable bifunctional electrocatalysts for overall water splitting at high current densities. *Nano Energy* 57: 1–13.

24 Qi, L., Su, Y.Q., Xu, Z. et al. (2020). Hierarchical 2D yarn-ball like metal–organic framework NiFe(dobpdc) as bifunctional electrocatalyst for efficient overall electrocatalytic water splitting. *J. Mater. Chem. A* 8 (43): 22974–22982.

25 Liu, T., Li, P., Yao, N. et al. (2019). CoP-doped MOF-based electrocatalyst for pH-universal hydrogen evolution reaction. *Angew. Chem. Int. Ed.* 58 (14): 4679–4684.

26 Tian, T., Huang, L., Ai, L., and Jiang, J. (2017). Surface anion-rich NiS_2 hollow microspheres derived from metal–organic frameworks as a robust electrocatalyst for the hydrogen evolution reaction. *J. Mater. Chem. A* 5 (39): 20985–20992.

27 Huang, Z., Xu, B., Li, Z. et al. (2020). Accurately regulating the electronic structure of Ni_xSe_Y@NC core-shell nanohybrids through controllable selenization of a Ni-MOF for pH-universal hydrogen evolution reaction. *Small* 16 (44): e2004231.

28 Zhu, B., Xia, D., and Zou, R. (2018). Metal-organic frameworks and their derivatives as bifunctional electrocatalysts. *Coordin. Chem. Rev.* 376: 430–448.

29 Liang, Z., Zhao, R., Qiu, T. et al. (2019). Metal-organic framework-derived materials for electrochemical energy applications. *EnergyChem* 1 (1): 100001.

30 Liang, Z., Qu, C., Xia, D. et al. (2018). Atomically dispersed metal sites in MOF-based materials for electrocatalytic and photocatalytic energy conversion. *Angew. Chem. Int. Ed.* 57 (31): 9604–9633.

31 Xiao, L., Wang, Z., and Guan, J. (2022). 2D MOFs and their derivatives for electrocatalytic applications: recent advances and new challenges. *Coordin. Chem. Rev.* 472: 214777.

32 Rui, K., Zhao, G., Lao, M. et al. (2019). Direct hybridization of noble metal nanostructures on 2D metal-organic framework nanosheets to catalyze hydrogen evolution. *Nano Lett.* 19 (12): 8447–8453.

33 Wang, M., Xu, Y., Peng, C.K. et al. (2021). Site-specified two-dimensional heterojunction of Pt nanoparticles/metal-organic frameworks for enhanced hydrogen evolution. *J. Am. Chem. Soc.* 143 (40): 16512–16518.

34 Koçyiğit, N., Özen, Ü.E., Özer, M. et al. (2017). Electrocatalytic activity of novel ball-type metallophthalocyanines with trifluoro methyl linkages in oxygen reduction reaction and application as Zn-air battery cathode catalyst. *Electrochim. Acta* 233: 237–248.

35 Monama, G.R., Mdluli, S.B., Mashao, G. et al. (2018). Palladium deposition on copper(II) phthalocyanine/metal organic framework composite and electrocatalytic activity of the modified electrode towards the hydrogen evolution reaction. *Renew. Energ.* 119: 62–72.

36 Wen, X. and Guan, J. (2019). Recent progress on MOF-derived electrocatalysts for hydrogen evolution reaction. *Appl. Mater. Today* 16: 146–168.

37 Chen, P., Zhou, T., Xing, L. et al. (2017). Atomically dispersed iron-nitrogen species as electrocatalysts for bifunctional oxygen evolution and reduction reactions. *Angew. Chem. Int. Ed.* 56 (2): 610–614.

38 Asefa, T. (2016). Metal-free and noble metal-free heteroatom-doped nanostructured carbons as prospective sustainable electrocatalysts. *Acc. Chem. Res.* 49 (9): 1873–1883.

39 Wang, Y., Liu, B., Shen, X. et al. (2021). Engineering the activity and stability of MOF-nanocomposites for efficient water oxidation. *Adv. Energy Mater.* 11 (16): 2003759.

40 Yang, D., Chen, Y., Su, Z. et al. (2021). Organic carboxylate-based MOFs and derivatives for electrocatalytic water oxidation. *Coordin. Chem. Rev.* 428: 213619.

41 Li, S., Gao, Y., Li, N. et al. (2021). Transition metal-based bimetallic MOFs and MOF-derived catalysts for electrochemical oxygen evolution reaction. *Energy Environ. Sci.* 14 (4): 1897–1927.

42 Gu, X., Ji, Y., Tian, J. et al. (2022). Combined MOF derivation and fluorination imparted efficient synergism of Fe-Co fluoride for oxygen evolution reaction. *Chem. Eng. J.* 427: 131576.

43 Qian, Q., Li, Y., Liu, Y. et al. (2019). Ambient fast synthesis and active sites deciphering of hierarchical foam-like trimetal-organic framework nanostructures as a platform for highly efficient oxygen evolution electrocatalysis. *Adv. Mater.* 31 (23): e1901139.

44 Li, J., Liu, P., Mao, J. et al. (2021). Structural and electronic modulation of conductive MOFs for efficient oxygen evolution reaction electrocatalysis. *J. Mater. Chem. A* 9 (18): 11248–11254.

45 Cao, C., Ma, D., Xu, Q. et al. (2018). Semisacrificial template growth of self-supporting MOF nanocomposite electrode for efficient electrocatalytic water oxidation. *Adv. Funct. Mater.* 29 (6): 1807418.

46 Liang, Q., Chen, J., Wang, F., and Li, Y. (2020). Transition metal-based metal-organic frameworks for oxygen evolution reaction. *Coordin. Chem. Rev.* 424: 213488.

47 Xia, Z., Fang, J., Zhang, X. et al. (2019). Pt nanoparticles embedded metal-organic framework nanosheets: a synergistic strategy towards bifunctional oxygen electrocatalysis. *Appl. Catal. B: Environ.* 245: 389–398.

48 Ye, B., Jiang, R., Yu, Z. et al. (2019). Pt (1 1 1) quantum dot engineered Fe-MOF nanosheet arrays with porous core-shell as an electrocatalyst for efficient overall water splitting. *J. Catal.* 380: 307–317.

49 Ma, X., Qi, K., Wei, S. et al. (2019). In situ encapsulated nickel-copper nanoparticles in metal-organic frameworks for oxygen evolution reaction. *J. Alloy Compd.* 770: 236–242.

50 Xie, X., Shang, L., Xiong, X. et al. (2021). Fe single-atom catalysts on MOF-5 derived carbon for efficient oxygen reduction reaction in proton exchange membrane fuel cells. *Adv. Energy Mater.* 12 (3): 2102688.

51 Kim, J., Yoo, J., Lee, H. et al. (2021). Single-atom M–N–C catalysts for oxygen reduction electrocatalysis. *Trends Chem.* 3 (9): 779–794.

52 Duan, X., Cao, F., Ding, R. et al. (2022). Cobalt-doping stabilized active and durable Sub-2 nm Pt nanoclusters for low-Pt-loading PEMFC cathode. *Adv. Energy Mater.* 12 (13): 2103144.

53 Wang, Y., Long, W., Wang, L. et al. (2018). Unlocking the door to highly active ORR catalysts for PEMFC applications: polyhedron-engineered Pt-based nanocrystals. *Energ. Environ. Sci.* 11 (2): 258–275.

54 Wu, Y., Yuan, M., Li, X. et al. (2022). Assistance of rearrangement of active sites in Fe/N/C catalyst for harvesting ultra-high power density PEMFCs. *Appl. Catal. B: Environ.* 312: 121365.

55 Liu, W., Zhang, C., Zhang, J. et al. (2022). Tuning the atomic configuration of Co-N-C electrocatalyst enables highly-selective H_2O_2 production in acidic media. *Appl. Catal. B: Environ.* 310: 121312.

56 Liu, M., Zhao, Z., Duan, X., and Huang, Y. (2019). Nanoscale structure design for high-performance Pt-based ORR catalysts. *Adv. Mater.* 31 (6): e1802234.

57 Deng, Y., Luo, J., Chi, B. et al. (2021). Advanced atomically dispersed metal–nitrogen–carbon catalysts toward cathodic oxygen reduction in PEM fuel cells. *Adv. Energy Mater.* 11 (37): 2101222.

58 Xiao, F., Wang, Y.C., Wu, Z.P. et al. (2021). Recent advances in electrocatalysts for proton exchange membrane fuel cells and alkaline membrane fuel cells. *Adv. Mater.* 33 (50): e2006292.

59 Yang, Y., Fu, J., Zhang, Y. et al. (2021). Molecular engineering for bottom-up construction of high-performance non-precious-metal electrocatalysts with well-defined active sites. *J. Phys. Chem. C* 125 (41): 22397–22420.

60 Fu, S., Zhu, C., Song, J. et al. (2017). Metal-organic framework-derived non-precious metal nanocatalysts for oxygen reduction reaction. *Adv. Energy Mater.* 7 (19): 1700363.

61 Singh, G., Lee, J., Karakoti, A. et al. (2020). Emerging trends in porous materials for CO_2 capture and conversion. *Chem. Soc. Rev.* 49 (13): 4360–4404.

62 Mu, Q., Zhu, W., Li, X. et al. (2020). Electrostatic charge transfer for boosting the photocatalytic CO_2 reduction on metal centers of 2D MOF/rGO heterostructure. *Appl. Catal. B: Environ.* 262: 118144.

63 Gulati, S., Vijayan, S., Mansi et al. (2023). Recent advances in the application of metal-organic frameworks (MOFs)-based nanocatalysts for direct conversion of carbon dioxide (CO_2) to value-added chemicals. *Coordin. Chem. Rev.* 474: 214853.

64 Younas, M., Rezakazemi, M., Daud, M. et al. (2020). Recent progress and remaining challenges in post-combustion CO_2 capture using metal-organic frameworks (MOFs). *Prog. Energ. Combust.* 80: 100849.

65 Wang, R., Haspel, H., Pustovarenko, A. et al. (2019). Maximizing Ag utilization in high-rate CO_2 electrochemical reduction with a coordination polymer-mediated gas diffusion electrode. *ACS Energy Lett.* 4 (8): 2024–2031.

66 Wu, Z.Z., Gao, F.Y., and Gao, M.R. (2021). Regulating the oxidation state of nanomaterials for electrocatalytic CO_2 reduction. *Energy Environ. Sci.* 14 (3): 1121–1139.

67 Wang, H., Liu, G., Chen, C. et al. (2021). Single-Ni sites embedded in multilayer nitrogen-doped graphene derived from amino-functionalized MOF for highly selective CO_2 electroreduction. *ACS Sustain. Chem. Eng.* 9 (10): 3792–3801.

68 Ju, W., Bagger, A., Hao, G.P. et al. (2017). Understanding activity and selectivity of metal-nitrogen-doped carbon catalysts for electrochemical reduction of CO_2. *Nat. Commun.* 8 (1): 944.

69 Meilikhov, M., Yusenko, K., Esken, D. et al. (2010). Metals@MOFs – loading MOFs with metal nanoparticles for hybrid functions. *Eur. J. Inorg. Chem.* 2010 (24): 3701–3714.

70 Moon, H.R., Lim, D.W., and Suh, M.P. (2013). Fabrication of metal nanoparticles in metal-organic frameworks. *Chem. Soc. Rev.* 42 (4): 1807–1824.

71 Liang, Z., Qiu, T., Gao, S. et al. (2021). Multi-scale design of metal–organic framework-derived materials for energy electrocatalysis. *Adv. Energy Mater.* 12 (4): 2003410.

72 Xin, Z., Liu, J., Wang, X. et al. (2021). Implanting polypyrrole in metal-porphyrin MOFs: enhanced electrocatalytic performance for CO_2RR. *ACS Appl. Mater. Interfaces* 13 (46): 54959–54966.

73 Hou, Y., Liang, Y.L., Shi, P.C. et al. (2020). Atomically dispersed Ni species on N-doped carbon nanotubes for electroreduction of CO_2 with nearly 100% CO selectivity. *Appl. Catal. B: Environ.* 271: 118929.

74 Hu, C., Wang, Y., Chen, J. et al. (2022). Main-group metal single-atomic regulators in dual-metal catalysts for enhanced electrochemical CO_2 reduction. *Small* 18 (22): e2201391.

75 Chen, C., Khosrowabadi Kotyk, J.F., and Sheehan, S.W. (2018). Progress toward commercial application of electrochemical carbon dioxide reduction. *Chem* 4 (11): 2571–2586.

76 Kornienko, N., Zhao, Y., Kley, C.S. et al. (2015). Metal-organic frameworks for electrocatalytic reduction of carbon dioxide. *J. Am. Chem. Soc.* 137 (44): 14129–14135.

77 Guntern, Y.T., Pankhurst, J.R., Vavra, J. et al. (2019). Nanocrystal/metal-organic framework hybrids as electrocatalytic platforms for CO_2 conversion. *Angew. Chem. Int. Ed.* 58 (36): 12632–12639.

78 Kung, C., Audu, C.O., Peters, A.W. et al. (2017). Copper nanoparticles installed in metal–organic framework thin films are electrocatalytically competent for CO_2 reduction. *ACS Energy Lett.* 2 (10): 2394–2401.

79 Xie, X., Zhang, X., Xie, M. et al. (2022). Au-activated N motifs in non-coherent cupric porphyrin metal organic frameworks for promoting and stabilizing ethylene production. *Nat. Commun.* 13 (1): 63.

80 Zhu, Z.H., Liang, Z., Hou, S. et al. (2021). Efficient CO_2 electroreduction coupled with semi-dehydrogenation of tetrahydroisoquinoline by MOFs modified electrodes. *J. Energy Chem.* 63: 328–335.

81 Yuan, W., Wu, J., Zhang, X. et al. (2020). In situ transformation of bismuth metal–organic frameworks for efficient selective electroreduction of CO_2 to formate. *J. Mater. Chem. A* 8 (46): 24486–24492.

82 Luo, H., Li, B., Ma, J.G., and Cheng, P. (2022). Surface modification of nano-Cu_2O for controlling CO_2 electrochemical reduction to ethylene and syngas. *Angew. Chem. Int. Ed.* 61 (11): e202116736.

83 Martín, N., Portillo, A., Ateka, A. et al. (2020). MOF-derived/zeolite hybrid catalyst for the production of light olefins from CO_2. *ChemCatChem* 12 (22): 5750–5758.

84 Lim, J., Fernández, C.A., Lee, S., and Hatzell, M.C. (2021). Ammonia and nitric acid demands for fertilizer use in 2050. *ACS Energy Lett.* 6 (10): 3676–3685.

85 Elishav, O., Mosevitzky Lis, B., Miller, E.M. et al. (2020). Progress and prospective of nitrogen-based alternative fuels. *Chem. Rev.* 120 (12): 5352–5436.

86 Deng, J., Iñiguez, J.A., and Liu, C. (2018). Electrocatalytic nitrogen reduction at low temperature. *Joule* 2 (5): 846–856.

87 Foster, S.L., Bakovic, S.I.P., Duda, R.D. et al. (2018). Catalysts for nitrogen reduction to ammonia. *Nat. Catal.* 1 (7): 490–500.

88 Kyriakou, V., Garagounis, I., Vourros, A. et al. (2020). An electrochemical Haber–Bosch process. *Joule* 4 (1): 142–158.

89 Smith, C., Hill, A.K., and Torrente-Murciano, L. (2020). Current and future role of Haber–Bosch ammonia in a carbon-free energy landscape. *Energy Environ. Sci.* 13 (2): 331–344.

90 Ren, Y., Yu, C., Tan, X. et al. (2021). Strategies to suppress hydrogen evolution for highly selective electrocatalytic nitrogen reduction: challenges and perspectives. *Energy Environ. Sci.* 14 (3): 1176–1193.

91 Chanda, D., Xing, R., Xu, T. et al. (2021). Electrochemical nitrogen reduction: recent progress and prospects. *Chem. Commun.* 57 (60): 7335–7349.

92 Guo, X., Du, H., Qu, F., and Li, J. (2019). Recent progress in electrocatalytic nitrogen reduction. *J. Mater. Chem. A* 7 (8): 3531–3543.

93 Cui, Q., Qin, G., Wang, W. et al. (2019). Mo-based 2D MOF as a highly efficient electrocatalyst for reduction of N_2 to NH_3: a density functional theory study. *J. Mater. Chem. A* 7 (24): 14510–14518.

94 Wan, Y., Wang, Z., Li, J., and Lv, R. (2021). Mo_2C-MoO_2 heterostructure quantum dots for enhanced electrocatalytic nitrogen reduction to ammonia. *ACS Nano* 16: 643–654.

95 Cai, W., Han, Y., Li, H. et al. (2020). Significantly enhanced electrocatalytic N_2 reduction to NH_3 by surface selenization with multiple functions. *J. Mater. Chem. A* 8 (39): 20331–20336.

96 Tang, C. and Qiao, S.Z. (2019). How to explore ambient electrocatalytic nitrogen reduction reliably and insightfully. *Chem. Soc. Rev.* 48 (12): 3166–3180.

97 He, H., Wen, H.M., Li, H.K., and Zhang, H.W. (2022). Recent advances in metal–organic frameworks and their derivatives for electrocatalytic nitrogen reduction to ammonia. *Coordin. Chem. Rev.* 471: 214761.

98 Lee, H., Koh, C.S.L., Lee, Y.H. et al. (2018). Favoring the unfavored: selective electrochemical nitrogen fixation using a reticular chemistry approach. *Sci. Adv.* 4 (3): eaar3208.

99 Sim, H.Y.F., Chen, J.R.T., Koh, C.S.L. et al. (2020). ZIF-induced d-band modification in a bimetallic nanocatalyst: achieving over 44% efficiency in the ambient nitrogen reduction reaction. *Angew. Chem. Int. Ed.* 59 (39): 16997–17003.

100 He, H., Zhu, Q.Q., Yan, Y. et al. (2022). Metal–organic framework supported Au nanoparticles with organosilicone coating for high-efficiency electrocatalytic N_2 reduction to NH_3. *Appl. Catal. B: Environ.* 302: 120840.

101 Wang, T., Cao, X., and Jiao, L. (2022). Progress in hydrogen production coupled with electrochemical oxidation of small molecules. *Angew. Chem. Int. Ed.* 61 (51): e202213328.

102 Bambagioni, V., Bevilacqua, M., Bianchini, C. et al. (2010). Self-sustainable production of hydrogen, chemicals, and energy from renewable alcohols by electrocatalysis. *ChemSusChem* 3 (7): 851–855.

103 Li, Y., Wei, X., Chen, L., and Shi, J. (2021). Electrocatalytic hydrogen production trilogy. *Angew. Chem. Int. Ed.* 60 (36): 19550–19571.

104 Wang, J., Li, W., Wen, Y. et al. (2015). Rh-Ni-B nanoparticles as highly efficient catalysts for hydrogen generation from hydrous hydrazine. *Adv. Energy Mater.* 5 (10): 1401879.

105 Rees, N.V. and Compton, R.G. (2011). Carbon-free energy: a review of ammonia- and hydrazine-based electrochemical fuel cells. *Energy Environ. Sci.* 4 (4): 1255–1260.

106 Hannauer, J., Akdim, O., Demirci, U.B. et al. (2011). High-extent dehydrogenation of hydrazine borane $N_2H_4BH_3$ by hydrolysis of BH_3 and decomposition of N_2H_4. *Energy Environ. Sci.* 4 (9): 3355–3558.

107 Kim, J.D., Choi, M.Y., and Choi, H.C. (2017). Graphene-oxide-supported Pt nanoparticles with high activity and stability for hydrazine electro-oxidation in a strong acidic solution. *Appl. Surf. Sci.* 420: 700–706.

108 Singh, S.K. and Xu, Q. (2009). Complete conversion of hydrous hydrazine to hydrogen at room temperature for chemical hydrogen storage. *J. Am. Chem. Soc.* 131 (50): 18032–18033.

109 Han, Y., Han, L., Zhang, L., and Dong, S. (2015). Ultrasonic synthesis of highly dispersed Au nanoparticles supported on Ti-based metal–organic frameworks for electrocatalytic oxidation of hydrazine. *J. Mater. Chem. A* 3 (28): 14669–14674.

110 Bie, C., Wang, L., and Yu, J. (2022). Challenges for photocatalytic overall water splitting. *Chem* 8 (6): 1567–1574.

111 Zou, Z., Ye, J., Sayama, K., and Arakawa, H. (2001). Direct splitting of water under visible light irradiation with an oxide semiconductor photocatalyst. *Nature* 414 (6864): 625–627.

112 Wang, C., Wang, J.L., and Lin, W. (2012). Elucidating molecular iridium water oxidation catalysts using metal-organic frameworks: a comprehensive structural, catalytic, spectroscopic, and kinetic study. *J. Am. Chem. Soc.* 134 (48): 19895–19908.

113 Li, Z., Wang, Q., Kong, C. et al. (2015). Interface charge transfer versus surface proton reduction: which is more pronounced on photoinduced hydrogen generation over sensitized Pt cocatalyst on RGO? *J. Phys. Chem. C* 119 (24): 13561–13568.

114 Dresselhaus, M.S. and Thomas, I.L. (2001). Alternative energy technologies. *Nature* 414 (6861): 332–337.

115 Rahman, M.Z., Kibria, M.G., and Mullins, C.B. (2020). Metal-free photocatalysts for hydrogen evolution. *Chem. Soc. Rev.* 49 (6): 1887–1931.

116 Shi, L., Li, P., Zhou, W. et al. (2016). n-type boron phosphide as a highly stable, metal-free, visible-light-active photocatalyst for hydrogen evolution. *Nano Energy* 28: 158–163.

117 Li, X., Kikugawa, N., and Ye, J. (2008). Nitrogen-doped lamellar niobic acid with visible light-responsive photocatalytic activity. *Adv. Mater.* 20 (20): 3816–3819.

118 Usman, M., Mendiratta, S., and Lu, K.L. (2017). Semiconductor metal-organic frameworks: future low-bandgap materials. *Adv. Mater.* 29 (6): 1605071.

119 Corma, A. and Garcia, H. (2004). Zeolite-based photocatalysts. *Chem. Commun.* 13: 1443–1459.

120 Alvaro, M., Carbonell, E., Ferrer, B. et al. (2007). Semiconductor behavior of a metal-organic framework (MOF). *Chemistry* 13 (18): 5106–5112.

121 Shi, Y., Yang, A.-F., Cao, C.-S., and Zhao, B. (2019). Applications of MOFs: recent advances in photocatalytic hydrogen production from water. *Coordin. Chem. Rev.* 390: 50–75.

122 Silva, C.G., Corma, A., and García, H. (2010). Metal–organic frameworks as semiconductors. *J. Mater. Chem.* 20 (16): 3141–3156.

123 Gomes Silva, C., Luz, I., Llabres i Xamena, F.X. et al. (2010). Water stable Zr-benzenedicarboxylate metal-organic frameworks as photocatalysts for hydrogen generation. *Chemistry* 16 (36): 11133–11138.

124 Wen, M., Mori, K., Kamegawa, T., and Yamashita, H. (2014). Amine-functionalized MIL-101(Cr) with imbedded platinum nanoparticles as a durable photocatalyst for hydrogen production from water. *Chem. Commun.* 50 (79): 11645–11648.

125 Yue, X., Yi, S., Wang, R. et al. (2018). Synergistic effect based Ni_xCo_{1-x} architected $Zn_{0.75}Cd_{0.25}S$ nanocrystals: an ultrahigh and stable photocatalysts for hydrogen evolution from water splitting. *Appl. Catal. B: Environ.* 224: 17–26.

126 Wang, M., Zhen, W., Tian, B. et al. (2018). The inhibition of hydrogen and oxygen recombination reaction by halogen atoms on over-all water splitting over Pt-TiO_2 photocatalyst. *Appl. Catal. B: Environ.* 236: 240–252.

127 Wang, C., deKrafft, K.E., and Lin, W. (2012). Pt nanoparticles@photoactive metal-organic frameworks: efficient hydrogen evolution via synergistic photoexcitation and electron injection. *J. Am. Chem. Soc.* 134 (17): 7211–7214.

128 Xiao, J., Shang, Q., Xiong, Y. et al. (2016). Boosting photocatalytic hydrogen production of a metal–organic framework decorated with platinum nanoparticles: the platinum location matters. *Angew. Chem., Int. Ed.* 55 (32): 9389–9393.

129 Huang, X., Li, X., Luan, Q. et al. (2021). Highly dispersed Pt clusters encapsulated in MIL-125-NH2 via in situ auto-reduction method for photocatalytic H_2 production under visible light. *Nano Res.* 14 (11): 4250–4257.

130 Guo, W., Lv, H., Chen, Z. et al. (2016). Self-assembly of polyoxometalates, Pt nanoparticles and metal–organic frameworks into a hybrid material for synergistic hydrogen evolution. *J. Mater. Chem. A* 4 (16): 5952–5957.

131 Xu, M., Li, D., Sun, K. et al. (2021). Interfacial microenvironment modulation boosting electron transfer between metal nanoparticles and MOFs for enhanced photocatalysis. *Angew. Chem. Int. Ed.* 60 (30): 16372–16376.

132 Zhen, W., Ma, J., and Lu, G. (2016). Small-sized Ni(1 1 1) particles in metal-organic frameworks with low over-potential for visible photocatalytic hydrogen generation. *Appl. Catal. B: Environ.* 190: 12–25.

133 Wang, G., Sun, Q., Liu, Y. et al. (2015). A bismuth-based metal-organic framework as an efficient visible-light-driven photocatalyst. *Chemistry* 21 (6): 2364–2367.

134 Lin, R., Shen, L., Ren, Z. et al. (2014). Enhanced photocatalytic hydrogen production activity via dual modification of MOF and reduced graphene oxide on CdS. *Chem. Commun.* 50 (62): 8533–8535.

135 Wang, Y., Zhang, Y., Jiang, Z. et al. (2016). Controlled fabrication and enhanced visible-light photocatalytic hydrogen production of Au@CdS/MIL-101 heterostructure. *Appl. Catal. B: Environ.* 185: 307–314.

136 Wang, Z., Li, C., and Domen, K. (2019). Recent developments in heterogeneous photocatalysts for solar-driven overall water splitting. *Chem. Soc. Rev.* 48 (7): 2109–2125.

137 Xiao, J.D., Han, L., Luo, J. et al. (2018). Integration of plasmonic effects and Schottky junctions into metal-organic framework composites: steering charge flow for enhanced visible-light photocatalysis. *Angew. Chem. Int. Ed.* 57 (4): 1103–1107.

138 Hou, C., Li, T., Chen, Y., and Fu, W. (2015). Improved photocurrents for water oxidation by using metal–organic framework derived hybrid porous Co_3O_4@carbon/$BiVO_4$ as a photoanode. *ChemPlusChem* 80 (9): 1465–1471.

139 Zhang, J., Bai, T., Huang, H. et al. (2020). Metal-organic-framework-based photocatalysts optimized by spatially separated cocatalysts for overall water splitting. *Adv. Mater.* 32 (49): e2004747.

140 Inoue, T., Fujishima, A., Konishi, S., and Honda, K. (1979). Photoelectrocatalytic reduction of carbon dioxide in aqueous suspensions of semiconductor powders. *Nature* 277 (5698): 637–638.

141 Wang, S. and Wang, X. (2016). Imidazolium ionic liquids, imidazolylidene heterocyclic carbenes, and zeolitic imidazolate frameworks for CO_2 capture and photochemical reduction. *Angew. Chem. Int. Ed.* 55 (7): 2308–2320.

142 Wang, J., Wang, C., and Lin, W. (2012). Metal–organic frameworks for light harvesting and photocatalysis. *ACS Catal.* 2 (12): 2630–2640.

143 Li, R., Zhang, W., and Zhou, K. (2018). Metal-organic-framework-based catalysts for photoreduction of CO_2. *Adv. Mater.* 30 (35): e1705512.

144 Li, D., Kassymova, M., Cai, X. et al. (2020). Photocatalytic CO_2 reduction over metal-organic framework-based materials. *Coordin. Chem. Rev.* 412: 213262.

145 He, H., Perman, J.A., Zhu, G., and Ma, S. (2016). Metal-organic frameworks for CO_2 chemical transformations. *Small* 12 (46): 6309–6324.

146 Wang, S. and Wang, X. (2015). Multifunctional metal-organic frameworks for photocatalysis. *Small* 11 (26): 3097–3112.

147 Zhu, Q.L. and Xu, Q. (2014). Metal-organic framework composites. *Chem. Soc. Rev.* 43 (16): 5468–5512.

148 Liang, Z., Qu, C., Guo, W. et al. (2018). Pristine metal-organic frameworks and their composites for energy storage and conversion. *Adv. Mater.* 30 (37): e1702891.

149 Zhang, W., Shi, W., Ji, W. et al. (2020). Microenvironment of MOF channel coordination with Pt NPs for selective hydrogenation of unsaturated aldehydes. *ACS Catal.* 10 (10): 5805–5813.

150 Guo, J., Wan, Y., Zhu, Y. et al. (2020). Advanced photocatalysts based on metal nanoparticle/metal-organic framework composites. *Nano Res.* 14 (7): 2037–2052.

151 Najafi, M., Abednatanzi, S., Yousefi, A., and Ghaedi, M. (2021). Photocatalytic activity of supported metal nanoparticles and single atoms. *Chemistry* 27 (72): 17999–18014.

152 Deng, X., Yang, L., Huang, H. et al. (2019). Shape-defined hollow structural Co-MOF-74 and metal nanoparticles@Co-MOF-74 composite through a transformation strategy for enhanced photocatalysis performance. *Small* 15 (35): e1902287.

153 Li, J., Cui, Z., Zheng, Y. et al. (2022). Atomic-layer Fe_2O_3-modified 2D porphyrinic metal-organic framework for enhanced photocatalytic disinfection through electron-withdrawing effect. *Appl. Catal. B: Environ.* 317: 121701.

154 Liu, L. and Corma, A. (2018). Metal catalysts for heterogeneous catalysis: from single atoms to nanoclusters and nanoparticles. *Chem. Rev.* 118 (10): 4981–5079.

155 Guo, F., Yang, S., Liu, Y. et al. (2019). Size engineering of metal–organic framework MIL-101(Cr)–Ag hybrids for photocatalytic CO_2 reduction. *ACS Catal.* 9 (9): 8464–8470.

156 Chen, L., Wang, Y., Yu, F. et al. (2019). A simple strategy for engineering heterostructures of Au nanoparticle-loaded metal–organic framework nanosheets to achieve plasmon-enhanced photocatalytic CO_2 conversion under visible light. *J. Mater. Chem. A* 7 (18): 11355–11361.

157 Guo, F., Wei, Y., Wang, S. et al. (2019). Pt nanoparticles embedded in flowerlike NH_2-UiO-68 for enhanced photocatalytic carbon dioxide reduction. *J. Mater. Chem. A* 7 (46): 26490–26495.

158 Chang, K., Wang, T., and Chen, J. (2017). Hydrogenation of CO_2 to methanol over CuCeTiO catalysts. *Appl. Catal. B: Environ.* 206: 704–711.

159 Onishi, N. and Himeda, Y. (2022). Homogeneous catalysts for CO_2 hydrogenation to methanol and methanol dehydrogenation to hydrogen generation. *Coordin. Chem. Rev.* 472: 214767.

160 Lais, A., Gondal, M.A., and Dastageer, M.A. (2017). Semiconducting oxide photocatalysts for reduction of CO_2 to methanol. *Environ. Chem. Lett.* 16 (1): 183–210.

161 Kar, S., Sen, R., Goeppert, A., and Prakash, G.K.S. (2018). Integrative CO_2 capture and hydrogenation to methanol with reusable catalyst and amine: toward a carbon neutral methanol economy. *J. Am. Chem. Soc.* 140 (5): 1580–1583.

162 Yang, Y., Pan, Y., Tu, X., and Liu, C. (2022). Nitrogen doping of indium oxide for enhanced photocatalytic reduction of CO_2 to methanol. *Nano Energy* 101: 107613.

163 Hu, J., Yu, L., Deng, J. et al. (2021). Sulfur vacancy-rich MoS_2 as a catalyst for the hydrogenation of CO_2 to methanol. *Nat. Catal.* 4 (3): 242–250.

164 Becerra, J., Nguyen, D., Gopalakrishnan, V., and Do, T. (2020). Plasmonic Au nanoparticles incorporated in the zeolitic imidazolate framework (ZIF-67) for the efficient sunlight-driven photoreduction of CO_2. *ACS Appl. Energ. Mater.* 3 (8): 7659–7665.

165 Izadpanah Ostad, M., Niknam Shahrak, M., and Galli, F. (2022). The effect of different reaction media on photocatalytic activity of Au- and Cu-decorated zeolitic imidazolate Framework-8 toward CO_2 photoreduction to methanol. *J. Solid State Chem.* 315: 123514.

166 Chen, D., Xing, H., Wang, C., and Su, Z. (2016). Highly efficient visible-light-driven CO_2 reduction to formate by a new anthracene-based zirconium MOF via dual catalytic routes. *J. Mater. Chem. A* 4 (7): 2657–2662.

167 Sun, D., Liu, W., Fu, Y. et al. (2014). Noble metals can have different effects on photocatalysis over metal-organic frameworks (MOFs): a case study on M/NH_2-MIL-125(Ti) (M=Pt and Au). *Chemistry* 20 (16): 4780–4788.

168 Liao, F., Huang, Y., Ge, J. et al. (2011). Morphology-dependent interactions of ZnO with Cu nanoparticles at the materials' interface in selective hydrogenation of CO_2 to CH_3OH. *Angew. Chem. Int. Ed.* 50 (9): 2162–2165.

169 Fang, F., Liu, Y., Sun, X. et al. (2021). TiO_2 Facet-dependent reconstruction and photocatalysis of CuO_x/TiO_2 photocatalysts in CO_2 photoreduction. *Appl. Surf. Sci.* 564: 150407.

170 Reske, R., Mistry, H., Behafarid, F. et al. (2014). Particle size effects in the catalytic electroreduction of CO_2 on Cu nanoparticles. *J. Am. Chem. Soc.* 136 (19): 6978–6986.

171 Dai, S., Kajiwara, T., Ikeda, M. et al. (2022). Ultrasmall copper nanoclusters in zirconium metal-organic frameworks for the photoreduction of CO_2. *Angew. Chem. Int. Ed.* 134: e202211848.

172 Xu, C., Pan, Y., Wan, G. et al. (2019). Turning on visible-light photocatalytic C-H oxidation over metal-organic frameworks by introducing metal-to-cluster charge transfer. *J. Am. Chem. Soc.* 141 (48): 19110–19117.

173 Xu, C., Liu, H., Li, D. et al. (2018). Direct evidence of charge separation in a metal-organic framework: efficient and selective photocatalytic oxidative coupling of amines via charge and energy transfer. *Chem. Sci.* 9 (12): 3152–3158.

174 Zhang, T. and Lin, W. (2014). Metal-organic frameworks for artificial photosynthesis and photocatalysis. *Chem. Soc. Rev.* 43 (16): 5982–5993.

175 Luo, T., Li, L., Chen, Y. et al. (2021). Construction of C-C bonds via photoreductive coupling of ketones and aldehydes in the metal-organic-framework MFM-300(Cr). *Nat. Commun.* 12 (1): 3583.

176 Li, X.C., Zhang, Z., Xiao, W.M. et al. (2019). Mechanochemistry-assisted encapsulation of metal nanoparticles in MOF matrices via a sacrificial strategy. *J. Mater. Chem. A* 7 (24): 14504–14509.

177 Wei, H.S., Liu, X.Y., Wang, A.Q. et al. (2014). FeO_x-supported platinum single-atom and pseudo-single-atom catalysts for chemoselective hydrogenation of functionalized nitroarenes. *Nat. Commun.* 5 (1): 5634.

178 Murugesan, K., Senthamarai, T., Chandrashekhar, V.G. et al. (2020). Catalytic reductive aminations using molecular hydrogen for synthesis of different kinds of amines. *Chem. Soc. Rev.* 49 (17): 6273–6328.

179 Li, B., Ma, J.G., and Cheng, P. (2018). Silica-protection-assisted encapsulation of Cu_2O nanocubes into a metal-organic framework (ZIF-8) to provide a composite catalyst. *Angew. Chem. Int. Ed.* 57 (23): 6834–6837.

180 Jagadeesh, R.V., Surkus, A.E., Junge, H. et al. (2013). Nanoscale Fe_2O_3-based catalysts for selective hydrogenation of nitroarenes to anilines. *Science* 342 (6162): 1073–1076.

181 Zhang, J., Wang, L., Shao, Y. et al. (2017). A Pd@zeolite catalyst for nitroarene hydrogenation with high product selectivity by sterically controlled adsorption in the zeolite micropores. *Angew. Chem. Int. Ed.* 56 (33): 9747–9751.

182 Wang, L., Guan, E., Zhang, J. et al. (2018). Single-site catalyst promoters accelerate metal-catalyzed nitroarene hydrogenation. *Nat. Commun.* 9 (1): 1362.

183 Furukawa, S., Yoshida, Y., and Komatsu, T. (2014). Chemoselective hydrogenation of nitrostyrene to aminostyrene over Pd- and Rh-based intermetallic compounds. *ACS Catal.* 4 (5): 1441–1450.

184 Wang, L., Zhang, J., Wang, H. et al. (2016). Activity and selectivity in nitroarene hydrogenation over Au nanoparticles on the edge/corner of anatase. *ACS Catal.* 6 (7): 4110–4116.

185 Li, L., Li, Y., Jiao, L. et al. (2022). Light-induced selective hydrogenation over PdAg nanocages in hollow MOF microenvironment. *J. Am. Chem. Soc.* 144 (37): 17075–17085.

186 Shen, L., Luo, M., Huang, L. et al. (2015). A clean and general strategy to decorate a titanium metal-organic framework with noble-metal nanoparticles for versatile photocatalytic applications. *Inorg. Chem.* 54 (4): 1191–1193.

187 Lou, Y., Zhang, Y., Cheng, L. et al. (2018). A stable plasmonic $Cu@Cu_2O/ZnO$ heterojunction for enhanced photocatalytic hydrogen generation. *ChemSusChem* 11 (9): 1505–1511.

188 Wang, Z., Song, H., Pang, H. et al. (2019). Photo-assisted methanol synthesis via CO_2 reduction under ambient pressure over plasmonic Cu/ZnO catalysts. *Appl. Catal. B: Environ.* 250: 10–16.

189 Xiao, L., Zhang, Q., Chen, P. et al. (2019). Copper-mediated metal-organic framework as efficient photocatalyst for the partial oxidation of aromatic alcohols under visible light irradiation: synergism of plasmonic effect and schottky junction. *Appl. Catal. B: Environ.* 248: 380–387.

190 Sun, Z.X., Sun, K., Gao, M.L. et al. (2022). Optimizing Pt electronic states through formation of a Schottky junction on non-reducible metal-organic frameworks for enhanced photocatalysis. *Angew. Chem. Int. Ed.* 61 (32): e202206108.

191 Sun, D. and Li, Z. (2016). Double-solvent method to Pd nanoclusters encapsulated inside the cavity of NH_2–UiO-66(Zr) for efficient visible-light-promoted Suzuki coupling reaction. *J. Phys. Chem. C* 120 (35): 19744–19750.

192 Sun, D., Xu, M., Jiang, Y. et al. (2018). Small-sized bimetallic CuPd nanoclusters encapsulated inside cavity of NH_2-UiO-66(Zr) with superior performance for light-induced Suzuki coupling reaction. *Small Methods* 2 (12): 1800164.

193 Guillena, G., Ramón, D.J., and Yus, M. (2010). Hydrogen autotransfer in the N-alkylation of amines and related compounds using alcohols and amines as electrophiles. *Chem. Rev.* 110 (3): 1611–1641.

194 Hamid, M.H.S.A., Allen, C.L., Lamb, G.W. et al. (2009). Ruthenium-catalyzed N-alkylation of amines and sulfonamides using borrowing hydrogen methodology. *J. Am. Chem. Soc.* 131 (5): 1766–1774.

195 Fertig, R., Irrgang, T., Freitag, F. et al. (2018). Manganese-catalyzed and base-switchable synthesis of amines or imines via borrowing hydrogen or dehydrogenative condensation. *ACS Catal.* 8 (9): 8525–8530.

196 Wang, D. and Li, Z. (2016). Coupling MOF-based photocatalysis with Pd catalysis over Pd@MIL-100(Fe) for efficient N-alkylation of amines with alcohols under visible light. *J. Catal.* 342: 151–157.

197 Ren, W., Gao, J., Lei, C. et al. (2018). Recyclable metal-organic framework/cellulose aerogels for activating peroxymonosulfate to degrade organic pollutants. *Chem. Eng. J.* 349: 766–774.

198 Hu, T., Tang, L., Feng, H. et al. (2022). Metal-organic frameworks (MOFs) and their derivatives as emerging catalysts for electro-Fenton process in water purification. *Coord. Chem. Rev.* 451: 214277.

199 Dapaah, M.F., Niu, Q., Yu, Y. et al. (2022). Efficient persistent organic pollutant removal in water using MIL-metal–organic framework driven Fenton-like reactions: a critical review. *Chem. Eng. J.* 431: 134182.

200 He, Y., Wang, Z., Wang, H. et al. (2022). Confinement of ZIF-derived copper-cobalt-zinc oxides in carbon framework for degradation of organic pollutants. *J. Hazard. Mater.* 440: 129811.

201 Pi, Y., Li, X., Xia, Q. et al. (2018). Adsorptive and photocatalytic removal of persistent organic pollutants (POPs) in water by metal-organic frameworks (MOFs). *Chem. Eng. J.* 337: 351–371.

202 Wang, C., Li, J., Lv, X. et al. (2014). Photocatalytic organic pollutants degradation in metal–organic frameworks. *Energy Environ. Sci.* 7 (9): 2831–2867.

203 Ding, J., Pan, Y., Li, L. et al. (2020). Synergetic adsorption and electrochemical classified recycling of Cr(VI) and dyes in synthetic dyeing wastewater. *Chem. Eng. J.* 384: 123232.

204 Zhu, Z.H., Liu, Y., Song, C. et al. (2021). Porphyrin-based two-dimensional layered metal-organic framework with sono-/photocatalytic activity for water decontamination. *ACS Nano* 16 (1): 1346–1357.

205 Ma, X., Zhao, S., Tian, Z. et al. (2022). MOFs meet wood: reusable magnetic hydrophilic composites toward efficient water treatment with super-high dye adsorption capacity at high dye concentration. *Chem. Eng. J.* 446: 136851.

206 Ali, S.A., Rachman, I.B., and Saleh, T.A. (2017). Simultaneous trapping of Cr(III) and organic dyes by a pH-responsive resin containing zwitterionic aminomethylphosphonate ligands and hydrophobic pendants. *Chem. Eng. J.* 330: 663–674.

207 Li, H., Su, L., Zheng, J. et al. (2023). MOFs derived carbon supporting CuCo nanospheres as efficient catalysts of peroxymonosulfate for rapid removal of organic pollutant. *Chem. Eng. J.* 451: 139114.

208 Liang, R., Jing, F., Shen, L. et al. (2015). M@MIL-100(Fe) (M = Au, Pd, Pt) nanocomposites fabricated by a facile photodeposition process: efficient visible-light photocatalysts for redox reactions in water. *Nano Res.* 8 (10): 3237–3249.

209 Liang, R., Luo, S., Jing, F. et al. (2015). A simple strategy for fabrication of Pd@MIL-100(Fe) nanocomposite as a visible-light-driven photocatalyst for the treatment of pharmaceuticals and personal care products (PPCPs). *Appl. Catal. B: Environ.* 176, 177: 240–248.

210 Chen, S., Huang, Y., Han, X. et al. (2018). Simultaneous and efficient removal of Cr(VI) and methyl orange on LDHs decorated porous carbons. *Chem. Eng. J.* 352: 306–315.

211 Shen, L., Wu, W., Liang, R. et al. (2013). Highly dispersed palladium nanoparticles anchored on UiO-66(NH_2) metal-organic framework as a reusable and dual functional visible-light-driven photocatalyst. *Nanoscale* 5 (19): 9374–9382.

212 Joseph, L., Jun, B., Jang, M. et al. (2019). Removal of contaminants of emerging concern by metal-organic framework nanoadsorbents: a review. *Chem. Eng. J.* 369: 928–946.

213 Yin, Y., Shi, M., Ren, Y. et al. (2020). Wrinkle structure on multifunctional MOFs to facilitate PPCPs adsorption in wastewater. *Chem. Eng. J.* 387: 124196.

214 Martínez, C., Canle, L.M., Fernández, M.I. et al. (2011). Aqueous degradation of diclofenac by heterogeneous photocatalysis using nanostructured materials. *Appl. Catal. B: Environ.* 107 (1-2): 110–118.

215 Xiao, W., Xiao, L., Yang, Y. et al. (2022). Catalytic degradation of organic pollutants for water remediation over Ag nanoparticles immobilized on amine-functionalized metal-organic frameworks. *Nano Res.* 15 (9): 7887–7895.

216 Guo, H., Guo, D., Zheng, Z. et al. (2015). Visible-light photocatalytic activity of Ag@MIL-125(Ti) microspheres. *Appl. Organomet. Chem.* 29 (9): 618–623.

217 Huang, Y., Zhang, Y., Chen, X. et al. (2014). Bimetallic alloy nanocrystals encapsulated in ZIF-8 for synergistic catalysis of ethylene oxidative degradation. *Chem. Commun.* 50 (70): 10115–10117.

218 Wang, X., Wang, Y., and Ying, Y. (2021). Recent advances in sensing applications of metal nanoparticle/metal–organic framework composites. *TrAC-Trend. Anal. Chem.* 143: 116395.

219 Yin, H., Wang, C., Zhu, H. et al. (2008). Colloidal deposition synthesis of supported gold nanocatalysts based on Au-Fe_3O_4 dumbbell nanoparticles. *Chem. Commun.* 36: 4357–4359.

220 Carrettin, S., Concepcion, P., Corma, A. et al. (2004). Nanocrystalline CeO_2 increases the activity of Au for CO oxidation by two orders of magnitude. *Angew. Chem. Int. Ed.* 43 (19): 2538–2540.

221 Dimitratos, N., Villa, A., Wang, D. et al. (2006). Pd and Pt catalysts modified by alloying with Au in the selective oxidation of alcohols. *J. Catal.* 244 (1): 113–121.

222 Dan, I.E., Jennifer, K.E., Philip, L. et al. (2006). Solvent-free oxidation of primary alcohols to aldehydes using Au-Pd/ TiO_2 catalysts. *Science* 5759: 362–365.

223 Jiang, H.L., Liu, B., Akita, T. et al. (2009). Au@ZIF-8: CO oxidation over gold nanoparticles deposited to metal-organic framework. *J. Am. Chem. Soc.* 131: 11302–11303.

224 Wu, R., Qian, X., Zhou, K. et al. (2013). Highly dispersed Au nanoparticles immobilized on Zr-based metal–organic frameworks as heterostructured catalyst for CO oxidation. *J. Mater. Chem. A* 1 (45): 14294–14299.

225 Carne-Sanchez, A., Imaz, I., Cano-Sarabia, M., and Maspoch, D. (2013). A spray-drying strategy for synthesis of nanoscale metal-organic frameworks and their assembly into hollow superstructures. *Nat. Chem.* 5 (3): 203–211.

226 Carne, A., Carbonell, C., Imaz, I., and Maspoch, D. (2011). Nanoscale metal-organic materials. *Chem. Soc. Rev.* 40 (1): 291–305.

227 Luan, Y., Qi, Y., Gao, H. et al. (2014). Synthesis of an amino-functionalized metal–organic framework at a nanoscale level for gold nanoparticle deposition and catalysis. *J. Mater. Chem. A* 2: 20588–20596.

228 Wu, F., Zhang, D., Peng, M. et al. (2016). Microfluidic synthesis enables dense and uniform loading of surfactant-free PtSn nanocrystals on carbon supports for enhanced ethanol oxidation. *Angew. Chem. Int. Ed.* 55 (16): 4952–4956.

229 Bai, L., Wang, X., Chen, Q. et al. (2016). Explaining the size dependence in platinum-nanoparticle-catalyzed hydrogenation reactions. *Angew. Chem. Int. Ed.* 55: 15656–15661.

230 Fan, Z. and Zhang, H. (2016). Template synthesis of noble metal nanocrystals with unusual crystal structures and their catalytic applications. *Acc. Chem. Res.* 49 (12): 2841–2850.

231 Li, G., Zhao, S., Zhang, Y., and Tang, Z. (2018). Metal-organic frameworks encapsulating active nanoparticles as emerging composites for catalysis: recent progress and perspectives. *Adv. Mater.* 30 (51): 1800702.

232 He, D., Zhang, L., He, D. et al. (2016). Amorphous nickel boride membrane on a platinum-nickel alloy surface for enhanced oxygen reduction reaction. *Nat. Commun.* 7: 12362.

233 Chen, Y.Z., Gu, B., Uchida, T. et al. (2019). Location determination of metal nanoparticles relative to a metal-organic framework. *Nat. Commun.* 10 (1): 3462.

234 Zhang, W., Zheng, B., Shi, W. et al. (2018). Site-selective catalysis of a multifunctional linear molecule: the steric hindrance of metal-organic framework channels. *Adv. Mater.* 30 (23): 1800643.

235 Delidovich, I., Hausoul, P.J., Deng, L. et al. (2016). Alternative monomers based on lignocellulose and their use for polymer production. *Chem. Rev.* 116 (3): 1540–1599.

236 Zhen, R., Yusen, Y., Wang, S. et al. (2021). Pt atomic clusters catalysts with local charge transfer towards selective oxidation of furfural. *Appl. Catal. B: Environ.* 295: 120290.

237 Kar, S., Zhou, Q.Q., Ben-David, Y., and Milstein, D. (2022). Catalytic furfural/5-hydroxymethyl furfural oxidation to furoic acid/furan-2,5-dicarboxylic acid with H_2 production using alkaline water as the formal oxidant. *J. Am. Chem. Soc.* 144 (3): 1288–1295.

238 Ning, L., Liao, S., Cui, H. et al. (2017). Selective conversion of renewable furfural with ethanol to produce furan-2-acrolein mediated by Pt@MOF-5. *ACS Sust. Chem. Eng.* 6 (1): 135–142.

239 Molnár, Á., Sárkány, A., and Varga, M. (2001). Hydrogenation of carbon–carbon multiple bonds: chemo-, regio- and stereo-selectivity. *J. Mol. Catal. A: Chem.* 173 (1-2): 185–221.

240 Liu, P., Zhao, Y., Qin, R. et al. (2016). Photochemical route for synthesizing atomically dispersed palladium catalysts. *Science* 352 (6287): 797–800.

241 Kar, A.K. and Srivastava, R. (2018). An efficient and sustainable catalytic reduction of carbon–carbon multiple bonds, aldehydes, and ketones using a Cu nanoparticle decorated metal organic framework. *New J. Chem.* 42 (12): 9557–9567.

242 Chen, W.-J., Cheng, B.-H., Sun, Q.-T., and Jiang, H. (2018). Preparation of MOF confined Ag nanoparticles for the highly active, size selective hydrogenation of olefins. *ChemCatChem* 10 (17): 3659–3665.

243 Zhang, W., Lu, G., Cui, C. et al. (2014). A family of metal-organic frameworks exhibiting size-selective catalysis with encapsulated noble-metal nanoparticles. *Adv. Mater.* 26 (24): 4056–4060.

244 Natesakhawat, S., Lekse, J.W., Baltrus, J.P. et al. (2012). Active sites and structure–activity relationships of copper-based catalysts for carbon dioxide hydrogenation to methanol. *ACS Catal.* 2 (8): 1667–1676.

245 Fang, R., Liu, H., Luque, R., and Li, Y. (2015). Efficient and selective hydrogenation of biomass-derived furfural to cyclopentanone using Ru catalysts. *Green Chem.* 17 (8): 4183–4188.

246 Xue, Y., Zhao, G., Yang, R. et al. (2021). 2D metal-organic framework-based materials for electrocatalytic, photocatalytic and thermocatalytic applications. *Nanoscale* 13 (7): 3911–3936.

247 Deng, Z., Yu, H., Wang, L. et al. (2019). Ferrocene-based metal–organic framework nanosheets loaded with palladium as a super-high active hydrogenation catalyst. *J. Mater. Chem. A* 7 (26): 15975–15980.

248 Yu, Q., Wu, X., Yao, X. et al. (2011). Mesoporous ceria–zirconia–alumina nanocomposite-supported copper as a superior catalyst for simultaneous catalytic elimination of NO–CO. *Catal. Commun.* 12 (14): 1311–1317.

249 Larmier, K., Liao, W.C., Tada, S. et al. (2017). CO_2 -to-methanol hydrogenation on zirconia supported copper nanoparticles: reaction intermediates and the role of the metal-support interface. *Angew. Chem. Int. Ed.* 56 (9): 2318–2323.

250 Redfern, L.R., Li, Z., Zhang, X., and Farha, O.K. (2018). Highly selective acetylene semihydrogenation catalyzed by Cu nanoparticles supported in a metal–organic framework. *ACS Appl. Nano Mater.* 1 (9): 4413–4417.

251 Jiang, H.L., Akita, T., Ishida, T. et al. (2011). Synergistic catalysis of Au@Ag core-shell nanoparticles stabilized on metal-organic framework. *J. Am. Chem. Soc.* 133 (5): 1304–1306.

252 Ke, F., Zhu, J., Qiu, L.G., and Jiang, X. (2013). Controlled synthesis of novel Au@MIL-100(Fe) core-shell nanoparticles with enhanced catalytic performance. *Chem. Commun.* 49 (13): 1267–1269.

253 Augustyniak, A.W., Sadakiyo, M., Navarro, J.A.R., and Trzeciak, A.M. (2018). Design of shape-palladium nanoparticles anchored on titanium(IV) metal-organic framework: highly active catalysts for reduction of p-nitrophenol in water. *ChemistrySelect* 3 (27): 7934–7939.

254 Goswami, S., Noh, H., Redfern, L.R. et al. (2019). Pore-templated growth of catalytically active gold nanoparticles within a metal–organic framework. *Chem. Mater.* 31 (5): 1485–1490.

255 Li, Z., Yu, R., Huang, J. et al. (2015). Platinum-nickel frame within metal-organic framework fabricated in situ for hydrogen enrichment and molecular sieving. *Nat. Commun.* 6: 8248.

256 Zhao, M., Huang, Y., Peng, Y. et al. (2018). Two-dimensional metal-organic framework nanosheets: synthesis and applications. *Chem. Soc. Rev.* 47 (16): 6267–6295.

257 Liu, X., Li, Y., Liu, Y. et al. (2022). Microwave-assisted synthesis of 2D Zr-MOF nanosheets supported gold nanocomposites as efficient catalysts for the reduction of 4-nitrophenol. *J. Alloys Compounds* 922: 165939.

258 Chen, J., Liu, R., Guo, Y. et al. (2014). Selective hydrogenation of biomass-based 5-hydroxymethylfurfural over catalyst of palladium immobilized on amine-functionalized metal–organic frameworks. *ACS Catal.* 5 (2): 722–733.

259 Yin, D., Ren, H., Li, C. et al. (2018). Highly selective hydrogenation of furfural to tetrahydrofurfuryl alcohol over MIL-101(Cr)-NH_2 supported Pd catalyst at low temperature. *Chin. J. Catal.* 39 (2): 319–326.

260 Yuan, Q., Zhang, D., van Haandel, L. et al. (2015). Selective liquid phase hydrogenation of furfural to furfuryl alcohol by Ru/Zr-MOFs. *J. Mol. Catal. A: Chem.* 406: 58–64.

261 Fang, R., Chen, L., Shen, Z., and Li, Y. (2021). Efficient hydrogenation of furfural to fufuryl alcohol over hierarchical MOF immobilized metal catalysts. *Catal. Today* 368: 217–223.

262 Zhang, X., Zhang, P., Chen, C. et al. (2019). Fabrication of 2D metal–organic framework nanosheets with tailorable thickness using bio-based surfactants and their application in catalysis. *Green Chem.* 21 (1): 54–58.

263 Chen, H., He, Y., Pfefferle, L.D. et al. (2018). Phenol catalytic hydrogenation over palladium nanoparticles supported on metal-organic frameworks in the aqueous phase. *ChemCatChem* 10 (12): 2558–2570.

264 Chen, D., Yang, W., Jiao, L. et al. (2020). Boosting Catalysis of Pd nanoparticles in MOFs by pore wall engineering: the roles of electron transfer and adsorption energy. *Adv. Mater.* 32 (30): e2000041.

265 Ruben, M. and Stephen, L.B. (2008). Palladium-catalyzed Suzuki–Miyaura cross-coupling reactions employing dialkylbiaryl phosphine ligands. *Acc. Chem. Res.* 41 (11): 1461–1473.

266 Nicolaou, K.C., Bulger, P.G., and Sarlah, D. (2005). Palladium-catalyzed cross-coupling reactions in total synthesis. *Angew. Chem. Int. Ed.* 44 (29): 4442–4489.

267 Adam, F.L. and Gregory, C.F. (2002). Palladium-catalyzed coupling reactions of aryl chlorides. *Angew. Chem. Int. Ed.* 41 (22): 4176–4211.

268 Miyaura, N. and Suzuki, A. (1995). Palladium-catalyzed cross-coupling reactions of organoboron compounds. *Chem. Rev.* 95: 2457–2483.

269 Yuan, B., Pan, Y., Li, Y. et al. (2010). A highly active heterogeneous palladium catalyst for the Suzuki-Miyaura and Ullmann coupling reactions of aryl chlorides in aqueous media. *Angew. Chem. Int. Ed.* 49 (24): 4054–4058.

270 Sonnauer, A., Hoffmann, F., Froba, M. et al. (2009). Giant pores in a chromium 2,6-naphthalenedicarboxylate open-framework structure with MIL-101 topology. *Angew. Chem. Int. Ed.* 48 (21): 3791–3794.

271 Huang, Y., Zheng, Z., Liu, T. et al. (2011). Palladium nanoparticles supported on amino functionalized metal-organic frameworks as highly active catalysts for the Suzuki–Miyaura cross-coupling reaction. *Catal. Commun.* 14 (1): 27–31.

272 Mishra, B., Ghosh, D., and Tripathi, B.P. (2022). Finely dispersed AgPd bimetallic nanoparticles on a polydopamine modified metal organic framework for diverse catalytic applications. *J. Catal.* 411: 1–14.

273 Otake, K., Ye, J., Mandal, M. et al. (2019). Enhanced activity of heterogeneous Pd(II) catalysts on acid-functionalized metal–organic frameworks. *ACS Catal.* 9 (6): 5383–5390.

274 Yang, H., Han, X., Li, G., and Wang, Y. (2009). N-Heterocyclic carbene palladium complex supported on ionic liquid-modified SBA-16: an efficient and highly recyclable catalyst for the Suzuki and Heck reactions. *Green Chem.* 11 (8): 1184–1193.

275 Gole, B., Sanyal, U., Banerjee, R., and Mukherjee, P.S. (2016). High loading of Pd nanoparticles by interior functionalization of MOFs for heterogeneous catalysis. *Inorg. Chem.* 55 (5): 2345–2354.

276 Chen, F., Shen, K., Chen, J. et al. (2019). General immobilization of ultrafine alloyed nanoparticles within metal-organic frameworks with high loadings for advanced synergetic catalysis. *ACS Cent. Sci.* 5 (1): 176–185.

277 Li, Y.A., Yang, S., Liu, Q.K. et al. (2016). Pd(0)@UiO-68-AP: chelation-directed bifunctional heterogeneous catalyst for stepwise organic transformations. *Chem. Commun.* 52 (39): 6517–6520.

278 Ning, L., Liao, S., Liu, X. et al. (2018). A regulatable oxidative valorization of furfural with aliphatic alcohols catalyzed by functionalized metal-organic frameworks-supported Au nanoparticles. *J. Catal.* 364: 1–13.

279 Datta, K.K., Reddy, B.V., Ariga, K., and Vinu, A. (2010). Gold nanoparticles embedded in a mesoporous carbon nitride stabilizer for highly efficient three-component coupling reaction. *Angew. Chem. Int. Ed.* 49 (34): 5961–5965.

280 Zhang, X. and Corma, A. (2008). Supported gold(III) catalysts for highly efficient three-component coupling reactions. *Angew. Chem. Int. Ed.* 47 (23): 4358–4361.

281 Grirrane, A., Alvarez, E., Garcia, H., and Corma, A. (2014). Deactivation of cationic Cu^I and Au^I catalysts for A^3 coupling by CH_2Cl_2 : mechanistic implications of the formation of neutral Cu^I and Au^I chlorides. *Angew. Chem. Int. Ed.* 53 (28): 7253–7258.

282 Yang, Q. and Jiang, H. (2018). Oxidation or reduction state of Au stabilized by an MOF: active site identification for the three-component coupling reaction. *Small Methods* 2 (12): 1800216.

283 Ji, X., Su, Z., Wang, P. et al. (2015). Tethering of nicotinamide adenine dinucleotide inside hollow nanofibers for high-yield synthesis of methano from carbon dioxide catalyzed by coencapsulated multienzymes. *ACS Nano* 9 (4): 4600–4610.

284 Arakawa, H., Aresta, M., Armor, J.N. et al. (2001). Catalysis research of relevance to carbon management: progress, challenges, and opportunities. *Chem. Rev.* 101: 953–996.

285 Zhou, Z., He, C., Yang, L. et al. (2017). Alkyne activation by a porous silver coordination polymer for heterogeneous catalysis of carbon dioxide cycloaddition. *ACS Catal.* 7 (3): 2248–2256.

286 Zhou, Z. and Gao, P. (2022). Direct carbon dioxide hydrogenation to produce bulk chemicals and liquid fuels via heterogeneous catalysis. *Chin. J. Catal.* 43 (8): 2045–2056.

287 Decortes, A., Castilla, A.M., and Kleij, A.W. (2010). Salen-complex-mediated formation of cyclic carbonates by cycloaddition of CO_2 to epoxides. *Angew. Chem. Int. Ed.* 49 (51): 9822–9837.

288 Tang, L., Zhang, S., Wu, Q. et al. (2018). Heterobimetallic metal–organic framework nanocages as highly efficient catalysts for CO_2 conversion under mild conditions. *J. Mater. Chem. A* 6 (7): 2964–2973.

289 Zhang, W., Ping, R., Lu, X. et al. (2023). Rational design of Lewis acid-base bifunctional nanopolymers with high performance on CO_2/epoxide cycloaddition without a cocatalyst. *Chem. Eng. J.* 451: 138715.

290 Guo, Y., Chen, W., Feng, L. et al. (2022). Greenery-inspired nanoengineering of bamboo-like hierarchical porous nanotubes with spatially organized bifunctionalities for synergistic photothermal catalytic CO_2 fixation. *J. Mater. Chem. A* 10: 12418.

291 Du, Y., Yang, H., Wan, S. et al. (2017). A titanium-based porous coordination polymer as a catalyst for chemical fixation of CO_2. *J. Mater. Chem. A* 5 (19): 9163–9168.

292 Abednatanzi, S., Gohari Derakhshandeh, P., Depauw, H. et al. (2019). Mixed-metal metal-organic frameworks. *Chem. Soc. Rev.* 48 (9): 2535–2565.

293 Lohr, T.L. and Marks, T.J. (2015). Orthogonal tandem catalysis. *Nat. Chem.* 7 (6): 477–482.

294 Chunlei, P. and Jinlong, G. (2021). Tandem catalysis at nanoscale. *Science* 371 (6535): 1203–1204.

295 Li, X., Guo, Z., Xiao, C. et al. (2014). Tandem catalysis by palladium nanoclusters encapsulated in metal–organic frameworks. *ACS Catal.* 4 (10): 3490–3497.

296 Zhao, M., Deng, K., He, L. et al. (2014). Core-shell palladium nanoparticle@metal-organic frameworks as multifunctional catalysts for cascade reactions. *J. Am. Chem. Soc.* 136 (5): 1738–1741.

297 Chen, Y., Zhou, Y., Wang, H. et al. (2015). Multifunctional PdAg@MIL-101 for one-pot cascade reactions: combination of host–guest cooperation and bimetallic synergy in catalysis. *ACS Catal.* 5 (4): 2062–2069.

298 Sun, J.L., Chen, Y.Z., Ge, B.D. et al. (2019). Three-shell Cu@Co@Ni nanoparticles stabilized with a metal-organic framework for enhanced tandem catalysis. *ACS Appl. Mater. Interfaces* 11 (1): 940–947.

299 Gremaud, R., Broedersz, C.P., Borsa, D.M. et al. (2007). Hydrogenography: an optical combinatorial method to find new light-weight hydrogen-storage materials. *Adv. Mater.* 19 (19): 2813–2817.

300 Niemann, M.U., Srinivasan, S.S., Phani, A.R. et al. (2008). Nanomaterials for hydrogen storage applications: a review. *J. Nanomater.* 2008: 1–9.

301 Zheng, D.Y., Zhou, X.M., Mutyala, S., and Huang, X.C. (2018). High catalytic activity of C60 Pdn encapsulated in metal-organic framework UiO-67, for tandem hydrogenation reaction. *Chemistry* 24 (72): 19141–19145.

8

Confinement Effects in Catalysis with Molecular Complexes Immobilized into Porous Materials

Maryse Gouygou, Philippe Serp, and Jérôme Durand

Université de Toulouse, LCC, CNRS-UPR 8241, ENSIACET, 4 allée Emile Monso BP44362, 31030 Toulouse Cedex 4, France

8.1 Introduction

Most catalytic reactions proceed on catalyst active sites/centers/species. The concept of active site, composed of an atom or an ensemble of atoms on heterogeneous catalysts, was introduced by H.S. Taylor in 1925 [1], and the importance of its design for optimizing catalyst performance was suggested a few years later by A.A. Balandin in its multiplet theory of heterogeneous catalysis [2]. The catalytic performance of the active site(s), defined as its stability, activity, and selectivity, is directly linked to its direct environment in the molecule for homogeneous catalysts or in the material for heterogeneous catalysts. This direct environment is composed of groups of atoms from: (i) ligand in homogeneous catalysts, (ii) the metallic surface or the support for supported catalysts, and (iii) ligand and support atoms for supported homogeneous catalysts (SHC). Thus, continuous and substantial efforts are dedicated to the rational design of new ligands [3, 4], to the rational construction of a metal/support interface [5] or of metallic nanoparticles [6] to fine-tuning the performance of catalysts. Identification and study of active sites remain however challenging [7], owing to their dynamic behaviors during catalysis process [8], and issues with current characterization techniques. It is however important to remember that placing the active site at the heart of the catalytic act makes sense only if the reaction is well under kinetic control. This is particularly relevant for supported catalysts, for which the reactions, occurring on the surface of porous materials, are accompanied by heat and mass transport in the catalyst porosity and across the boundary layer at its external surface. If these heat and mass fluxes cause finite gradients of concentration and temperature in the material as well as in the film on the external surface, the resulting reaction rate is governed by both the kinetics of the reaction and the transport process(es) giving rise to the gradient. An appreciation of the regime in which a reaction is occurring is thus of great importance in interpreting catalytic data and designing a catalyst [9]; and there is a real need to understand and decouple the underlying roles of active site modification from other factors, such as diffusional limitations

Catalysis in Confined Frameworks: Synthesis, Characterization, and Applications, First Edition.
Edited by Hermenegildo Garcia and Amarajothi Dhakshinamoorthy.

[10], since both can impact catalyst performances. In that context, understanding the physics of small-scale confined fluid flow will bring important information for catalysis [11]; and the use of multiscale molecular simulation should be encouraged [12, 13]. It therefore appears clear that it is important to look beyond the active site and its immediate environment when seeking to understand catalysis, in particular with catalysts supported on porous materials.

The connection to enzymatic catalysis then becomes evident. With enzymes, understanding the effects from the molecular to macroscopic scale is central due to the complex effect of the protein environment. Indeed, this microenvironment will make that any nonchemical or chemical step can determine the enzyme activity. In this case, it is also essential to take into account several weak interactions between the active site and the substrate to be transformed, involving the active site itself but also its microenvironment, to be able to evaluate the activation energy in a reliable way [14]. Finally, we can also mention that both surface- or interface-confined enzymatic reactions can be considered [15]. Indeed, most *in vivo* enzyme-catalyzed reactions occur in a confined environment, which differs from one of isolated enzymes in a buffered aqueous solution. Thus, it has been proposed that protein nano-compartment can modulate the transport of small molecules, and this effect could increase the local concentration of intermediates [16].

Synergism between a catalyst's active site and its particular local environment is at the origin of the concept of confinement in catalysis, which refers to the spatial restriction of active site, reactants, and/or intermediates within structures of nano- or subnano-scale dimensions. Confinement effect in zeolites is a representative example, with seminal contributions from E.G. Derouane [17, 18], A. Corma [19, 20], and E. Iglesias [21] explaining catalytic activity and selectivity in systems where confinement influences transition state stability as well as the flux of products and intermediates. Thus, Eric Derouane established in the 1980s that confined reactants and intermediates tend to optimize their van der Waals interactions with the pore walls, and experience enhanced physisorption energetics due to the curvature inside the micropores [17]. Such effects modify the diffusion of reactants and intermediates inside the pores and directly influence reaction rates.

The idea to encapsulate metal complexes or clusters in the cavities of zeolites dates back from the early 70's [22], and the term "ship in the bottle" synthesis (Figure 8.1a,b) or species was tackled in 1985 by du Pont de Nemours scientists [27]. The possibility to use the restrictive dimensions of the zeolite pores: (i) to increase the catalyst selectivity by molecular sieving and orientating effects on the reactant that must diffuse to the active sites; and/or (ii) to force unusual coordination geometry or coordinative unsaturation upon the entrapped metal catalyst, has been the subject of numerous studies [28–31]. Besides zeolites, immobilization of coordination complexes in other microporous and mesoporous materials has been studied, with a particular attention being paid on mesoporous silica [32–34], clays [35, 36], metal–organic frameworks (MOFs) [37–41], covalent organic frameworks (COFs) [42], and polymers [43]. Worth to mention are the seminal works in asymmetric catalysis of J.M. Thomas using mesoporous silica MCM-41 as a support or tether for new single-site catalysts [44, 45], and of C. Li in confinement effects in chiral synthesis [46, 47].

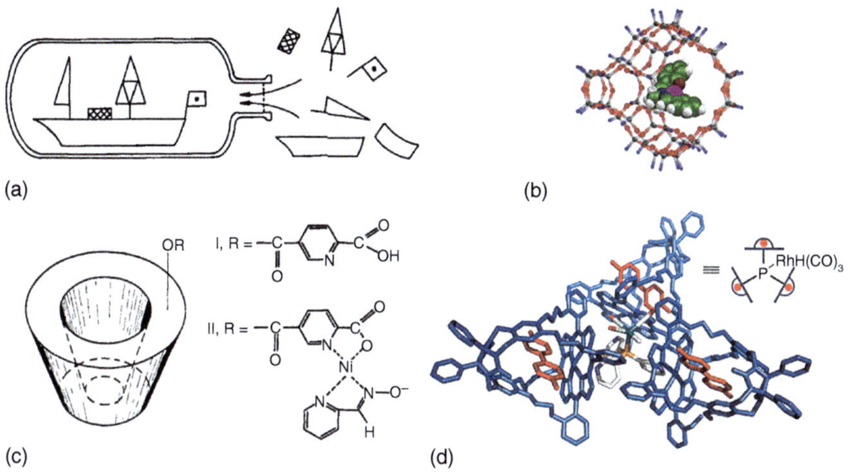

Figure 8.1 Representative examples of confined metal complexes. (a) Ship in the bottle synthesis/species concept. Source: Reproduced with permission from Ichikawa [23] © 2000 Johnson Johnson Matthey plc. (b) Encapsulated Schiff base nickel complex in the cage of a zeolite Y. Source: Reproduced with permission from Choudhary et al. [24] © 2017 Elsevier. (c) Artificial enzyme combining a metal complex and a hydrophobic binding cavity. Source: Reproduced with permission from Breslow and Overman [25] © 1970 American Chemical Society. (d) Cofactor-controlled encapsulated rhodium hydroformylation catalyst. Source: Reproduced with permission from Jongkind et al. [26] © 2019 Wiley.

Inspired by enzymes, it has also been early recognized that the microenvironment is as equally important as the active site for homogeneous catalysts grafted to molecular cages. The first attempt to immobilize a complex on a molecular cage (cyclohexaamylose, the α-cyclodextrin) in order to take advantage of cyclodextrin binding ability was published in 1970 (Figure 8.1c) [25]. This supramolecular approach, which has been widened to more complex supramolecular capsules (Figure 8.1d) [26], offers the possibility to combine molecular recognition, phase transfer catalysis, and transition metal catalysis on the same catalytic system [48]. With the aim to develop molecular catalysts in confined spaces, for which both activity and selectivity can be switched, many supramolecular catalytic systems have been investigated. These systems, which refer to homogeneous catalysis and fall outside the subject of the present chapter, have been discussed in details in review articles [49–57], and books edited by S. Sadjadi [58] and R. Poli [59].

It therefore appears that over the last fifty years, many proof of principle examples have been published for confined molecular complexes, where the confinement occurs within pores of solid porous materials or within self-assembled molecular cavities or cages. While we will see later on that the nature of the confinement effects varies between these diverse systems, all demonstrate the importance of nano-interfaces and co-location of reactants, catalysts, and intermediates in governing improvements in catalytic performance. Such improvements require an in-depth knowledge of diffusion phenomena at the nanometric scale, as well as the possibility of studying the energetics of transition states in a confined environment.

Table 8.1 Comparison between the structure, physical, and chemical properties of zeolites, CNTs, and MOFs.

Properties	Zeolites	CNTs	MOFs
Surface area ($m^2\ g^{-1}$)	<700	50–500	10^3–10^4
Pore size (nm)	≤1	1–100	0.5–9.8
Thermal stability	High	High	Low to medium
Chemical stability	High	High	Low to medium
Pore hydrophilicity	Tunable	Very low	Tunable
Flexibility	Low	Medium	Medium
Tailorability	Low	High	High
Gas diffusivity	Low	High	Low to medium
Ordering of site distribution	Medium	Low	High

Source: Adapted from Dai and Zhang [60].

For porous materials, which are the supports discussed for coordination complex confinement in the present chapter, the confinement effects, which will be discussed hereafter, are closely correlated to the pore features of the material, such as their size/shape, surface chemistry, and potential dynamics. The structural, physical, and chemical properties of three supports discussed in this chapter (zeolites, MOFs, and carbon nanotubes [CNTs]) are summarized in Table 8.1. Albeit they all have some common characteristics, such as a relatively high surface area and uniform pore size distributions, they differ in many significant aspects, which can affect both the choice of the immobilization method for the coordination complex, and the extent/type of confinement effects. We can mention that the influence of pore features such as pore diameter [61], pore shape [62], pore hydrophobicity [63, 64], pore surface chemistry [65], or pore flexibility [66, 67] has been evidenced for the adsorption, dynamics, or freezing/melting of confined water molecules. For immobilization of coordination complexes, an obvious drawback of CNTs (or silica) compared to MOFs is that it is difficult to synthesize single-site catalysts (i.e. sites structurally well-characterized presenting the same energy of interaction with a reactant [68]) due to the poor ordering of anchoring site distribution. Besides spectroscopy and microscopy, which can be used to evidence the uniformity of metal sites on supports [69, 70], some computational works, including *ab initio* computational studies, have started to address the role of active site heterogeneity for SHC [71–74].

Several factors can influence catalytic performance in porous materials, often in a convoluted way; as the active site, the adsorbed intermediates and transition states may be strongly influenced by the confining environment, with a resulting influence on reaction rate and catalyst stability. For some reactions, bulky transition states may be stabilized within pores to a different degree than others, leading to selectivity shifts. For other reactions, it is often challenging to distinguish these intrinsic differences in rate from diffusion-limited rates within the catalyst pore. Nevertheless, it is now well established that the active site microenvironment is as important as the

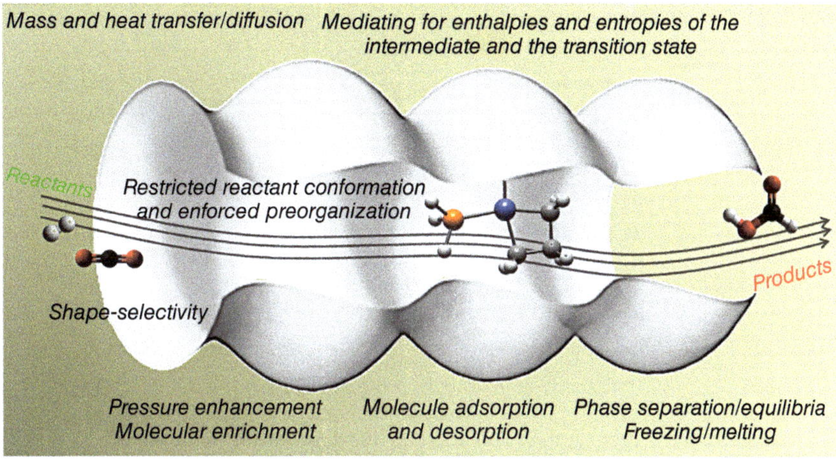

Figure 8.2 Different types of effects can contribute to catalytic performance of confined coordination complexes. Source: Philippe Serp.

active site itself for confined catalysts. Figure 8.2 shows the different types of effects that can contribute to catalytic performance of confined coordination complexes. The confinement effect can be the collective result of several effects affecting reaction energetics, mechanisms, and dynamics. Consequently, it is a key factor affecting the reaction on porous materials. We can classify these effects in long-range-mesoscopic scale effects (physical or "soft" effects) and short-range-microscopic scale effects (chemical or "hard" effects). Table 8.2 gives a classification of these effects and provides to the reader useful bibliographic references.

The long-range-mesoscopic scale effects include changes upon confinement in: (i) mass transfer and diffusion of reactants and products; (ii) heat transfer; (iii) shape selectivity; (iv) molecular enrichment; (v) pressure enhancement; (vi) reactant pre-organization; and (vii) phase separation or equilibria. The short range-microscopic scale effects include changes upon confinement in: (i) the adsorption and desorption steps; and (ii) the transition state energy.

Despite numerous research works in the field of confinement effects in catalysis, the elucidation of the relative importance of each of these factors on a given reaction remains extremely challenging. The main reason for these difficulties is that experimental data often reflect the influence of several effects at once, and it is expensive and time-consuming to design experiments to study the influence of each of these effects separately. Moreover, the relative importance of each effect will depend on the reaction and the catalyst.

Due to the difficulties and costs associated with a purely experimental approach, the study of confinement effects in catalysis is a good candidate for molecular simulation studies that can be assisted by machine learning [98]; and such approaches will play an invaluable role in the future of confined catalysis.

The intention of the authors of this chapter is to provide the reader with an overview of the different stages in the study of the effects of confinement with molecular complexes immobilized into porous materials. This involves the preparation

Table 8.2 Individual effects contributing to the confinement effect.

Effects	Comment	References
Mass transfer	The relative contribution of the surface resistance to the overall transport resistance varied with both the material and the guest molecule	[75]
Diffusion	In addition to the diffusional resistance of the pore space itself, molecular transport is affected also by transport resistances in the intra-crystalline space and on the external surface	[76]
Heat transfer	Small pores lead to better heat transfer in MOFs	[77]
Shape selectivity	The ability of microporous materials to control the access and egress of some reactants/products and to preferentially contain some transition states, while excluding others based on pore size constitutes the shape selectivity concept. Basic mechanisms giving rise to shape selectivity are: (i) reactant shape selectivity (molecules that are too large to enter the pores cannot reach the active sites); (ii) product shape selectivity (new molecules are formed but are too large to desorb as a product); and (iii) transition state shape selectivity (transition states that are too large to fit inside a pore do not form)	[78]
Molecular enrichment (pore enrichment effect)	Some molecules can be enriched on the concave surface of nanotubes/spheres. With increasing pore size, the reactant concentration inside the pore decreases while the diffusion flux increases	[79–82]
Pressure enhancement	Phenomena that occur only at high pressures in bulk phases are often observed in micropores, suggesting that the pressure in such confined phases is large. High-pressure effect in the micropores deform the solids and affect properties of the fluids themselves	[83–87]
Reactant preorganization	Supramolecular organization can occur in confined nano-spaces. Such organization stems from the space restrictions imposed by the pore to the guest species and often implies a change in their physico-chemical properties and release profile	[88–90]
Adsorption/desorption	Desorption can be slower than adsorption because of guest-guest interaction (molecular cluster formation)	[91]
Transition state energy	Tighter confinement is preferred at low temperatures because the functional form of Gibbs free energies causes enthalpic stabilization to prevail over entropy losses caused by confinement at modest temperatures	[92–94]
Phase separation/equilibria	The introduction of pore wall forces and the competition between fluid-wall and fluid-fluid forces can lead to surface-driven phase changes. These include new kinds of phase transitions not found in the bulk phase, as well as shifts in transitions (freezing, gas-liquid, liquid-liquid)	[95–97]

and characterization of confined catalysts, and then the study of catalysis in a confined environment. The objective is not an exhaustive review of the literature but rather to provide, *via* relevant representative examples, a methodology adapted to the study of complex systems and phenomena in chemistry.

8.2 Immobilization of Molecular Complexes into Porous Materials

The strategies described in the literature to confine molecular coordination catalysts into porous materials can be divided into two categories. The first one consists in approaches that have been applied to several kinds of materials, while the second concerns routes that apply to one kind of specific material, in close relationship with both its morphology, in particular the pore size, and the chemical nature of its surface. The selective confinement of the catalytic phase inside the pores is still a challenge, as some of the strategies described below do not exclude the possibility of species located on the outer surface of the porous material. We will describe below representative strategies for confined metal complexes in various porous materials.

8.2.1 Confinement of Molecular Complexes in Mesoporous Silica

Silica has been widely used as support for the confinement of molecular catalysts, taking advantage of the reactive silanol groups, as described in a review published in 2021 on silica-supported metal complexes for catalysis [33]. The covalent anchoring routes for the silica functionalization with transition metal complexes can be divided into two categories. The surface organometallic chemistry (SOMC, Figure 8.3a), involving a direct interaction of the surface oxygen atoms with the metallic center, or the SHC (Figure 8.3b), where the surface hydroxyl functions are exploited to achieve covalent grafting of a ligand. In the latter case, depending on the length and flexibility of the linker, the catalytic active center can be located more or less far away from the silica surface, unlike what was observed for the SOMC approach (Figure 8.3). In all cases, if no preliminary passivation of the external surface is performed, the chemical grafting does not occur selectively into the pores, and a small amount of the catalyst can also be located outside.

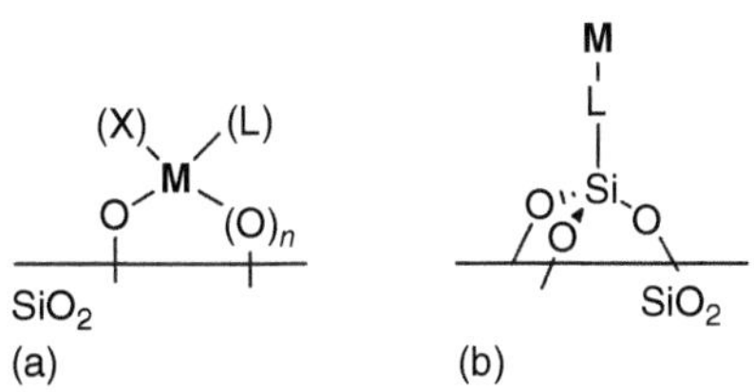

Figure 8.3 Covalent anchoring of a metal complex onto a silica surface. (a) SOMC by direct attachment of the metallic center onto the silica surface. Source: Adapted with permission from Motokura et al. [33] © 2021 American Chemical Society; and (b) SHC through a linker (L) grafted through coupling with the surface Si-OH.

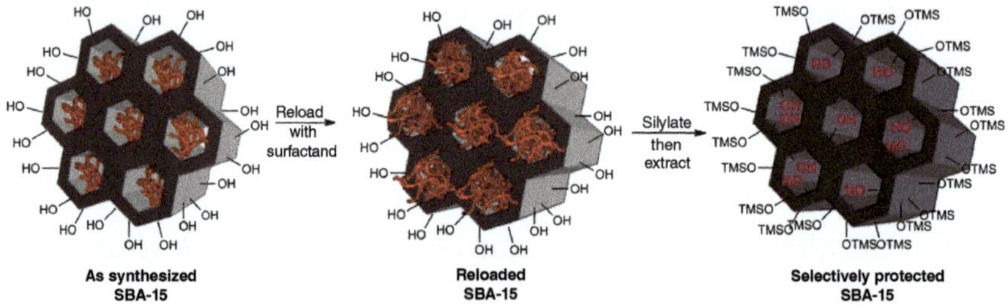

Figure 8.4 Multistep modification of SBA-15 (ordered mesoporous silica, OMS) for the pore-selective functionalization inside the mesopores. Source: Reproduced with permission from Webb et al. [101] © 2015 Elsevier.

The SOMC approach was applied to various catalytic species, like zirconocene derivatives for ethylene polymerization into the pores of molecular sieves [99]. Several mesoporous silica were treated at 500 °C under high vacuum to generate isolated silanols prior to the immobilization of WMe_5 fragments by SOMC to generate catalysts, which were successfully applied to cyclooctane metathesis [100].

Using a methodology described in 2015 [101] for the selective functionalization of the mesopores of SBA-15 with a silylation Rh catalyst [102], Mo and Ru ring closure metathesis catalysts [103, 104] were selectively confined into SBA 15 by prior passivation of the external surface. This step was achieved by filling the pores with Pluronic P123 (a triblock copolymer), followed by selective reaction of the outer silanol groups with hexamethyldisilazane, and removal of the Pluronic by extraction, prior to catalyst anchoring by SOMC (Figure 8.4).

The SHC approach was used: (i) to immobilize an Ir chiral catalytic species into SBA-15 for the enantioselective ketone hydrogenation reaction starting from a support previously functionalized by a diamine moiety [105], (ii) for the confinement of Pd catalyst in various mesoporous silica for allylation reactions [106, 107], (iii) for immobilization of a second-generation Hoveyda–Grubbs type catalyst for ring-closing metathesis [104, 108], and (iv) for grafting a Ru complex for dynamic kinetic resolution of secondary alcohols [109].

As a variation of the SHC method, covalent immobilization can also be achieved by elaboration of the mesoporous silica by co-condensation of TEOS (tetraethyl orthosilicate) with a ligand containing a pending $-Si(OMe)_3$ function, resulting in, after reaction with a Rh precursor, a core–shell structured catalyst applied to the asymmetric transfer hydrogenation reaction [110].

A similar approach was used to elaborate mesoporous silica aerogels covalently functionalized with Cu(II) phenol oxidation catalysts, from alkoxysilane-anchored macrocyclic complexes and TMOS (tetramethyl orthosilicate) [111].

Alternative routes also involve non-covalent interaction between the surface and the catalytic species, namely, hydrogen bonds and electrostatic interactions. This was reported for instance for a Rh cationic hydroformylation catalyst, immobilized on MCM-41 and SBA-15 through the triflate counter anion in interaction with the silica surface by the formation of hydrogen bonds [112].

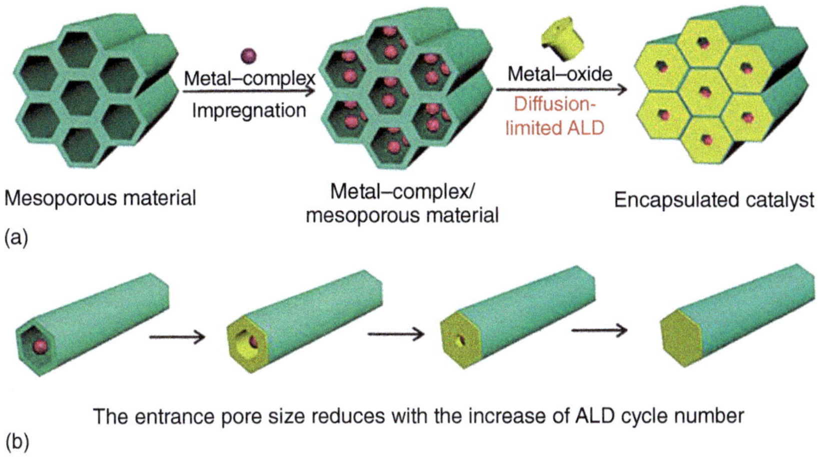

Figure 8.5 Illustration of (a) the encapsulation process into ordered mesoporous materials; and (b) the influence of ALD cycle number on the pore size. Source: Reproduced with permission from Zhang et al. [114] © 2018 Wiley.

The ship-in-a-bottle approach can also be applied to mesoporous silica. This was described for instance for a chiral Fe(salen) catalyst for the asymmetric sulfoxidation reaction, which was confined into SBA-16 and modified mesocellular foams, characterized by large cage size (15–22 nm) but small side windows (1.8–3.7 nm), particularly adapted to this approach. The authors reported that modulation of the acidity, the hydrophobicity, and the pore size of the support can be achieved by post-treatment with organosilanes containing alkyl chains of various sizes [113].

The encapsulation of metal complexes into nano-channels of mesoporous materials can also be achieved by atomic layer deposition (ALD) to generate a metal oxide coating near the pore entrance. The pore size of the hollow plug is precisely controlled on the sub-nanometer scale by the number of ALD cycles to fit various metal complexes with different molecular sizes. Typically, Co or Ti molecular catalysts, respectively, for the hydrolytic kinetic resolution of epoxides and asymmetric cyanosilylation of carbonyl compounds, were successfully encapsulated into the nanochannels of SBA-15, SBA-16, and MCM-41. A schematic illustration of this encapsulation strategy is depicted in Figure 8.5.

Excellent reusability of these catalysts was reported by the authors, evidencing the efficiency of this approach to avoid leaching of the active species [114]. This contrasts strongly for instance with the report of a Mo catalyst supported on MCM-41 by impregnation, where an important leaching was observed after three runs of the catalyzed olefin epoxidation reaction [115].

8.2.2 Confinement of Molecular Complexes in Zeolites

The synthesis of zeolite-encapsulated transition metal complexes was reviewed in 2014 [28]. Concerning the confinement of coordination catalysts, the main strategies reported are either dealing with the synthesis of the catalyst inside the cavity (the ship-in-a-bottle or flexible ligand method [FLM]) or the covalent anchoring through the –OH surface groups as described for mesoporous silica.

The FLM is based on the use of a ligand flexible enough to enter the cavity of the zeolite containing a metal ion. Upon complexation into the pore, the reduced flexibility of the molecular structure of the coordination compounds and its size ensures its confinement. This method was applied to various molecular catalytic systems, most of the time confined into faujasite zeolite Y, whose pore size of 7.4 Å is particularly adapted to this strategy. In particular, in the case of different Fe, Ni, Pd, and Cu catalysts, confinement achieved through FLM was reported to have, depending of the size, geometries showing distortion due to the reduced cavity size, which can influence the catalytic performances [24, 116–123].

Alternatively, another "ship-in-a-bottle" approach involves the diffusion of ligand precursors into the zeolite pores, where they assemble around a templating metallic center. This method was successfully applied for the encapsulation of V, Ni, and Cu Schiff base catalysts, including chiral ones; where the ligand is generated inside the pores by reaction of a diamine and 2 equivalents of a 2-hydroxy-benzaldehyde derivative [124–126]. These catalytic systems were then applied to the asymmetric nitraldol reaction (Cu), the asymmetric Henry reaction (Ni), and styrene epoxidation (V).

Surface chemistry can also be used to covalently immobilize coordination catalysts on zeolite. In the case of a Cr oligomerization catalyst, this was achieved by a reaction between the surface group and a ligand precursor containing an $-Si(OMe)_3$ group, followed by metalation. Depending on the nature of the zeolite employed, it was reported that the resulting catalyst can be grafted either in the pore channel or both in and outside the pores. This observation was attributed to a different acidity of the studied supports [127]. A similar observation was reported in the case of a preformed Cr catalyst, immobilized on a zeolite that was previously treated with methylaluminoxane. According to the nature of the zeolite, the resulting catalytic phase can be located either mainly outside or within the pores [128].

8.2.3 Confinement of Molecular Complexes in Covalent Organic Frameworks (COF)

Principle and challenges in the use of COFs in heterogeneous catalysis was recently reviewed [129]. Confinement of molecular metallic catalysts into the pores of COFs can be achieved through molecular engineering. This can involve the elaboration of a framework containing functional groups with coordination ability into the network, affording the possibility to covalently anchor molecular complexes on the side walls by treatment with a metallic precursor. An alternative strategy consists of a condensation reaction between appropriate pending functions on both the COF and a preformed metallic species.

The first example using COFs to confine a molecular catalyst was reported in 2011 [130]. The authors used an imine-containing covalent framework to incorporate inside its crystalline porous structure palladium centers by a simple

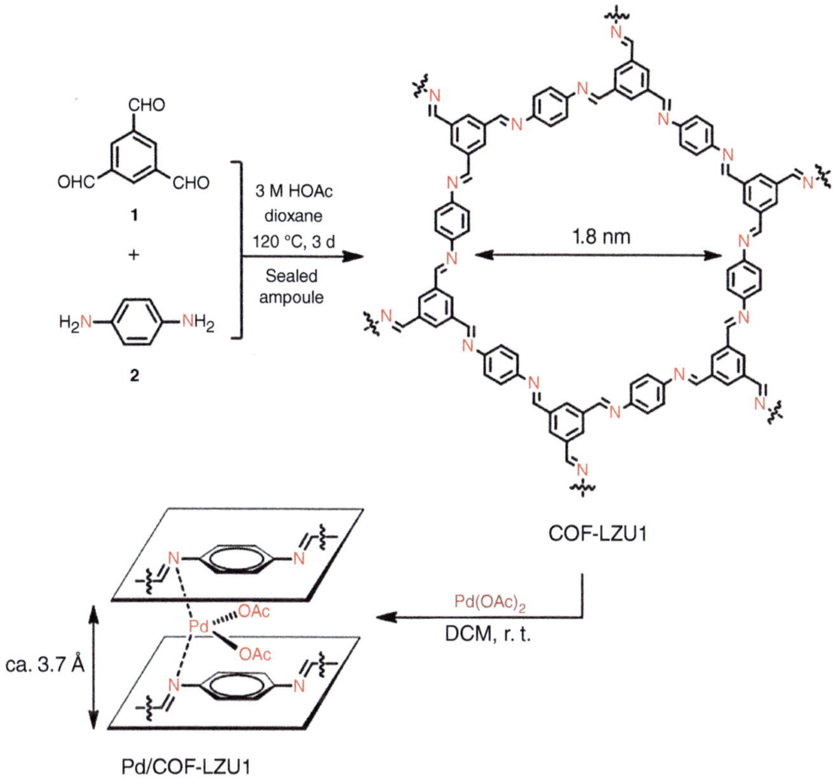

Figure 8.6 Schematic representation for the synthesis of a Pd/COF material. Source: Reprinted with permission from Ding et al. [130] © 2011 American Chemical Society.

post-treatment from Pd acetate (Figure 8.6), and the resulting catalyst was applied to the catalytic Suzuki–Miyaura reaction.

This approach was then successfully applied to other catalytic systems based on different metals for various catalytic applications [131, 132]. This also includes some examples in enantioselective catalysis, based on the specific design of COFs containing chiral moiety with coordinating ability incorporated into their structure [133, 134].

Recently, polyoxometalates were also homogeneously confined into COFs through the condensation of aldehyde functions present on the walls of the framework and pending amino groups on the clusters [135]. A similar approach was reported for a Ru-carbene catalyst containing an $-Si(OMe)_3$ function allowing condensation with COF hydroxyl groups [136].

8.2.4 Confinement of Molecular Complexes in Metal–Organic Frameworks (MOFs)

The incorporation of molecular catalysts in MOFs for heterogeneous catalytic applications was recently reviewed [37, 41, 137–139]. In MOF, the catalytically active

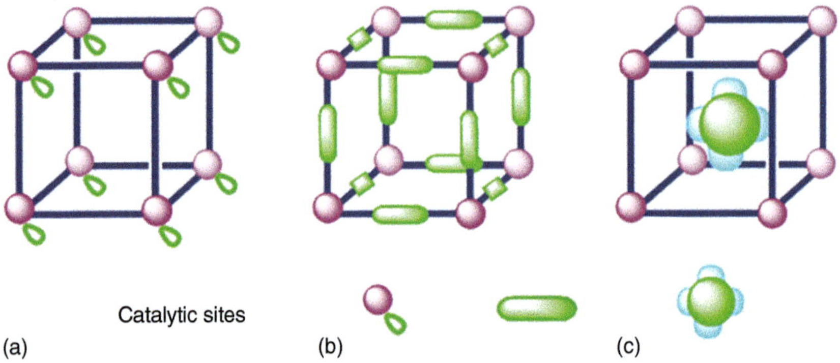

Figure 8.7 Possible catalytic site location in a MOF structure: (a) at the node; (b) anchored to the organic linker; or (c) encapsulated into the porous cavities. Source: Reproduced with permission from Wu and Zhao [37] © 2017 Wiley.

center can be either located at the node, anchored to the organic linker, or encapsulated into the porous cavities as illustrated in Figure 8.7.

The use of the metallic nodes as active centers (Figure 8.7a) for catalytic applications can present some drawbacks, such as limited accessibility or collapse of the framework due to changes of coordination, geometry, and/or oxidation state of the metal during the catalytic cycle.

Post-synthetic metalation of a linker possessing coordinating ability, an approach referred to as "linker design" [140], has been reported using in particular bipyridine [141], phenanthroline dicarboxylic acid [142], or porphyrin [143]. This approach was also successfully applied to MOFs built with chiral coordinating linkers derived from binaphtol [144], biphenol [145], diphosphine [146], or phosphoramidite [147] for asymmetric transformations.

To a lesser extent, a similar approach was also reported but starting from organic linkers [148–150] metalated prior to the MOF synthesis. Examples include Rh or Ir-preformed porphyrin, and applications in cyclopropanation of addition of CO_2 to epoxides, or Pd-containing pincer diphosphine ligands for the catalytic hydrogen transfer [151].

Apart from the situations where the catalytic site is directly the metallic node of the MOF structure or anchored to the organic linkers (Figure 8.7b), recent works appeared dealing with post-synthetic strategies for molecular catalysts' confinement into the cages of the MOF (Figure 8.7c). The ship-in-a-bottle technique was applied: (i) to the encapsulation of a Cu-phenanthroline catalyst achieved by preabsorption of a Cu precursor followed by the addition of a solution of the ligand [152], or (ii) for the encapsulation of a Co carbene catalyst [153]. Confinement of a Co phtalocyanine into the pores of a MOF was also achieved by treatment with a cobalt precursor followed by the addition of 1,2-dicyanobenzene and formamide [154]. Another approach described in the case of MOF is their elaboration in the presence of the molecular catalyst in the solution that ends confined into the pores of the structure, as reported for the encapsulation of a Ni-phenanthroline-based catalyst for the synthesis of benzimidazoles [155].

Immobilization of a molecular catalyst can also be efficiently achieved by electrostatic interaction between the MOF side walls and the catalyst. The charge can be located on the ligand, like in the case of a protonated amine function of the ligand in interaction with an anionic (Cr)MIL-101-SO_3^- for a Ru olefin metathesis catalyst [156]. Another possibility is the interaction with a charged complex as reported for the encapsulation of an anionic Rh catalyst for methanol carbonylation into a MOF-containing pending imidazolium groups on the linkers [157]. Metalated phthalocyanines were also encapsulated into a MIL-101 MOF by simple wet infiltration, by solution mixing followed by solvent evaporation and washing of the solid obtained [158]. This approach was also used for the sorption of various Ru olefin metathesis catalyst [159], or a bimetallic Ni–Fe hydrogenase mimic [160]. Recently, effective encapsulation of a W catalyst into a MOF structure was achieved by impregnation, followed by sealing of the surface by coating with a thin cross-linked polymer. This coating was reported to have no detrimental effect on the catalytic activity and to allow avoiding leaching even after 10 cycles. As this approach is not dependent on the type of MOF and/or the catalytic species confined, it could easily be applied to other systems [161].

8.2.5 Confinement of Molecular Complexes in Carbon Materials

As far as carbon materials are concerned, most of the strategies developed for the immobilization of molecular catalysts are based on the specific surface chemistry of the graphitic surface. In particular, the preliminary oxidation of the surface leads to the formation of, among others, carboxylic functional groups that can be further used to covalently graft the desired catalysts partially confined into the carbon material [162].

Covalent immobilization can also be achieved using phenolic functions of the surface, also created during surface oxidation, for the condensation with a Rh complex containing an –$Si(OMe)_3$ moiety. The authors reported that when CNTs are used as support, the catalyst is preferentially located in the inner cavity, as evidenced by the absence of a signal for Rh in the XPS spectrum, unlike what was observed for other supports like activated carbon [163]. Alternatively, deprotonated carboxylic and phenolic groups of CNTs have also been used to interact with a Pd catalyst. Unexpectedly, the authors reported that in the resulting immobilized catalyst, the palladium is located in the inner cavity of the CNTs [164]. Self-assembling Nd/Na heterobimetallic asymmetric catalyst was also successfully confined into multi-walled CNTs, leading to smaller catalyst clusters with increased catalytic performances compared to their unsupported analog [165]. This approach was then extended by the authors to other rare-earth-based catalysts like Er [166].

8.3 Characterization of Molecular Complexes Immobilized into Porous Materials

Detailed characterization of coordination complexes confined in porous materials is by no means an easy task. First, structural, textural, and physicochemical

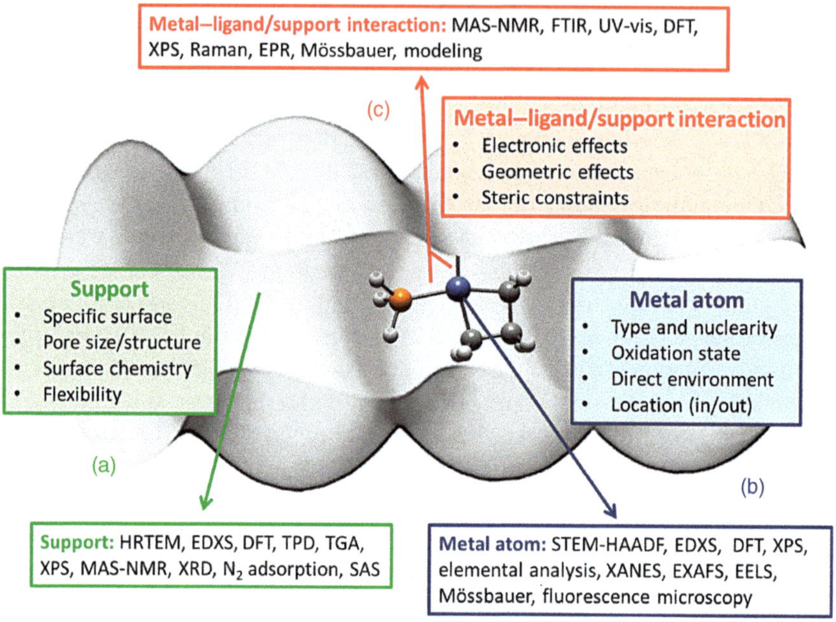

Figure 8.8 Overview of the complementary characterization methods required to assess the structure and composition of molecular complexes immobilized into porous materials. Source: Philippe Serp.

characterization of the porous support should be performed by conventional techniques of porous material characterization. These information are crucial for the study of confinement effects since these properties of the support will affect the long-range-mesoscopic scale effects. A peculiar attention should be devoted to the elucidation of: (i) the pore structure by nitrogen adsorption methods[167] or molecular simulation [168], and other methods such as small angle X-ray and neutron scattering [169, 170]; and (ii) the surface chemistry of the support by various techniques that will depend on the type of support (Figure 8.8a). The reader is encouraged to look at review articles for zeolite [171], carbon materials [172], or MOF characterization [173]. In addition, the investigations of dynamic structural changes in flexible supports, which can be relevant to confinement effect, require *ex-situ* diffraction and/or spectroscopic techniques, or specific *in-situ* analytical techniques (^{129}Xe NMR, Raman, IR, scanning transmission electron microscopy [STEM], or X-ray Absorption Spectroscopy [XAS]) together with molecular modeling [174–178].

Detailed characterization of the anchored complex passed by the characterization of the metal atom (Figure 8.8b), and its coordination sphere (Figure 8.8c).

It goes without saying that the structural uniformity of the system (i.e. single-site catalysts) will be a major asset for easier characterization of the material. Assessing uniformity of metal species and their surroundings can be performed by using different strategies and a variety of techniques, as reviewed by B.C. Gates et al. [179]. For a detailed understanding of the structural features and catalytic properties of such materials, a combination of several techniques should be applied. For information related to the metal, the reader can profitably look at the rich literature on single-atom catalyst characterization [180–184]. Basically, information can be obtained concerning: (i) the metal nuclearity by STEM-HAADF [185, 186]; (ii) the nature of the metal and its oxidation state by conventional techniques such as XPS or XAS; (iii) the coordination numbers by XANES [187]; and (iv) the spatial distribution of the metal atoms. There is no specific method for quantitatively determining the spatial distribution of the active sites; and distinguishing between locations inside the pores and on the external surface is challenging. Besides, high-resolution STEM analyses, which can be time-consuming and only offer a very local analysis of the sample [186, 188], solid-state NMR techniques [189–192], and fluorescence microscopy paired with spectral probes [193], constitute useful tools for determining the spatial distribution of the active sites.

The metal–ligand but also metal-support [194] interactions have also to be characterized into details, since they will affect the short-range-microscopic scale effects. The structural characterization of tethered homogeneous catalysts onto solid surfaces is challenging due to the low density of active sites, and the presence of several species resulting from the often complex surface chemistry of the support. To achieve a good understanding of metal–ligand and metal-support interaction, the surface complexes need to be characterized through a toolbox of advanced spectroscopic techniques [195], such as Fourier transform infrared (FT-IR), extended X-ray absorption fine structure (EXAFS) [187, 196] or multi-dimensional solid-state NMR spectroscopy [197, 198]. Extraction methods have also been proposed since the eventually extracted species can be easily characterized by classical methods of molecular chemistry [199]. Finally, computational chemistry is an essential complementary tool to confirm or rule out possible structures based on thermodynamics or spectroscopic signatures of surface [200].

8.4 Catalysis with Molecular Complexes Immobilized into Porous Materials and Evidences of Confinement Effects

Much work has been done in recent years to understand the effects of confinement in catalysis, in particular with molecular complexes immobilized into porous materials. The concept of confinement has now expanded beyond zeolites, and it has also been observed inside the channels of mesoporous silica, CNTs, and MOFs. The investigated catalytic reactions include hydrogenation, hydroformylation, oxidation, oligomerization and polymerization, metathesis, and miscellaneous reactions. In addition, several attempts to perform asymmetric catalysis in confined spaces have also been reported. We will detail below these different reactions by taking representative examples of the work carried out, focusing on those where the effects of confinements have been studied in detail.

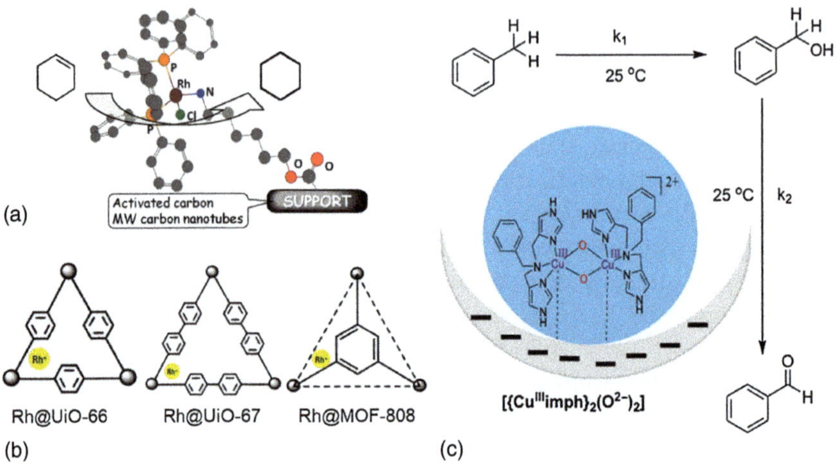

Figure 8.9 (a) A Rh complex, derived from Wilkinson's catalyst immobilized on activated carbon and CNTs through a linear organic molecule (6-amino-1-hexanol), covalently bonded to the support. Source: Reproduced with permission from Pérez-Cadenas et al. [162] © 2011 Wiley. (b) Schematic drawing of the pore windows of Rh@UiO-66, Rh@UiO-67, and Rh@MOF-808. Source: Reproduced with permission from [201] © 2022 Elsevier. (c) Oxidation reaction of toluene with a tripodal tridentate copper(II) complex immobilized in the nanochannels of functionalized mesoporous silica nanoparticles. Source: Reproduced with permission from Liu et al. [202] © 2015 Elsevier.

8.4.1 Hydrogenation Reactions

M.C. Roman-Martinez et al. used CNTs to covalently immobilize a rhodium complex derived from the Wilkinson's catalyst through an organic linker (6-amino-1-hexanol) in order to keep the complex apart of the support surface, and to mimic its environment in homogeneous phase (Figure 8.9a) [162, 203]. In the hydrogenation of cyclohexene, a noticeably enhanced catalytic activity was observed with the catalyst prepared on CNTs (TON = 3564) compare to a catalyst prepared on the activated carbon (TON = 1404), which is as active as the homogeneous Wilkinson's catalyst. Negligible leaching was observed (lower than 4%), and the catalyst was recyclable. The positive effect of the support has been attributed to the confinement of the active species inside the CNT pores, and it was suggested that the structure and pore size of the support plays a remarkable role in the catalytic behavior.

The same authors have highlighted the influence of the CNT diameter on the catalytic properties in the hydrogenation of cyclohexene and carvone. With Rh-catalysts prepared with small CNTs with inner diameter (inner $\Phi \sim 4$ nm), the support has no positive effect, whereas catalysts prepared with CNTs with an internal diameter around 7 nm give better results. Although the location of the complex with respect to the porosity of the support could not be established precisely, this increased catalytic activity was considered as an indirect proof of the confinement of the complex (or at least of a part of it) inside pores.

On the other hand, Pd-tridecilamine complex was immobilized by impregnation only in the internal cavity of CNTs open at both ends, as shown by XPS analysis [164].

A significant enhancement of the catalytic activity was observed in hydrogenation of cyclohexene (TON = 1696), compared to another Pd-catalyst (TON = 58) located both at the outer and at the inner surface of CNTs, with similar BET surface area and surface chemistry.

8.4.2 Hydroformylation Reactions

Hydroformylation of olefins is recognized as one of the most momentous and atom-economical access to aldehydes, which are widely applied as intermediates for the synthesis of esters, alcohols, carboxylic acids, aliphatic amines, and other fine chemicals [204]. Controlling the regioselectivity to the linear or branched aldehyde is a long-standing challenge in hydroformylation. Several groups have studied the confinement of rhodium(I) species in mesoporous silica (MCM-41 and SBA-15) by covalent or non-covalent immobilization of Rh-complexes [112, 205, 206]. For instance, the strategy involving the use of the cationic electrostatic interaction for anchoring of rhodium complex onto mesoporous materials resulted in a catalyst system active, regioselective, and recyclable for hydroformylation of octene toward *n*-nonanal. The effect of pore size on the hydroformylation reaction has been observed, SBA-15-Rh catalysts being a better catalyst than the MCM-41-Rh catalysts due to larger pore diameter of SBA-15 support (4.0–13.6 nm) compared to MCM-41 (pore diameters in the range of 2.0–3.2 nm) [112]. When [$RhCl(CO)(PPh_3)_2$] was anchored to mesoporous silicas SBA-15 and MCM-41 through *n*-alkyl as spacer (carbon number n = 1, 3, 5, 8, 11), the specific activity and stability of the immobilized catalyst were both increased when increasing the alkyl spacer length [206].

For Co-catalyzed hydroformylation of olefins without directing groups, M. Ranocchiari et al. [207] described in confined MOF systems (Co@MOF), an unexpected product formation and branched aldehyde selectivity up to 90% that is not achievable with existing catalysts. To investigate the cause of this change in selectivity, the interaction energies of the active pre-catalyst [$HCo(CO)_4$] and the active catalyst [$HCo(CO)_3$] in the pores of the MOF were modeled by DFT. The results suggest that a bond between the unsaturated cobalt complex and the support is unlikely to form under catalytic conditions. Based on Monte Carlo simulations combined with kinetic models, the enhanced selectivity for branched products with Co@MOF was ascribed, in part, to adsorption/partition driven increases in the local concentration of olefin in the micropores, relative to the olefin concentration near the free catalyst in homogeneous solution. The MOF micropores create the right conditions to tune the concentration of the reactants and push the kinetic limit of the hydroformylation reaction, favoring the formation of the branched aldehydes.

Schulz et al. also studied catalysts in MOF matrices [201]. They incorporated [Rh(acac)(cod)] complex in highly porous Zr-MOF of UiO-66, UiO-67, and MOF-808 (Figure 8.9b). The Zr-MOF-based catalysts were very active at low Rh loading: less than 1 Rh atom per 100 unit cells (Zr metal nodes). They showed high aldehyde selectivity in the hydroformylation of linear *n*-alkenes of different chain lengths, such as *n*-hexene, *n*-octene, *n*-decene, *n*-dodecene, and *n*-hexadecene. The superior catalytic properties were assigned to the presence of highly dispersed active Rh sites located at the isolated zirconia moieties, and the enhanced accessibility of sites via the large pores.

One can also note the work of Cui and Shi [208] who described the confinement of Rh-atom catalysts within porous monophosphine polymers by one-pot condensation of 4,4,4″-phosphanetriyltribenzaldehyde (PTBA), hydrazine, and $[Rh(CO)_2(acac)]$. In the hydroformylation of 1-octene, the resulting Rh-catalysts were found to be highly active (TON up to 60 000) and regioselective, with the linear selectivity of aldehydes remarkably improved (92% linear aldehyde selectivity) compared to a Rh catalyst supported on porous monophosphine polymers, in which the Rh species were located on the support surface (51% linear aldehyde selectivity). The authors attributed the high regioselectivity to the encapsulation of atomically dispersed Rh species within the porous monophosphine polymers. In addition, the encirclement of the active Rh species by an abundant but dispersive phosphine ligand improved the stability of the catalysts through multisite coordination, preventing leaching of Rh species during the catalytic reaction.

8.4.3 Oxidation Reactions

The selective catalytic oxidation of C—H bonds of hydrocarbons to specific oxygenated products under mild conditions is a major challenge. The work of Liu et al. highlights the oxidation of hydrocarbons using copper complexes immobilized on mesoporous silica [202, 209]. They reported on a $Cu^{III}(\mu\text{-}O)_2Cu^{III}$ dicopper complex immobilized in mesoporous silica nanoparticles (MSNs) that is capable of catalyzing aerobic oxidation of the aliphatic C—H bonds of toluene, to yield benzyl alcohol and further benzaldehyde at ambient temperature under mild conditions (Figure 8.9c). The dicopper complex is formed by immobilizing the Cu(II) complex of a tripodal tridentate histidine-like ligand, Cu^{III}mph (Imph = bis(4-imidazolyl methyl)- benzylamine), in the nano-channels of the MSNs, followed by reduction of the Cu(II) complexes, and activation by O_2. The complex, with only O_2 as the bridging ligand, is apparently stabilized by confinement within the pores of the MSNs, in which the oxidizing power of both O atoms is harnessed for catalytic turnover. This is the first example of a $Cu^{III}(\mu\text{-}O)_2Cu^{III}$ complex capable of functioning like a "dioxygenase" in hydrocarbon oxidation.

In addition, Alkordi et al. [210] described the oxidation of cyclohexane to cyclohexanol/cyclohexanone with 91.5% yield (based on the oxidant) and 23.5 TON catalyzed with a Mn-porphyrin encapsulated in the pores of an indium imidazoledicarboxylate-based *rho*-ZMOF (Mn-TMPyP@*rho*-ZMOF). Another interesting example is the work of Larsen et al. [211] on the encapsulation of catalytically active Fe(III)-tetrakis(4-sulphonatophenyl)porphyrin (Fe4SP), by the "ship-in-a-bottle" strategy, within one of the three cages that exist in the platform of HKUST-1. Fe4SP@HKUST-1 was applied to mimic the catalytic behaviors of peroxidase enzyme by using 2,2′-azinodi(3-ethylbenzthiazoline)-6-sulfonate (ABTS) as a redox indicator to the addition of H_2O_2. The catalytic efficiency of Fe4SP@HKUST-1 was found to be comparable to that of microperoxidase-11 enzyme (MP-11) and homogeneous molecular metalloporphyrin Fe4SP.

8.4.4 Ethylene Oligomerization and Polymerization Reactions

The oligomerization of ethylene to linear α-olefins stands as one of the largest scale homogeneously catalyzed reactions, as short linear α-olefins (C_4–C_8) are used as co-monomers in the industrial synthesis of linear low-density polyethylene, and higher molecular weight oligomers are important intermediates in the production of plasticizers (C_6–C_{10}), lubricants (C_{10}–C_{12}), and detergents (C_{12}–C_{16}).

In 2011, Toulhoat et al. [212] reported a computational study of the effect of confinement within microporous structures of zeolitic frameworks on the activity and selectivity of titanocene catalysts for ethylene oligomerization. In this theoretical study, they have shown that the catalyst activity and selectivity can be significantly altered by performing the catalysis in hosts with an adequate pore size, thereby optimizing host-guest van der Waals interactions, with respect to performing the reaction in a "traditional" homogeneous solution. By the use of Grand Canonical Monte Carlo (GCMC) calculations, the adsorption energies and entropies in the Henry's regime were calculated for the identified reaction intermediates, for oligomer formation using Ti-based catalysts, in representative microporous structures with increasing pore sizes from 10.1 to 15.4 Å. They developed a microkinetic model to account for confinement effects on the activity and selectivity. Using the thermodynamic parameters characterizing the confinement effect determined for a choice of microporous guests, they predicted the optimal window and temperature to shift the selectivity from α-hexene to α-octene. For instance, when the pore size becomes too small (10.1 Å), the key transition state structure to form α-octene cannot enter anymore, since its molecular volume is too large with respect to the available volume of the host. These theoretical results have been supported by experimental studies on ethylene oligomerization catalyzed by a Cr(III)-diphosphinoamine catalyst supported on various zeolites [127, 128], and different mesoporous silicas [213].

8.4.5 Metathesis Reactions

Selectivity in olefin metathesis is an important issue in the pharmaceutical industry that has also been addressed using catalysis in confined spaces. Various olefin metathesis reactions, such as ring-closing metathesis (RCM) [159, 214], cross metathesis (CM) [214], ring-opening metathesis polymerization, and ring closing metathesis based macrocylization, were studied. The first and second-generation Grubbs–Hoveyda catalyst covalently immobilized inside the pores of micro- and mesoporous silica supports have been particularly investigated [103, 104, 214–216]. Most of these confined catalysts showed good to excellent activity, improved selectivity, and good stability in olefin metathesis reactions. Some of them have been used to perform reactions under continuous-flow conditions [104].

Interesting examples come from the work of Buchmeiser's group, which performed macrocyclization *via* olefin RCM using Grubbs–Hoveyda type catalysts or

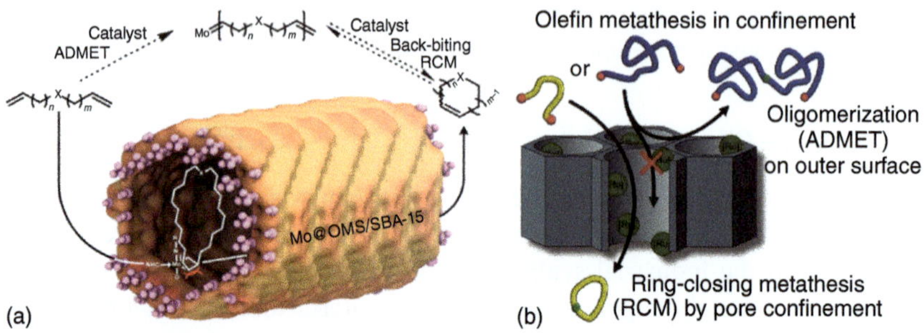

Figure 8.10 (a) Homogeneous ADMET reaction to the oligomer followed by back-biting RCM to the desired macrocycle, heterogeneous reaction with the intermediary metallacyclobutane inside the pore, and the confinement effect by the pore to obtain the desirable macrocycle. Source: Reproduced with permission from Ziegler et al. [103] © 2021 American Chemical Society. (b) COF for molecular heterogeneous olefin metathesis that utilizes spatial confinement of substrates in its pores for a substrate-size dependent and increased macro(mono)cyclization selectivity. Source: Reproduced with permission from Emmerling et al. [136] © 2022 Wiley.

a cationic molybdenum imido alkylidene *N*-heterocyclic carbene (NHC) catalyst confined inside the pores of mesoporous silicas (SBA-15 and an OMS) [103, 104]. Selective macro(mono)cyclization is difficult to obtain, since this reaction competes with acyclic diene metathesis (ADMET) oligomerization for entropic reasons. However, confinement limits ADMET oligomerization and macro(mono)cyclization reactions can be performed with selectivity up to 98% at elevated substrate concentrations (up to 100 mM) and moderate catalyst loadings (1 mol%), using tailored cationic molybdenum imido alkylidene NHC catalysts immobilized inside the pores of mesoporous silicas without any tethering using SOMC (Figure 8.10a) [103, 104].

The best selectivity in macro(mono)cyclization was obtained with molybdenum complexes immobilized in small diameter OMS pores (average pore diameter 2.5 ± 0.75 nm). In addition, the higher Z/E ratios obtained (up to 1.1) show that the confinement influences the stereoselectivity of the reaction, allowing to transiently overcome the thermodynamic. The increase in Z/E ratio with decreasing pore diameter can be rationalized by a confinement effect on the intermediary molybdacyclobutane, which pushes the substituents into the same direction and requires less space than a molybdacyclobutane with two substituents trans to each other. Molecular dynamics simulations (*vide infra*) have been performed to obtain a molecular picture of the distribution and mobility of substrates and products at the single pore level, using cylindrical model pore geometries of 2.55, 5.03, and 6.24 nm diameter, respectively, and 10 nm length. These simulations show that the high conversions are a direct result of the proximity between the surface-bound catalyst, analyzed by extended X-ray absorption spectroscopy, and the surface-located substrates. In addition, simulations strongly suggest that diffusion of the macrocycle out of the pore is much faster than back-diffusion into the pore, while all pores are well accessible for all substrates. This is in line with the finding that substrate diffusion into a pore is hindered if its hydrodynamic diameter exceeds one-third of the pore diameter.

MOFs [159] and COFs [136] are also very useful confinement platforms for Grubbs–Hoveyda-type catalysts. For instance, a Ru@MOF catalyst is active in olefin RCM of various substrates with TONs up to 8900 (in batch conditions) or 4700 (in continuous flow [159]). In α,ω-diene RCM, increased macrocyclization selectivity over oligomerization (ADMET) was observed for the Ru@COF catalyst (Figure 8.10b) compared to the homogeneous catalyst [136].

8.4.6 Miscellaneous Reactions on Various Supports

Several other catalytic applications of metal-based catalysts confined in zeolites, mesoporous silica, and MOFs for instance are reported in this section. In view of the variety of reactions studied, we preferred to make a classification by type of support used.

8.4.6.1 Zeolites

Zeolites have been used for the encapsulation of transition metal Schiff bases complexes. Ray et al. [24, 118] reported Heck coupling reaction and styrene oxidation catalyzed by Schiff-base palladium- and nickel-complexes encapsulated in zeolite-Y supercage, respectively. The encapsulated palladium complexes showed a remarkable improvement in reactivity during the formation of E-stilbene by C–C coupling of bromobenzene and styrene. The nickel complexes improved both the catalytic activity and the selectivity for the oxidation of styrene to benzaldehyde using H_2O_2 as the oxidant. According to theoretical and experimental studies, encapsulation leads to a substantial change in structural parameters of the square planar metal-based Schiff complexes, which undergoes a distortion in geometry around the Ni center. Consequently, these steric and electrostatic constraints are responsible for the modifications of the electronic properties of the complexes and their reinforced catalytic activities.

8.4.6.2 Mesoporous Silica

Combining two or more catalytic centers on a support to work cooperatively in reactions is considered as the state of the art in heterogeneous catalysis. Active sites are usually spatially separated, and cooperativity is difficult, making the combination a great challenge. The efficiency of accumulation of two or more active species with respect to catalysis can be significantly affected by the confined mesoporous environment. An interesting example is the work of Motokura et al. [106, 107, 217] using palladium catalysts immobilized on a sophisticated multifunctionalized mesoporous silica support for C—C bond reaction (Figure 8.11). A bisphosphine Pd complex homogeneously dispersed inside the mesopores of various silica supports (pore size: 16–31 Å) was found to be an excellent catalyst for the allylation reaction of nucleophiles (acyclic, cyclic, aromatic ketoesters, and diketones) with allylic alcohols as alkylating agents, in presence of specific base additives, such as 1,8-diazabicyclo[5.4.0]undec-7-ene (DBU). The catalytic activity of the supported Pd complex was *c.* 9 times higher than that of its homogeneous precursor Pd complex (TON up to 4500). The accelerating effect strongly depends on the size of

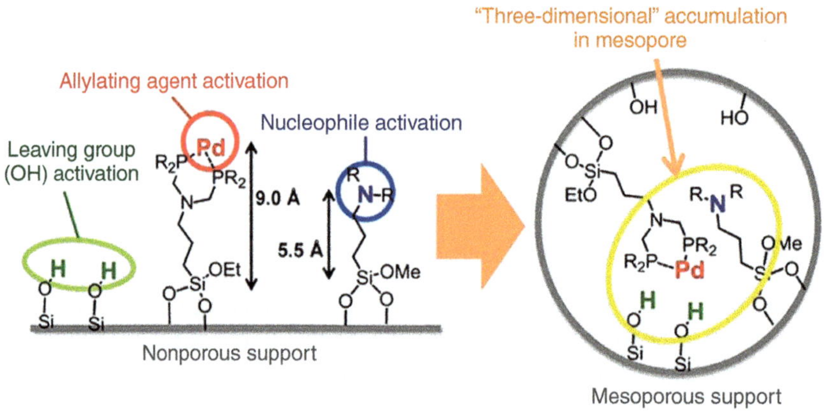

Figure 8.11 Concept of mesoporous silica-supported catalyst design for Pd-catalyzed allylation reactions. Source: Reprinted with permission from Motokura et al. [106] © 2017 Wiley.

the mesopores; the catalyst with a pore size of ~20 Å showing the highest reaction rate compared to other catalysts with smaller or larger pores. Kinetic studies imply that the maximum pre-exponential factor (A) is reached at a size of 23 Å, whereas the activation energy insignificantly changes (E_a = 66.1–73.3 kJ mol^{-1}), suggesting an accumulation efficiency of DBU near the Pd complex in the silica mesopores as shown in Figure 8.11. These results indicate that the confinement in an appropriate pore enhances the accumulation of active catalytic centers (DBU and the Pd complex), resulting in a high catalytic performance.

Other transition metal complexes confined in mesoporous silica have been reported as efficient heterogeneous catalysts for hydrosilylation of alkynes such as Rh(I)/(III)-*N*-heterocyclic carbene complexes on SBA-15 [102]. Notably, the SBA-15 immobilized Rh(I) catalysts, in which the pore walls provide an additional confinement, showed excellent β-(Z)-selectivity in the hydrosilylation of aliphatic alkynes.

8.4.6.3 MOFs

NH_2-MIL-125(Ti) is a visible-region photo-active MOF, which, in principle, can be coupled to molecular catalysts to accomplish light-to-chemical energy conversion. Nasalevich et al. [218] showed that a cobaloxime complex could be incorporated by first infiltrating the MOF pores with a flexible free ligand, and then providing cobalt ions for the ligand to wrap in a pseudo-macrocycle fashion. This "ship in a bottle" assembly strategy yields site-isolated, permanently pore confined, but fully functional molecular catalysts, positioned immediately proximal to two photo-active octa-titanium/oxy structures. The resulting **Co@MOF** system proved to be a robust, efficient, and fully recyclable photocatalytic production of H_2 from water under visible light illumination.

On the other hand, the large-pores of MIL-101-NH_2(Al) [219] were able to immobilize a molecular Rh catalyst [Cp*Rh(4,4′-bpydc)]$^{2+}$ and a molecular photosensitizer $[Ru(bpy)_2(4,4'\text{-bpydc})]^{2+}$ (bpydc = bipyridinedicarboxylic acid) upon easy post-synthetic impregnation. This photosystem allows the reduction of CO_2 under similar conditions to those for homogeneous catalysis. The exclusive production of formates with the suppression of H_2 production pointed toward a remarkable change in the selectivity of the Rh-Ru photosystem when immobilized in the Rh-Ru@MIL-101-NH_2 composite.

Zhang et al. [142] exploited cavity-imparted steric limitations of UiO-67 functionalized with an iridium(III) complex to accomplish mono-functionalization of methane using B_2Pin_2 (bis(pinacolato)diboron). The classic problem with selective activation/functionalization of methane (for example, to form methanol, rather than formaldehyde, formic acid, or CO_2) is that the first C—H bond is the most energetically difficult to activate. With subsequent C—H bonds being energetically easier to activate, it is challenging to halt functionalization after activating just one C—H bond. UiO-67 features 8 Å apertures and an octahedral cavity of volume 1.1 nm^3, as well as a tetrahedral cavities that are too small to participate in the catalytic reaction. In the absence of cavity for the confinement of reactants, both mono- and di-borylation of methane are readily feasible, with the doubly functionalized product being thermodynamically favored. By requiring methane and B_2Pin_2 to permeate UiO-67 in order to encounter an iridium catalyst, and by employing dodecane as a solvent to intercept a portion of the activated co-reactant, chemoselectivity for the desired mono-borylated product can be made to exceed 99%. Computational studies by Truhlar et al. [220] further revealed that the overwhelming preference for mono-borylation can be largely understood in terms of how local confinement preferentially attenuates the rate of transport of mono-borylated methane relative to CH_4. Related findings regarding selective mono-borylation, based on Zr–P1–Ir, have been described by Lin et al. [221].

Other transition metal complexes encapsulated in MOFs have been reported as efficient heterogeneous catalysts, among which we can cite as representative examples:

(i) a Fe phorphyrin MOF [143], for its ability for carbene transfer reactions to olefins, leading to the formation of pyrazoline with an unusual large increase in selectivity upon recycling;
(ii) a Co-porphyrin MOF [222], as an efficient electrocatalyst for oxygen evolution reaction (OER) at a wide pH range (from acidic to neutral); and
(iii) a Ni-phenanthroline complex [155], for the dehydrogenative coupling of alcohols with aromatic diamines, leading to the selective synthesis of mono-benzimidazoles and 1,2-disubstituted benzimidazoles.

8.4.7 Asymmetric Catalysis Reactions

Numerous examples exist where asymmetric catalysis occurs when appropriately selected chiral active centers are immobilized within the pores of mesoporous materials. Comprehensive review articles [44, 46, 47, 223] confirm the merit of exploiting nanospace in the manner adumbrated above. The most important reactions surveyed for the asymmetric synthesis in porous materials include epoxidation, hydrogenation, hydroformylation, C—X, and C—C bond-forming reactions. The confinement effect originated from the surfaces and the pores turns out to be a general phenomenon, which may make the enantioselectivity to increase (positive effect) or to decrease (negative effect). The effect becomes more pronounced particularly when the bonding between the catalyst and the surface is more rigid, and the pore size is tuned to a suitable range. It is proposed that the confinement in chiral synthesis is essentially a consequence of subtle change in transition states, induced by weak interaction in pores or on surfaces, such as hydrogen bonding, physical adsorption, and electronic interaction, in addition to the chemical bonding between the catalysts and the surface. In addition, advances have been shed on precisely how enantioselectivity is enhanced thanks to molecular dynamics simulations [224]. This also embraces the fruitful notion of continuous chirality measurement (CCM) introduced by Alvarez and Avnir [225], which provides a quantitative parameter that evaluates the degree of chirality of a given molecule, and allows to associate the chirality measures with enantioselectivity in asymmetric catalysis.

For example, the factors influencing the chiral catalysis in nanopores have been studied in detail for the asymmetric epoxidation of unfunctionalized olefins on Mn(salen) catalysts immobilized in the pores of silica-based mesoporous materials. It is found that chiral Mn(salen) catalysts immobilized in the pores generally exhibit higher chemical selectivity and enantioselectivity than those anchored on the external surface of the support. This is particularly true for pores with the optimized pore size [226], suggesting that the confinement effect from the pores can improve the chiral induction for the asymmetric catalysis. The stereo properties of the linkages of the chiral catalysts also affect the catalytic performance [227]. The enantiomeric excess (ee) values are increased with increasing the linkage lengths [228, 229] for the asymmetric epoxidation on Mn(salen) catalysts immobilized in nanopores (Figure 8.12), but the ee values are not affected by the linkage lengths for the catalysts anchored on the external surface.

Theoretical insights into epoxidation of cis- and trans-methylstyrene, catalyzed by anchored oxo-Mn(V)-salen into MCM-41 channels, were investigated by Malek et al. [224, 230]. A molecular dynamics simulation has led to a deeper understanding of the detailed perturbations that contribute to enhanced enantioselectivities [224]. This study shows that the immobilized linker influences the enantioselectivity of the catalyst due to the increasing chirality content of the Mn-salen complex. Avnir's CCM method was used to quantify the chirality content of the optimized geometries of the Mn-salen complex. Simulations with docked olefin (β-methyl styrene) suggest that cis- and trans-substrates have different levels of asymmetric induction to the Mn-salen catalyst, the trans-substrate inducing higher chirality to the immobilized Mn-salen complex than the *cis*-olefin. The importance of immobilization related to the steric communication between the substrate and the Mn-salen complex improves the chiral recognition of the catalyst, due to the increasing chirality content of the Mn-salen complex. This fact enables to interpret the significant differences in ee obtained when the chiral confined catalyst is

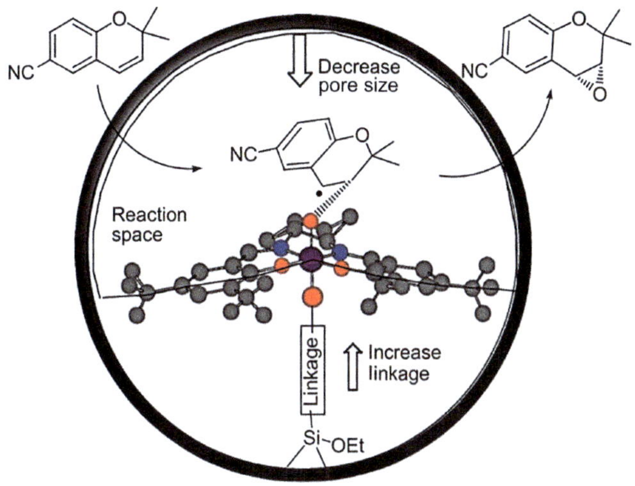

Figure 8.12 A schematic description of the heterogeneous asymmetric epoxidation in nanopores of mesoporous supports with decrease in pore sizes and increase in linkage lengths. Source: Reprinted with permission from Zhang and Li [229] © 2017 Wiley.

compared to an unconfined (but immobilized) catalyst, or the same homogeneous catalyst. The results are consistent with the experimental studies of epoxidation reactions using heterogenized Mn-salen complexes in nanoporous materials [229, 231].

The study of asymmetric catalytic reactions in porous materials has expanded in recent years. Some representative examples are given below. The asymmetric Henry's reaction was performed with chiral Schiff base complexes encapsulated inside the cavity of zeolite-NaY via a "ship in a bottle" synthesis method [124, 125, 232]. The Rh-catalyzed 1,2-addition of phenylboroxine to *N*-tosylimine was conducted with confined catalysts into the OMS material SBA-15 [233]. Preliminary results in enantioselective hydrogenation of olefins catalyzed by a chiral complex of rhodium encapsulated in MOFs have been reported by Ranocchiari and van Bokhoven [234]. Using a chiral [Rh(Me-BPE)(cod)]OTf complex (Me-BPE = 1,2-bis(2,5-dimethylphospholano)ethane) prepared *in situ* by absorption into different MOFs (Figure 8.13), a substantially enhanced reactivity (increase in the ee by 4–13%) was observed depending on the material.

Another interesting example is the confinement of multi-active sites to perform tandem catalysis. Liu et al. [235] immobilized a Lewis base and a chiral ruthenium/diamine dual species into the inner surface of Me-FDU-12 *via* a BF_4^- anion–bonding approach, to perform one-pot enantioselective aza-Michael addition/asymmetric transfer hydrogenation. They report a synergistic contribution of both functionalities that boost greatly one-pot enantioselective synthesis of a series of chiral g-secondary amino alcohols from enones in 92% yield with 96% ee. Such a yield was markedly higher than that of 79% obtained by the use of mixed homogeneous (DABCO)BF_4 and homogeneous (Mesitylene Ru TsDPEN)BF_4 as dual catalysts.

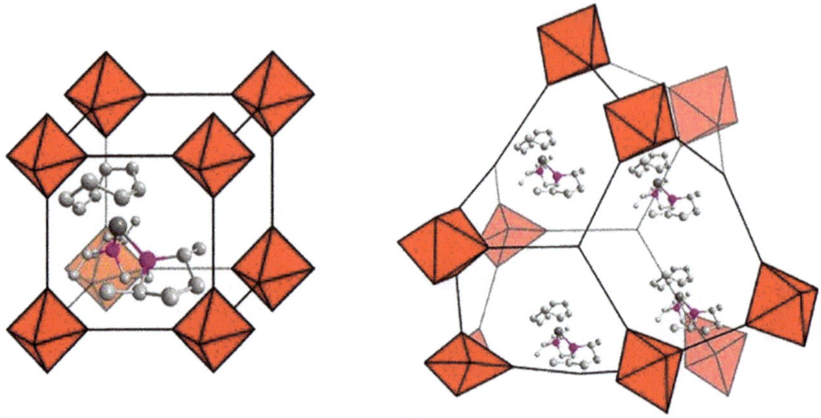

(a) (b)

Figure 8.13 Schematic representation of [Rh(Me-BPE)(cod)]OTf complex adsorbed in (a) IRMOF-3 or MIXMOF-5-NH_2; and (b) UMCM-1 or UMCM-1-NH_2 materials. Source: Reproduced with permission from Vilhanová et al. [234] © 2016 Wiley.

8.5 Conclusion

A rich literature reports confinement effects in catalysis using molecular complexes immobilized into porous materials. Many reactions have been investigated, mainly related to fine chemical synthesis including asymmetric catalysis, and promising results have been obtained in most cases. Improvement of catalyst activity, selectivity and in some cases stability have been reported upon confinement using mono- and in some cases dual active centers.

To be carried out successfully, such studies first need the synthesis of a catalytic system in which all or almost all of the active phase is confined. This requires the development of a synthetic route, which can be complex, and which will have an impact on the cost of the catalyst. It is therefore important that the added benefit of confinement on catalytic performance is significant. This also implies being able to compare the confined catalyst with an unconfined counterpart. For a supported complex, this would require being able to prepare a catalyst on the same support, but with the complex grafted onto the outer surface. This is generally not done and the comparison is limited to a comparison with the homogeneous species. A fine characterization of the catalyst is also necessary not only to ensure confinement but also to obtain the most precise image possible of the environment close to the active site. To do so, a combination of several techniques will be necessary (Figure 8.8), often combined with modeling. Finally, if a positive effect on catalysis is noted (activity, selectivity, or even stability), the study of confinement effects strictly speaking may require sustained efforts. Indeed, several effects can be combined (Table 8.2) and distinguishing between each effect on catalysis can prove to be a complicated task. In this field, modeling is an essential tool, since it offers a series of implements that can provide insight into crucial steps of catalysis including the understanding of catalysis inside the pores at a molecular level, as well as the analysis of the reactant-active center interactions that are at the heart of these processes.

References

1 Taylor, H.S. and Armstrong, E.F. (1925). A theory of the catalytic surface. *Proc. R. Soc. London, Ser. A* 108 (745): 105–111.

2 Balandin, A.A. (1929). Zur Theorie der heterogenen katalytischen Reaktionen. Multipletthypothese. Modell der Dehydrierungskatalyse. *Z. Phys. Chem.* 2B (1): 289–316.

3 Lundgren, R.J. and Stradiotto, M. (2016). Key concepts in ligand design. In: *Ligand Design in Metal Chemistry* (ed. M. Stradiotto and R.J. Lundgren), 1–14.

4 Durand, D.J. and Fey, N. (2019). Computational ligand descriptors for catalyst design. *Chem. Rev.* 119 (11): 6561–6594.

5 van Deelen, T.W., Hernández, M.C., and de Jong, K.P. (2019). Control of metal-support interactions in heterogeneous catalysts to enhance activity and selectivity. *Nat. Catal.* 2 (11): 955–970.

6 Friend, C.M. and Xu, F. (2018). Perspectives on the design of nanoparticle systems for catalysis. *Faraday Discuss.* 208: 595–607.

7 Maganas, D., Trunschke, A., Schlögl, R., and Neese, F. (2016). A unified view on heterogeneous and homogeneous catalysts through a combination of spectroscopy and quantum chemistry. *Faraday Discuss.* 188: 181–197.

8 Eremin, D.B. and Ananikov, V.P. (2017). Understanding active species in catalytic transformations: from molecular catalysis to nanoparticles, leaching, "cocktails" of catalysts and dynamic systems. *Coord. Chem. Rev.* 346: 2–19.

9 Rajadhyaksha, R.A. and Doraiswamy, L.K. (1976). Falsification of kinetic parameters by transport limitations and its role in discerning the controlling regime. *Catal. Rev.* 13 (1): 209–258.

10 Crossley, S.P., Resasco, D.E., and Haller, G.L. (2019). Clarifying the multiple roles of confinement in zeolites: from stabilization of transition states to modification of internal diffusion rates. *J. Catal.* 372: 382–387.

11 Nazari, M., Davoodabadi, A., Huang, D. et al. (2020). Transport phenomena in nano/molecular confinements. *ACS Nano* 14 (12): 16348–16391.

12 Keil, F.J. (2012). Multiscale modelling in computational heterogeneous catalysis. In: *Multiscale Molecular Methods in Applied Chemistry* (ed. B. Kirchner and J. Vrabec), 69–107. Berlin, Heidelberg: Springer Berlin Heidelberg.

13 Bhatt, J.S., Chutia, A., Catlow, C.R.A., and Coppens, M.-O. (2017). Quantum and statistical mechanical simulations for porous catalyst modelling. In: *Modern Developments in Catalysis* (ed. G. Hutchings, M. Davidson, R. Catlow, et al.), 253–288.

14 Warshel, A., Sharma, P.K., Kato, M. et al. (2006). Electrostatic basis for enzyme catalysis. *Chem. Rev.* 106 (8): 3210–3235.

15 Küchler, A., Yoshimoto, M., Luginbühl, S. et al. (2016). Enzymatic reactions in confined environments. *Nat. Nanotechnol.* 11 (5): 409–420.

16 Glasgow, J.E., Asensio, M.A., Jakobson, C.M. et al. (2015). Influence of electrostatics on small molecule flux through a protein nanoreactor. *ACS Synth. Biol.* 4 (9): 1011–1019.

17 Derouane, E.G., Andre, J.-M., and Lucas, A.A. (1988). Surface curvature effects in physisorption and catalysis by microporous solids and molecular sieves. *J. Catal.* 110 (1): 58–73.

18 Sastre, G. and Corma, A. (2009). The confinement effect in zeolites. *J. Mol. Catal. A: Chem.* 305 (1): 3–7.

19 Márquez, F., García, H., Palomares, E. et al. (2000). Spectroscopic evidence in support of the molecular orbital confinement concept: case of anthracene incorporated in zeolites. *J. Am. Chem. Soc.* 122 (27): 6520–6521.

20 del Campo, P., Martínez, C., and Corma, A. (2021). Activation and conversion of alkanes in the confined space of zeolite-type materials. *Chem. Soc. Rev.* 50 (15): 8511–8595.

21 Noh, G., Shi, Z., Zones, S.I., and Iglesia, E. (2018). Isomerization and β-scission reactions of alkanes on bifunctional metal-acid catalysts: consequences of confinement and diffusional constraints on reactivity and selectivity. *J. Catal.* 368: 389–410.

22 Kllerman, R. and Klier, K. (1975). Intrazeolitic transition-metal ion complexes. In: *Surface and Defect Properties of Solids: Volume 4*, vol. 4 (ed. M.W. Roberts and J.M. Thomas), 1–33. The Royal Society of Chemistry.

23 Ichikawa, M. (2000). "Ship-in-bottle" catalyst technology. *Platinum Metals Rev.* 44 (1): 3–14.

24 Choudhary, A., Das, B., and Ray, S. (2017). Encapsulated Schiff base nickel complex in zeolite Y: correlation between catalytic activities and extent of distortion supported by experimental and DFT studies. *Inorg. Chim. Acta* 462: 256–265.

25 Breslow, R. and Overman, L.E. (1970). "Artificial enzyme" combining a metal catalytic group and a hydrophobic binding cavity. *J. Am. Chem. Soc.* 92 (4): 1075–1077.

26 Jongkind, L.J., Elemans, J.A.A.W., and Reek, J.N.H. (2019). Cofactor controlled encapsulation of a rhodium hydroformylation catalyst. *Angew. Chem. Int. Ed.* 58 (9): 2696–2699.

27 Herron, N., Stucky, G.D., and Tolman, C.A. (1985). The reactivity of tetracarbonylnickel encapsulated in zeolite X. A case history of intrazeolite coordination chemistry. *Inorg. Chim. Acta* 100 (1): 135–140.

28 Ahmed, A.H. (2014). Zeolite-encapsulated transition metal chelates: synthesis and characterization. *Rev. Inorg. Chem.* 34 (3): 153–175.

29 Kosinov, N., Liu, C., Hensen, E.J.M., and Pidko, E.A. (2018). Engineering of transition metal catalysts confined in zeolites. *Chem. Mater.* 30 (10): 3177–3198.

30 Dai, C., Zhang, A., Song, C., and Guo, X. (2018). Chapter Two – Advances in the synthesis and catalysis of solid and hollow zeolite-encapsulated metal catalysts. In: *Advances in Catalysis*, vol. 63 (ed. C. Song), 75–115. Academic Press.

31 Ogino, I. (2015). Chapter 2 Zeolite-supported molecular metal complex catalysts. In: *Atomically-Precise Methods for Synthesis of Solid Catalysts*, 27–54. The Royal Society of Chemistry.

32 Liang, Y. (2021). Recent advanced development of metal-loaded mesoporous organosilicas as catalytic nanoreactors. *Nanoscale Adv.* 3 (24): 6827–6868.

33 Motokura, K., Ding, S., Usui, K., and Kong, Y. (2021). Enhanced catalysis based on the surface environment of the silica-supported metal complex. *ACS Catal.* 11 (19): 11985–12018.

34 Anwander, R. (2001). SOMC@PMS. Surface organometallic chemistry at periodic mesoporous silica. *Chem. Mater.* 13 (12): 4419–4438.

35 Zhou, C.H. (2011). An overview on strategies towards clay-based designer catalysts for green and sustainable catalysis. *Appl. Clay Sci.* 53 (2): 87–96.

36 Bedioui, F. (1995). Zeolite-encapsulated and clay-intercalated metal porphyrin, phthalocyanine and Schiff-base complexes as models for biomimetic oxidation catalysts: an overview. *Coord. Chem. Rev.* 144: 39–68.

37 Wu, C.-D. and Zhao, M. (2017). Incorporation of molecular catalysts in metal–organic frameworks for highly efficient heterogeneous catalysis. *Adv. Mater.* 29 (14): 1605446.

38 Dhakshinamoorthy, A., Asiri, A.M., and Garcia, H. (2020). Catalysis in confined spaces of metal organic frameworks. *ChemCatChem* 12 (19): 4732–4753.

39 Hemmer, K., Cokoja, M., and Fischer, R.A. (2021). Exploitation of intrinsic confinement effects of MOFs in catalysis. *ChemCatChem* 13 (7): 1683–1691.

40 Liu, J., Goetjen, T.A., Wang, Q. et al. (2022). MOF-enabled confinement and related effects for chemical catalyst presentation and utilization. *Chem. Soc. Rev.* 51 (3): 1045–1097.

41 Guo, J., Qin, Y., Zhu, Y. et al. (2021). Metal–organic frameworks as catalytic selectivity regulators for organic transformations. *Chem. Soc. Rev.* 50 (9): 5366–5396.

42 Dong, J., Han, X., Liu, Y. et al. (2020). Metal–covalent organic frameworks (MCOFs): a bridge between metal–organic frameworks and covalent organic frameworks. *Angew. Chem. Int. Ed.* 59 (33): 13722–13733.

43 Nghiem, T.-L., Coban, D., Tjaberings, S., and Gröschel, A.H. (2020). Recent advances in the synthesis and application of polymer compartments for catalysis. *Polymers* 12 (10): 2190.

44 Thomas, J.M. and Raja, R. (2008). Exploiting nanospace for asymmetric catalysis: confinement of immobilized, single-site chiral catalysts enhances enantioselectivity. *Acc. Chem. Res.* 41 (6): 708–720.

45 Maschmeyer, T., Rey, F., Sankar, G., and Thomas, J.M. (1995). Heterogeneous catalysts obtained by grafting metallocene complexes onto mesoporous silica. *Nature* 378: 159–162.

46 Li, C., Zhang, H., Jiang, D., and Yang, Q. (2007). Chiral catalysis in nanopores of mesoporous materials. *Chem. Commun.* 6: 547–558.

47 Li, C. (2004). Chiral synthesis on catalysts immobilized in microporous and mesoporous materials. *Catal. Rev.* 46 (3–4): 419–492.

48 Reetz, M.T. and Waldvogel, S.R. (1997). β-Cyclodextrin-modified diphosphanes as ligands for supramolecular rhodium catalysts. *Angew. Chem. Int. Ed.* 36 (8): 865–867.

49 Leenders, S.H.A.M., Gramage-Doria, R., de Bruin, B., and Reek, J.N.H. (2015). Transition metal catalysis in confined spaces. *Chem. Soc. Rev.* 44 (2): 433–448.

50 Roland, S., Suarez, J.M., and Sollogoub, M. (2018). Confinement of metal–N-heterocyclic carbene complexes to control reactivity in catalytic reactions. *Chemistry, A. Eur. J.* 24 (48): 12464–12473.

51 Mouarrawis, V., Plessius, R., van der Vlugt, J.I., and Reek, J.N.H. (2018). Confinement effects in catalysis using well-defined materials and cages. *Frontiers in Chemistry* 6: 1.

52 Takezawa, H. and Fujita, M. (2021). Molecular confinement effects by self-assembled coordination cages. *Bull. Chem. Soc. Jpn.* 94 (10): 2351–2369.

53 Gaeta, C., La Manna, P., De Rosa, M. et al. (2021). Supramolecular catalysis with self-assembled capsules and cages: what happens in confined spaces. *ChemCatChem* 13 (7): 1638–1658.

54 Matt, D. and Harrowfield, J. (2021). Phosphines and other P(III)-derivatives with cavity-shaped subunits: valuable ligands for supramolecular metal catalysis, metal confinement and subtle steric control. *ChemCatChem* 13 (1): 153–168.

55 Jin, Y., Zhang, Q., Zhang, Y., and Duan, C. (2020). Electron transfer in the confined environments of metal–organic coordination supramolecular systems. *Chem. Soc. Rev.* 49 (15): 5561–5600.

56 Wang, K., Jordan, J.H., Hu, X.-Y., and Wang, L. (2020). Supramolecular strategies for controlling reactivity within confined nanospaces. *Angew. Chem. Int. Ed.* 59 (33): 13712–13721.

57 Yang, J., Chatelet, B., Hérault, D. et al. (2018). Covalent cages with inwardly directed reactive centers as confined metal and organocatalysts. *Eur. J. Org. Chem.* 2018 (41): 5618–5628.

58 Sadjadi, S. (ed.) (2016). *Organic Nanoreactors*. Elsevier.

59 Poli, R. (ed.) (2017). *Effects of Nanoconfinement on Catalysis*. New York: Springer.

60 Dai, J. and Zhang, H. (2021). Recent advances in catalytic confinement effect within micro/meso-porous crystalline materials. *Small* 17 (22): 2005334.

61 Morishige, K. (2018). Influence of pore wall hydrophobicity on freezing and melting of confined water. *J. Phys. Chem. C* 122 (9): 5013–5019.

62 Morishige, K., Yasunaga, H., and Matsutani, Y. (2010). Effect of pore shape on freezing and melting temperatures of water. *J. Phys. Chem. C* 114 (9): 4028–4035.

63 Malani, A., Ayappa, K.G., and Murad, S. (2009). Influence of hydrophilic surface specificity on the structural properties of confined water. *J. Phys. Chem. B* 113 (42): 13825–13839.

64 Salles, F., Bourrelly, S., Jobic, H. et al. (2011). Molecular insight into the adsorption and diffusion of water in the versatile hydrophilic/hydrophobic flexible MIL-53(Cr) MOF. *J. Phys. Chem. C* 115 (21): 10764–10776.

65 Malfait, B., Jani, A., Mietner, J.B. et al. (2021). Influence of pore surface chemistry on the rotational dynamics of nanoconfined water. *J. Phys. Chem. C* 125 (30): 16864–16874.

66 Coudert, F.-X. (2020). Water adsorption in soft and heterogeneous nanopores. *Acc. Chem. Res.* 53 (7): 1342–1350.

67 Altabet, Y.E., Haji-Akbari, A., and Debenedetti, P.G. (2017). Effect of material flexibility on the thermodynamics and kinetics of hydrophobically induced evaporation of water. *Proc. Natl. Acad. Sci.* 114 (13): E2548–E2555.

68 Pham, M.D. and Serp, P. (2022). Introduction to supported metal single atom catalysis. In: *Supported Metal Single Atom Catalysis* (ed. P. Serp and D.P. Minh), 1–49.

69 Serna, P. and Gates, B.C. (2014). Molecular metal catalysts on supports: organometallic chemistry meets surface science. *Acc. Chem. Res.* 47 (8): 2612–2620.

70 Ehresmann, J.O., Kletnieks, P.W., Liang, A. et al. (2006). Evidence from NMR and EXAFS studies of a dynamically uniform mononuclear single-site zeolite-supported rhodium catalyst. *Angew. Chem. Int. Ed.* 45 (4): 574–576.

71 Khan, S.A., Vandervelden, C.A., Scott, S.L., and Peters, B. (2020). Grafting metal complexes onto amorphous supports: from elementary steps to catalyst site populations via kernel regression. *React. Chem. Eng.* 5 (1): 66–76.

72 Vandervelden, C.A., Khan, S.A., Scott, S.L., and Peters, B. (2020). Site-averaged kinetics for catalysts on amorphous supports: an importance learning algorithm. *React. Chem. Eng.* 5 (1): 77–86.

73 Patel, P., Wells, R.H., Kaphan, D.M. et al. (2021). Computational investigation of the role of active site heterogeneity for a supported organovanadium(III) hydrogenation catalyst. *ACS Catal.* 11 (12): 7257–7269.

74 Vandervelden, C.A., Khan, S.A., and Peters, B. (2020). Importance learning estimator for the site-averaged turnover frequency of a disordered solid catalyst. *J. Chem. Phys.* 153 (24): 244120.

75 Remi, J.C.S., Lauerer, A., Chmelik, C. et al. (2016). The role of crystal diversity in understanding mass transfer in nanoporous materials. *Nat. Mater.* 15 (4): 401–406.

76 Kärger, J. (2015). Transport phenomena in nanoporous materials. *ChemPhysChem* 16 (1): 24–51.

77 Babaei, H., McGaughey, A.J.H., and Wilmer, C.E. (2017). Effect of pore size and shape on the thermal conductivity of metal-organic frameworks. *Chem. Sci.* 8 (1): 583–589.

78 Smit, B. and Maesen, T.L.M. (2008). Towards a molecular understanding of shape selectivity. *Nature* 451 (7179): 671–678.

79 Guan, J., Pan, X., Liu, X., and Bao, X. (2009). Syngas segregation induced by confinement in carbon nanotubes: a combined first-principles and monte carlo study. *J. Phys. Chem. C* 113 (52): 21687–21692.

80 Yao, D., Wang, Y., Li, Y. et al. (2018). A high-performance nanoreactor for carbon–oxygen bond hydrogenation reactions achieved by the morphology of nanotube-assembled hollow spheres. *ACS Catal.* 8 (2): 1218–1226.

81 Zhou, W., Deng, Q.-W., Ren, G.-Q. et al. (2020). Enhanced carbon dioxide conversion at ambient conditions via a pore enrichment effect. *Nat. Commun.* 11 (1): 4481.

82 Yao, D., Wang, Y., Hassan-Legault, K. et al. (2019). Balancing effect between adsorption and diffusion on catalytic performance inside hollow nanostructured catalyst. *ACS Catal.* 9 (4): 2969–2976.

83 Long, Y., Palmer, J.C., Coasne, B. et al. (2011). Pressure enhancement in carbon nanopores: a major confinement effect. *Phys. Chem. Chem. Phys.* 13 (38): 17163–17170.

84 Long, Y., Śliwińska-Bartkowiak, M., Drozdowski, H. et al. (2013). High pressure effect in nanoporous carbon materials: effects of pore geometry. *Colloids Surf., A* 437: 33–41.

85 Srivastava, D., Santiso, E.E., and Gubbins, K.E. (2017). Pressure enhancement in confined fluids: effect of molecular shape and fluid–wall interactions. *Langmuir* 33 (42): 11231–11245.

86 Long, Y., Palmer, J.C., Coasne, B. et al. (2012). Under pressure: quasi-high pressure effects in nanopores. *Microporous Mesoporous Mater.* 154: 19–23.

87 Emelianova, A., Maximov, M.A., and Gor, G.Y. (2021). Solvation pressure in spherical mesopores: macroscopic theory and molecular simulations. *AlChE J.* 67 (3): e16542.

88 Tabacchi, G. (2018). Supramolecular organization in confined nanospaces. *ChemPhysChem* 19 (11): 1249–1297.

89 Balestri, D., Mazzeo, P.P., Perrone, R. et al. (2021). Deciphering the supramolecular organization of multiple guests inside a microporous MOF to understand their release profile. *Angew. Chem. Int. Ed.* 60 (18): 10194–10202.

90 Di Iorio, J.R., Johnson, B.A., and Román-Leshkov, Y. (2020). Ordered hydrogen-bonded alcohol networks confined in lewis acid zeolites accelerate transfer hydrogenation turnover rates. *J. Am. Chem. Soc.* 142 (45): 19379–19392.

91 Li, C., Zhang, Z., and Heinke, L. (2022). Mass transfer of toluene in a series of metal–organic frameworks: molecular clusters inside the nanopores cause slow and step-like release. *Phys. Chem. Chem. Phys.* 24 (6): 3994–4001.

92 Gounder, R. and Iglesia, E. (2013). The catalytic diversity of zeolites: confinement and solvation effects within voids of molecular dimensions. *Chem. Commun.* 49 (34): 3491–3509.

93 Artioli, N., Lobo, R.F., and Iglesia, E. (2013). Catalysis by confinement: enthalpic stabilization of NO oxidation transition states by micropororous and mesoporous siliceous materials. *J. Phys. Chem. C* 117 (40): 20666–20674.

94 Jones, A.J., Zones, S.I., and Iglesia, E. (2014). Implications of transition state confinement within small voids for acid catalysis. *J. Phys. Chem. C* 118 (31): 17787–17800.

95 Gelb, L.D., Gubbins, K.E., Radhakrishnan, R., and Sliwinska-Bartkowiak, M. (1999). Phase separation in confined systems. *Rep. Prog. Phys.* 62 (12): 1573–1659.

96 Binder, K., Puri, S., Das, S.K., and Horbach, J. (2010). Phase separation in confined geometries. *J. Stat. Phys.* 138 (1): 51–84.

97 Alba-Simionesco, C., Coasne, B., Dosseh, G. et al. (2006). Effects of confinement on freezing and melting. *J. Phys. Condens. Matter* 18 (6): R15–R68.

98 Qiao, C., Yu, X., Song, X. et al. (2020). Enhancing gas solubility in nanopores: a combined study using classical density functional theory and machine learning. *Langmuir* 36 (29): 8527–8536.

99 Xu, H. and Guo, C.-Y. (2015). Polymerization in the confinement of molecular sieves: facile preparation of high performance polyethylene. *Eur. Polym. J.* 65: 15–32.

100 Pump, E., Cao, Z., Samantaray, M.K. et al. (2017). Exploiting confinement effects to tune selectivity in cyclooctane metathesis. *ACS Catal.* 7 (10): 6581–6586.

101 Webb, J.D., Seki, T., Goldston, J.F. et al. (2015). Selective functionalization of the mesopores of SBA-15. *Microporous Mesoporous Mater.* 203: 123–131.

102 Panyam, P.K.R., Atwi, B., Ziegler, F. et al. (2021). Rh(I)/(III)-N-heterocyclic carbene complexes: effect of steric confinement upon immobilization on regio- and stereoselectivity in the hydrosilylation of alkynes. *Chem. Eur. J.* 27 (68): 17220–17229.

103 Ziegler, F., Kraus, H., Benedikter, M.J. et al. (2021). Confinement effects for efficient macrocyclization reactions with supported cationic molybdenum imido alkylidene N-heterocyclic carbene complexes. *ACS Catal.* 11 (18): 11570–11578.

104 Ziegler, F., Roider, T., Pyschik, M. et al. (2021). Olefin ring-closing metathesis under spatial confinement and continuous flow. *ChemCatChem* 13 (9): 2234–2241.

105 Liu, G., Wang, J., Huang, T. et al. (2010). Mesoporous silica-supported iridium catalysts for asymmetric hydrogenation reactions. *J. Mater. Chem.* 20 (10): 1970–1975.

106 Motokura, K., Ikeda, M., Nambo, M. et al. (2017). Concerted catalysis in tight spaces: palladium-catalyzed allylation reactions accelerated by accumulated active sites in mesoporous silica. *ChemCatChem* 9 (15): 2924–2929.

107 Motokura, K., Kawashima, S., Nambo, M. et al. (2020). Accumulation of active species in silica mesopore: effect of the pore size and free base additives on Pd-catalyzed allylation using allylic alcohol. *ChemCatChem* 12 (10): 2783–2791.

108 Tallarek, U., Hochstrasser, J., Ziegler, F. et al. (2020). Olefin ring-closing metathesis under spatial confinement: morphology–transport relationships. *ChemCatChem* 13 (1): 281–292.

109 Cao, H., Zhu, X.-H., Wang, D. et al. (2014). Selectivity enhancement in dynamic kinetic resolution of secondary alcohols through adjusting the micro-environment of metal complex confined in nanochannels: a promising strategy for tandem reactions. *ACS Catal.* 5 (1): 27–33.

110 Zhang, H., Jin, R., Yao, H. et al. (2012). Core-shell structured mesoporous silica: a new immobilized strategy for rhodium catalyzed asymmetric transfer hydrogenation. *Chem. Commun.* 48 (63): 7874–7876.

111 Bereczki, H.F., Daróczi, L., Fábián, I., and Lázár, I. (2016). Sol-gel synthesis, characterization and catalytic activity of silica aerogels functionalized with copper(II) complexes of cyclen and cyclam. *Microporous Mesoporous Mater.* 234: 392–400.

112 Vunain, E., Ncube, P., Jalama, K., and Meijboom, R. (2017). Confinement effect of rhodium(I) complex species on mesoporous MCM-41 and SBA-15: effect of pore size on the hydroformylation of 1-octene. *J. Porous Mater.* 25 (1): 303–320.

113 Oliveira, R.L., Nijholt, T., Shakeri, M. et al. (2016). Encapsulation of chiral Fe(salen) in mesoporous silica structures for use as catalysts to produce optically active sulfoxides. *Cataly. Sci. Technol.* 6 (13): 5124–5133.

114 Zhang, S., Zhang, B., Liang, H. et al. (2018). Encapsulation of homogeneous catalysts in mesoporous materials using diffusion-limited atomic layer deposition. *Angew. Chem. Int. Ed.* 57 (4): 1091–1095.

115 Fernandes, C.I., Saraiva, M.S., Nunes, T.G. et al. (2014). Highly enantioselective olefin epoxidation controlled by helical confined environments. *J. Catal.* 309: 21–32.

116 Mori, K., Kagohara, K., and Yamashita, H. (2008). Synthesis of tris(2,2′-bipyridine)iron(II) complexes in zeolite y cages: influence of exchanged alkali metal cations on physicochemical properties and catalytic activity. *J. Phys. Chem. C* 112 (7): 2593–2600.

117 Ma, Y., He, Y., Zhang, Q. et al. (2010). Self-assembly of ionic liquids and metal complexes in super-cages of NaY: integration of free catalysts and solvent molecules into confined catalytic sites. *Chin. J. Catal.* 31 (8): 933–937.

118 Kumari, S., Maddipoti, K., Das, B., and Ray, S. (2019). Palladium-schiff base complexes encapsulated in zeolite-Y host: functionality controlled by the structure of a guest complex. *Inorg. Chem.* 58 (2): 1527–1540.

119 Li, S., Liu, M., Liu, Q. et al. (2022). Zeolite encapsulated Cu(II)-salen complexes for the catalytic degradation of dyes in a neutral condition. *Colloids Surf., A* 648: 129153.

120 Kumari, S. and Ray, S. (2020). Zeolite encapsulated Ni(ii) Schiff-base complexes: improved catalysis and site isolation. *New J. Chem.* 44 (35): 14953–14963.

121 Choudhary, A., Das, B., and Ray, S. (2015). Encapsulation of a Ni salen complex in zeolite Y: an experimental and DFT study. *Dalton Trans.* 44 (8): 3753–3763.

122 Choudhary, A., Kumari, S., and Ray, S. (2017). Tuning of catalytic property controlled by the molecular dimension of palladium–schiff base complexes encapsulated in zeolite Y. *ACS Omega* 2 (10): 6636–6645.

123 Choudhary, A., Das, B., and Ray, S. (2016). Enhanced catalytic activity and magnetization of encapsulated nickel Schiff-base complexes in zeolite-Y: a correlation with the adopted non-planar geometry. *Dalton Trans.* 45 (47): 18967–18976.

124 Bania, K.K., Karunakar, G.V., Sarma, B., and Deka, R.C. (2014). Effect of nanospace confinement on the catalytic activity and stability of a chiral Schiff base complex (CuL; L=$C_{22}H_{24}N_2O_4$): a ombined experimental and theoretical study. *Chempluschem* 79 (3): 427–438.

125 Sharma, M., Das, B., Karunakar, G.V. et al. (2016). Chiral Ni-schiff base complexes inside zeolite-Y and their application in asymmetric henry reaction: effect of initial activation with microwave irradiation. *J. Phys. Chem. C* 120 (25): 13563–13573.

126 Mandal, M., Nagaraju, V., Karunakar, G.V. et al. (2015). Electronic, conjugation, and confinement effects on structure, redox, and catalytic behavior of oxido-vanadium(IV) and -(V) chiral schiff base complexes. *J. Phys. Chem. C* 119 (52): 28854–28870.

127 Shao, H., Wang, J., Wang, R. et al. (2017). Ethylene oligomerization in zeolite-grafted Cr(III)-diphosphinoamine catalysts using triisobutylaluminium as cocatalyst: change from dimerization to trimerization due to confinement effect. *Appl. Catal., A* 544: 154–160.

128 Shao, H., Li, Y., Gao, X. et al. (2014). Microporous zeolite supported $Cr(acac)_3$/PNP catalysts for ethylene tetramerization: influence of supported patterns and confinement on reaction performance. *J. Mol. Catal. A: Chem.* 390: 152–158.

129 Guo, J. and Jiang, D. (2020). Covalent organic frameworks for heterogeneous catalysis: principle, current status, and challenges. *ACS Cent. Sci.* 6 (6): 869–879.

130 Ding, S.Y., Gao, J., Wang, Q. et al. (2011). Construction of covalent organic framework for catalysis: Pd/COF-LZU1 in Suzuki-Miyaura coupling reaction. *J. Am. Chem. Soc.* 133 (49): 19816–19822.

131 Cai, M., Ding, S., Gibbons, B. et al. (2020). Nickel(II)-modified covalent-organic framework film for electrocatalytic oxidation of 5-hydroxymethylfurfural (HMF). *Chem. Commun.* 56 (92): 14361–14364.

132 Li, L.-H., Feng, X.-L., Cui, X.-H. et al. (2017). Salen-based covalent organic framework. *J. Am. Chem. Soc.* 139 (17): 6042–6045.

133 Wang, X., Han, X., Zhang, J. et al. (2016). Homochiral 2D porous covalent organic frameworks for heterogeneous asymmetric catalysis. *J. Am. Chem. Soc.* 138 (38): 12332–12335.

134 Han, X., Xia, Q., Huang, J. et al. (2017). Chiral covalent organic frameworks with high chemical stability for heterogeneous asymmetric catalysis. *J. Am. Chem. Soc.* 139 (25): 8693–8697.

135 Lu, M., Zhang, M., Liu, J. et al. (2022). Confining and highly dispersing single polyoxometalate clusters in covalent organic frameworks by covalent linkages for CO_2 photoreduction. *J. Am. Chem. Soc.* 144 (4): 1861–1871.

136 Emmerling, S.T., Ziegler, F., Fischer, F.R. et al. (2022). Olefin metathesis in confinement: towards covalent organic framework scaffolds for increased macrocyclization selectivity. *Chem. Eur. J.* 28 (8): e202104108.

137 Ranocchiari, M. (2021). Design of molecular heterogeneous catalysts with metal–organic frameworks. In: *Heterogeneous Catalysts* (ed. W.Y. Teoh, A. Urakawa, Y.H. Ng, and P. Sit), 147–160. WILEY-VCH GmbH, Weinheim, Germany.

138 Wei, Y.-S., Zhang, M., Zou, R., and Xu, Q. (2020). Metal–organic framework-based catalysts with single metal sites. *Chem. Rev.* 120 (21): 12089–12174.

139 Kumar, S., Mohan, B., Tao, Z. et al. (2021). Incorporation of homogeneous organometallic catalysts into metal–organic frameworks for advanced heterogenization: a review. *Cataly. Sci. Technol.* 11 (17): 5734–5771.

140 Zhang, Y., Yang, X., and Zhou, H.-C. (2018). Synthesis of MOFs for heterogeneous catalysis via linker design. *Polyhedron* 154: 189–201.

141 Gonzalez, M.I., Bloch, E.D., Mason, J.A. et al. (2015). Single-crystal-to-single-crystal metalation of a metal–organic framework: a route toward structurally well-defined catalysts. *Inorg. Chem.* 54 (6): 2995–3005.

142 Zhang, X., Huang, Z., Ferrandon, M. et al. (2018). Catalytic chemoselective functionalization of methane in a metal–organic framework. *Nat. Catal.* 1 (5): 356–362.

143 Abeykoon, B., Devic, T., Greneche, J.M. et al. (2018). Confinement of Fe-Al-PMOF catalytic sites favours the formation of pyrazoline from ethyl diazoacetate with an unusual sharp increase of selectivity upon recycling. *Chem. Commun.* 54 (73): 10308–10311.

144 Ma, L., Falkowski, J.M., Abney, C., and Lin, W. (2010). A series of isoreticular chiral metal–organic frameworks as a tunable platform for asymmetric catalysis. *Nat. Chem.* 2 (10): 838–846.

145 Chen, X., Qiao, Z., Hou, B. et al. (2021). Chiral metal-organic frameworks with tunable catalytic selectivity in asymmetric transfer hydrogenation reactions. *Nano Res.* 14 (2): 466–472.

146 Falkowski, J.M., Sawano, T., Zhang, T. et al. (2014). Privileged phosphine-based metal-organic frameworks for broad-scope asymmetric catalysis. *J. Am. Chem. Soc.* 136 (14): 5213–5216.

147 Gong, W., Chen, X., Zhang, W. et al. (2022). Leveraging chiral Zr(IV)-based metal–organic frameworks to elucidate catalytically active Rh species in asymmetric hydrogenation reactions. *J. Am. Chem. Soc.* 144 (7): 3117–3126.

148 Liu, J., Fan, Y.Z., Li, X. et al. (2018). Catalytic space engineering of porphyrin metal-organic frameworks for combined CO_2 capture and conversion at a low concentration. *ChemSusChem* 11 (14): 2340–2347.

149 Epp, K., Bueken, B., Hofmann, B.J. et al. (2019). Network topology and cavity confinement-controlled diastereoselectivity in cyclopropanation reactions catalyzed by porphyrin-based MOFs. *Cataly. Sci. Technol.* 9 (22): 6452–6459.

150 Wang, Y., Cui, H., Wei, Z.W. et al. (2017). Engineering catalytic coordination space in a chemically stable Ir-porphyrin MOF with a confinement effect inverting conventional Si-H insertion chemoselectivity. *Chem. Sci.* 8 (1): 775–780.

151 Burgess, S.A., Kassie, A., Baranowski, S.A. et al. (2016). Improved catalytic activity and stability of a palladium pincer complex by incorporation into a metal–organic framework. *J. Am. Chem. Soc.* 138 (6): 1780–1783.

152 Chen, Y., Fan, B., Lu, N., and Li, R. (2015). Aluminum metal–organic framework as a new host for preparation of encapsulated metal complex catalysts. *Catal. Commun.* 64: 91–95.

153 Zhou, Z., Liu, X., Ma, J.-G., and Cheng, P. (2022). MOF-incorporated binuclear N-heterocyclic carbene-cobalt catalyst for efficient conversion of CO_2 to formamides. *ChemSusChem* 15 (19): e202201386.

154 Li, B., Zhang, Y., Ma, D. et al. (2014). Metal-cation-directed de novo assembly of a functionalized guest molecule in the nanospace of a metal–organic framework. *J. Am. Chem. Soc.* 136 (4): 1202–1205.

155 Shadab, D.G., Sk, M., Banerjee, D., and Aijaz, A. (2021). Heterogenizing a homogeneous nickel catalyst using nanoconfined strategy for selective synthesis of mono- and 1,2-disubstituted benzimidazoles. *Inorg. Chem.* 60 (21): 16042–16047.

156 Chołuj, A., Karczykowski, R., and Chmielewski, M.J. (2019). Simple and robust immobilization of a ruthenium olefin metathesis catalyst inside MOFs by acid–base reaction. *Organometallics* 38 (18): 3392–3396.

157 Ivko, S.A., Bailey, T., Brammer, L., and Haynes, A. (2022). Ionic encapsulation of a methanol carbonylation catalyst in a microporous metal–organic framework. *Chem. Commun.* 58 (80): 11252–11255.

158 Kockrick, E., Lescouet, T., Kudrik, E.V. et al. (2011). Synergistic effects of encapsulated phthalocyanine complexes in MIL-101 for the selective aerobic oxidation of tetralin. *Chem. Commun.* 47 (5): 1562–1564.

159 Chołuj, A., Zieliński, A., Grela, K., and Chmielewski, M.J. (2016). Metathesis@MOF: simple and robust immobilization of olefin metathesis catalysts inside (Al)MIL-101-NH2. *ACS Catal.* 6 (10): 6343–6349.

160 Balestri, D., Roux, Y., Mattarozzi, M. et al. (2017). Heterogenization of a [NiFe] hydrogenase mimic through simple and efficient encapsulation into a mesoporous MOF. *Inorg. Chem.* 56 (24): 14801–14808.

161 Wang, D., Li, S., Wu, C., and Li, T. (2022). Surface-seal encapsulation of a homogeneous catalyst in a mesoporous metal–organic framework. *J. Am. Chem. Soc.* 144 (2): 685–689.

162 Pérez-Cadenas, M., Lemus-Yegres, L.J., Román-Martínez, M.C., and Salinas-Martínez de Lecea, C. (2011). Immobilization of a Rh complex derived from the Wilkinson's catalyst on activated carbon and carbon nanotubes. *Appl. Catal., A* 402 (1–2): 132–138.

163 Lemus-Yegres, L.J., Román-Martínez, M.C., Such-Basáñez, I., and Salinas-Martínez de Lecea, C. (2008). Effects of confinement in hybrid diamine-Rh complex-carbon catalysts used for hydrogenation reactions. *Microporous Mesoporous Mater.* 109 (1–3): 305–316.

164 Such-Basáñez, I., Román-Martínez, M.C., and Serp, P. (2016). Enhancement of the hydrogenation activity of a Pd-tridecilamine (TDA) complex by confinement in carbon nanotubes. *Microporous Mesoporous Mater.* 225: 378–384.

165 Ogawa, T., Kumagai, N., and Shibasaki, M. (2013). Self-assembling neodymium/sodium heterobimetallic asymmetric catalyst confined in a carbon nanotube network. *Angew. Chem. Int. Ed.* 52 (24): 6196–6201.

166 Hashimoto, K., Kumagai, N., and Shibasaki, M. (2015). A carbon nanotube confinement strategy to implement homogeneous asymmetric catalysis in the solid phase. *Chemistry* 21 (11): 4262–4266.

167 Sing, K.S.W. (2004). Characterization of porous materials: past, present and future. *Colloids Surf., A* 241 (1): 3–7.

168 Kraus, H., Rybka, J., Höltzel, A. et al. (2021). PoreMS: a software tool for generating silica pore models with user-defined surface functionalisation and pore dimensions. *Mol. Simul.* 47 (4): 306–316.

169 Rouquerol, J., Avnir, D., Fairbridge, C.W. et al. (1994). Recommendations for the characterization of porous solids (Technical Report). *Pure Appl. Chem.* 66 (8): 1739–1758.

170 Melnichenko, Y.B. (2016). Structural characterization of porous materials using SAS. In: *Small-Angle Scattering from Confined and Interfacial Fluids: Applications to Energy Storage and Environmental Science* (ed. Y.B. Melnichenko), 139–171. Cham: Springer International Publishing.

171 van Hooff, J.H.C. and Roelofsen, J.W. (1991). Chapter 7 Techniques of zeolite characterization. In: *Studies in Surface Science and Catalysis*, vol. 58 (ed. H. van Bekkum, E.M. Flanigen, and J.C. Jansen), 241–283. Elsevier.

172 Iwanow, M., Gärtner, T., Sieber, V., and König, B. (2020). Activated carbon as catalyst support: precursors, preparation, modification and characterization. *Beilstein J. Org. Chem.* 16: 1188–1202.

173 Howarth, A.J., Peters, A.W., Vermeulen, N.A. et al. (2017). Best practices for the synthesis, activation, and characterization of metal–organic frameworks. *Chem. Mater.* 29 (1): 26–39.

174 Schneemann, A., Bon, V., Schwedler, I. et al. (2014). Flexible metal–organic frameworks. *Chem. Soc. Rev.* 43 (16): 6062–6096.

175 Ilić, B. and Wettstein, S.G. (2017). A review of adsorbate and temperature-induced zeolite framework flexibility. *Microporous Mesoporous Mater.* 239: 221–234.

176 Kapko, V., Dawson, C., Treacy, M.M.J., and Thorpe, M.F. (2010). Flexibility of ideal zeolite frameworks. *Phys. Chem. Chem. Phys.* 12 (30): 8531–8541.

177 Treacy, M.M.J., Dawson, C.J., Kapko, V., and Rivin, I. (2014). Flexibility mechanisms in ideal zeolite frameworks. *Philos. Trans. R. Soc. London, Ser. A* 372 (2008): 20120036.

178 Xiong, H., Liu, Z., Chen, X. et al. (2022). In situ imaging of the sorption-induced subcell topological flexibility of a rigid zeolite framework. *Science* 376 (6592): 491–496.

179 Babucci, M., Guntida, A., and Gates, B.C. (2020). Atomically dispersed metals on well-defined supports including zeolites and metal–organic frameworks: structure, bonding, reactivity, and catalysis. *Chem. Rev.* 120 (21): 11956–11985.

180 Kottwitz, M., Li, Y., Wang, H. et al. (2021). Single atom catalysts: a review of characterization methods. *Chem. Methods* 1 (6): 278–294.

181 Qi, P., Wang, J., Djitcheu, X. et al. (2022). Techniques for the characterization of single atom catalysts. *RSC Adv.* 12 (2): 1216–1227.

182 Li, X., Yang, X., Zhang, J. et al. (2019). In situ/operando techniques for characterization of single-atom catalysts. *ACS Catal.* 9 (3): 2521–2531.

183 Zhang, L., Doyle-Davis, K., and Sun, X. (2022). Characterization of supported metal single-atom catalysts. In: *Supported Metal Single Atom Catalysis* (ed. P. Serp and D.P. Minh), 169–198. WILEY-VCH GmbH, Weinheim, Germany.

184 Casu, A., Ould-Chikh, S., Mountjoy, G. et al. (2022). In situ/operando techniques for characterization of supported metal single-atom catalysts. In: *Supported Metal Single Atom Catalysis* (ed. P. Serp and D.P. Minh), 199–239. WILEY-VCH GmbH, Weinheim, Germany.

185 Browning, N.D., Aydin, C., Lu, J. et al. (2013). Quantitative Z-contrast imaging of supported metal complexes and clusters—a gateway to understanding catalysis on the atomic scale. *ChemCatChem* 5 (9): 2673–2683.

186 Yang, D., Xu, P., Browning, N.D., and Gates, B.C. (2016). Tracking Rh atoms in zeolite HY: first steps of metal cluster formation and influence of metal nuclearity on catalysis of ethylene hydrogenation and ethylene dimerization. *J. Phys. Chem. Lett.* 7 (13): 2537–2543.

187 Alexeev, O. and Gates, B.C. (2000). EXAFS characterization of supported metal-complex and metal-cluster catalysts made from organometallic precursors. *Top. Catal.* 10 (3): 273–293.

188 Lu, J., Aydin, C., Liang, A.J. et al. (2012). Site-isolated molecular iridium complex catalyst supported in the 1-dimensional channels of zeolite HSSZ-53: characterization by spectroscopy and aberration-corrected scanning transmission electron microscopy. *ACS Catal.* 2 (6): 1002–1012.

189 Werner, M., Rothermel, N., Breitzke, H. et al. (2014). Recent advances in solid state NMR of small molecules in confinement. *Isr. J. Chem.* 54 (1–2): 60–73.

190 Rieg, C., Dittmann, D., Li, Z. et al. (2021). Quantitative distinction between noble metals located in mesopores from those on the external surface. *Chem. A. Eur. J.* 27 (68): 17012–17023.

191 Grosso-Giordano, N.A., Schroeder, C., Xu, L. et al. (2021). Characterization of a molecule partially confined at the pore mouth of a zeotype. *Angew. Chem. Int. Ed.* 60 (18): 10239 10246.

192 Lou, C., Zhang, W., Ma, C. et al. (2021). Revealing the specific spatial confinement in 8-membered ring cage-type molecular sieves via solid-state NMR and theoretical calculations. *ChemCatChem* 13 (5): 1299–1305.

193 Dong, B., Mansour, N., Huang, T.-X. et al. (2021). Single molecule fluorescence imaging of nanoconfinement in porous materials. *Chem. Soc. Rev.* 50 (11): 6483–6506.

194 Basset, J.-M., Lefebvre, F., and Santini, C. (1998). Surface organometallic chemistry: some fundamental features including the coordination effects of the support. *Coord. Chem. Rev.* 178–180: 1703–1723.

195 Copéret, C., Comas-Vives, A., Conley, M.P. et al. (2016). Surface organometallic and coordination chemistry toward single-site heterogeneous catalysts: strategies, methods, structures, and activities. *Chem. Rev.* 116 (2): 323–421.

196 Lee, J.S. and Park, E.D. (2002). In situ XAFS characterization of supported homogeneous catalysts. *Top. Catal.* 18 (1): 67–72.

197 Blümel, J. (2008). Linkers and catalysts immobilized on oxide supports: new insights by solid-state NMR spectroscopy. *Coord. Chem. Rev.* 252 (21): 2410–2423.

198 Pump, E., Viger-Gravel, J., Abou-Hamad, E. et al. (2017). Reactive surface organometallic complexes observed using dynamic nuclear polarization surface enhanced NMR spectroscopy. *Chem. Sci.* 8 (1): 284–290.

199 Roberto, D., Cariati, E., Pizzotti, M., and Psaro, R. (1996). Extraction methods and surface reactions as a convenient methodology for the characterization of surface organometallic species. *J. Mol. Catal. A: Chem.* 111 (1): 97–108.

200 Sautet, P. and Delbecq, F. (2010). Catalysis and surface organometallic chemistry: a view from theory and simulations. *Chem. Rev.* 110 (3): 1788–1806.

201 Sun, Y., Harloff, J., Kosslick, H. et al. (2022). Influence of the framework on the catalytic performance of Rh-supported Zr-MOFs in the hydroformylation of n-alkenes. *Mol. Catal.* 517: 112005.

202 Liu, C.-C., Lin, T.-S., Chan, S.I., and Mou, C.-Y. (2015). A room temperature catalyst for toluene aliphatic C–H bond oxidation: tripodal tridentate copper complex immobilized in mesoporous silica. *J. Catal.* 322: 139–151.

203 Lemus-Yegres, L.J., Pérez-Cadenas, M., Román-Martínez, M.C., and de Lecea, C.S.-M. (2011). Hybrid Rh catalysts prepared with carbon nanotubes of different inner diameter. *Microporous Mesoporous Mater.* 139 (1-3): 164–172.

204 Franke, R., Selent, D., and Börner, A. (2012). Applied hydroformylation. *Chem. Rev.* 112 (11): 5675–5732.

205 Jeon, J.-K., Ko, Y.S., and Yim, J.-H. (2013). A selective functionalized mesoporous silica-supported rh catalyst for effective 1-octene hydroformylation. *J. Nanosci. Nanotechnol.* 13 (4): 3087–3095.

206 Zhou, W. and He, D. (2009). A facile method for promoting activities of ordered mesoporous silica-anchored Rh–P complex catalysts in 1-octene hydroformylation. *Green Chem.* 11 (8): 1146–1154.

207 Bauer, G., Ongari, D., Tiana, D. et al. (2020). Metal-organic frameworks as kinetic modulators for branched selectivity in hydroformylation. *Nat. Commun.* 11 (1): 1059.

208 Zhao, K., Wang, H., Wang, X. et al. (2021). Confinement of atomically dispersed Rh catalysts within porous monophosphine polymers for regioselective hydroformylation of alkenes. *J. Catal.* 401: 321–330.

209 Liu, C.-C., Tsai, Y.-F., Mou, C.-Y. et al. (2019). Dicopper dioxygenase model immobilized in mesoporous silica nanoparticles for toluene oxidation: a mechanism to harness both O atoms of O_2 for catalysis. *J. Phys. Chem. C* 123 (17): 11032–11043.

210 Alkordi, M.H., Liu, Y., Larsen, R.W. et al. (2008). Zeolite-like metal–organic frameworks as platforms for applications: on metalloporphyrin-based catalysts. *J. Am. Chem. Soc.* 130 (38): 12639–12641.

211 Larsen, R.W., Wojtas, L., Perman, J. et al. (2011). Mimicking heme enzymes in the solid state: metal–organic materials with selectively encapsulated heme. *J. Am. Chem. Soc.* 133 (27): 10356–10359.

212 Toulhoat, H., Lontsi, F.M., and de Bruin, T. (2011). Computational study of the effect of confinement within microporous structures on the activity and selectivity of metallocene catalysts for ethylene oligomerization. *J. Am. Chem. Soc.* 133 (8): 2481–2491.

213 Shao, H., Zhou, H., Guo, X. et al. (2015). Chromium catalysts supported on mesoporous silica for ethylene tetramerization: effect of the porous structure of the supports. *Catal. Commun.* 60: 14–18.

214 Yang, H., Ma, Z., Zhou, T. et al. (2013). Encapsulation of an olefin metathesis catalyst in the nanocages of SBA-1: facile preparation, high encapsulation efficiency, and high activity. *ChemCatChem* 5 (8): 2278–2287.

215 Ziegler, F., Teske, J., Elser, I. et al. (2019). Olefin metathesis in confined geometries: a biomimetic approach toward selective macrocyclization. *J. Am. Chem. Soc.* 141 (48): 19014–19022.

216 Dewaele, A., Van Berlo, B., Dijkmans, J. et al. (2016). Immobilized Grubbs catalysts on mesoporous silica materials: insight into support characteristics and their impact on catalytic activity and product selectivity. *Cataly. Sci. Technol.* 6 (8): 2580–2597.

217 Motokura, K., Ikeda, M., Kim, M. et al. (2018). Silica support-enhanced Pd-catalyzed allylation using allylic alcohols. *ChemCatChem* 10 (20): 4536–4544.

218 Nasalevich, M.A., Becker, R., Ramos-Fernandez, E.V. et al. (2015). Co@NH_2-MIL-125(Ti): cobaloxime-derived metal–organic framework-based composite for light-driven H_2 production. *Energy Environ. Sci.* 8 (1): 364–375.

219 Wang, X., Wisser, F.M., Canivet, J. et al. (2018). Immobilization of a full photosystem in the large-pore MIL-101 metal–organic framework for CO_2 reduction. *ChemSusChem* 11 (18): 3315–3322.

220 Yang, B., Wu, X.-P., Gagliardi, L., and Truhlar, D.G. (2019). Methane functionalization by an Ir(III) catalyst supported on a metal–organic framework: an alternative explanation of steric confinement effects. *Theor. Chem. Acc.* 138 (9): 107.

221 Feng, X., Song, Y., Li, Z. et al. (2019). Metal–organic framework stabilizes a low-coordinate iridium complex for catalytic methane borylation. *J. Am. Chem. Soc.* 141 (28): 11196–11203.

222 Mukhopadhyay, S., Basu, O., and Das, S.K. (2020). ZIF-8 MOF encapsulated co-porphyrin, an efficient electrocatalyst for water oxidation in a wide pH range: works better at neutral pH. *ChemCatChem* 12 (21): 5430–5438.

223 Song, C.E., Kim, D.H., and Choi, D.S. (2006). Chiral organometallic catalysts in confined nanospaces: significantly enhanced enantioselectivity and stability. *Eur. J. Inorg. Chem.* 2006 (15): 2927–2935.

224 Malek, K., Jansen, A.P.J., Li, C., and van Santen, R.A. (2007). Enantioselectivity of immobilized Mn-salen complexes: a computational study. *J. Catal.* 246 (1): 127–135.

225 Alvarez, S., Alemany, P., and Avnir, D. (2005). Continuous chirality measures in transition metal chemistry. *Chem. Soc. Rev.* 34 (4): 313–326.

226 Yu, K., Gu, Z., Ji, R. et al. (2007). Effect of pore size on the performance of mesoporous material supported chiral Mn(III) salen complex for the epoxidation of unfunctionalized olefins. *J. Catal.* 252 (2): 312–320.

227 Xiang, S., Zhang, Y., Xin, Q., and Li, C. (2002). Enantioselective epoxidation of olefins catalyzed by Mn (salen)/MCM-41 synthesized with a new anchoring method. *Chem. Commun.* 22: 2696–2697.

228 Zhang, H., Xiang, S., and Li, C. (2005). Enantioselective epoxidation of unfunctionalised olefins catalyzed by Mn(salen) complexes immobilized in porous materials via phenyl sulfonic group. *Chem. Commun.* 9: 1209–1211.

229 Zhang, H. and Li, C. (2006). Asymmetric epoxidation of 6-cyano-2,2-dimethylchromene on Mn(salen) catalyst immobilized in mesoporous materials. *Tetrahedron* 62 (28): 6640–6649.

230 Malek, K., Li, C., and van Santen, R.A. (2007). New theoretical insights into epoxidation of alkenes by immobilized Mn-salen complexes in mesopores: effects of substrate, linker and confinement. *J. Mol. Catal. A: Chem.* 271 (1): 98–104.

231 Zhang, H., Zhang, Y., and Li, C. (2006). Enantioselective epoxidation of unfunctionalized olefins catalyzed by the Mn(salen) catalysts immobilized in the nanopores of mesoporous materials. *J. Catal.* 238 (2): 369–381.

232 Bania, K.K., Karunakar, G.V., Goutham, K., and Deka, R.C. (2013). Enantioselective henry reaction catalyzed by "ship in a bottle" complexes. *Inorg. Chem.* 52 (14): 8017–8029.

233 Beurer, A.-K., Kirchhof, M., Bruckner, J.R. et al. (2021). Efficient and spatially controlled functionalization of SBA-15 and initial results in asymmetric Rh-catalyzed 1,2-additions under confinement. *ChemCatChem* 13 (10): 2407–2419.

234 Vilhanová, B., Ranocchiari, M., and van Bokhoven, J.A. (2016). Enantioselective hydrogenation of olefins enhanced by metal–organic framework additives. *ChemCatChem* 8 (2): 308–312.

235 Li, L., Yang, D., Zhao, Z. et al. (2018). Boron tetrafluoride anion bonding dual active species within a large-pore mesoporous silica for two-step successive organic transformaion to prepare optically pure amino alcohols. *Frontiers in Chemistry* 6: 272.

9

Size-Selective Catalysis by Metal–Organic Frameworks

Amarajothi Dhakshinamoorthy[1,2] *and Hermenegildo García*[3]

[1] *Universitat Politècnica de València, Departamento de Quimica, Av. De los Naranjos s/n, Valencia 46022, Spain*
[2] *Madurai Kamaraj University, School of Chemistry, Madurai 625021, Tamil Nadu, India*
[3] *Instituto Universitario de Tecnología Química, Consejo Superior de Investigaciones Científicas-Universitat Politecnica de Valencia, Av. De los Naranjos s/n, Valencia 46022, Spain*

9.1 Introduction

Metal–organic frameworks (MOFs) are a class of porous, crystalline solids, whose structure is built of nodes of metal ions or metal oxo clusters held in lattice by coordination with bi- or multipodal rigid organic linkers (Figure 9.1) [1–4]. The forces responsible for maintaining the crystal structure are coordinative metal-ligand bonds with a defined directionality of the interaction and, very often also, Coulombic forces lacking directionality, but providing considerable additional robustness to the structure [4–6]. MOFs are among the solids with the largest surface area [7], highest pore volume, and lowest framework density with a considerable percentage of empty space within the unit cell.

The high weight percentage of metal ions in MOFs and the fact that some of the coordination positions are not compromised with the construction of the crystal lattice determine that these materials can act as solid Lewis acids [8–11]. Exchangeable coordination positions around metal ions, normally occupied by solvent molecules introduced during the synthesis of these materials, can interact with substrates and reagents, promoting a chemical reaction (Figure 9.2) [12–14]. Structural defects and missing linkers can also generate unsaturated coordination positions around metal ions [15]. MOFs offer a large compositional flexibility in the synthesis regarding the metal, normally di-, tri- or tetravalent, and the structure of the organic ligand. There are by now over 90 000 MOFs reported and this number is growing due to the current interest in these materials.

The ability of metal ions to act as Lewis acid sites is the most common origin of the catalytic activity of MOFs, ligand substituents, and the presence of adsorbed guests being other possibilities for which MOFs can perform as solid catalysts (Figure 9.3) [16–19]. In fact, the study of the activity of MOFs is currently one of the most active fronts in catalysis and there are considerable number of reviews covering this broad

Catalysis in Confined Frameworks: Synthesis, Characterization, and Applications, First Edition.
Edited by Hermenegildo Garcia and Amarajothi Dhakshinamoorthy.

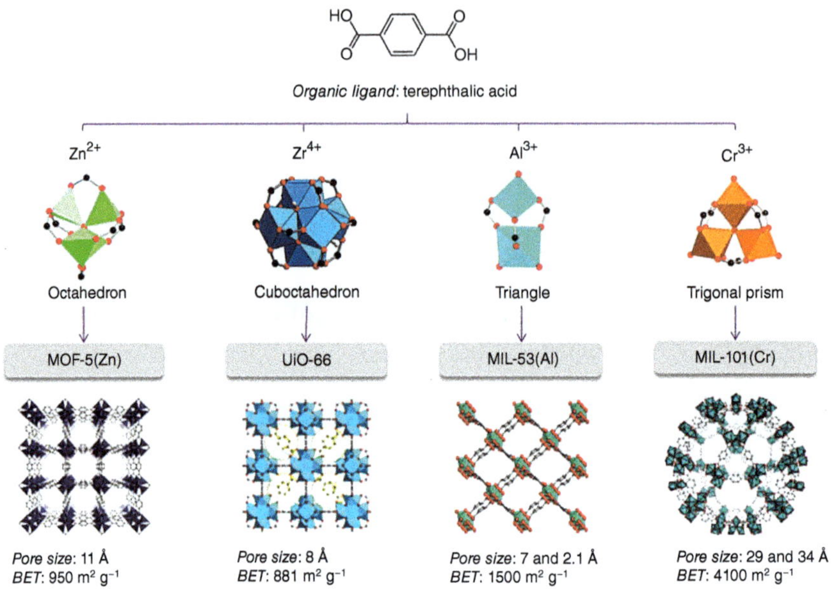

Figure 9.1 Some representative MOFs with their metal cluster and a ligand. Each framework topology has its characteristic pore size and available surface area.

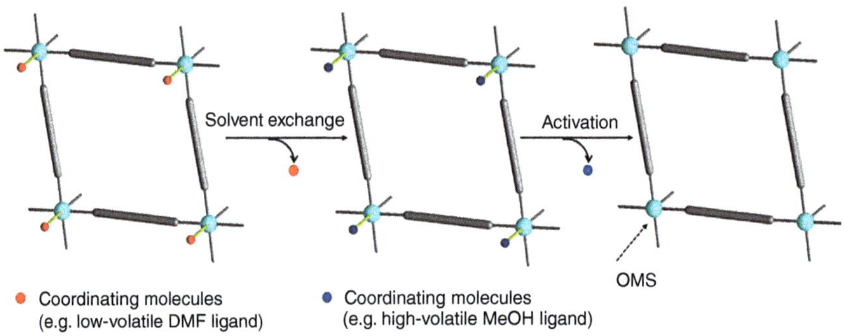

Figure 9.2 Schematic presentation of the generation of unsaturated coordination positions on metal atoms in MOFs by replacing a low-volatile ligand (e.g. DMF) by a higher-volatile ligand (e.g. MeOH) through "solvent exchange" for the subsequent activation, that is, solvent removal.

topic [20–24]. In this chapter, we will focus on the specific subject of shape-selectivity in MOFs and the reader is referred to other relevant works in this area for an exhaustive coverage of other aspects in catalysis [25–30].

Catalysis is continuously searching for increasing substrate conversion at the largest possible rate and for increasing selectivity to the wanted products. Reaction rate and selectivity, together with stability, are the most important features to be considered in a catalyst. Product selectivity is most frequently achieved by using sites that preferentially form certain intermediates among other possible alternatives. Tuning the nature, electronic density and coordination sphere of the

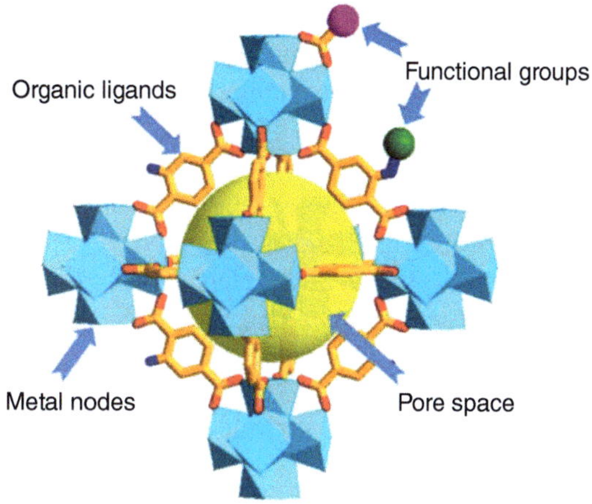

Figure 9.3 Possible active sites in MOFs beyond Lewis acid sites at the metal nodes.

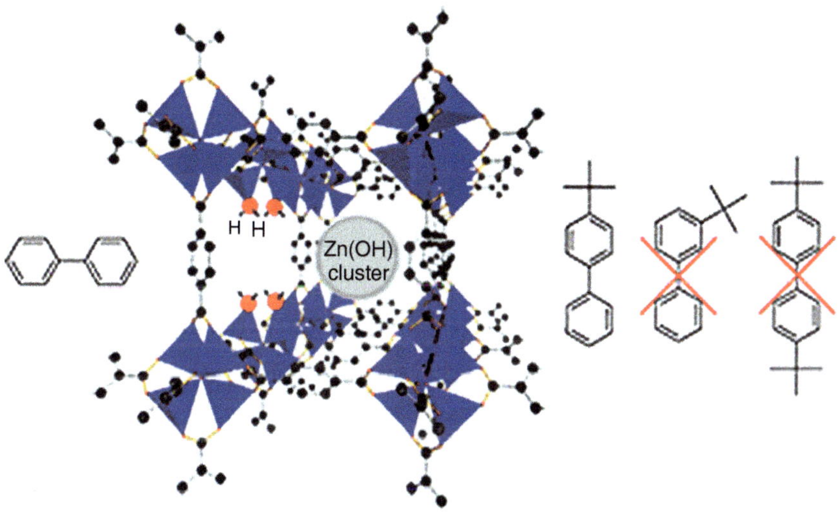

Figure 9.4 *tert-Butylation* reaction of biphenyl using Zn-based MOFs to illustrate the concept of shape selectivity arising from dimension constraints.

active site are tools to increase selectivity. However, in the case of porous materials, in which the reaction is taking place in a reaction cavity in which the catalytic center is located, geometrical constraints can also modify and alter product selectivity. When the main reason to explain product selectivity is the dimensions of substrates, products, or transition states, then, it is accepted that *shape-selectivity* is operating (Figure 9.4).

The best examples of shape selectivity were reported to explain the reaction selectivity in gas phase reactions of the petrochemical industry catalyzed by zeolites [31–33]. Zeolites are microporous crystalline aluminosilicates having Brönsted and Lewis acid sites within the pores [34]. For ZSM-5, a medium pore zeolite with two perpendicular channel systems (straight of 5.1×5.5 Å or sinusoidal 5.3×5.6 Å), it was observed that gas-phase methylation of toluene [35] or xylene isomerization gives the preferential formation of the *para* isomer, much above what is expected based on the thermodynamic equilibrium for the positional *ortho-*, *meta-*, and *para*-xylene isomers. Other examples also refer to the preferential cracking reactivity of linear paraffins with higher diffusion coefficients over the branched isomers with more impeded diffusion throughout the channels [36].

In these classical studies, three types of shape selectivity were established. *Substrate/reactant selectivity* refers to the case in which some substrates react faster than others for the only reason of their higher diffusion and smaller molecular dimensions [37]. This is the case of preferential linear cracking. Product and transition state shape selectivity refer to the case in which the selectivity is due to the molecular product dimensions as it could be the case of *p*-xylene formation.

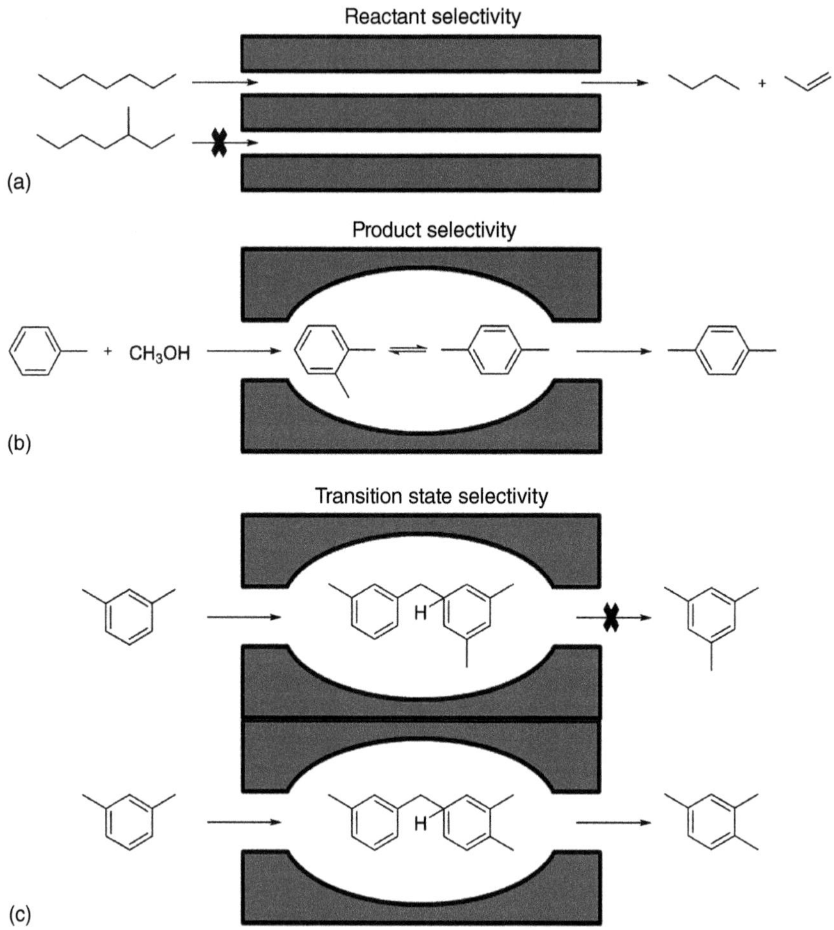

Figure 9.5 Illustration of shape selectivity using zeolites as solid catalysts in (a) reactant selectivity in the cleavage of linear hydrocarbons, (b) product selectivity through the methylation of toluene with methanol, and (c) transition state selectivity in the disproportionation reaction of *m*-xylene.

The dimensions of *p*-xylene allow its escape from the zeolite pores by a much faster diffusion, while the *ortho* or *meta* isomers cannot easily escape from the active center and they undergo consecutive isomerizations, until the *p*-xylene isomer is formed. Transition state selectivity occurs when the intermediate leading to a product is not formed because the formation of this intermediate requires larger space than that available in the reaction cavity. This is shown again operating in the case of *m*-xylene disproportionation in medium pore zeolites in which bimolecular transition states should fit within the pores (Figure 9.5).

Since similarly to zeolites, MOFs are porous materials, the catalytic reaction taking place inside a confined space defined by the lattice, shape-selectivity effects can also be expected to be observed for these catalysts. The larger empty volume of MOFs in comparison to zeolites should establish some differences between them, as well as the use of MOFs in liquid phase reactions. This chapter intends to provide a general view of the current state of the art in *shape catalysis* in MOF catalysis by presenting selected examples to illustrate the potential that MOFs offer as size-selective catalysts to achieve high selectivity of the product taking advantage of the confinement of the reaction cavity. These examples are presented based on the type of reactions. Emphasis is made on the influence of porosity of MOFs in controlling the reactivity of substrates depending on their molecular dimensions.

9.2 Friedel–Crafts Alkylation

Rod-like crystals of $[Zn_3(pbdc)_2)\ 2H_2O]_n$, a phosphonate MOF, were prepared by reaction of 5-phosphonobenzene-1,3-dicarboxylic acid (pbdc) and $Zn(OAc)_2\ 2H_2O$ in the presence of triethylamine [38]. The size-selective catalysis by this MOF was examined in the Friedel–Crafts benzylation of toluene with a series of benzyl bromides (4-chlorobenzyl bromide [CBB], 3-biphenylmethyl bromide [BPMB] and 3,5-di-tert-butylbenzyl bromide [TBBB]) of different dimensions (Scheme 9.1). While the reaction of CBB and toluene reached complete conversion, the reaction of toluene with BPMB and TBBB showed 89.8% and 0% conversions, respectively, under identical conditions. Since the expected reactivity of these primary benzyl bromides based on inductive and resonance effects should be very similar, these catalytic data can be taken as an indirect evidence that the active sites are located within the MOF channels. The two bulky *tert*-butyl groups impede TBBB to enter

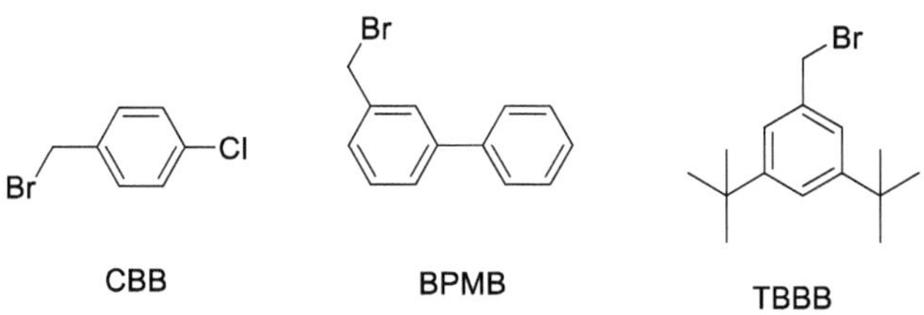

Scheme 9.1 Structure of benzyl bromides used for benzylation of toluene with $[(Zn_3(pbdc)_2)\ 2H_2O]_n$ catalyst.

into the channels and access to the active sites. Thus, TBBB with a large molecular dimension was used to determine the absence of active sites on the external surface of the crystals, thus convincingly demonstrating the confinement effect of MOFs that is the origin of the shape-selective catalysis in MOFs.

Urea-containing porous MOF (NU-601) was synthesized by the combination of a symmetrical urea tetracarboxylate strut, 4,4′-bipyridine, and $Zn(NO_3)_2\ 6H_2O$ under solvothermal conditions [39]. The activity of this solid was tested as an effective hydrogen-bond-donor catalyst in the Friedel–Crafts reactions between pyrroles and nitroalkenes. Activity data indicate that NU-601 is superior (100% conversion) than dissolved urea (80% conversion) considered as the active sites for the room-temperature reaction between *N*-methylpyrrole and (E)-1-nitroprop-1-ene (Scheme 9.2) in toluene. Among the various reasons for the lower activity of the homogeneous sites, self-aggregation is among the major ones. In comparison, immobilization in the porous lattice of the MOF makes persistent isolated and accessible single sites. Observation of a higher reaction rate for small substrates relative to larger substrates (Scheme 9.2) using NU-601 suggests that the reaction occurs preferentially within the MOF pores. Furthermore, the negligible conversion of bulky substrates (see yields in Scheme 9.3), probing exclusively the sites on the external surface, indicates that the alkylation reaction does not occur on the external surface. These results further indicate the potential of shape selectivity offered by MOFs to control the product distribution in otherwise poorly selective Friedel–Crafts alkylation.

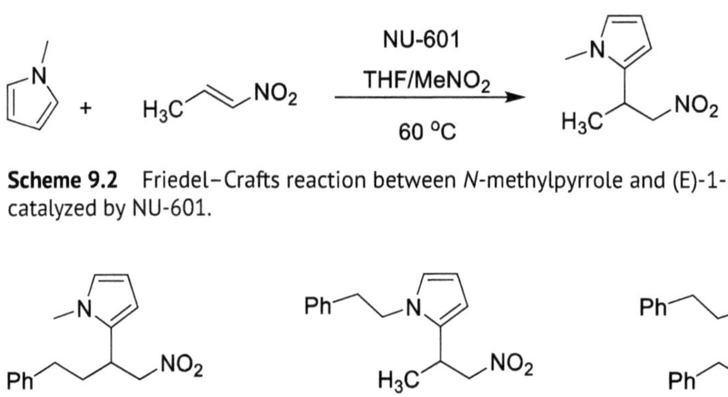

Scheme 9.2 Friedel–Crafts reaction between *N*-methylpyrrole and (E)-1-nitroprop-1-ene catalyzed by NU-601.

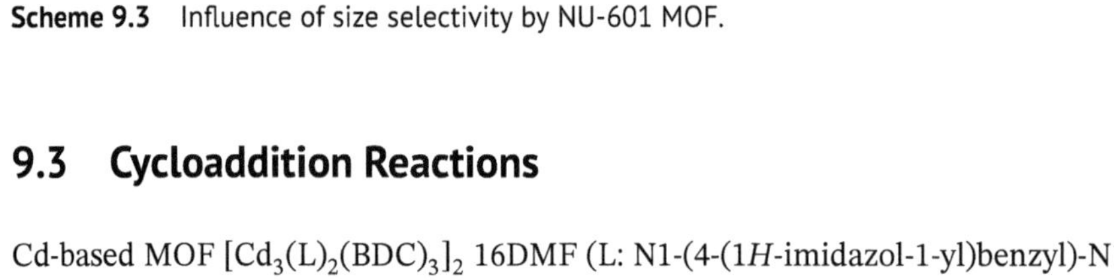

Scheme 9.3 Influence of size selectivity by NU-601 MOF.

9.3 Cycloaddition Reactions

Cd-based MOF $[Cd_3(L)_2(BDC)_3]_2$ 16DMF (L: N1-(4-(1*H*-imidazol-1-yl)benzyl)-N1-(2-aminoethyl)ethane-1,2-diamine, BDC: 1,4-benzenedicarboxylic) was synthesized by reacting cadmium salt with amino tripodal imidazole ligand L and BDC [40].

Due to the presence of amino groups, this Cd MOF is a good CO_2 adsorbent and promotes CO_2 fixation reactions with propargylamines. Interestingly, the reaction in acetonitrile between *N*-methylpropargylamine with CO_2 catalyzed by Cd MOF afforded the corresponding oxazolidinone (Scheme 9.4) in >99% yield with a TON value of 9300. Furthermore, propargylamines with larger kinetic dimensions (Scheme 9.5) were also tested using Cd-MOF as a heterogeneous catalyst, but very low yields of 5 (for the substrate shown in Scheme 9.5, left) or 0% yield (for the substrate shown in Scheme 9.5, right) were achieved for bulky substrates. As in the previously commented examples of shape-selective catalysis, this catalytic performance was considered as indicating the occurrence of the reaction within the MOF pores, since the yield of the oxazolidinone product correlates with substrate dimension. Thus, these substrates with larger dimensions experience diffusion limitations which prevent reaching the active sites located inside the pores.

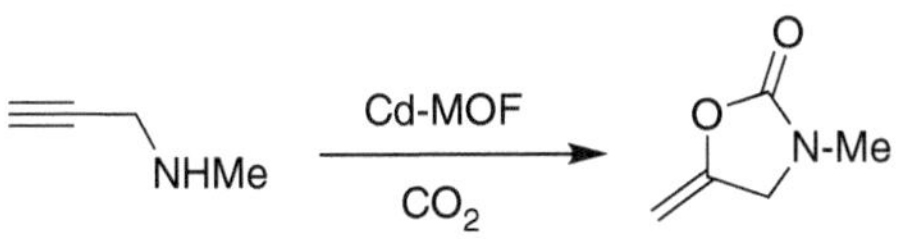

Scheme 9.4 Cd-MOF catalyzed in the synthesis of oxazolidinone from *N*-methylpropargylamine and CO_2.

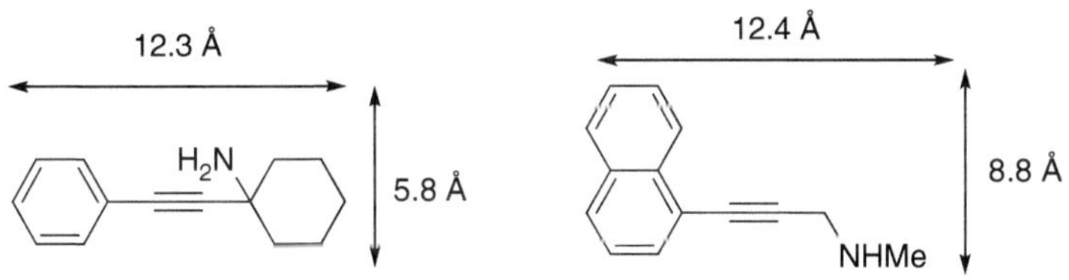

Scheme 9.5 Structures of bulky propargylamines used to test site accessibility in Cd-MOF.

In a recent study, a new Zr-MOF, namely, PCN-625 was synthesized with an identical topology to a well-known PCN-222 and NU-1000 MOFs (Figure 9.6). However, the main difference between these two MOFs is the conformation of porphyrin rings, which are vertical to the pore surfaces rather than in parallel alignment. Furthermore, PCN-625 possesses two types of one-dimensional channels with diameters of 2.03 and 0.43 nm [41]. The catalytic performance of PCN-625(Fe) was studied as heterogeneous catalyst for the size-selective [4 + 2] hetero-Diels–Alder cycloaddition reaction in toluene using $AgBF_4$ as cocatalyst. The reaction between benzaldehyde and 2,3-dimethyl-1,3-butadiene at 80 °C (Scheme 9.6) with PCN-625(Fe) as catalyst reached a 99% yield. Similarly, aliphatic aldehydes possessing smaller dimensions like acetaldehyde (6.2 × 3.6 Å [2]) and cyclopropanecarboxaldehyde (6.1 × 3.7 Å [2]) reacted with 2,3-dimethyl-1,3-butadiene facilely to give 90% and 92% yields, respectively, of the cycloadduct, without experiencing diffusion limitations. On the other hand, the reaction of 2,3-dimethyl-1,3-butadiene with 2-naphthaldehyde (11.2 × 7.1 Å [2]) was 56% after 5 hours and 92% yield after 24 hours. In addition,

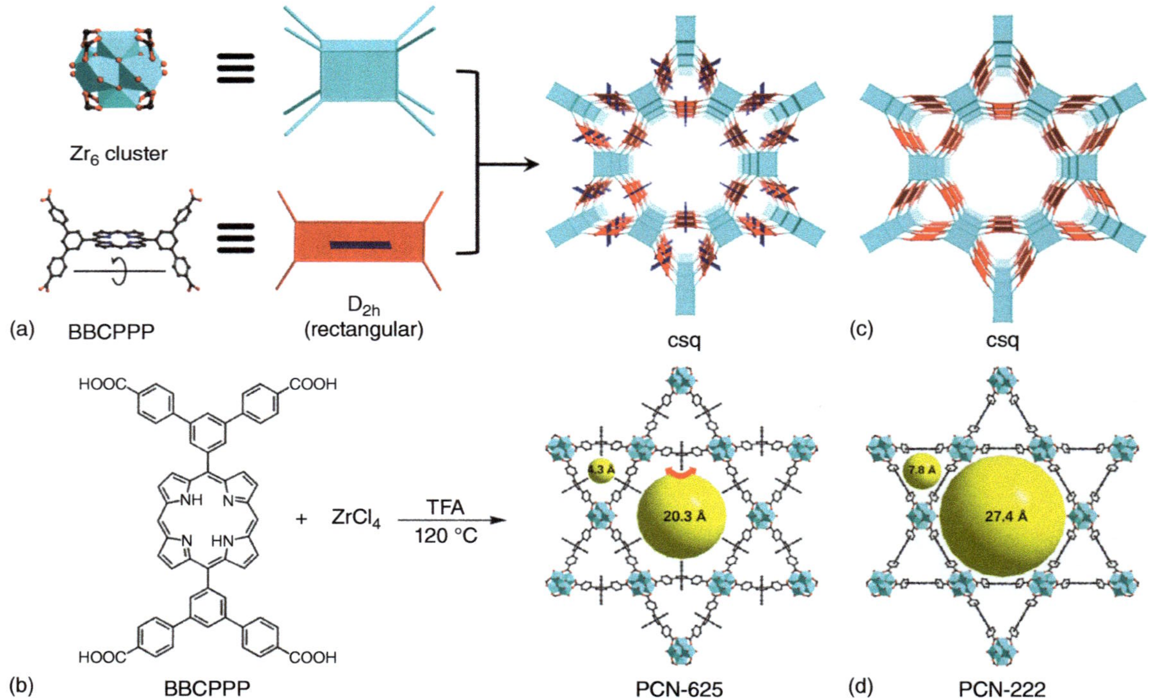

Figure 9.6 Schematic representation of the building units of PCN-625 and PCN-222. (a) Construction of PCN-625 using an 8-connected Zr_6 cluster and rectangular, 4-connected BBCPPP ligand. (b) Synthetic route for the construction of PCN-625 under solvothermal conditions. (c) The csq topology diagram of PCN-222. (d) Crystal structure of PCN-222 viewed along the *c* axis; BBCPPP: bis[3,5-bis(4-carboxyphenyl)phenyl]porphyrin; TFA: trifluoracetic acid. Source: Reproduced with permission from Yang et al. [41]/American Chemical Society.

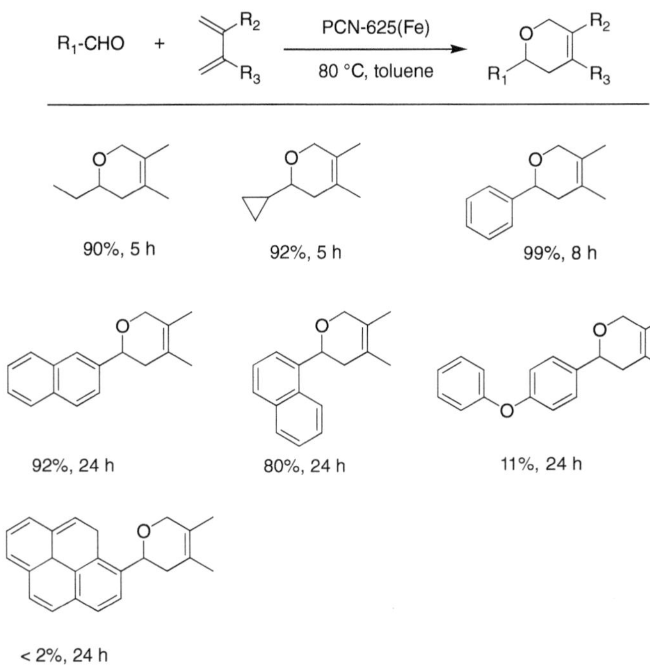

Scheme 9.6 Catalytic performance of PCN-625(Fe) in the [4 + 2] hetero-Diels–Alder cycloaddition between 2,3-dimethyl-1,3-butadiene and various aldehydes.

86% yield was achieved for 1-naphthaldehyde (10.2 × 7.6 Å [2]) under identical conditions. Interestingly, 11% and 2% yields were reached in the cycloaddition between 2,3-dimethyl-1,3-butadiene and the bulky 4-phenoxybenzaldehyde (14.7 × 7.1 Å [2]) and 1-pyrenecarboxaldehyde (13.3 × 7.3 Å [2]), respectively, after 24 hours. These results again can easily be rationalized assuming that the reaction takes place within the MOF pores and that site accessibility is a prerequisite in order for the catalytic reaction to occur.

9.4 Oxidation of Olefins

Besides MOFs, related porous materials such as covalent organic frameworks (COFs) can exhibit similar shape selectivity behavior derived from the confinement of the active sites. In report of shape selectivity in COFs, 3D microporous aromatic frameworks (PAF-76) were prepared by cross-coupling polymerization between 5,10,15,20-tetrakis-(4-bromophenyl)-porphyrin (TBPP) and tetrakis[4-(4′,4′,5′,5′-tetramethyl-1′,3′,2′-dioxaborolane-phenyl)]methane (TTBPM). PAF-76 exhibits a conjugated network linked by intensive and newly created C–C bonds between the two building units [42]. Gas adsorption measurements indicated that PAF-76 has permanent porosity, high surface area, and large pore sizes. This COF was then subjected to metalation and the catalytic activity of metalloporphyrin-functionalized PAF-76 was studied in the oxidation of styrene to phenylacetaldehyde at 80 °C using isobutyraldehyde. PAF 76 afforded 27% conversion of styrene with 99% selectivity to phenylacetaldehyde while no activity is observed with TBPP under these conditions. In contrast, PAF-76-Fe significantly enhanced the conversion of styrene to 80.7% with phenylacetaldehyde selectivity to 89.3% and 10.7% benzaldehyde. On the other hand, Fe-TBPP (homogeneous molecular catalyst analog) showed 64.1% conversion with 75.5% and 24.5% selectivity to phenylacetaldehyde and benzaldehyde, respectively. These catalytic results indicate that the conversion of styrene is highly influenced by the nature of catalyst as well as the uniform distribution of active sites. Furthermore, the enhanced selectivity to the wanted phenylacetaldehyde offered by PAF-76-Fe in comparison to Fe-TBPP is attributed to the porosity of PAF-76-Fe and the special polarity of the reaction cavity.

In another work, the influence of MOF crystal morphology was studied in achieving high selectivity. In this context, two HKUST-1 samples, namely, nano-HKUST-1 and bulk HKUST-1, with different particle sizes were prepared by performing the HKUST synthesis in the presence of 2-methylimidazole (2-MI) [43]. These solids were characterized by a series of techniques (PXRD, FT-IR, TGA, and EDS) to confirm their particle size, morphology, their crystallinity, and HKUST-1 structure. Furthermore, SEM and TEM images show the differences in morphology and size, thus confirming the influence of the presence of 2-MI in the formation of the HKUST-1 particles. The bulk HKUST-1 obtained in the absence of 2-MI has a crystal size of around 10 μm, while the synthesis in the presence of 2-MI afforded nano HKUST-1 with a particle size of 10–20 nm. The BET-specific surface areas of HKUST-1 and nano HKUST-1 were 1092 and 1915 $m^2 g^{-1}$, respectively. The activity of these solids was tested in the oxidation of styrene with t-butylhydroperoxide (TBHP) as an oxidant at 80 °C under solvent-free conditions. Scheme 9.7 summarizes the catalytic data with these two solids. While benzaldehyde is the predominant product and benzoic acid being the main side-product after 50 minutes, benzoic acid is the major product after six hours, indicating a consecutive benzaldehyde oxidation to benzoic acid.

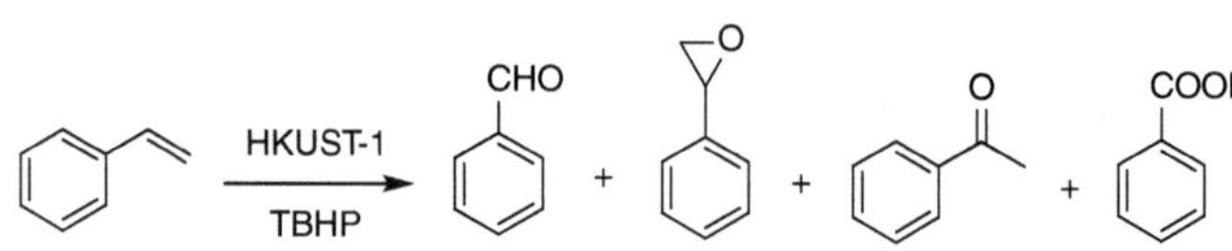

Catalyst	Time	Con (%)	Sel.(%)	Sel.(%)	Sel.(%)	Sel.(%)
HKUST-1	50 min	72	43	6	9	14
HKUST-1	6 h	99	21	7	10	61
nHKUST-1	50 min	92	72	4	7	9
nHKUST-1	6 h	99	14	6	8	71

Scheme 9.7 The catalytic activity of HKUST-1 in the oxidation of styrene with TBHP.

9.5 Hydrogenation Reactions

A bimetallic CuRhBTC (BTC: 1,3,5-benzenetricarboxylate) MOF was prepared by partial substitution of Cu^{2+} in preformed CuBTC by Rh^{2+}. The room temperature activity of CuRhBTC was studied in propylene hydrogenation [44]. In contrast, CuBTC itself and CuMBTC (M = Co, Ir, Ni, Ru ions) showed no activity in propylene hydrogenation at room temperature, indicating that Rh^{2+} are the active sites for H_2 activation under these mild conditions. Importantly, Rh^{2+} ions in CuRhBTC do not leach and the activity is stable under these conditions for at least 24 hours. It was claimed that the specific role of Cu^{2+} ions in CuRhBTC is to stabilize the MOF structure, preventing the release of Rh^{2+} ions and further disfavoring the reduction of Rh^{2+} to metallic Rh. This study illustrates the design of stable heterogeneous MOF catalysts by combining two or more metallic ions in the composition with different roles, including at least stabilization of the lattice and activity as catalytic centers in a chemical reaction.

In another recent work, Cu^{2+} ions were doped within the framework of ZIF-8 (ZIFs: Zeolitic Imidazolate Frameworks) to obtain Cu@ZIF-8. ZIF-8 with a formula of $Zn(2\text{-MI})_2$ is a MOF in which Zn^{2+} ions are tetrahedrally coordinated to four 2-MI and each 2-MI bridges two Zn^{2+} tetrahedra. For comparison, Cu^{2+} ions were also adsorbed over ZIF-8 at identical copper loading as Cu@ZIF-8 to obtain Cu/ZIF-8 [45]. The difference between Cu@ZIF-8 and Cu/ZIF-8 is that Cu^{2+} ions are doped within the framework occupying the nodal positions in ZIF-8 in the former solid, while Cu^{2+} ions (and the corresponding charge-compensating anion) are just adsorbed within the pores, as well as on the outer surface of ZIF-8 crystals. The catalytic performance of these solids was tested in the hydrogenation of olefins using hydrazine hydrate at room temperature in ethanol. The hydrogenation of 1-hexene and cyclohexene to their corresponding products was 84% and 80%, respectively, due to the accessibility of these two linear and cyclic alkenes to the Cu sites. Similarly, the conversion of 1-octene under same conditions was 92% using Cu@ZIF-8 as catalyst due to the linear structure of this alkene which can easily diffuse within the pore aperture of Cu@ZIF-8 (3.4 Å). In contrast, the conversion of cyclooctene was lower (71%), a fact that was interpreted as derived from the molecular size that is bigger than the pore size of Cu@ZIF-8. Interestingly, the hydrogenation of t-stilbene was significantly reduced to 12% even after a longer reaction time due to its molecular size (5.6 Å) again larger than the pores of Cu@ZIF-8 solid. This difference in the activity data indicates that the reactivity of olefins decreases upon increasing their size and that the reaction occurs within the pores of Cu@ZIF-8. In comparison Cu/ZIF-8 exhibits higher activity in *t*-stilbene hydrogenation (36%), a fact that should reflect the easier accessibility of t-stilbene to the Cu^{2+} ions adsorbed on the outer surface without the need to diffuse inside the pores. Scheme 9.8 summarizes the catalytic data for the hydrogenation of these olefins.

Substrate	Conversion (%) Cu@ZIF-8	Conversion (%) Cu/ZIF-8
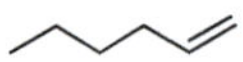	84	99
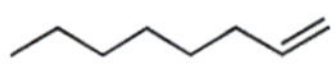	92	99
	80	96
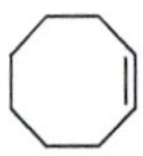	71	100
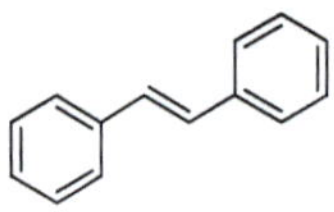	12	36

Scheme 9.8 Size-selective hydrogenation of various olefins with different dimensions using Cu@ZIF-8 (internal Cu^{2+} sites) compared to Cu/ZIF-8 (external Cu^{2+} sites).

9.6 Aldehyde Cyanosilylation

In one of the seminal works, a sodalite-type of MOF, $Mn_3[(Mn_4Cl)_3(BTT)_8(CH_3OH)_{10}]_2$ (BTT: 1,3,5-benzenetristetrazol-5-yl), having a uniform distribution of Mn^{2+} ions, serving as potential Lewis acid sites, was prepared (Figure 9.7) [46]. This solid MOF was thermally stable with two micropore sizes of 7 and 10 Å and a BET surface area of 2100 m^2 g^{-1}. The catalytic performance of this solid was

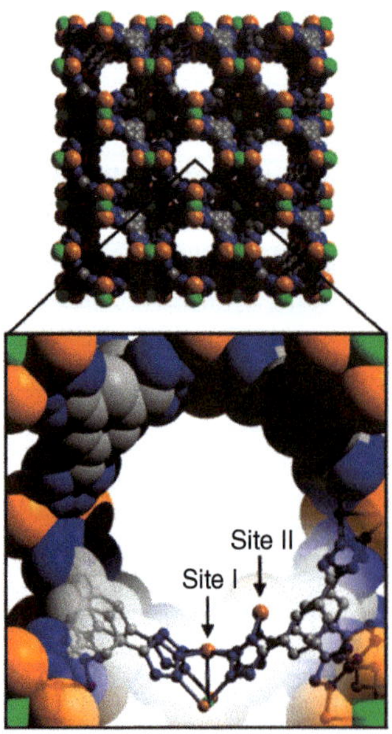

Figure 9.7 A portion of the crystal structure of $Mn_3[(Mn_4Cl)_3(BTT)_8(CH_3OH)_{10}]_2$ showing the two different types of Mn^{II} sites exposed within its three-dimensional pore system of 10 Å wide channels. Source: Reproduced with permission from Horike et al. [46]/American Chemical Society.

studied in the room-temperature cyanosilylation of carbonyl compounds with Me_3SiCN in dichloromethane. The molecular dimension of 1-naphthaldehyde (9.7 × 8.4 Å) permits this molecule to penetrate within the MOF pores without strong diffusion limitation, thereby reaching 90% conversion. In contrast, the conversion values of 4-phenoxybenzaldehyde (13.3 × 7.3 Å) and 4-phenylbenzaldehyde (13.1 × 6.7 Å) were below 20% under identical conditions. These experimental data were rationalized based on the relative molecular dimensions of the latter two substrates that are too big to penetrate within the MOF pores, thus explaining the low conversions. On the other hand, the conversion of acetophenone to its respective cyanosilylated product was only 28% after 24 hours. In contrast, the conversion of 4-acetylbiphenyl (13.9 × 6.7 Å) with a larger dimension and similar reactivity was below 1% conversion after 24 hours. These catalytic data unambiguously prove the influence of developing porous solids for achieving size-selective catalysis.

Two Zn-MOFs have been prepared and characterized with 1D channels [$Zn_3(TCPB)_2 \cdot 2DEF$]·3DEF and [$Zn_3(TCPB)_2$ $2H_2O$]·$2H_2O \cdot 4DMF$ (TCPB: (1,3,5-tri(4-carboxyphenoxy)benzene, DEF: *N*,*N*-diethylformamide, DMF: *N*,*N*-dimethylformamide). Also, these two MOFs exhibit size-selective catalysis for the room-temperature cyanosilylation of aldehydes by Me_3SiCN in hexane as solvent (Scheme 9.9) [47]. The reaction of Me_3SiCN with benzaldehyde (8.9 × 6.7 Å) reached 100% conversion. In contrast, the conversions of 1-naphthaldehyde (9.6 × 8.6 Å) and 4-phenylbenzaldehyde (13.1 × 6.7 Å) were, respectively, 85% and 65% under identical conditions. These catalytic data agree with the reaction taking place within the channels of these MOFs, since the conversion of aldehyde decreases upon increasing its molecular size. Furthermore, these catalytic data also infer that tuning the pore size of MOFs is one of the facile strategies to achieve high selectivity.

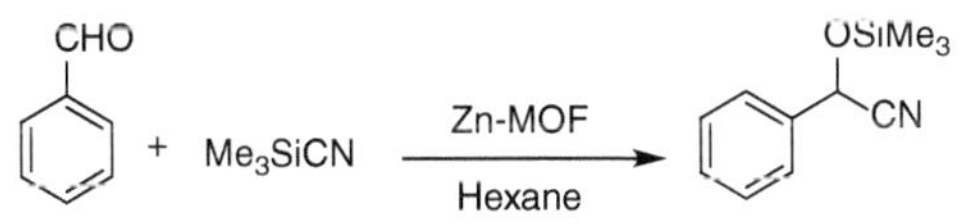

Scheme 9.9 Cyanosilylation of benzaldehyde by Me_3SiCN promoted by Zn-MOF as a solid catalyst.

In another example of shape-selective cyanosilylation of aldehydes, a dual-functional MOF $Cd_2(L)(DMF)_2(H_2O)_2$ (L: 2,3′,5,5′-biphenyl tetracarboxylic acid) was synthesized by solvothermal method and its catalytic activity for the reaction of aldehydes with Me_3SiCN in dichloromethane screened at 40 °C [48]. Under these conditions, the reaction between benzaldehyde and Me_3SiCN afforded the respective adduct in 97% yield with a TOF value of 64.96 h^{-1}. Other aldehydes, such as *n*-heptaldehyde and *p*-anisaldehyde, showed 96% and 94% yields, respectively. In contrast, the reaction of 1-naphthaldehyde and 4-benzyloxybenzaldehyde with Me_3SiCN reached 81% and 34% yields, respectively, after double reaction time. Since it can be expected that the reactivity of the aldehyde functional groups should be very similar in all these compounds, these catalytic data are interpreted considering that the reactivity order decreases as the dimension of the substrate increases, thus confirming the diffusion limitations experienced by the larger substrates and resulting in a shape selectivity.

9.7 Knoevenagel Condensation

Shape selectivity has been claimed in some cases in which the experimental evidence is not fully convincing. One of these examples is the use of two mesoporous MOFs having large cavities, namely, PCN-100 and PCN-101, whose structure is built by $Zn_4O(CO_2)_6$ as secondary building units with 4,4′,4″-s-triazine-1,3,5-triyltri-*p*-aminobenzoate (TATAB) and 4,4′,4″-(benzene-1,3,5-triyltris(azanediyl) tribenzoate) (BTATB), were synthesized. Gas adsorption measurements indicated that the dimension of the cavities is up to 2.73 nm [49]. The shape selectivity of these two solids was studied in the Knoevenagel condensation between aldehydes and butyl cyanoacetate in DMSO at 50 °C (Scheme 9.10). The reaction between benzaldehyde (0.87 x 0.61 nm) and butyl cyanoacetate using PCN-100 and PCN-101 afforded 93% and 96% conversions, respectively. In comparison, bulkier 4-phenylbenzaldehyde (1.33 nm x 0.61 nm) reacted with butyl cyanoacetate in the presence of PCN-100 and PCN-101 reaching 58% and 65% conversions, respectively. Interestingly, both solids did not show any conversion for the reaction between benzophenone (1.14 x 0.66 nm) with butyl cyanoacetate under similar conditions. While benzaldehyde and 4-phenylbenzaldehyde are linear-shaped molecules that can diffuse relatively easier than benzophenone (carbonyl group at the center of the bent molecule), which should have lower diffusion coefficient within the pores of PCN-100 and PCN-101. However, as commented earlier in the Introduction, the concept of shape selectivity should only be used only in those cases in which the reactivity of two molecules are identical and the only reason to justify differences in reactivity not observed in homogeneous catalysis is the confinement of the catalytic sites in a restricted reaction cavity imposing geometric restriction. This is not the case here since aldehydes are far more reactive than diaryl ketones due to the operation of electronic conjugation effects and molecular encumbrance. Thus, molecular factors, rather than catalytic site confinement, can contribute partially or totally to explain this contrasting reactivity between aldehydes and benzophenone.

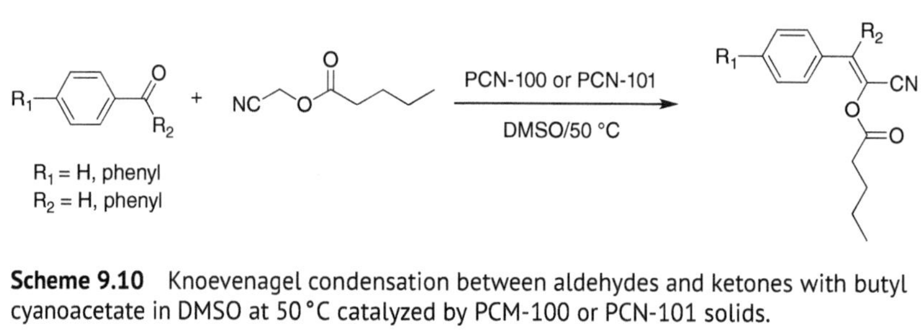

Scheme 9.10 Knoevenagel condensation between aldehydes and ketones with butyl cyanoacetate in DMSO at 50 °C catalyzed by PCM-100 or PCN-101 solids.

In another study, a porous bimetallic double-walled MOF material based on unique Cd(II)–K(I) octanuclear clusters with the molecular formula $[Cd_4K_4(TADA)_4(H_2O)_{12}]\cdot 6DMF$ (TADA: 3,3′-((6-hydroxy-1,3,5-triazine-2,4-diyl)bis(azanediyl))dibenzoate) was prepared and its activity tested as a size-selective catalyst in the Knoevenagel condensation in toluene at 80 °C for three hours [50]. This MOF promoted the reaction between benzaldehyde and malononitrile to give 2-benzylidenemalononitrile (Scheme 9.11) in 100% yield. Similarly, *para*-substituted benzaldehydes with -F, $-NO_2$, and phenyl groups also afforded 100% yield of the respective products. In contrast, the activity of this catalyst decreased for the reaction between benzaldehyde with ethyl cyanoacetate with 85% yield of the condensation product and even further decrease with bulkier diethyl malonate for which the Knoevenagel condensation product is formed in 23% yield under the same conditions (Scheme 9.11). These catalytic data prove that the relative pore dimension of the MOF with respect to the reaction products can play a vital role in determining the reactivity of substrates. The catalytic activity correlates well with the kinetic dimensions of the corresponding substrates.

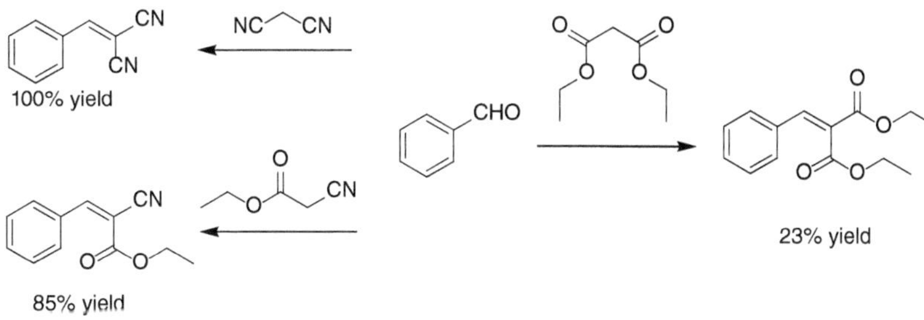

Scheme 9.11 Size-selective catalysis in the Knoevenagel condensation by $[Cd_4K_4(TADA)_4(H_2O)_{12}]\cdot 6DMF$.

9.8 Conclusions

The above Sections (9.2–9.7) have summarized the current state of the art of shape-selective catalysis in MOFs, by presenting selected examples of different reaction types. The general conclusion is that the shape selectivity in MOFs has been considerably less exploited and less systematically studied than in zeolites. Most of the examples reported so far refer to substrate selectivity, with only a few that can be considered to refer to product selectivity as in the case of benzylation of toluene by $[(Zn_3(pbdc)_2)\ 2H_2O]_n$ or Knoevenagel condensation. There are almost no examples of *transition state* shape selectivity, since strong evidence on the operation of this type of shape selectivity requires a deep understanding of the reaction mechanism. This present situation indicates that the field of shape-selectivity in MOFs has still to develop much further and has not reached the maturity that has got for zeolites. In part, this situation can be explained considering that MOFs are stable and better suited for liquid-phase reactions, examples of gas-phase reactions catalyzed by MOFs being less common. In liquid-phase reaction, products formed inside the

MOF pores, but not able to diffuse outside, will block the internal space resulting in poisoning and catalyst deactivation. In comparison, in the gas-phase reactions occurring at much higher temperature, the trapped products will continue reacting until they form the less encumbered isomer, finding the way out of the pores after isomerization. These isomerizations do not happen at low temperatures due to their high activation barrier. In any case, the differences between liquid and gas phase reactions regarding shape selectivity still need to be better delineated.

Another rather surprising fact in the current state of the art is that the most robust and widely used MOFs, such as those of the MIL or UiO series have not been much studied with respect to their potential shape selectivity. As in the case of zeolites, Friedel–Crafts alkylations is a rather unselective reaction resulting in positional isomers and polyalkylation products that would be worth to explore in terms of selectivity with structurally stable and robust MOFs.

It should be commented that there are claims in the literature proposing the operation of shape-selectivity, such as in the case of alkene hydrogenation, that may deserve to be critically revisited. Substrate shape selectivity requires that the substrates should exhibit equal reactivity in a homogeneous phase, but they have to present different molecular sizes. Then, the only reason why differences are observed in substrate reactivity with a porous catalyst is their kinetic diameter. However, it is well known that primary, secondary, or cyclic alkenes exhibit different chemical reactivity as a consequence of the operation of inductive effects. Then, if the substrates have different chemical reactivity, claims on shape selectivity require careful comparisons on reaction rates with homogeneous catalysts to discriminate the contribution of all the factors to the final product selectivity.

Overall, a goal of this chapter is to trigger further interest and research in exploiting spatial confinement in MOFs as a valid tool to gain selectivity. It is proposed that, at least, the field should follow the research already well-established in the case of zeolites and related aluminosilicates.

References

1 Kitagawa, S., Kitaura, R., and Noro, S.-I. (2004). *Angew. Chem. Int. Ed.* 43: 2334–2375.

2 Furukawa, H., Cordova, K.E., O'Keeffe, M., and Yaghi, O.M. (2013). *Science.* 341: 1230444.

3 Li, H., Eddaoudi, M., O'Keeffe, M., and Yaghi, O.M. (1999). *Nature.* 402: 276–279.

4 Stock, N. and Biswas, S. (2012). *Chem. Rev.* 112: 933–969.

5 James, S.L. (2003). *Chem. Soc. Rev.* 32: 276–288.

6 Rowsell, J.L.C. and Yaghi, O.M. (2004). *Microporous Mesoporous Mater.* 73: 3–14.

7 Hong, D.-Y., Hwang, Y.K., Serre, C. et al. (2009). *Adv. Funct. Mater.* 19: 1537–1552.

8 Kökçam-Demir, U., Goldman, A., Esrafili, L. et al. (2020). *Chem. Soc. Rev.* 49: 2751–2798.

9 Dietzel, P.D.C., Besikiotis, V., and Blom, R. (2009). *J. Mater. Chem.* 19: 7362–7370.
10 Wei, Y.-S., Zhang, M., Zou, R., and Xu, Q. (2020). *Chem. Rev.* 120: 12089–12174.
11 Wang, D., Huang, R., Liu, W. et al. (2014). *ACS Catal.* 4: 4254–4260.
12 Zhou, W., Wu, H., and Yildirim, T. (2008). *J. Am. Chem. Soc.* 130: 15268–15269.
13 Hu, Z. and Zhao, D. (2017). *CrystEngComm.* 19: 4066–4081.
14 Bhattacharjee, S., Chen, C., and Ahn, W.-S. (2014). *RSC Adv.* 4: 52500–52525.
15 López-Maya, E., Montoro, C., Colombo, V. et al. (2014). *Adv. Funct. Mater.* 24: 6130–6135.
16 Hartmann, M. and Fischer, M. (2012). *Microporous Mesoporous Mater.* 164: 38–43.
17 Valvekens, P., Vandichel, M., Waroquier, M. et al. (2014). *J. Catal.* 317: 1–10.
18 Yang, Q., Xu, Q., and Jiang, H.-L. (2017). *Chem. Soc. Rev.* 46: 4774–4808.
19 Dhakshinamoorthy, A. and Garcia, H. (2012). *Chem. Soc. Rev.* 41: 5262–5284.
20 Zhu, L., Liu, X.-Q., Jiang, H.-L., and Sun, L.-B. (2017). *Chem. Rev.* 117: 8129–8176.
21 Corma, A., García, H., and Xamena, F.X.L.I. (2010). *Chem. Rev.* 110: 4606–4655.
22 Czaja, A.U., Trukhan, N., and Müller, U. (2009). *Chem. Soc. Rev.* 38: 1284–1293.
23 Farrusseng, D., Aguado, S., and Pinel, C. (2009). *Angew. Chem. Int. Ed.* 48: 7502–7513.
24 Huang, Y.-B., Liang, J., Wang, X.-S., and Cao, R. (2017). *Chem. Soc. Rev.* 46: 126–157.
25 Llabrés i Xamena, F.X., Abad, A., Corma, A., and Garcia, H. (2007). *J. Catal.* 250: 294–298.
26 Ravon, U., Domine, M.E., Gaudillère, C. et al. (2008). *New J. Chem.* 32: 937–940.
27 Khajavi, H., Stil, H.A., Kuipers, H.P.C.E. et al. (2013). *ACS Catal.* 3: 2617–2626.
28 Wang, B., Liu, W., Zhang, W., and Liu, J. (2017). *Nano Res.* 10: 3826–3835.
29 Liu, X., Oh, M., and Lah, M.S. (2011). *Inorg. Chem.* 50: 5044–5053.
30 Qi, B., Zhang, T., Li, M. et al. (2017). *Catal. Sci. Technol.* 7: 5872–5881.
31 Csicsery, S.M. (1984). *Zeolites.* 4: 202–213.
32 Marcilly, C.R. (2000). *Topics Catal.* 13: 357–366.
33 Degnan, T.F. Jr. (2003). *J. Catal.* 216: 32–46.
34 Valtchev, V., Majano, G., Mintova, S., and Pérez-Ramírez, J. (2013). *Chem. Soc. Rev.* 42: 263–290.
35 Kaeding, W.W., Chu, C., Young, L.B. et al. (1981). *J. Catal.* 67: 159–174.
36 Odedairo, T., Balasamy, R.J., and Al-Khattaf, S. (2011). *J. Mol. Catal. A Chem.* 345: 21–36.
37 Kim, J.-H., Kunieda, T., and Niwa, M. (1998). *J. Catal.* 173: 433–439.
38 Liao, T.-B., Ling, Y., Chen, Z.-X. et al. (2010). *Commun.* 46: 1100–1102.
39 Roberts, J.M., Fini, B.M., Sarjeant, A.A. et al. (2012). *J. Am. Chem. Soc.* 134: 3334–3337.
40 Zhao, D., Liu, X.-H., Zhu, C. et al. (2017). *ChemCatChem.* 9: 4598–4606.
41 Yang, L., Cai, P., Zhang, L. et al. (2021). *J. Am. Chem. Soc.* 143: 12129–12137.
42 Meng, S., Zou, X., Liu, C. et al. (2016). *ChemCatChem.* 8: 2393–2400.
43 Guo, C., Zhang, Y., Zhang, L. et al. (2018). *ACS Appl. Nano Mater.* 1: 5289–5296.

44 Shakya, D.M., Ejegbavwo, O.A., Rajeshkumar, T. et al. (2019). *Angew. Chem. Int. Ed.* 58: 16533–16537.

45 Nagarjun, N., Arthy, K., and Dhakshinamoorthy, A. (2021). *Eur. J. Inorg. Chem.* 2021: 2108–2119.

46 Horike, S., Dinca, M., Tamaki, K., and Long, J.R. (2008). *J. Am. Chem. Soc.* 130: 5854–5855.

47 Lin, X.-M., Li, T.-T., Wang, Y.-W. et al. (2012). *Chem. Asian J.* 7: 2796–2804.

48 Zhang, X.-N., Liu, L., Han, Z.-B. et al. (2015). *RSC Adv.* 5: 10119–10124.

49 Fang, Q.-R., Yuan, D.-Q., Sculley, J. et al. (2010). *Inorg. Chem.* 49: 11637–11642.

50 He, H., Xue, Y.-Q., Wang, S.-Q. et al. (2018). *Inorg. Chem.* 57: 15062–15068.

10

Selective Oxidations in Confined Environment

Oxana A. Kholdeeva

Boreskov Institute of Catalysis, Lavrentieva 5, Novosibirsk 630090, Russia

10.1 Introduction

The role of liquid-phase selective oxidation in the production of valuable oxygen-containing products and intermediates of bulk and fine organic synthesis is continuously increasing [1–4]. The development of environmentally benign and cost-effective oxidation processes that employ safe, readily available, and atom-efficient oxidants is a demanding task of the modern chemical industry. Two oxidants, molecular oxygen (air) and hydrogen peroxide (atom efficiency 50% and 47%, respectively), rank first in terms of environmental acceptability, since both produce water as the only by-product [5]. Although molecular oxygen is cheaper than hydrogen peroxide, the latter does not require special equipment for operation under high pressure–high-temperature conditions and thus can be advantageous for small-scale manufacturers [6]. Considerable progress made toward the direct synthesis of H_2O_2 from H_2 and O_2 contributes to the growth of interest in this oxidant [7–9]. When a nonaqueous medium is preferred, the use of alkyl hydroperoxides (HPs), e.g. *tert*-butyl hydroperoxide (TBHP, atom efficiency 22%) or cumyl hydroperoxide (CHP, atom efficiency 12%), as oxidant can be reasonable. In this case, alcohols formed as by-products can be easily separated by distillation and reconverted to HPs or used in other processes [5, 9].

Since the environmentally friendly oxidants are inert to most organic substrates, catalysts are required to use them in the organic synthesis. Heterogeneous catalysts offer the evident advantages of easy recovery and recycling, as well as suitability for continuous fixed-bed operation and thus they meet well the requirements of sustainable chemistry, provided that the solid catalyst is resistant to leaching of active species under the conditions of liquid-phase oxidation [1–3, 10]. In the past years, much research efforts were directed at designing active catalytic sites within pores of various solids. Ideally, such sites should be uniform in their nature, spatially separated, and surrounded by a uniform environment to ensure high catalytic activity and selectivity of the oxidation process (the concept of single-site heterogeneous catalysts [11, 12]). Furthermore, porous frameworks may

Catalysis in Confined Frameworks: Synthesis, Characterization, and Applications, First Edition.
Edited by Hermenegildo Garcia and Amarajothi Dhakshinamoorthy.

endow confinement effects and provide reaction spaces different from solutions, resulting in the creation of unique catalytic performances, in particular, significant acceleration of the target oxidation reaction and suppression of overoxidation, appearance of shape selectivity, and unusual chemo- and regioselectivity, increasing enantioselectivity, and even enhancing stability and productivity of active species.

Several reviews devoted to various aspects of confinement effects in catalysis have been recently published [13–17]. The aim of this chapter is to give a brief overview of some key advances in the field of liquid-phase selective oxidation with environmentally benign oxidants using heterogeneous porous catalysts. We restrict ourselves here to consideration of such catalysts as transition-metal-substituted molecular sieves (zeolites and aluminophosphates), mesoporous metal–silicates, metal–organic frameworks (MOFs), and polyoxometalates (POMs) confined within various porous matrixes. A selection of the results reported in the literature was made to demonstrate how spatial confinement can affect the catalyst activity, selectivity, productivity, and stability of the active species. The scope and limitations of such catalysts for the synthesis of oxygen-containing products will be also discussed. When appropriate, special attention will be paid to mechanistic aspects of the oxidation reactions.

10.2 Transition-Metal-Substituted Molecular Sieves

A constrained environment for a catalytic reaction can be easily realized by using microporous crystalline materials, such as transition-metal-substituted zeolites (zeotypes) and aluminophosphates.

10.2.1 Ti-Substituted Zeolites and H_2O_2

A real breakthrough in the field of heterogeneous liquid-phase oxidation was the invention of titanium silicalite-1 (shortly TS-1) by ENI researchers in the early 1980s [18]. TS-1 has the MFI structure and possesses a 3D-pore system with straight channels (0.53 × 0.56 nm) where 1–2% of Si atoms are isomorphically substituted for Ti(IV) atoms, which possess a unique ability to oxidize selectively small organic substrates using dilute H_2O_2 as the oxidant [19–22]. Nowadays TS-1 is employed in three large-scale industrial processes, such as hydroxylation of phenol, epoxidation of propene, and ammoximation of cyclohexanone [2, 9, 20, 21] (Figure 10.1).

The hydrophobicity of TS-1 is one of the main characteristics, which endows it with high catalytic activity in the oxidation of hydrocarbons. The hydrophobic environment of Ti sites favors adsorption of nonpolar hydrocarbons and pushes out polar oxidation products from the reaction zone, thereby preventing overoxidation. Steric restrictions to the diffusion of organic molecules in the micropores (reactant shape selectivity) or to the attack of the substrate on the active species (transition state shape selectivity) usually offset electronic factors in TS-1-catalyzed oxidations, leading to unprecedented reactivities (for reviews, see [20–22]). For linear alkenes, in particular propene, epoxidation selectivity can reach 97% based on the oxidant, with no formation of allylic oxidation products [23, 24]. Once the kinetic

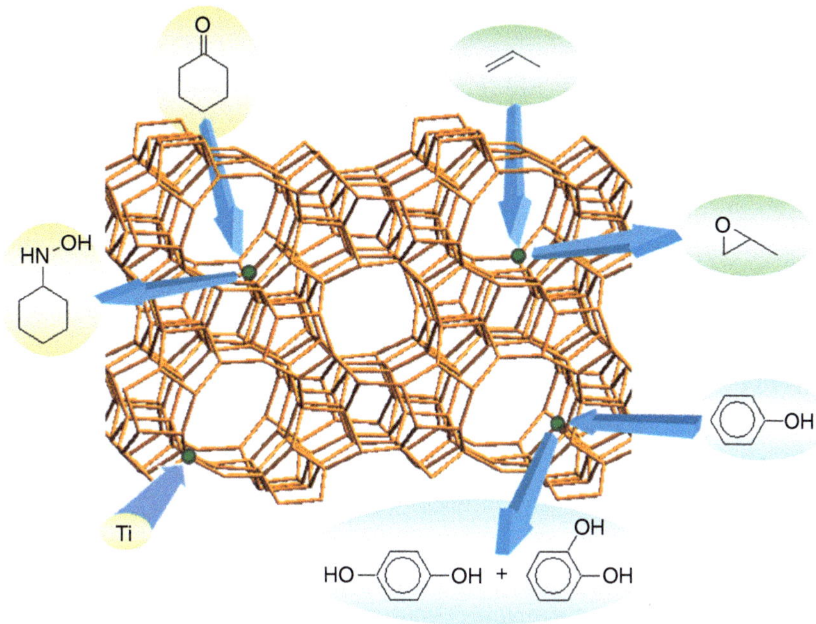

Figure 10.1 Three industrial processes based on TS-1 and H_2O_2.

diameter of organic substrate approaches 0.6 nm, oxidation becomes very slow and product yields decrease dramatically. On the one hand, the small pore size of TS-1 is perceived as a drawback that restricts the scope of potential substrates, but on the other hand, it is also an important factor that ensures unique catalytic performances [20–22]. In addition to shape selectivity effects, the concentration of hydrocarbon molecules near the active Ti sites increases greatly due to the combined effect of microporosity and hydrophobicity [21, 22, 24–26]. High affinity of the internal surface of TS-1 to nonpolar substrates makes possible hydroxylation of alkanes at temperatures below 55 °C [26].

In the TS-1-based phenol hydroxylation process, conversion and product (catechol and hydroquinone) selectivity are significantly higher than these parameters attained with conventional homogeneous catalysts, because overoxidation is suppressed in the confined zeolite space [20, 21, 27]. Moreover, the catechol/hydroquinone ratio (0.5–1.3) differs significantly from statistical distribution, reflecting shape selectivity effects, which can result from steric restrictions in the transition state and/or different diffusivity of the two products. The smaller kinetic diameter of hydroquinone favors the hydroxylation at the less hindered *p*-position of phenol.

It has been widely accepted (and supported by various spectroscopic techniques) that active centers in TS-1 are tetrahedrally coordinated site-isolated Ti(IV) atoms, which activate H_2O_2 through the formation of a hydroperoxo titanium species "TiOOH" capable of electrophilic oxygen transfer to organic substrates ([21, 22, 28, 29] and references therein).

The question on the coordination mode of the hydroperoxo group (end-on or side-on, Scheme 10.1) remains open. Model studies using Ti-substituted polyoxometalates (Ti-POMs) revealed that both configurations can be, in principle, realized for single-site Ti(IV), depending on its coordination environment [30–33].

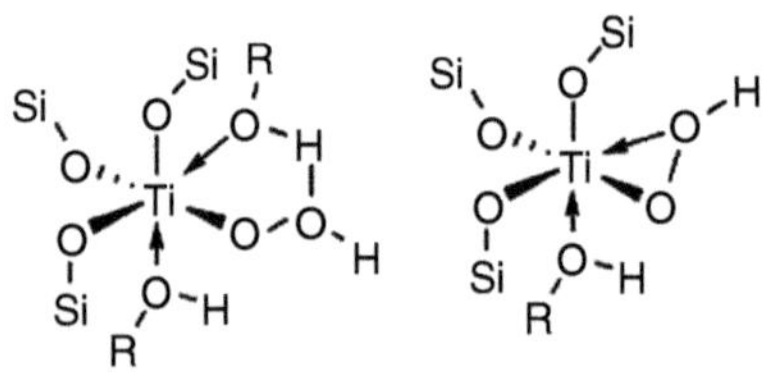

Scheme 10.1 Conventional structures of Ti-hydroperoxo species operated in TS-1: left – end-on Ti(η^1-OOH) and right – side-on Ti(η^2-OOH).

Recent studies of Copéret and coworkers using ^{17}O NMR spectroscopy, electron microscopy, and DFT calculations implicated that dinuclear Ti species are present in the industrially used TS-1 epoxidation catalyst and these species enable the low-energy epoxidation of propylene via an oxygen transfer [34]. On contact with labeled $H_2{}^{17}O_2$, the catalyst samples exhibited a characteristic solid-state ^{17}O NMR signature indicative of the formation of a bridging peroxo species on a dinuclear titanium site. Transfer of β-oxygen atom from this di-Ti-peroxo species to C=C bond to form epoxide was suggested on the basis of DFT calculations [34]. Previously, Carbó and coworkers demonstrated that, for rigid and hindered Ti centers in Ti-POMs, β-oxygen transfer from Ti(η^1-OOH) becomes favored over the α-oxygen transfer from Ti(η^2-OOH) [30].

The tremendous success of TS-1 stimulated the development of works on the synthesis of large-pore zeolites and hierarchical micro-mesoporous zeolite structures (see for reviews [21, 35, 36]). While cyclohexene and other cyclic alkenes are too large to penetrate easily into the pores of TS-1, their epoxidation can be accomplished using larger pore zeolites, e.g. Ti-β and ITQ-17 [35, 37]. However, stability of the large-pore zeolites is inferior to that of TS-1 [21, 38].

Cycloalkenes (along with linear olefins) can be also epoxidized using Ti-MWW zeolite the channel system of which comprises supercages ($0.7 \times 0.7 \times 1.8$ nm) [39–41]. Interestingly, Ti-MWW behaved differently from TS-1 and Ti-β in the epoxidation of *cis*- and *trans*-olefins. In the oxidation of hex-2-enes (*cis*/*trans* = 41/59) with equimolar amount of H_2O_2, the ratio of *trans*-/*cis*-epoxides was c. 80/20 while *cis*-isomer dominated with TS-1 and Ti-β [39]. This difference was rationalized by a better fitting of *trans*-configuration to the tortuosity of the 10MR channel system of Ti-MWW. In contrast to TS-1, bulky alkyl HPs can be employed as oxidant with the large-pore zeolites. The supercages within the Ti–MWW crystals and the exterior pockets directly related to the crystal form of the thin platelets make the Ti species in Ti-MWW highly accessible to both bulky substrates (cycloalkenes) and bulky oxidants (TBHP). While the intralayer sinusoid 10-membered ring channels are not accessible to the bulky molecules, they can accomplish epoxidation of linear olefins and allyl alcohol with H_2O_2 [40, 41].

The effects of hydrophobicity are not always easily distinguishable from other surface effects [22]. The paramount importance of surface hydrophobicity in H_2O_2-based oxidations of hydrocarbons was demonstrated in the epoxidation of methyl oleate with hydrogen peroxide over Ti-β zeolites. A more hydrophobic Al-free Ti-β prepared in a fluoride medium, Ti-β(F), despite larger size of crystallites, showed higher substrate conversion and epoxide yield compared to conventional Ti-β(OH) [42]. More recently, Tatsumi and coworkers have shown that Ti-MWW containing the smallest amount of silanols and fewer defect sites showed the best hydrophobicity and the highest catalytic activity in the oxidation of 1-hexene [43]. Surprisingly, Flaherty and coworkers revealed that turnover rates for 1-octene epoxidation increased monotonically with the density of silanol nests $(SiOH)_4$ in Ti-BEA zeolite [44]. Materials containing ~5 $(SiOH)_4$ per unit cell revealed 1-octene epoxidation rates 100 times greater than nearly defect-free Ti-BEA. Rates of H_2O_2 decomposition were, however, insensitive to the number of silanol nests in Ti-BEA. More recently, the same group reported on the synthesis of transition metal-substituted (Ti, Nb, Ta, and Sn) zeolite FAU (pores 1.2 nm) with ultralow Al contents (Si : Al > 900) [45]. Turnover rates for styrene epoxidation with Ti-FAU were two- and seven-fold higher than with Ti-BEA and Ti-grafted SiO_2, respectively, while rates of H_2O_2 decomposition were similar for all three materials. The improved performance of Ti-FAU was suggested to reflect differences in activation-free energies for epoxidation that show an enthalpic preference in Ti-FAU relative to Ti–SiO_2 and an entropic gain relative to Ti-BEA [45]. However, since the size of crystallites was not provided for the zeolite materials used, it is difficult to understand whether the observed activity was measured in the kinetic regime, or it could reflect diffusion limitations within the micropores.

10.2.2 Co-Substituted Aluminophosphates and O_2

The benefits that might be given by confinement of micropores have been nicely realized in the oxyfunctionalization of linear alkanes at the terminal positions – one of the major challenging transformations in oxidation catalysis. Thomas and coworkers have designed a molecular sieve catalyst, a Co(III)-enriched aluminophosphate CoAPO-18 (Co:P = 0.1), which demonstrated unprecedented terminal selectivity for the oxidation of *n*-hexane to adipic acid (AA) using molecular oxygen (air) [46, 47]. In this catalyst, two tetrahedrally coordinated Co(III) ions occupy the opposite vertices of the chabazitic cage, with a separation distance (ca. 0.76 nm) close to the distance between two CH_3 groups in *n*-hexane (Figure 10.2). Using the CoAPO-18 catalyst, AA was obtained with 34% selectivity at 9.5% conversion of *n*-hexane after 24 hours at 100 °C and 1.5 MPa of air. The major by-products were 2-hexanone and hexanoic acid, and the overall terminal selectivity attained 65%. The shape and size of the catalyst pores were crucial for the catalytic performance. The larger pore CoAPO-36 revealed drastically different selectivity, with no AA among the oxidation products. The retention of the structural integrity of CoAPO-18 and the absence of cobalt leaching were thoroughly confirmed.

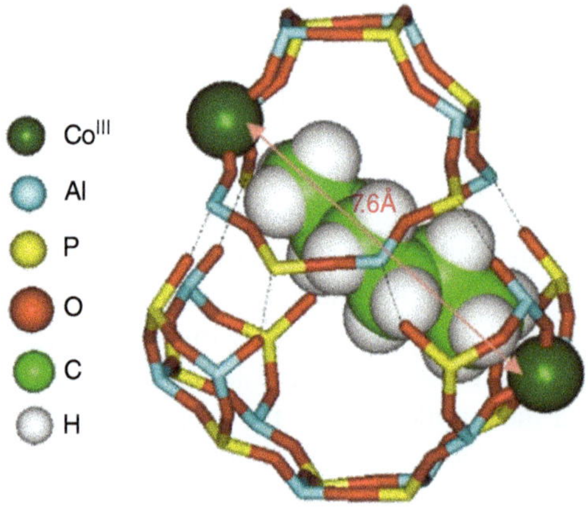

Figure 10.2 The chabazitic cage of Co-enriched CoAlPO-18 with *n*-hexane between two Co(III) ions. Source: Reproduced from Thomas and Raja [47] by permission of The Royal Society of Chemistry.

While microporous molecular sieves can be unique catalysts for transformations of small or linear molecules, they become inapplicable when reactants with sizes comparable with the micropore dimensions have to be converted.

10.3 Mesoporous Metal–Silicates

On the border of the twentieth and twenty-first centuries, a large scientific effort was devoted to the elaboration of methods for the synthesis and characterization of mesoporous materials (pore dimensions 2–50 nm) [49]. Since the discovery of the MCM-41 family of molecular sieves by Mobil Oil Corporation researchers [50], the surfactant-templating strategy was widely employed for the preparation of ordered transition-metal-containing mesoporous silicates of the MCM, HMS, SBA, MSU, TUD, MMM, KIT, FDU, and other families. A characteristic structural feature of all these materials is a long-range ordering of mesopores channels and lack of short-range ordering (at the atomic level). This distinguishes them from zeolites, which possess both types of ordering.

The great interest of the catalytic community in mesoporous materials was caused by their potential advantage in transformations of large molecules, meeting the demands of the fine chemicals industry [1, 51], and transformation of biomass [52, 53]. Following the target of developing a mesoporous analog of TS-1, much work has been devoted to the preparation of mesoporous titanium-silicates using direct hydrothermal synthesis or post-synthesis approaches and assessment of their catalytic performance. Significant progress has been also achieved in the inclusion of other metals (M = Zr, Ce, Nb, Cr, W, Fe, Co, Al, Sn) into mesoporous silicates. Comprehensive reviews on liquid-phase selective oxidations over mesoporous metal–silicates can be found elsewhere [48, 54]. Some representative oxidation reactions that have been successfully realized over mesoporous metal–silicates are presented in Figure 10.3.

Below we provide a few examples to demonstrate some specific features of this class of catalysts, with particular attention to the role of spatial confinement for the oxidative transformations.

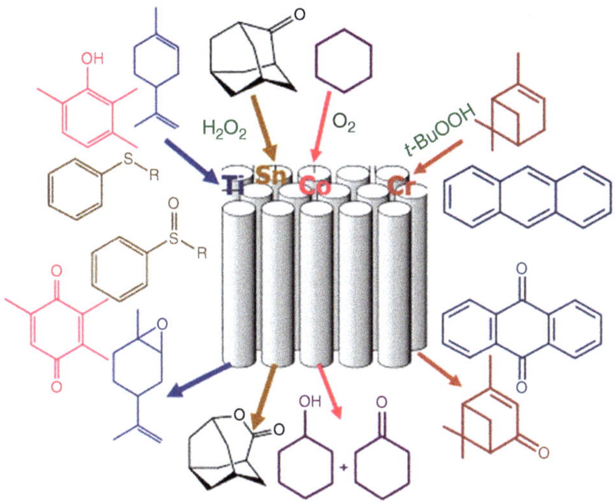

Figure 10.3 Representative selective oxidations realized over mesoporous metal–silicates. Source: Kholdeeva [48]/Royal Society of Chemistry.

10.3.1 Mesoporous Ti-Silicates in Oxidation of Hydrocarbons

Along with the large pore size, one of the principal features of mesopores silicates is a large density of surface Si–OH groups and, as a result, much greater affinity for water than for hydrocarbon molecules [55]. Because of the intrinsic hydrophilicity, mesoporous Ti-silicates are usually poor catalysts for oxidation of hydrocarbons with aqueous H_2O_2. Preferable adsorption of water and oxidant favors unproductive decomposition of H_2O_2 and enhances side reactions caused by radical species. While TS-1 produces mostly *o*- and *p*-cresols in the oxidation of toluene, oxidation of the side chain prevails in case of mesoporous titanium-silicates [56].

A hydrothermally stable catalyst Ti-MMM-2 catalyzed the oxidation of a range of alkanes (cyclooctane, *n*-heptane, *n*-octane, *iso*-octane, methylcyclohexane, and *cis*- and *trans*-1,2-dimethylcyclohexanes) by H_2O_2 in acetonitrile [57]. The selectivity parameters C(1):C(2):C(3):C(4), which reflect the relative reactivities of hydrogens at carbons atoms of *n*-heptane or *n*-octane, differed significantly for Ti-MMM-2 and TS-1 (1 : 5.2 : 4.5 : 4.4 vs. 1 : 80 : 62 : 59 for *n*-octane) whereas the bond-selectivity parameters 1°:2° in the oxidation of 2,2,4-trimetylpentane and methylcyclohexane over Ti-MMM-2 were close to the parameters reported for homogeneous catalysts which generate hydroxyl radicals [57].

In general, mesoporous Ti-silicates are less active than TS-1 in epoxidation of linear olefins with H_2O_2 (e.g. 9% vs. 49% conversion of hexene-1 [58]), but they enable epoxidation of bulky alkene molecules. The oxidation of cyclohexene, the well-known test substrate that allows an easy discrimination between homo- and heterolytic oxidation mechanisms, shows a significant contribution of allylic oxidation products over mesoporous Ti-silicates; however, a controlled, drop-wise addition of H_2O_2 makes possible minimization of its unproductive decomposition and increases greatly epoxidation selectivity [59]. For cyclic olefins, epoxidation

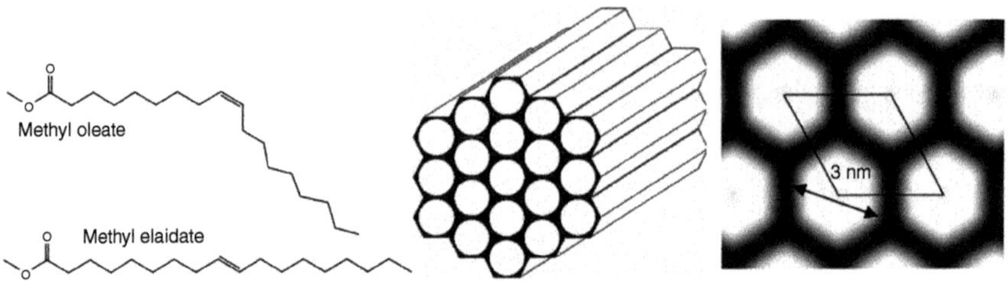

Figure 10.4 Structures of methyl oleate and methyl elaidate (left) and Ti/MCM-41 catalyst (right).

becomes predominating reaction as the olefin ring size increases (*cis*-cyclooctene, cyclododecene, caryophyllene, etc.). For example, (−)-caryophyllene, a component of glove oil, can be converted to 4,5-monoepoxide (a food and cosmetics stabilizer approved by FDA) with 70–80% selectivity at 75% substrate conversion, while negligible conversion of this substrate was observed with TS-1 [60].

Irrespective of the alkene nature, high epoxide yields can be reached over mesoporous titanium-silicates using alkyl HPs as oxidants [58, 61, 62]. Natural terpenes (*R*)-limonene and α-pinene produced epoxides with 97% and 91% yields and 93% and 96% HP efficiencies, respectively [62]. Grafted Ti-silicates showed promising results in the epoxidation of C_{18} unsaturated fatty acid methyl esters (FAMEs) and their mixtures obtained from high-oleic sunflower, castor, coriander, and soy-bean oils using TBHP as oxidant [63]. The effect of confinement was manifested in the stereospecific oxidation of methyl oleate and methyl elaidate (see Figure 10.4 for the structures) over grafted Ti catalysts. When Ti(IV) was grafted on a nonporous commercial silica, the initial epoxidation rate was 4.3 times higher for the *cis*-isomer, methyl oleate, but the difference in the initial rates for the two isomers decreased to 2.2 and 1.4 over mesoporous Ti/SiO_2 and Ti/MCM-41, respectively, indicating that mesopores affect the epoxidation process and that the straight aliphatic chains of the *trans*-isomer, methyl elaidate, match better inside the ordered cylindrical pores of Ti/MCM-41 (Figure 10.4) than the bent chains of methyl oleate [63].

Surface hydrophobization by silylation or other approaches allows significant improvement of epoxide selectivity and substrate conversion to be achieved [64, 65]. With this in mind, in 2003, Sumitomo commercialized the process for the production of propylene oxide by propene oxidation with CHP obtained by cumene autoxidation and using a proprietary silylated mesoporous catalyst obtained by grafting titanium onto a silica [9, 66].

10.3.2 Mesoporous Ti-Silicates in Oxidation of Bulky Phenols

If polar bulky substrates have to be oxidized, the situation changes drastically and hydrophilic Ti-silicates can show advantages over hydrophobic catalysts in terms of the reaction rates and selectivity. A clear demonstration of this is the oxidation of sterically hindered alkylphenols to corresponding *p*-benzoquinones (*p*-BQs) and diphenoquinones (DPQs) with aqueous H_2O_2, which was first demonstrated by Pinnavaia and coworkers using mesoporous catalysts Ti-HMS and Ti-MCM-41 [67, 68]. After optimization of reaction conditions, nearly quantitative yields of trimethyl-*p*-benzoquinone (TMBQ, vitamin E key intermediate) could be obtained in H_2O_2-based oxidation of 2,3,6-trimethylphenol (TMP) using diverse mesoporous titanium-silicates, including TiO_2–SiO_2 aerogels [69], SBA-15-supported TiO_2 nanoparticles [70], Ti(IV) grafted on commercial mesoporous silica [71, 72], and ordered titanium-silicates prepared by evaporation-induced self-assembly (EISA) [73]. Mesoporous silicates (both ordered and amorphous) with Ti surface concentration in the range of 0.7–1.0 Ti nm^{-2} have been identified as optimal catalysts for the selective transformation of alkylphenols to *p*-BQs [72, 74]. The key point to achieve superior *p*-BQ selectivity is the presence of Ti(IV) dimers or small oligomers on the surface of silica mesopores (maximum at 230–240 nm in DRS UV) to ensure fast oxidation of intermediate phenoxyl radicals, thereby preventing them from coupling to dimeric by-products (Scheme 10.2) [72, 74].

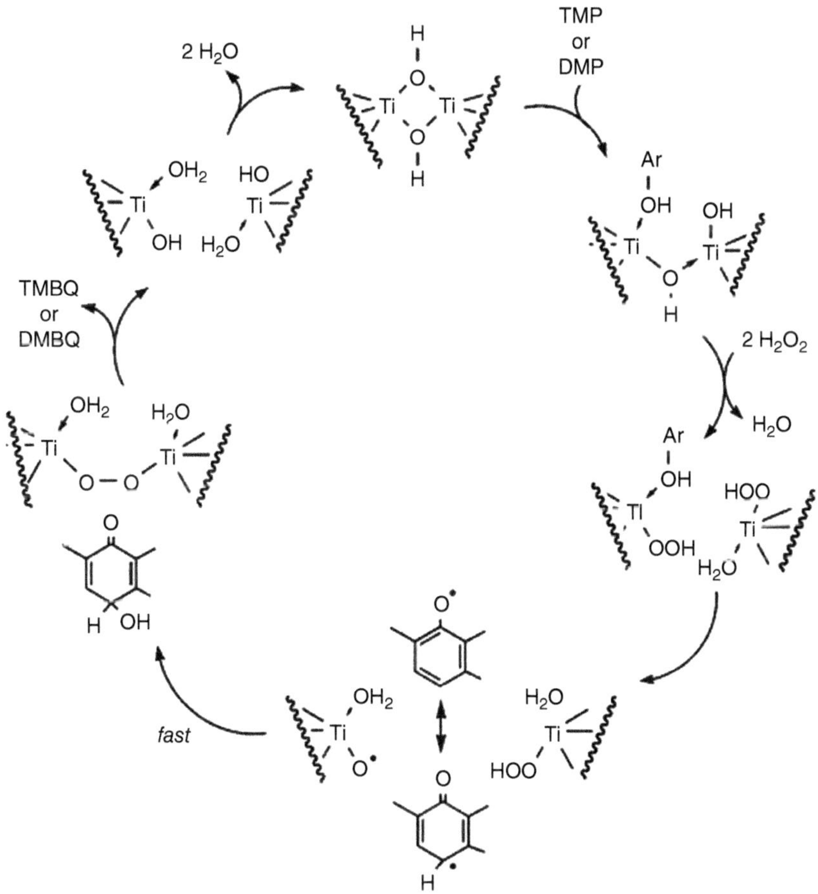

Scheme 10.2 Plausible mechanism of alkylphenol oxidation to *p*-BQ over a dimeric Ti site within a mesopore. Source: Reproduced with permission from Kholdeeva et al. [72]. Copyright 2019 Wiley.

It is noteworthy that with Ti grafted onto nonporous Aerosil, TMBQ selectivity reached only 47%, although the catalyst endowed the optimal density of Ti sites (0.85 Ti nm^{-2}) [74]. This example clearly demonstrates the role of constrained environment of the mesopores in the formation of TMBQ via the pathway shown in Scheme 10.2. When a dimeric Ti(IV) precursor was used for grafting, high yields of TMBQ were attained even at a rather low surface density of the Ti sites [72].

10.3.3 Alkene Epoxidation over Mesoporous Nb-Silicates

The long-range ordering of mesoporous silicates and a relative flexibility of their network make possible incorporation of transition metals with the radius much larger than that of Si, which is practically impossible for microporous crystalline zeolite frameworks because the size incompatibility between M and Si atoms causes longer M–O–Si bonds and more strained bond angles [54]. Significant progress was achieved in the preparation of mesoporous Nb-silicates via direct templated synthesis [75–79] or post-modification [80–82]. Several groups demonstrated that mesoporous Nb-silicates possess better hydrolytic stability than Ti counterparts and are often superior catalysts for the epoxidation of bulky alkenes with H_2O_2 [80–84]. The high stability has been ascribed to the optimal geometry of Nb–O–Si bond angles and relatively low surface mobility of niobium atoms [75]. The presence of surface Nb–OH groups capable of the formation of active Nb-hydroperoxo species was suggested to be crucial for high catalytic activity and epoxidation selectivity [83]. Low activity of Nb catalysts in H_2O_2 degradation is an additional factor that favors their high heterolytic pathway selectivity [84].

A remarkable feature of Nb-silicates is their unusual regioselectivity in the epoxidation of terpene molecules and dependence of the regioselectivity on the solvent employed [80, 81, 83]. While epoxidation of less nucleophilic exocyclic double bond of limonene occurred first in acetonitrile [79, 81, 84], more nucleophilic endocyclic epoxide preferably formed in alcoholic solvents [77, 84] (Scheme 10.3). With Ti-silicates, endocyclic double bond was first epoxidized in both solvents (Scheme 10.3), indicating electrophilic oxygen transfer mechanism and lack of steric hindrances.

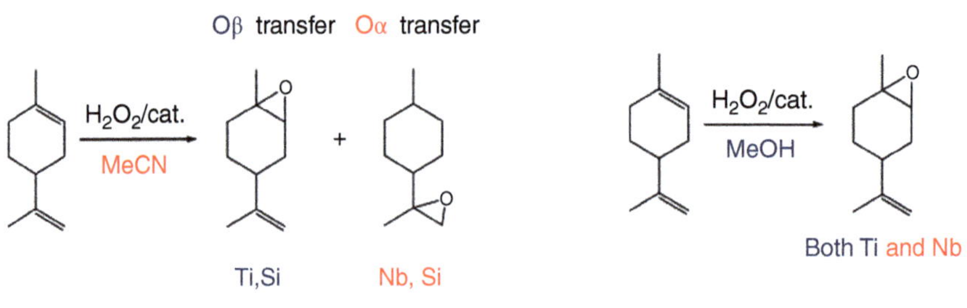

Scheme 10.3 Effect of solvent on limonene epoxidation over mesoporous Nb- and Ti-silicates.

The superior catalytic performance of Nb-silicates in alkene epoxidation and different regioselectivity observed for the limonene epoxidation over Nb- and

Ti-silicates in CH_3CN have become more clear in view of the recent comparative model studies using Nb- and Ti-POMs of the Lindquist structure [31, 85]. These studies revealed that the main reason for the superior catalytic performance of Nb(V) over Ti(IV) in alkene epoxidation is the lower free energy barrier for the oxygen transfer from Nb(η^2–OOH) due to the higher electrophilicity of niobium. While transference of the more electrophilic α-oxygen is favored for larger Nb(V) endowed with a more flexible coordination sphere, preferential transfer of the undistorted β-oxygen from Ti(η^1-OOH) becomes more favorable for smaller Ti(IV), for which extension of the coordination sphere in the transition state can be critical [31]. The steric hindrance around the α-oxygen in Nb(η^2–OOH) may account for the epoxidation of less electrophilic but more accessible C=C bonds in terpene molecules.

10.4 Metal–Organic Frameworks

MOFs are a unique class of materials composed of metal clusters (the so-called nodes) connected by multidentate organic ligands into a regular porous structure. The specific construction of MOFs endows the materials with intrinsic hybrid nature, tunable functionality, extremely high surface areas, and porosity [86–91]. A homogeneous distribution and spatial isolation of the metal nodes over the framework render the metal sites highly accessible for organic molecules, provided that the pore entrances ensure their penetration within the crystallites. The amount of active metal (20–40 wt%) that can be accommodated within MOF structure without aggregation into extra-framework species is an order of magnitude larger than that in zeolite and zeotype materials. Such structural features potentially meet all the requirements of the concept of single-site heterogeneous catalysts [12] and attract a growing interest in this class of materials from the catalytic community, as evidenced by numerous review papers [92–101].

The intrinsic hybrid nature of MOFs imposes constraints on their thermal and hydrothermal stability. The majority of them decompose below 350–400 °C in inert atmosphere, which certainly makes problematic use of MOFs in gas-phase catalytic processes. On the other hand, this problem is not so critical for catalysis in the liquid phase, including selective oxidation, where relatively mild conditions are employed. In view of the numerous review literature on this topic [102–108], below we just give selected examples to define some milestones in selective oxidation catalysis by MOFs and to demonstrate how confinement of active catalytic sites within MOFs can lead to new or improved catalytic performances.

10.4.1 Selective Oxidations over Cr- and Fe-Based MOFs

The presence of mesopores is crucial for liquid-phase processes if transformation of large organic molecules is required. The discovery of the mesoporous chromium carboxylates MIL-100 ($[Cr_3X(H_2O)_3O(BTC)_2]$; X = F, OH; BTC = 1,3,5-benzenetricarboxylate) [109] and MIL-101 ($[Cr_3X(H_2O)_2O(BDC)_3]$; X = F, OH; BDC = 1,4-benzenedicarboxylate) [110] (see Figure 10.5 for the

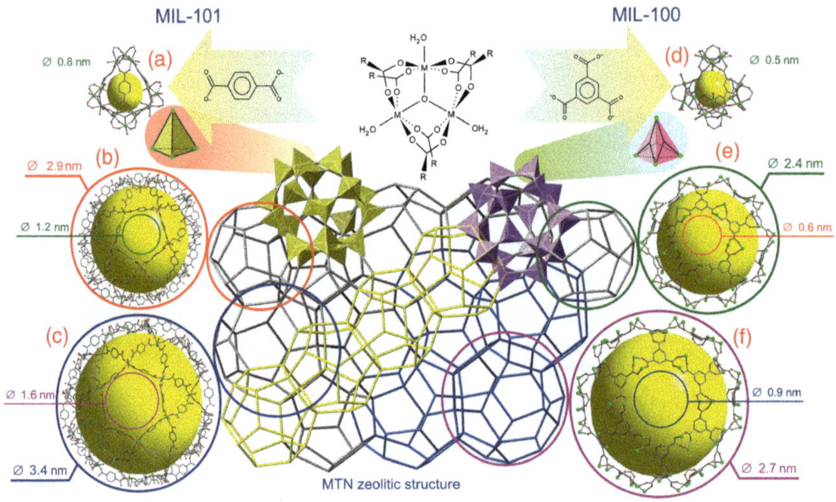

Figure 10.5 Schematic representation of the MIL-101 and MIL-100 structures: M_3O-carboxylate trimer – primary building unit, a and d – supertetrahedra, secondary building units, b and e – small cages, c and f – large cages of MIL-101 and MIL-100, respectively. Source: Reproduced with permission from Kholdeeva et al. [111]. Copyright 2014 Elsevier.

structures) by the Férey group (MIL stands for Matérial Institut Lavoisier) has marked the beginning of the use of MOFs in liquid-phase oxidations.

Cr(III)-based MIL-100 and MIL-101 possess relatively good thermal stability (250–275 °C in air), are stable to common solvents, and easily withstand the removal of guest molecules. Additional advantage for catalysis arises from the possibility of creating coordinatively unsaturated sites (CUSs) by removal of terminal water molecules from the metal nodes [102, 112]. This feature was first realized in the selective oxidation of thioethers with hydrogen peroxide [112]. It was demonstrated that catalyst activity could be increased by enlarging the CUS number with various treatments of the material. Since the pioneering work of Chang and coworkers [112], a range of selective oxidations has been accomplished using the larger pore MIL-101 [111, 113–117] (some key reactions are shown in Figure 10.6). TBHP or its combinations with molecular oxygen were most often used as oxidants because of the MOF stability issue.

Alkene oxidation with TBHP over Cr-MIL-101 produced unsaturated ketones with yields and conversions superior to the ones acquired using mesoporous chromium-silicates Cr-MCM-41 and Cr-MCM-48 [118], which demonstrates advantages of the confined space of the MIL-101 cages which prevents overoxidation [114].

Fe-based MOFs are less stable than the Cr-based counterparts, and a caution should be taken in estimation of their catalytic activity [111, 116]. It is noteworthy that a pronounced prevalence of allylic alcohols over ketones, not typical of autoxidation processes, was observed in the alkene oxidation with molecular oxygen over Fe-MIL-101, Fe-MIL-100, and Basolite F300, indicating that these MOFs can reveal a biomimetic, hydroxylase-like activity [111, 116]. A plausible mechanism of

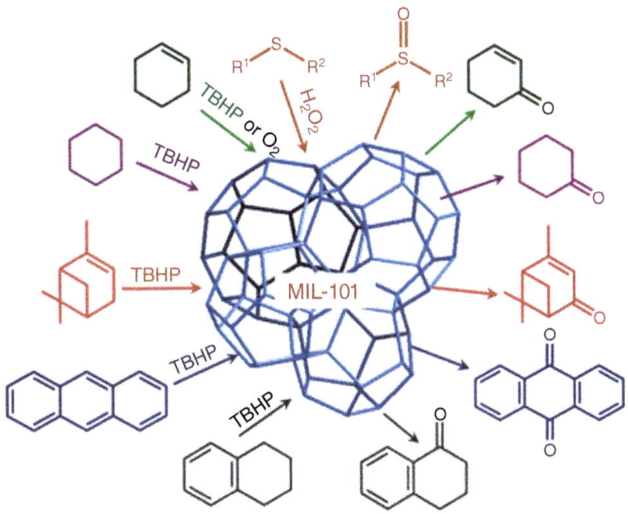

Figure 10.6 Summary on selective oxidations over Cr-MIL-101. Source: Reproduced with permission from Kholdeeva[107]. Copyright 2016 Elsevier.

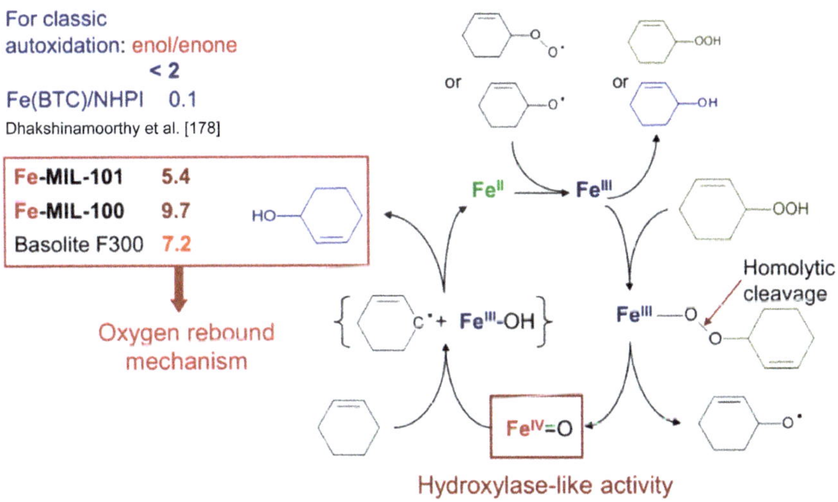

Figure 10.7 Tentative mechanism of the formation of allylic alcohols over Fe-based MOFs.

the transformation of the primary oxidation product cyclohexenyl HP over Fe sites within the MOF cages is shown in Figure 10.7.

More recently, Gascon and coworkers reported on the synthesis of a Fe-containing MOF, MIL-53(Al, Fe) that comprised a spatially isolated oxo-bridged Fe_2 unit in a coordination environment resembling that of the carboxylate-bridging diiron active species in methane monooxygenase (MMO) [119]. They demonstrated catalytic activity of this hybrid material in methane oxidation with H_2O_2 in water, producing methanol as the principal oxidation product (selectivity to oxygenates ca. 80% and turnover number (TON) up to 350). The catalyst contained no extra-framework iron oxide species and was stable in aqueous H_2O_2 at temperatures below 60 °C. A pure phase MIL-53(Fe) revealed substantial iron leaching under the reaction conditions and showed a typical Fenton-type behavior instead of MMO simulation, which indicates that the use of a nonredox scaffold, MIL-53(Al), is vital for both the catalyst stability and catalytic performance [119].

Another important organic transformation realized using MIL-101 is the oxidation of cyclohexane (CH) to a mixture of cyclohexanone (K) and cyclohexanol (A) [115], the so-called K-A oil, which has particular significance for the manufacture of Nylon-6 and Nylon-6,6. In industry, K-A oil is produced via CH autoxidation using homogeneous cobalt or manganese catalysts [120, 121]. Since both A and K are much more reactive than CH, high selectivity to K-A can be achieved only at substrate conversions below <5%. The development of a heterogeneous catalyst for this important reaction remains a challenging goal. Both Cr- and Fe-based MIL-101 are active catalysts for CH oxidation with TBHP, but the product selectivity is determined by the nature of active metal [115]. While Cr-MIL-101 produces cyclohexanone as the main product (82% selectivity at 25% CH conversion; total selectivity to K-A oil 92%), Fe-MIL-101 gives cyclohexyl HP as the major product, the yield of which strongly depends on the presence of air (oxygen), indicating significant contribution of autoxidation. The total selectivity to the sum of HP and K-A was as high as 99% at 38% substrate conversion [115]. If molecular oxygen (1 atm) was employed as oxidant (with small additives of TBHP as initiator), HP predominated among the oxidation products for both Cr- and Fe-MIL-101. Covering autoxidation processes promoted by MOFs is beyond the scope of this chapter, so we refer the reader to the recent review of Dhakshinamoorthy et al. devoted to this particular class of oxidation reactions [108].

A comparison of catalytic performances of various heterogeneous single-site catalysts for CH oxidation with TBHP is given in Figure 10.8. Clearly, Cr-MIL-101 shows advantages in terms of substrate conversion, K-A oil selectivity, catalyst productivity, and stability to active metal leaching.

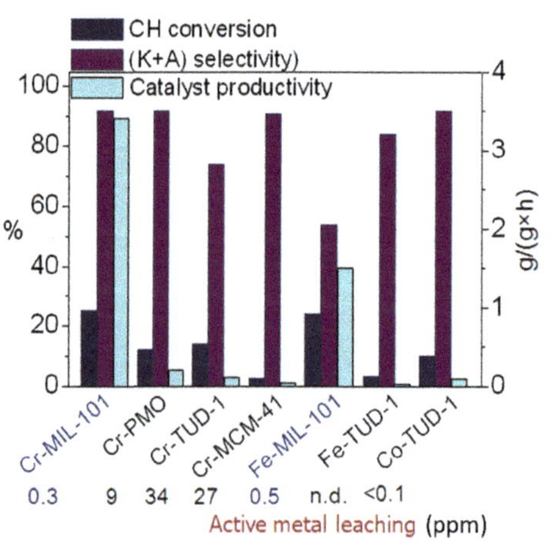

Figure 10.8 CH oxidation with TBHP at 70 °C using various solid catalysts (solvent-free conditions for MIL-101 [115] and TUD-1 [122] and acetone as solvent for Cr-PMO and Cr-MCM-41 [123]). Source: Reproduced with permission from Kholdeeva [107]. Copyright 2016 Elsevier.

More information about CH oxidation in the presence of MOFs can be found in the recent review of Garcia and coworkers [124]. In addition to systems where the oxidation reaction is due to the metal nodes, the authors consider systems in which the linker has been modified for the purpose of controlling the target reaction and those in which MOFs hosted metal nanoparticles inside the micropores.

10.4.2 Selective Oxidations with H_2O_2 over Zr- and Ti-Based MOFs

Use of hydrogen peroxide as oxidant with most MOFs is problematic because aqueous H_2O_2 possesses strong hydrolyzing and complexing ability, leading to the MOF structure damage, especially at elevated temperatures. However, the discovery of a range of microporous Zr-based MOFs with outstanding chemical and thermal stability (400–520 °C) [125–130] has made possible employment of H_2O_2 for selective oxidations. A recent review paper surveyed on the application of MOFs as heterogeneous catalysts for H_2O_2-based oxidations, with special focus on the mechanistic aspects of these reactions and catalyst stability/reusability issues [131].

While Zr-containing silicates did not show high activity in selective oxidations, at the expense of a few specific substrates [132, 133], Zr-based MOFs revealed interesting catalytic performances. First of all, Zr-MOFs (UiO-66, UiO-67, MOF-801, Zr-abtc, MIP-200, and others), regardless of their structural and morphological characteristics, unveil unprecedentedly high selectivity toward sulfones in the oxidation of thioethers in CH_3CN (Table 10.1) [131, 134–136]. Under the same conditions, a Ti-based MOF, MIL-125, gave mostly sulfoxide (see Table 10.1) [131]. The superior sulfone selectivity is an intrinsic feature of the Zr nodes, and open

Table 10.1 Oxidation of methyl phenyl sulfide (MPS) with H_2O_2 over various Zr-MOFs and Ti-MOF MIL-125.

MOF	Average particle size (nm)	Time (h)	MPS conv. (%)	Product selectivity (%)	
				MPSO	$MPSO_2$
—		24	14	79	21
UiO-66	20	0.25	49	1	99
UiO-66	200	0.5	51	8	92
UiO-67	200	0.5	51	4	96
MOF-801	n.d.	2.5	47	4	96
Zr-abtc	200	2	52	8	92
Zr-abtc	50	2	46	9	91
MIP-200	400	4	51	10	90
MIL-125	500	2	69	80	20

Reaction conditions: MPS 0.1 mmol, H_2O_2 0.1 mmol, MOF 1.6–2.0 mg, 27 °C, CH_3CN 1 ml.
n.d. = Not determined.
Source: Adapted from [131, 134, 135].

defect sites comprising a pair of Zr(OH)/Zr(OH_2) play a crucial role in the fast transformation of the primary oxidation product sulfoxide to sulfone [134, 136].

Zalomaeva et al. investigated the reaction mechanism using specific test substrates, kinetic, adsorption, ^{18}O labeling, and spectroscopic methods and came to the conclusion that weak basic sites present in the Zr-MOF defects (most likely, terminal Zr–OH groups) ensure nucleophilic activation of H_2O_2 and thus favor fast oxidation of sulfoxide to sulfone [134] (Scheme 10.4). Coordination of sulfoxide on the second Zr site also facilitates its oxidation [134, 136].

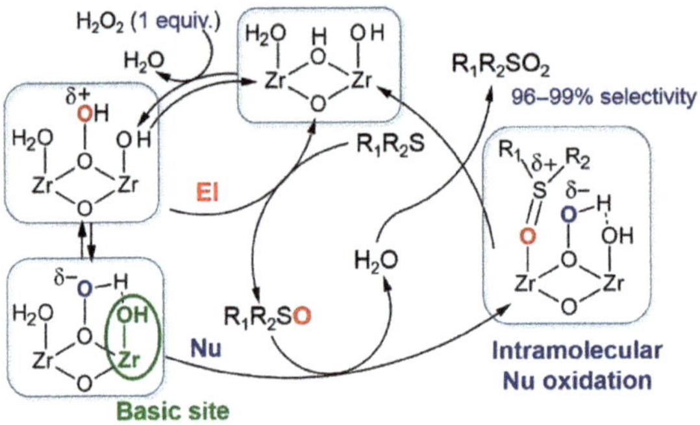

Scheme 10.4 Plausible mechanism of H_2O_2 activation and thioether oxidation over Zr-MOF defects. Source: Reprinted with permission from Zalomaeva et al. [134]. Copyright 2020 American Chemical Society.

In contrast to selectivity, activity strongly depends on both the specific Zr-MOF nature and sample characteristics, such as average particle size and defectiveness. In general, activity decreased in the order UiO-66/67 > Zr-abtc > MOF-801 > MIP-200 (Table 10.1) [131, 135]. While the average particle size had a strong impact on the activity of UiO-66, no effect was observed for larger-pore Zr-abtc (Table 10.1). Interestingly, MIP-200 which has the largest pore entrance (1.3 nm vs. 0.6 and 0.7 nm in UiO-66 and Zr-abtc, respectively) and high amount of terminal Zr-OH groups showed the lowest activity, which was explained by its specific water sorption properties (high H_2O uptake at low P/P_0 [135]) disfavoring adsorption of organic substrates [135]. On the other hand, the superior activity of UiO-66/67 may reflect a favorable, for adsorption and chemical transformation, environment of the active Zr sites within the cages of these isoreticular MOFs.

Nucleophilic activation of H_2O_2 makes also possible epoxidation of electron-deficient C=C bonds in α,β-unsaturated carbonyl compounds [131, 134, 135]. In this case, the highest product yields were obtained using the 8-coordinated MOF Zr-abtc [135]. The superior catalytic performance of Zr-abtc correlated with a larger amount of weak basic sites in this MOF relative to other MOFs studied, as shown using adsorption of isobutyric acid.

Importantly, the mode of H_2O_2 activation over UiO-66 and other Zr-MOFs can be tuned by the addition of a source of protons [135, 137]. Since Zr-MOFs possess significant activity in H_2O_2 dismutation to water and molecular oxygen, oxidation of alkenes with highly reactive allylic hydrogen atoms (e.g. cyclohexene)

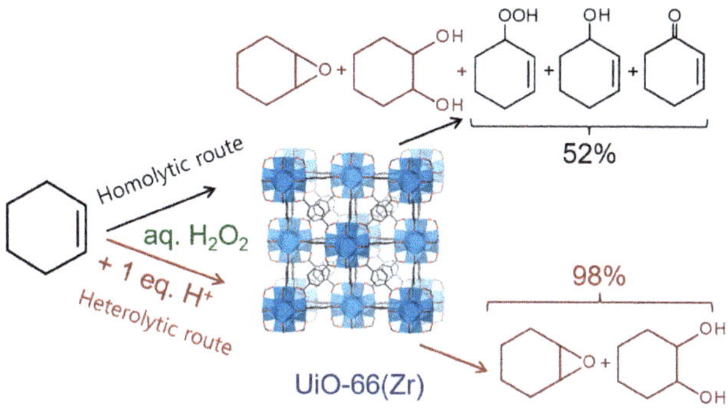

Figure 10.9 Effect of protons on cyclohexene oxidation with H_2O_2 over UiO-66. Source: Reproduced with permission from Maksimchuk et al. [137]. Copyright 2019 American Chemical Society.

leads to a mixture of products typical of radical oxidations, among which allylic HP, alcohol, and ketone are formed along with epoxide and diol (Figure 10.9) [137, 138]. The formation of allylic oxidation products could not be completely suppressed through the modification of Zr-MOFs with other transition metals (Ti [138] or Nb [139]). Maksimchuk et al. have found that small additives of a mineral acid (1 equiv. to Zr) into the reaction mixture drastically change selectivity in the oxidation of cyclohexene over UiO-66 and other Zr-MOFs, favoring the formation of heterolytic oxidation products, epoxide, and diol (Figure 10.9) [135, 137]. Even minor additives of acid (only 0.1 equiv. to Zr) resulted in a significant suppression of H_2O_2 decomposition [131, 137].

The enhancement of the heterolytic oxidation pathway by the addition of acid could also be realized for the Ti-based MOF MIL-125 [140]. Protons, most likely, favor the formation of active zirconium (titanium) hydroperoxo species, which are responsible for the alkene epoxidation.

10.5 Polyoxometalates in Confined Environment

Early transition-metal oxygen-anion nano-size clusters, usually called POMs, have attracted much attention in oxidation catalysis due to a unique combination of properties, including inorganic nature, metal oxide-like structure, thermodynamic stability to oxidation, good thermal and hydro(solvo)lytic stability, and tunable functionality [141–150]. Solubility, redox, and acid–base properties of POMs can be regulated by choosing the specific POM structure and chemical composition of both polyanion and its countercations. Of particular interest are heterometallic POMs which can activate various oxidants and, sometimes, organic substrates.

The obvious structural analogy between POMs and surface of metal oxides makes it possible to consider POMs as discrete soluble molecular models of heterogeneous catalysts, which can be systematically investigated at the atomic level by experimental and computational techniques [30–33, 85, 151–158]. A few examples demonstrating how useful such studies can be have been provided in Sections 10.2.1 and 10.3.3.

Below we give some examples to illustrate how POM encapsulation with various porous matrices may lead to interesting confinement effects and enables the construction of efficient heterogeneous selective oxidation catalysts.

10.5.1 Silica-Encapsulated POM

Encapsulation of heteropolyacid $H_5PW_{11}TiO_{40}$ within silica by a sol–gel technique has led to a leaching-tolerant and highly active catalyst for α-pinene oxidation to a mixture of verbenol and verbenone using aqueous H_2O_2 [159]. The silica-encapsulated catalyst revealed activity similar to that of homogeneous Ti-POM while selectivity to the allylic oxidation products markedly increased, most likely, thanks to suppression of overoxidation, polymerization, and rearrangement processes within the solid catalyst. Microporocity of POM/SiO_2 composites is crucial for their stability toward POM leaching in the liquid phase.

A promising POM immobilization strategy based on both covalent and non-covalent supramolecular interactions has been developed by Wu and coworkers [160–163]. This approach involves surfactant encapsulation of a POM via replacing counter ions with quaternary ammonium cations, e.g. di(11-hydroxyundecyl) dimethylammonium (DOHDA), thus forming a structure similar to that of a reverse micelle, followed by covalent anchoring within silica through sol–gel condensation with tetraethyl orthosilicate. The hydrophobic nano-environments around the POM provided by the surfactant chains favor adsorption of substrates of low polarity (e.g. sulfides or olefins) and release of polar oxidation products [161]. Following a similar approach, they prepared chiral POM complexes via electrostatic interaction of a catalytically active sandwich POM, $[WZn_3(H_2O)_2(ZnW_9O_{34})_2]^{12-}$ (Zn_5W_{19}), and a cationic surfactant with a chiral head, which was then covalently attached to the silica matrix using a sol–gel procedure (Figure 10.10) [163]. The supramolecular chiral environment around the encapsulated POM enabled kinetic resolution of racemic alcohols through catalytic oxidation with hydrogen peroxide [163].

10.5.2 MOF-Incorporated POM

In 2005, Férey et al. first demonstrated the successful inclusion of a Keggin-type heteropolytungstate within the cages of the chromium terephthalate MIL-101 [110]. Then a range of POMs, including $[PW_{11}CoO_{39}]^{5-}$ ($PW_{11}Co$), $[PW_{11}TiO_{40}]^{5-}$ ($PW_{11}Ti$) [164], $[PMo_{10}V_2O_{40}]^{5-}$ ($PMo_{10}V_2$) [165], $[PW_{12}O_{40}]^{3-}$ (PW_{12}), and $[PO_4\{WO(O_2)_2\}_4]^{3-}$ (PW_4, also called Venturello complex) [166], and many others, were incorporated within MIL-101 cages and evaluated as heterogeneous catalysts for liquid-phase selective oxidations (for recent reviews, see [150, 167]). Adsorption studies revealed strong, irreversible adsorption of about 1 POM anion per cage of MIL-101 (Figure 10.11), which corresponds to ca. 10 wt% POM [164–166].

Interesting confinement effects have been found for MIL-101-included catalytically active POMs [164, 166]. Hybrid materials PW_{12}/MIL-101 and PW_4/MIL-101 containing 5–10 wt% of POM were efficient catalysts for the epoxidation of various alkenes (3-carene, limonene, α-pinene, cyclohexene, cyclooctene, and 1-octene)

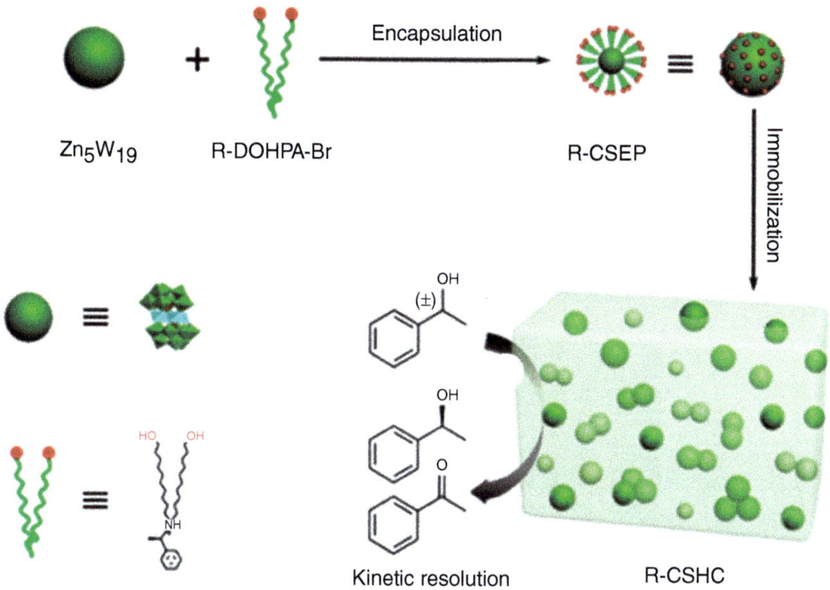

Figure 10.10 Schematic representation for the preparation of chiral surfactant-encapsulated POM complexes (R-CSEP) immobilized within silica for kinetic resolution of racemic secondary alcohols. Source: Reproduced from Shi et al. [163] by permission of The Royal Society of Chemistry.

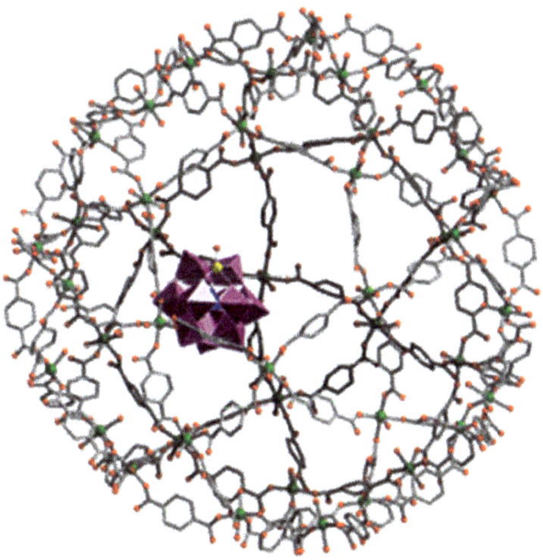

Figure 10.11 Keggin-type heterometallic POM incorporated within a large cage of MIL-101. Source: Reproduced with permission from Kholdeeva [107]. Copyright 2016 Elsevier.

using aqueous H_2O_2 in MeCN [166]. In sharp contrast to the homogeneous catalyst, increase in H_2O_2 concentration allowed increasing simultaneously both the alkene conversion and epoxide selectivity despite the augmenting amount of water in the system [107, 166]. A similar effect was observed for caryophyllene epoxidation over $PW_{11}Ti$/MIL-101 (Figure 10.12) [164].

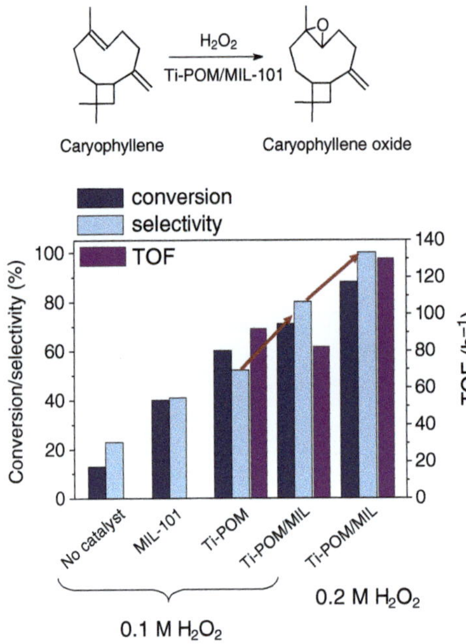

Figure 10.12 Caryophyllene epoxidation over Ti-POM/MIL-101 in comparison with MIL-101 and homogeneous Ti-POM.

Such unusual behavior of the POM/MIL-101 catalysts could be explained by the specific sorption properties of MIL-101. The hydrophobic part of terephthalate linkers favors adsorption of nonpolar hydrocarbons and, oppositely, disfavors adsorption of water. This suggestion is supported by the character of the water sorption isotherm reported for MIL-101, which shows that water uptake starts at high relative pressures ($p/p_0 = 0.4$) [168]. This implies that the MIL-101 surface reveals hydrophobic character when the concentration of water in the organic solvent is rather low.

Importantly, POM inclusion within MIL-101 enhanced the thermal stability of the MOF itself [110] and also improved stability and productivity of the POM [166]. The overall TON reached without evident degradation of PW_{12} was 770 and 155 for PW_{12}/MIL-101 and homogeneous PW_{12}, respectively [166].

10.5.3 POMs Supported on Carbon Nanotubes

Carbon nanotubes (CNTs) have attracted significant attention as supports for various active complexes, including POMs [167, 169–173]. Evtushok et al. first used N-doped CNTs for immobilization of a di-V-substituted γ-Keggin phosphotungstate $[\gamma\text{-}PW_{10}O_{38}V_2(\mu\text{-}O)(\mu\text{-}OH)]^{4-}$ (PV_2W_{10}) [174]. N-doping enabled a quasi-molecular dispersion of PV_2W_{10} on the carbon surface and superior catalytic performance in the selective oxidation of alkylphenols. With an optimal catalyst, $PW_{10}V_2$/N-CNTs (15 wt% POM, 1.8 at% N), TMBQ was produced in a nearly quantitative yield and 80% H_2O_2 utilization efficiency. The confined space inside nanotube tangles endowed the catalyst with unprecedented high activity (TOF = 500 h^{-1}) and space-time yield (450 $g\,l^{-1}\,h^{-1}$) [174, 175]. A blend of electrostatic forces and hydrogen bonding ensured excellent stability and recyclability of the catalyst

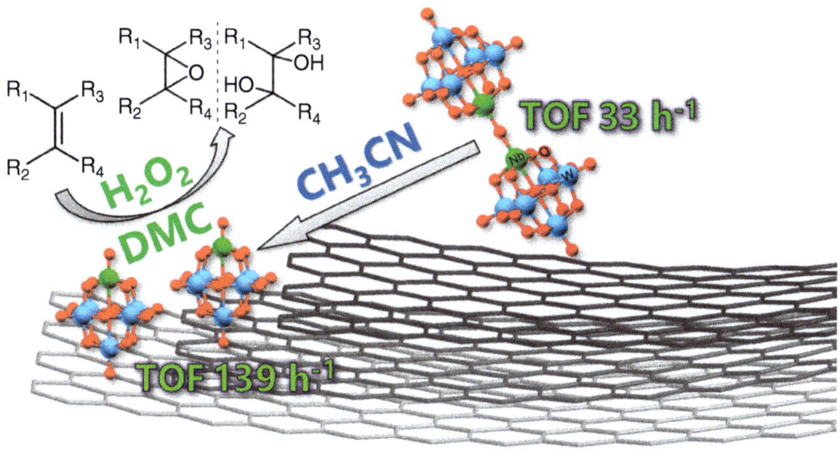

Figure 10.13 Stabilization of NbW_5 on CNTs for efficient alkene epoxidation with H_2O_2. Source: Evtushok et al. [177]/Royal Society of Chemistry.

without loss of its activity and selectivity. Immobilization on other supports, e.g. amine-modified SiO_2 or MIL-101, led to a significant deterioration of the catalytic performance [174].

N-free CNTs can also serve as supports for efficient immobilization of small POMs, in particular, the Venturello complex PW_4 [176] or heterometallic POM of the Lindquist structure [177], leading to efficient epoxidation catalysts. Stabilization of highly reactive monomeric forms, e.g. $[Nb(OH)W_5O_{19}]^{2-}$ (NbW_5) formed upon dissociation of the dimer $[(NbW_5O_{18})_2O]^{4-}$, on the surface of CNTs resulted in a pronounced enhancement of catalytic activity of the Nb-POM (Figure 10.13) [177].

10.6 Conclusion and Outlook

To summarize, a range of efficient heterogeneous catalysts have been elaborated for the selective oxidation of organic compounds to produce important products and intermediates for the base and fine organic synthesis. While large-scale processes using TS-1 have been already commercialized, other types of porous catalysts that have been touched upon in this chapter (mesoporous metal-silicates, MOFs, supported POMs, and others) have also shown promising results, although they still require some improvements. Specifically, oxidant utilization efficiency, catalyst productivity, and lifetime have to be increased to make them attractive for industry. Meanwhile, the remarkable effects of confined space on activity, selectivity (chemo/regio/enantio), and stability revealed so far for the new types of porous catalysts allow expectations that further developments of the confinement concept would certainly lead to new "smart" and efficient catalytic materials and make possible significant progress on the way toward more selective, safe and environmentally benign catalytic oxidation processes. Detailed studies on the structure/activity/selectivity relationships along with mechanistic investigations are indispensable for the rational design of novel generations of catalysts and catalytic processes.

Acknowledgments

The author thanks all coauthors of the joint publications cited in this chapter. Fruitful discussions of M.G. Clerici on Ti-catalyzed oxidations are greatly appreciated. This work was partially supported by the Ministry of Science and Higher Education of the Russian Federation within the governmental order for the Boreskov Institute of Catalysis (project AAAA-A21-121011390008-4).

References

1 Sheldon, R.A. and van Bekkum, H. (ed.) (2001). *Fine Chemicals Through Heterogeneous Catalysis*. Weinheim: Wiley-VCH.
2 Clerici, M.G. and Kholdeeva, O.A. (ed.) (2013). *Liquid Phase Oxidation via Heterogeneous Catalysis: Organic Synthesis and Industrial Applications*. Hoboken, NJ, USA: Wiley.
3 Duprez, D. and Cavani, F. (ed.) (2014). *Handbook of Advanced Methods and Processes in Oxidation Catalysis*. London, UK: Imperial College Press.
4 Jiao, N. and Stahl, S.S. (ed.) (2019). *Green Oxidation in Organic Synthesis*. Wiley.
5 Strukul, G. and Scarso, A. (2013). Environmentally benign oxidants. In: *Liquid Phase Oxidation via Heterogeneous Catalysis: Organic Synthesis and Industrial Applications* (ed. M.G. Clerici and O.A. Kholdeeva), 1–20. Hoboken, NJ, USA: Wiley.
6 Sheldon, R.A. and Dakka, J. (1994). *Catal. Today* 19: 215–246.
7 Edwards, J.K., Freakley, S.J., Lewis, R.J. et al. (2015). *Catal. Today* 248: 3–9.
8 Special issue on direct synthesis of hydrogen peroxide, *Catalysts* 2019. https://www.mdpi.com/si/catalysts/hydrogen_peroxide.
9 Cavani, F. and Teles, J.H. (2009). *ChemSusChem* 2: 508–534.
10 Sheldon, R.A., Wallau, M., Arends, I.W.C.E., and Schuchardt, U. (1998). *Acc. Chem. Res.* 31: 485–493.
11 Thomas, J.M., Raja, R., and Lewis, D.W. (2005). *Angew. Chem. Int. Ed.* 44: 6456–6482.
12 Thomas, J.M. (ed.) (2012). Thematic issue on single-site heterogeneous catalysis. *Proc. R. Soc. A* 468: 468.
13 Fu, Q., Yang, F., and Bao, X. (2013). *Acc. Chem. Res.* 46: 1692–1701.
14 Mouarrawis, V., Plessius, R., van der Vlugt, J.I., and Reek, J.N.H. (2018). *Front. Chem.* 6: 623.
15 Liu, L. and Corma, A. (2020). *Nat. Rev. Mater.* 6: 244–263.
16 Grommet, A.B., Feller, M., and Klajn, R. (2020). *Nat. Nanotechnol.* 15: 256–271.

17 Liu, L., Lopez-Haro, M., Perez-Omil, J.A. et al. (2022). *Nat. Commun.* 13: 821.

18 Taramasso, M., Perego, G., and Notari, B. (1983). US Patent 4 410 501, 21 December 1979 Granted: 18 October 1983.

19 Notari, B. (1996). Microporous crystalline titanium silicates. In: *Advances in Catalysis*, vol. 41 (ed. D.D. Eley, W.O. Haag, and B.C. Gates), 253–334. San Diego: Academic Press.

20 Perego, C., Carati, A., Ingallina, P. et al. (2001). *Appl. Catal., A* 221: 63–72.

21 Clerici, M.G. and Domine, M.E. (2013). Oxidation reactions catalyzed by transition metal-substituted zeolites. In: *Liquid Phase Oxidation via Heterogeneous Catalysis: Organic Synthesis and Industrial Applications* (ed. M.G. Clerici and O.A. Kholdeeva), 21–93. Hoboken, NJ, USA: Wiley.

22 Clerici, M.G. (2015). *Kinet. Catal.* 56: 450–455.

23 Clerici, M.G., Bellussi, G., and Romano, U. (1991). *J. Catal.* 129: 159–167.

24 Clerici, M.G. and Ingallina, P. (1993). *J. Catal.* 140: 71–83.

25 Tatsumi, T., Nakamura, M., Negishi, S., and Tominaga, H. (1990). *J. Chem. Soc., Chem. Commun.* 476–477.

26 Clerici, M.G. (1991). *Appl. Catal.* 68: 249–261.

27 Romano, U. and Ricci, M. (2013). In: *Liquid Phase Oxidation via Heterogeneous Catalysis: Organic Synthesis and Industrial Applications* (ed. M.G. Clerici and O.A. Kholdeeva), 451–462. Hoboken, NJ, USA: Wiley.

28 Bordiga, S., Bonino, F., Damin, A., and Lamberti, C. (2007). *Phys. Chem. Chem. Phys.* 9: 4854–4878.

29 Bordiga, S., Groppo, E., Agostini, G. et al. (2013). *Chem. Rev.* 113: 1736–1850.

30 Jiménez-Lozano, P., Skobelev, I.Y., Kholdeeva, O.A. et al. (2016). *Inorg. Chem.* 55: 6080–6084.

31 Maksimchuk, N.V., Ivanchikova, I.D., Maksimov, G.M. et al. (2019). *ACS Catal.* 9: 6262–6275.

32 Zhang, T., Mazaud, L., Chamoreau, L.-M. et al. (2018). *ACS Catal.* 8: 2330–2342.

33 Solé-Daura, A., Zhang, T., Fouilloux, H. et al. (2020). *ACS Catal.* 10: 4737–4750.

34 Gordon, C.P., Engler, H., Tragl, A.S. et al. (2020). *Nature* 586: 708–713.

35 Moliner, M. and Corma, A. (2014). *Microporous Mesoporous Mater.* 189: 31–40.

36 Bellussi, G., Carati, A., Rizzo, C., and Millini, R. (2013). *Catal. Sci. Technol.* 3: 833–857.

37 Corma, A., Esteve, P., Martínez, A., and Valencia, S. (1995). *J. Catal.* 152: 18–24.

38 Carati, A., Flego, C., Previde Massara, E. et al. (1999). *Microporous Mesoporous Mater.* 30: 137–144.

39 Wu, P. and Tatsumi, T. (2001). *Chem. Commun.* 897–898.

40 Tatsumi, T., Komatsu, T., and Yashima, T. (2001). *J. Catal.* 202: 245–255.

41 Wu, P. and Tatsumi, T. (2003). *J. Catal.* 214: 317–326.

42 Blasco, T., Camblor, M.A., Corma, A. et al. (1996). *Chem. Commun.* 2368.

43 Zhao, H., Yokoi, T., Kondo, J.N., and Tatsumi, T. (2015). *Appl. Catal., A* 503: 156–164.

44 Bregante, D.T., Johnson, A.M., Patel, A.Y. et al. (2019). *J. Am. Chem. Soc.* 141 (18): 7302–7319.

45 Bregante, D.T., Tan, J.Z., Sutrisno, A., and Flaherty, D.W. (2020). *Catal. Sci. Technol.* 10: 635–647.
46 Raja, R., Sankar, G., and Thomas, J.M. (2000). *Angew. Chem. Int. Ed.* 39: 2313–2316.
47 Thomas, J.M. and Raja, R. (2001). *Chem. Commun.* 675–687.
48 Kholdeeva, O.A. (2014). *Catal. Sci. Technol.* 4: 1869–1889.
49 Galarneau, A., Renzo, D.F., Fajula, F., and Vedrine, J. (ed.) (2001). Zeolites and mesoporous materials at the dawn of the 21st century. In: *Studies in Surface Science and Catalysis*, vol. 135. Amsterdam: Elsevier.
50 Kresge, C.T., Leonowicz, M.E., Roth, W.J. et al. (1992). *Nature* 359: 710–712.
51 Corma, A. (1997). *Chem. Rev.* 97: 2373–2419.
52 Perego, C. and Bosetti, A. (2011). *Microporous Mesoporous Mater.* 144: 28–39.
53 Dapsens, P.Y., Mondelli, C., and Perez-Ramírez, J. (2012). *ACS Catal.* 2: 1487–1499.
54 Kholdeeva, O.A. (2013). In: *Liquid Phase Oxidation via Heterogeneous Catalysis: Organic Synthesis and Industrial Applications* (ed. M.G. Clerici and O.A. Kholdeeva), 127–219. Hoboken, NJ, USA: Wiley.
55 Di Renzo, F., Galarneau, A., Trens, P., and Fajula, F. (2002). *Handbook of Porous Solids*, vol. 3 (ed. F. Schüth, K.S.W. Sing, and J. Weitkamp), 1311–1395. Weinheim: Wiley-VCH.
56 Ratnasamy, P., Srinivas, D., and Knozinger, H. (2004). *Adv. Catal.* 48: 1–169.
57 Bonon, A.J., Mandelli, D., Kholdeeva, O.A. et al. (2009). *Appl. Catal., A* 365: 96–104.
58 Blasco, T., Corma, A., Navarro, M.T., and Pérez-Pariente, J. (1995). *J. Catal.* 156: 65–74.
59 Guidotti, M., Pirovano, C., Ravasio, N. et al. (2009). *Green Chem.* 11: 1421–1427.
60 Kholdeeva, O.A., Melgunov, M.S., Shmakov, A.N. et al. (2004). *Catal. Today* 91, 92: 205–209.
61 Shan, Z., Jansen, J.C., Marchese, L., and Maschmeyer, T. (2001). *Microporous Mesoporous Mater.* 48: 181–187.
62 Chiker, F., Launay, F., Nogier, J.P., and Bonardet, J.L. (2003). *Green Chem.* 5: 318–322.
63 Guidotti, M., Ravasio, N., Psaro, R. et al. (2003). *Green Chem.* 5: 421–424.
64 Wu, P., Tatsumi, T., Komatsu, T., and Yashima, T. (2002). *Chem. Mater.* 14: 1657–1664.
65 Corma, A., Domine, M., Gaona, J.A. et al. (1998). *Chem. Commun.* 2211–2212.
66 Cavani, F. and Gaffney, A.M. (2009). *Sustainable Industrial Processes* (ed. F. Cavani, G. Centi, S. Perathoner, and F. Trifiro), 287–337. Wiley-VCH.
67 Tanev, P.T., Chibwe, M., and Pinnavaia, T. (1994). *Nature* 368: 321–323.
68 Zhang, W., Fröba, M., Wang, J. et al. (1996). *J. Am. Chem. Soc.* 118: 9164–9171.
69 Mrowiec-Bialon, J., Jarzebski, A.B., Kholdeeva, O.A. et al. (2004). *Appl. Catal., A* 273: 47–53.
70 Tuel, A. and Hubert-Pfalzgraf, L.G. (2003). *J. Catal.* 217: 343–353.
71 Kholdeeva, O.A., Ivanchikova, I.D., Guidotti, M., and Ravasio, N. (2007). *Green Chem.* 9: 731–733.

72 Kholdeeva, O.A., Ivanchikova, I.D., Guidotti, M. et al. (2009). *Adv. Synth. Catal.* 351: 1877–1889.

73 Ivanchikova, I.D., Kovalev, M.K., Mel'gunov, M.S. et al. (2014). *Catal. Sci. Technol.* 4: 200–207.

74 Kholdeeva, O.A., Ivanchikova, I.D., Guidotti, M. et al. (2009). *Catal. Today* 141: 330–336.

75 Ziolek, M. (2003). *Catal. Today* 78: 47–64.

76 Selvaraj, M., Kawi, S., Park, D.W., and Ha, C.S. (2009). *J. Phys. Chem. C* 113: 7743–7749.

77 Feliczak-Guzik, A. and Nowak, I. (2009). *Catal. Today* 142: 288–292.

78 Ramanathan, A., Masheswari, R., and Subramanian, B. (2015). *Top. Catal.* 58: 314–324.

79 Ivanchikova, I.D., Maksimchuk, N.V., Skobelev, I.Y. et al. (2015). *J. Catal.* 332: 138–148.

80 Gallo, C., Tiozzo, R., Psaro, F. et al. (2013). *J. Catal.* 298: 77–83.

81 Tiozzo, C., Bisio, C., Carniato, F., and Guidotti, M. (2014). *Catal. Today* 235: 49–57.

82 Thornburg, N.E., Thompson, A.B., and Notestein, J.M. (2015). *ACS Catal.* 5: 5077–5088.

83 Ivanchikova, I.D., Skobelev, I.Y., Maksimchuk, N.V. et al. (2017). *J. Catal.* 356C: 85–99.

84 Kholdeeva, O.A., Ivanchikova, I.D., Maksimchuk, N.V., and Skobelev, I.Y. (2019). *Catal. Today* 333: 63–70.

85 Maksimchuk, N.V., Ivanchikova, I.D., Maksimov, G.M. et al. (2018). *ACS Catal.* 8: 9722–9737.

86 Férey, G. (2008). *Chem. Soc. Rev.* 37: 191–214.

87 Thematic issue on metal-organic frameworks, H.-C. Zhou, J. R. Long, O. M. Yaghi (Guest Editors), *Chem. Rev.* 2012, 112 (2).

88 Zhou, H.-C. and Kitagawa, S. (ed.) (2014). Thematic issue on metal-organic frameworks. *Chem. Soc. Rev.* 43: 5415–5418. (Guest Editors).

89 Maurin, G., Serre, C., Cooper, A., and Férey, G. (ed.) (2017). Themed issue metal-organic frameworks and porous polymers – current and future challenges. *Chem. Soc. Rev.* 46: 3104–3107. (Guest Editors)

90 Schröder, M. (ed.) (2010). *Functional Metal-Organic Frameworks: Gas Storage, Separation and Catalysis.* Heidelberg, Germany: Springer-Verlag.

91 Levason, B. and Bradshaw, D. (ed.) (2016). Special issue on chemistry and applications of metal organic frameworks. *Coord. Chem Rev.* 307: 105–424. (Guest Editors)

92 Corma, A., García, H., and Llabres i Xamena, F.X. (2010). *Chem. Rev.* 110: 4606–4655.

93 Llabrés i Xamena, F. and Gascon, J. (ed.) (2013). *Metal–Organic Frameworks as Heterogeneous Catalysts.* Cambridge, UK: RCS Publishing.

94 García, H. and Navalón, S. (ed.) (2018). *Metal-Organic Frameworks: Applications in Separations and Catalysis.* Weinheim, Germany: Wiley-VCH.

95 Chughtai, A.H., Ahmad, N., Younus, H.A. et al. (2015). *Chem. Soc. Rev.* 44: 6804–6849.

96 Gascon, J., Corma, A., Kapteijn, F., and Llabrés i Xamena, F.X. (2014). *ACS Catal.* 4: 361–378.

97 Rogge, S.M.J., Bavykina, A., Hajek, J. et al. (2017). *Chem. Soc. Rev.* 46: 3134–3184.

98 Xu, W., Thapa, K.B., Ju, Q. et al. (2018). *Coord. Chem. Rev.* 373: 199–232.

99 Dhakshinamoorthy, A., Li, Z., and García, H. (2018). *Chem. Soc. Rev.* 47: 8134–8172.

100 Yang, D. and Gates, B.C. (2019). *ACS Catal.* 9: 1779–1798.

101 Bavykina, A., Kolobov, N., Khan, I.S. et al. (2020). *Chem. Rev.* 120: 8468–8535.

102 Hwang, Y.K., Férey, G., Lee, U.-H., and Chang, J.-S. (2013). In: *Liquid Phase Oxidation via Heterogeneous Catalysis: Organic Synthesis and Industrial Applications* (ed. M.G. Clerici and O.A. Kholdeeva), 371–409. Hoboken, NJ, USA: Wiley.

103 Dhakshinamoorthy, A., Alvaro, M., and García, H. (2011). *Catal. Sci. Technol.* 1: 856–867.

104 Maksimchuk, N.V., Zalomaeva, O.V., Skobelev, I.Y. et al. (2012). *Proc. R Soc. A* 468: 2017–2034.

105 Leus, K., Liu, Y.-Y., and Van Der Voort, P. (2014). *Catal Rev.* 56: 1–56.

106 Dhakshinamoorthy, A., Asiri, A.M., and García, H. (2016). *Chem. Eur. J.* 22: 8012–8024.

107 Kholdeeva, O.A. (2016). *Catal. Today* 278: 22–29.

108 Dhakshinamoorthy, A., Asiric, A.M., Herance, J.R., and García, H. (2018). *Catal. Today* 306: 2–8.

109 Ferey, G., Serre, C., Mellot-Draznieks, C. et al. (2004). *Angew. Chem. Int. Ed.* 116: 6456–6461.

110 Férey, G., Mellot-Draznieks, C., Serre, C. et al. (2005). *Science* 309: 2040–2042.

111 Kholdeeva, O.A., Skobelev, I.Y., Ivanchikova, I.D. et al. (2014). *Catal. Today* 238: 54–61.

112 Hwang, Y.K., Hong, D.-Y., Chang, J.-S. et al. (2009). *Appl. Catal., A* 358: 249–253.

113 Kim, J., Bhattacharjee, S., Jeong, K.-E. et al. (2009). *Chem. Commun.* 3904–3906.

114 Maksimchuk, N.V., Kovalenko, K.A., Fedin, V.P., and Kholdeeva, O.A. (2010). *Adv. Synth. Catal.* 352: 2943–2948.

115 Maksimchuk, N.V., Kovalenko, K.A., Fedin, V.P., and Kholdeeva, O.A. (2012). *Chem. Commun.* 48: 6812–6814.

116 Skobelev, I.Y., Sorokin, A.B., Kovalenko, K.A. et al. (2013). *J. Catal.* 298: 61–69.

117 Ivanchikova, I.D., Skobelev, I.Y., and Kholdeeva, O.A. (2015). *J. Organomet. Chem.* 793: 175–181.

118 Sakthivel, A., Dapurkar, S.E., and Selvam, P. (2003). *Appl. Catal., A* 246: 283–293.

119 Osadchii, D.Y., Olivos-Suarez, A.I., Szécsényi, A. et al. (2018). *ACS Catal.* 8: 5542–5548.

120 Musser, M.T. (2011). *Ullmann's Encyclopedia of Industrial Chemistry*, Electronic edition. Wiley-VCH.
121 Cavani, F. and Alini, S. (2009). In: *Sustainable Industrial Processes* (ed. F. Cavani, G. Centi, S. Perathoner, and F. Trifiro), 331–391. Wiley-VCH.
122 Anand, R., Hamdy, M.S., Gkourgkoulas, P. et al. (2006). *Catal. Today* 117: 279–283.
123 Shylesh, S., Srilakshmi, C., Singh, A.P., and Anderson, B.G. (2007). *Microporous Mesoporous Mater.* 99: 334–344.
124 Dhakshinamoorthy, A., López-Francés, A., Navalon, S., and Garcia, H. (2022). *ChemCatChem* 14: e202201036. https://doi.org/10.1002/cctc.202201036.
125 Cavka, J.H., Jakobsen, S., Olsbye, U. et al. (2008). *J. Am. Chem. Soc.* 130: 13850–13851.
126 Wißmann, G., Schaate, A., Lilienthal, S. et al. (2012). *Microporous Mesoporous Mater.* 152: 64–70.
127 Bai, Y., Dou, Y., Xie, L.-H. et al. (2016). *Chem. Soc. Rev.* 45: 2327–2367.
128 Yuan, S., Feng, L., Wang, K. et al. (2018). *Adv. Mater.* 30: 1704303.
129 Wang, H., Dong, X., Lin, J. et al. (2018). *Nat. Commun.* 9: 1745.
130 Wang, S., Lee, J.S., Wahiduzzaman, M. et al. (2018). *Nat. Energy* 3: 985–993.
131 Kholdeeva, O.A. and Maksimchuk, N.V. (2021). *Catalysts* 11: 283.
132 Gontier, S. and Tuel, A. (1996). *Appl. Catal., A* 143: 125–135.
133 Ivanchikova, I.D., Zalomaeva, O.V., Maksimchuk, N.V. et al. (2022). *Catalysts* 12: 742.
134 Zalomaeva, O.V., Evtushok, V.Y., Ivanchikova, I.D. et al. (2020). *Inorg. Chem.* 59: 10634–10649.
135 Maksimchuk, N.V., Ivanchikova, I.D., Cho, K.H. et al. (2021). *Chem. Eur. J.* 27: 1–9.
136 Limvorapitux, R., Chen, H., Mendonca, M.L. et al. (2019). *Catal. Sci. Technol.* 9: 327–335.
137 Maksimchuk, N.V., Lee, J.S., Solovyeva, M.V. et al. (2019). *ACS Catal.* 9: 9699–9704.
138 Nguyen, H.G.T., Mao, L., Peters, A.W. et al. (2015). *Catal. Sci. Technol.* 5: 4444–4451.
139 Ahn, S., Thornburg, N.E., Li, Z. et al. (2016). *Inorg. Chem.* 55: 11954–11961.
140 Maksimchuk, N., Lee, J.S., Ayupov, A. et al. (2019). *Catalysts* 9: 324.
141 Hill, C.L. and Prosser-McCartha, C.M. (1995). *Coord. Chem. Rev.* 143: 407–455.
142 Neumann, R. (2004). *Transition Metals for Organic Synthesis*, 2e, vol. 2 (ed. M. Beller and C. Bolm), 415–426. Weinheim, Germany: Wiley-VCH.
143 Pope, M.T. (2004). *Comprehensive Coordination Chemistry II*, vol. 4 (ed. A.G. Wedd), 635–678. NY, USA: Elsevier Science.
144 Hill, C.L. (2004). *Comprehensive Coordination Chemistry II*, vol. 4 (ed. A.G. Wedd), 679–759. NY, USA: Elsevier Science.
145 Mizuno, N., Kamata, K., Uchida, S., and Yamaguchi, K. (2009). *Modern Heterogeneous Oxidation Catalysis: Design, Reactions and Characterization* (ed. N. Mizuno), 185–216. Weinheim, Germany: Wiley-VCH.

146 Hill, C.L. and Kholdeeva, O.A. (2013). In: *Liquid Phase Oxidation via Heterogeneous Catalysis: Organic Synthesis and Industrial Applications* (ed. M.G. Clerici and O.A. Kholdeeva), 263–319. Hoboken, NJ, USA: Wiley.

147 Sun, M., Zhang, J., Putaj, P. et al. (2014). *Chem. Rev.* 114: 981–1019.

148 Wang, S.-S. and Yang, G.-Y. (2015). *Chem. Rev.* 115: 4893–4962.

149 Weinstock, I.A., Schreiber, R.E., and Neumann, R. (2018). *Chem. Rev.* 118: 2680–2717.

150 Kholdeeva, O.A. (2019). In: *Frontiers of Green Catalytic Selective Oxidations. Green Chemistry and Sustainable Technology* (ed. K.P. Bryliakov), 61–91. Singapore: Springer.

151 Kholdeeva, O.A. (2006). *Top. Catal.* 40: 229–243.

152 Carabineiro, H., Villanneau, R., Carrier, X. et al. (2006). *Inorg. Chem.* 45: 1915–1923.

153 Kholdeeva, O.A. and Maksimovskaya, R.I. (2007). *J. Mol. Catal. A: Chem.* 262: 7–24.

154 Kholdeeva, O.A. (2013). *Eur. J. Inorg. Chem.* 1595–1605.

155 Antonova, N.S., Carbó, J.J., Kortz, U. et al. (2010). *J. Am. Chem. Soc.* 132: 7488–7497.

156 Jiménez-Lozano, P., Ivanchikova, I.D., Kholdeeva, O.A. et al. (2012). *Chem. Commun.* 48: 9266–9268.

157 Zhang, T., Solé-Daura, A., Fouilloux, H. et al. (2021). *ChemCatChem* 13: 1220–1229.

158 Maksimchuk, N.V., Evtushok, V.Y., Zalomaeva, O.V. et al. (2021). *ACS Catal.* 11: 10589–10603.

159 Maksimchuk, N.V., Melgunov, M.S., Mrowiec-Białoń, J. et al. (2005). *J. Catal.* 235: 175–183.

160 Wu, L. (2017). In: *Encapsulated Catalysts* (ed. S. Sadjadi), 1–33. Elsevier: Academic Press.

161 Qi, W., Wang, Y., Li, W., and Wu, L. (2010). *Chem. Eur. J.* 16: 1068–1078.

162 Yan, Y. and Wu, L. (2011). *Isr. J. Chem.* 2: 181–190.

163 Shi, L., Wang, Y.Z., Li, B., and Wu, L. (2014). *Dalton Trans.* 43: 9177–9188.

164 Maksimchuk, N.V., Timofeeva, M.N., Melgunov, M.S. et al. (2008). *J. Catal.* 257: 315–323.

165 Maksimchuk, N.V., Kholdeeva, O.A., Kovalenko, K.A., and Fedin, V.P. (2011). *Isr. J. Chem.* 2: 281–289.

166 Maksimchuk, N.V., Kovalenko, K.A., Arzumanov, S.S. et al. (2010). *Inorg. Chem.* 49: 2920–2930.

167 Maksimchuk, N.V. and Kholdeeva, O.A. (2023). H_2O_2-based selective oxidations catalyzed by supported polyoxometalates: recent advances. *Catalysts* 13: 360. https://doi.org/10.3390/catal13020360.

168 Küsgens, P., Rose, M., Senkovska, I. et al. (2009). *Microporous Mesoporous Mater.* 120: 325–330.

169 Ji, Y., Huang, L., Hu, J. et al. (2015). *Energy Environ. Sci.* 8: 776–789.

170 Toma, F.M., Sartorel, A., Iurlo, M. et al. (2010). *Nat. Chem.* 2: 826–831.

171 Zhu, J., Holmen, A., and Chen, D. (2013). *ChemCatChem* 5: 378–401.

172 Evtushok, V.Y., Podyacheva, O.Y., Suboch, A.N. et al. (2020). *Catal. Today* 354: 196–203.
173 Evtushok, V.Y., Lopatkin, V.A., Podyacheva, O.Y., and Kholdeeva, O.A. (2022). *Catalysts* 12: 472.
174 Evtushok, V.Y., Suboch, A.N., Podyacheva, O.Y. et al. (2018). *ACS Catal.* 8: 1297–1307.
175 Kholdeeva, O.A. and Zalomaeva, O.V. (2016). *Coord. Chem. Rev.* 306: 302–330.
176 Evtushok, V.Y., Ivanchikova, I.D., Podyacheva, O.Y. et al. (2019). *Front. Chem.* 7: 858.
177 Evtushok, V.Y., Ivanchikova, I.D., Lopatkin, V.A. et al. (2021). *Catal. Sci. Technol.* 11: 3198–3207.
178 Dhakshinamoorthy, A., Alvaro, M., and Garcia, H. (2011). *Chem. Eur. J.* 17: 6256–6262.

11

Tailoring the Porosity and Active Sites in Silicoaluminophosphate Zeolites and Their Catalytic Applications

Jacky H. Advani, Abhinav Kumar, and Rajendra Srivastava

Indian Institute of Technology Ropar, Catalysis Research Laboratory, Department of Chemistry, Rupnagar, 140001, Punjab, India

11.1 Introduction

Since the discovery of aluminophosphates (AlPOs) in the late seventies, much research has been dedicated to their discovery with microporous open-framework structures. By far, 312 open-framework AlPO structures have been reported [1]. Although alike, the synthetic chemistry of zeolite and AlPO is very different. The structure of AlPO comprises PO_4 tetrahedra (associated with bridging O atoms), and the Al site has variable coordination numbers (4 to 6) depending on the AlPO conformation [2]. In AlPOs, octahedral Al sites are found dominant. However, the tetrahedral Al sites are found to dominate when the O/P ratio is high. The Al polyhedra form Al–O–P bonds. Compared to aluminosilicates, AlPO-based materials are available in a large variety of topologies due to the ability of Al to attain different coordination sites. Incorporating Si/metal, via substitution of Al/P atoms, in AlPOs can aid in generating chemical activity [1]. For instance, the silicoaluminophosphate (SAPO) molecular sieves possess Brønsted and Lewis acid sites which aid in catalyzing different organic transformations while the metal-substituted AlPOs can catalyze the redox-based reactions.

The acidic SAPO-based molecular sieves have been extensively reported for acid-catalyzed reactions like methanol to olefin synthesis, *n*-alkane cracking/hydrocracking, alkane dehydrogenation, and various other organic transformations. The acidity arises from three different mechanisms as illustrated in Figure 11.1 [3]. The first mechanism involves the Si insertion at the P site, which results in the formation of a negatively charged framework. The protons attached to the Al–O–Si bridges counter this negative charge. Another mechanism involves the Al substitution by Si. However, such substitution forms a highly energetically unfavorable Si–O–P bond (not observed experimentally) [4]. The third mechanism, which is very popular, involves the simultaneous substitution of P and Al by Si atoms which results in the formation of Si islands within the AlPO framework. The bond lengths, bond angles, and coulombic potential within zeolite cages and around acid centers

Catalysis in Confined Frameworks: Synthesis, Characterization, and Applications, First Edition.
Edited by Hermenegildo Garcia and Amarajothi Dhakshinamoorthy.

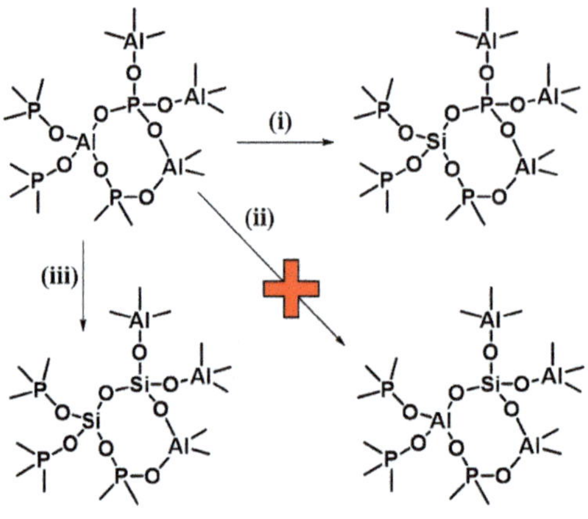

Figure 11.1 Different mechanisms of Si insertion in AlPO.

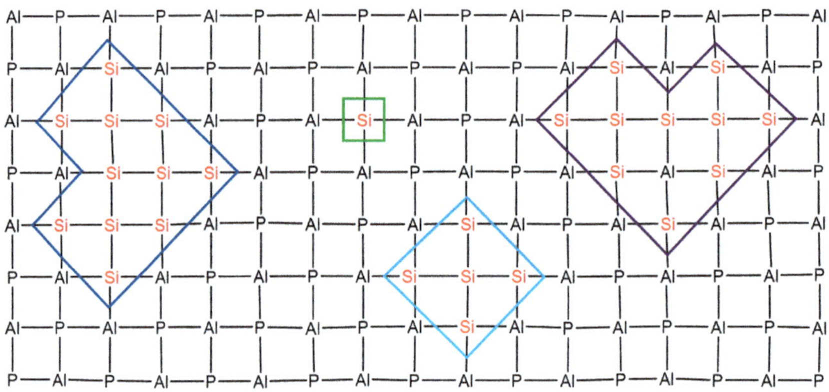

Figure 11.2 Distribution of Al, Si, and P in SAPO framework with isolated island (marked in green), 5Si island (marked in cyan), 11Si island (marked in blue), and Si/Al phase with one Al in Si island (marked in violet) [5, 6].

control the acidity of these materials. Figure 11.2 illustrates the distribution of Si, Al, and P in SAPO framework. The Si islands generated provides a high number of acid sites (due to the formation of Si(nAl) species where $n < 4$ on the border of islands) in the SAPOs, contrary to other aluminosilicates [7, 8]. The density of these islands is directly proportional to Si atoms along with the topology largely governing the acidity of these SAPOs.

The Si distribution in such materials is dependent on the number of framework charges which is again affected by the number of SDA molecules within cages and Si atoms introduced in the framework. The organic structure directing agents "SDA" not only acts as porogen but also influences the crystal size, chemical composition, and crystallization evolution of the SAPOs. Thus, the type of SDA used for the synthesis of SAPOs also plays a vital role in their physicochemical properties. Apart from the SDA used for preparing the SAPOs, various additives such as surfactants, ionic liquids, and silanes have been utilized to generate mesoporosity and synthesize hierarchical SAPOs. The physicochemical properties of the SAPOs are also influenced by the synthetic methodology. For instance, the microwave-assisted synthesis of the SAPOs shows high crystallization speed, high crystal purity, smaller particle diameter, and uniform grain distribution than that synthesized via conventional hydrothermal treatment.

This chapter discusses the methods to tailor the porosity and/or acidity of different SAPOs and their catalytic applications in various organic transformations. A brief background of these phosphate materials, different synthetic methodologies, and characterizations are provided. The catalytic applications of these SAPOs are discussed, particularly in acid catalysis and reductive transformations such as selective catalytic reduction (using ammonia/urea and hydrocarbon), hydroisomerization, and hydroprocessing. The catalytic activity of different SAPO-based composites in the CO_2 hydrogenation reaction is also covered in this chapter.

11.2 Synthesis of SAPO-*n* Zeolites

Conventionally, the synthesis of SAPOs has been reported using various methods such as hydrothermal synthesis, two-step crystallization, ionothermal synthesis, dry gel method, extremely dense system, ultrasonic synthesis, and microwave synthesis (Figure 11.3) [9–14]. Each method involves using Al source, Si source, and P source for synthesizing SAPOs.

Hydrothermal synthesis, widely used for the synthesis of SAPOs, typically involves structure directing agents (also known as templating agents) and deionized water along with the Si, Al, and P source. The general steps of the synthesis include (a) preparation of the initial gel using Si/Al/P/SDA/H_2O in a particular composition, (b) aging of the gel, (c) hydrothermal treatment at the desired temperature, and (d) removal of the template by calcination at a desired temperature. Table 11.1 shows the formation of various SAPO molecular sieves with their topology/framework and the type of SDAs used for respective synthesis. The two-stage crystallization method is an extension of hydrothermal synthesis. In this synthesis, the initial gel is aged and then hydrothermally treated at a lower temperature for a certain time. After completing this hydrothermal treatment, the gel is again diluted with a certain amount of water and hydrothermally treated at a higher temperature, resulting in the formation of uniform and smaller-sized SAPOs. The modified hydrothermal synthesis utilizing the use of eutectic mixtures or ionic liquids as reactants/solvents are termed ionothermal synthesis. The ionic liquids/eutectic solvents aid in the synthesis of SAPOs with a varied structure providing a tunability to produce a specific type of pores.

As the name suggests, the dry gel method involves mixing Si/Al/P sources and H_2O at a specific temperature to form a gel that is dried and powdered using milling. The dry powder is then placed between the crystallizing autoclave, while the aqueous solution is poured at the bottom. After a particular time of crystallization, SAPO molecular sieves are obtained. Next, the extremely dense system involves the synthesis of SAPOs by the addition of raw materials without extra water. The raw materials are mixed in a particular order and proportion followed by crystallization in the

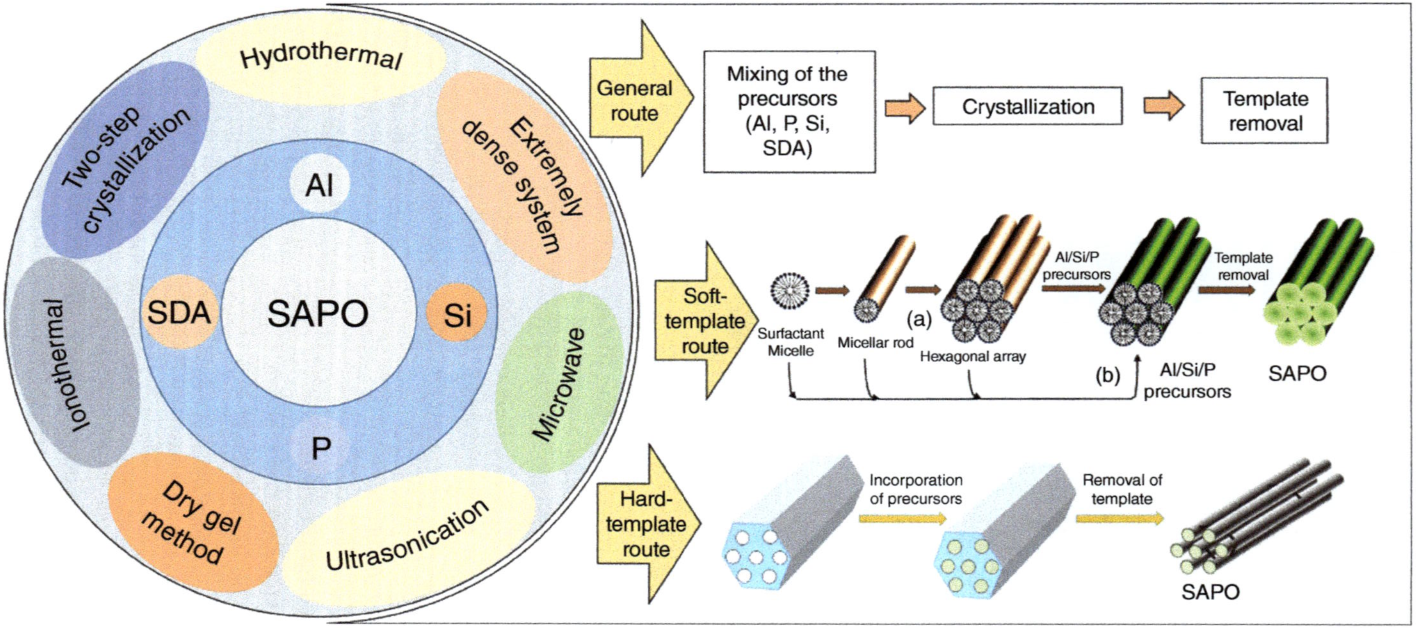

Figure 11.3 General routes for preparing various SAPO molecular sieves.

Table 11.1 Family of SAPOs with diverse framework structures synthesized by several SDAs.

E. no.	SAPO-*n*	Framework/ topology	SDA	References
1.	SAPO-5	AFI	DEA	[15]
			TEA	[16, 17]
			DIPA	[18]
			HMI	[19]
			DBA	[20]
			TPA	[17]
			MDEA	[21]
			DPA	[22]
			TEAOH	[23, 24]
			MDCHA	[25]
			TEED	[26]
2.	SAPO-11	AEL	DEA	[27]
			DIPA	[27]
			DBA	[28]
			DPA	[16, 17, 27, 28]
3.	SAPO-16	AST	HMI	[19]
			Qui	[29]
4.	SAPO 17	ERI	CHA	[30]
			Qui	[29]
			TMHD	[31]
5.	SAPO-18	AEI	TEAOH	[24]
			TMPOH	[32]
			DIPEA	[33, 34]
6.	SAPO-20	SOD	TMAOH	[29]
7.	SAPO-31	ATO	DBA	[35]
			DHA	[35]
			DPA	[17]
			DPeA	[35]
8.	SAPO-34	CHA	nBA	[18]
			DEA	[36]
			TEA	[18, 36, 37]
			DIPA	[18]
			Morp	[18, 36, 37]

(Continued)

Table 11.1 (Continued)

E. no.	SAPO-*n*	Framework/ topology	SDA	References
			DPA	[18]
			TEAOH	[16, 36, 37]
			TMAdaOH	[38]
			Pip	[39]
			TEED	[26]
			MDEA	[21]
9.	SAPO-35	LEV	HMI	[40]
			MDEA	[21]
			CHA	[41]
			Qui	[29]
10.	SAPO-37	FAU	TMAOH/TPAOH	[42]
11.	SAPO-39	ATN	DPA	[43]
12.	SAPO-40	AFR	TPAOH	[44]
13.	SAPO-41	AFO	DEA	[45]
			DIPA	[46]
			DBA	[47]
			DPA	[17]
14.	SAPO-42	LTA	TMAOH	[40]
			MTPQ	[48]
			DDBQ	[48]
15.	SAPO-43	—	iPA	[49]
16.	SAPO-44	CHA	DEA	[15]
			HMI	[50]
			MDEA	[21]
			CHA	[51]
17.	SAPO-46	AFS	DPA	[52]
18.	SAPO-47	CHA	nPA	[53]
			DEA	[10]
			TEA	[54]
			Morp	[54]
			NMBA	[55]
			DPA	[52]
			sBA	[56]
19.	SAPO-56	AFX	TMHD	[31, 40]
20.	SAPO-57	AFV	DEDMAOH	[57]

(Continued)

Table 11.1 (Continued)

E. no.	SAPO-*n*	Framework/ topology	SDA	References
21.	SAPO-59	AVL	ETMAOH	[57, 58]
22.	SAPO-67	—	ETMAOH	[58]
23.	SAPO-69	—	ETMAOH	[59]
24.	SAPO-79	—	DEDMAOH	[59]

DEA: Diethylamine, CHA: Cyclohexylamine, DBA: Dibutylamine, DEDMAOH: diethyldimethylammonium hydroxide, DDBQ: 2,2-dimethyl-2,3-dihydro-1*H*-benzo[de]isoquinoline-2-ium, DHA: Dihexylamine, DIPA: Diisopropylamine, DIPEA: Diisopropylethylamine, DPA: Dipropylamine, DPeA: Dipentylamine, ETMAOH: ethyltrimethylammonium hydroxide, HMI: Hexamethyleneimine, iPA: iso-Propylaimine, MDCHA: *N,N*-methyldicyclohexylamine, MDEA: Methyldiethanolamine, Morp: Morpholine, MTPQ: 4-methyl-2,3,6,7-tetrahydro-1*H*,5*H*-pyrido [3.2.1-ij] quinolinium, nBA: *n*-butylamine, NMBA: *N*-Methylbutylamine, nPA: *n*-Propylamine, Pip: Piperidine, Qui: Quinuclidine, sBA: sec-Butylamine, TEA: Triethylamine, TEAOH: Tetraethylammonium hydroxide, TEED: N1, N1, N2, N2-tetraethylethane-1,2-diamine, TMAOH: Tetramethylammonium hydroxide, TMAdaOH: *N,N,N*-trimethyl-1-adamantylammonium hydroxide, TMHD: N1, N1, N6, N6-tetramethylhexane-1,6-diamine, TMPOH: 1,1,3,5-tetramethyl piperidin-1-ium hydroxide, TPA: Tripropylamine, TPAOH: Tetrapropyl ammonium hydroxide.

autoclave to produce SAPOs. Furthermore, the use of ultrasonic waves to mix the initial gel composition gives rise to the ultrasonic method of SAPO synthesis. In this method, the ultrasonic waves are transmitted to produce cavitation, facilitating the mass transfer, and making it uniform in a shorter time. The gel after ultrasonic treatment is then crystallized to produce SAPO molecular sieves. Similarly, in microwave synthesis, microwaves are used to provide the necessary temperature via collision of polar molecular sieves producing evenly dispersed gel with a dense crystal nucleus. This method has added the advantage of high crystallization speed, high crystal purity, smaller particle diameter, and uniform grain distribution.

Apart from the conventional SAPO molecular sieves, several methods have been developed for synthesizing hierarchical SAPOs. Such materials possess high surface area and microporous structure which can aid in several organic transformations. The synthesis of hierarchical pores can be realized via a template or posttreatment method. The template method can be further sub-classified into two categories (i) in-situ synthesis method and (ii) nano-assembly. The in-situ method utilizes a hard or soft template for the generation of medium and large pores along with the simultaneous growth of microporous SAPOs. However, in the nano-assembly method, the nanocrystals of the SAPO are initially grown followed by their arrangement around the template in an orderly fashion. This method involves the formation of SAPO crystals/seeds that arrange around the template to form a multilevel pore structure. The posttreatment methods for synthesizing hierarchical zeolites involve various routes of modifications such as acid treatment, alkali treatment, high-temperature steam treatment, and fluoride etching [60–64]. The acid treatment method and the high-temperature treatment method remove the aluminum from the SAPO framework producing mesopores while the alkali treatment removes the framework Si, producing hierarchical pores. In fluoride etching, HF or NH_4F solutions are used to preferentially etch the defect area of the crystal, penetrating deeply to form a mesoporous zeolite structure. Among the various post-treatment methods, the generation of the mesopores in the SAPO can be realized via the acid treatment method, high-temperature treatment, and fluoride etching method.

11.3 Characterization of SAPO Zeolites

The introduction of the mesopores largely addresses the diffusion limitations and hence the ease of access to the acid sites in the micropores of the zeolite. The shape, size, and mesopores connectivity can be understood as the effect of textural properties on catalytic performance. The bulk textural properties of the zeolites are generally studied using gas adsorption [65], mercury porosimetry [66], thermoporometry [67], and NMR [68]. These physicochemical characterizations provide information relating to the shape, size, and ease of access of the pores. However, the catalytic activity is largely dependent on the interconnectivity of the various type of pores. The complex network of the micro-, meso-, and macropores facilitate the reactants/products diffusion toward/from the active catalytic sites. Hence, advanced image and microscopy analytical techniques have been used to understand the pore network in the catalyst. These techniques include simple confocal microscopy [69], electron microscopy (SEM [70], FIB-SEM [71], TEM [72]), electron tomography [73], and X-ray electron tomography. The probe used in these microscopy techniques helps in visualizing the porosity of the SAPOs from microscopic to macroscopic scales. Temperature-programmed desorption (TPD) using ammonia as the probe molecules and the pyridine- Fourier transform infrared spectroscopy (FTIR) techniques provide information relating to the acidity of the catalysts [74].

11.4 SAPO-Based Catalysts in Organic Transformations

SAPOs are of high importance, both scientifically and technologically, owing to the tunable acidity, high surface area, and porosity which aids in improved diffusion and superior catalytic performances in organic transformations.

11.4.1 Acid Catalysis

The direct utilization of high-viscosity oils is not suitable for conventional combustion engines. Hence upgradation of these free fatty acids (FFAs) is necessary. Transesterification of these FFAs has been identified as one of the potential

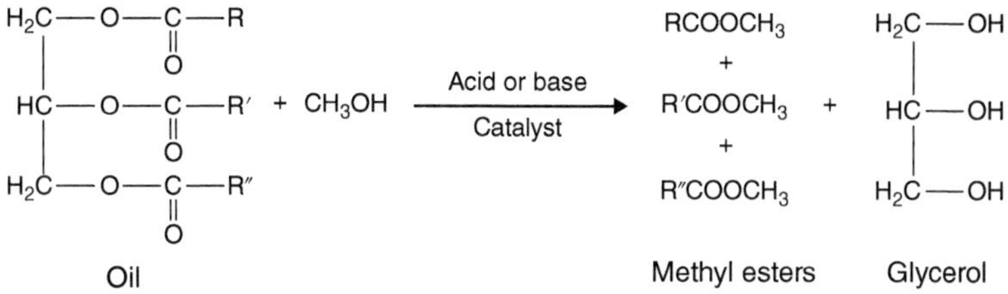

Scheme 11.1 Transesterification of oil to produce fuels.

methods to produce such high-end products to be used as fuel (Scheme 11.1) [75]. The synthesis of SAPO-34 was demonstrated by utilizing carbon-based secondary hard templates via ultrasonic-assisted synthesis and the catalyst was evaluated for the transformation of triglyceride to biodiesel [76]. It was observed that the use of a secondary template gave an increment of the SAPO pore size. These large pores aid in overcoming the diffusion barriers for the large FFA molecules. The catalyst synthesized using ultrasound-assisted synthesis showed larger porosity and hence higher biodiesel formation (83% oleic acid conversion) than the traditional SAPO-34. To increase the acidic strength of the catalyst, 5% CeO_2 was accommodated on the SAPO-34 catalyst. This composite catalyst gave a 94% conversion. The reaction rate constant was found to increase by elevating the pore radius and catalyst acidity, which resulted from the ease of penetration of the large oil molecules. The FFA transesterification over the SAPO-34 took place in seven sequential steps (Figure 11.4a). The first step involves the diffusion of reactant toward the pores of the catalyst, followed by diffusion in the pores. The reactants are then absorbed on the catalyst surface in the third step. These adsorbed reactants react on the active catalytic sites, and the resulting products desorb from the surface. The products then diffuse from the pore toward the surface in the last two steps. The mechanism of the catalytic reaction involves the formation of a carbocation via protonation of the carbonyl group from the FFA, followed by an attack by alcohol on the protonated carbon over the acid sites of the catalyst (Figure 11.4b,c).

The production of platform chemicals such as 5-HMF and furfural, precursors to several important fuels and chemicals, from cellulose/hemicellulose has been a central point of interest due to limited fossil fuel resources [77]. Wang and coworkers developed a series of Sn/SAPO-34 catalysts for glucose dehydration to 5-HMF (Scheme 11.2a) [78]. The report demonstrated that the Sn loading variation facilitated acid strength tuning of the Sn/SAPO-34 composite. The mechanistic pathway over the Sn/SAPO-34 is shown in Figure 11.5. The hydroxyl group on glucose is adsorbed on the –Cl group on the catalyst via hydrogen bonding. The adsorbed glucose is isomerized over tetrahedrally coordinated Sn^{4+} (generated via Sn incorporation) and Al^{3+} (available in the SAPO-34 support) Lewis acid sites to fructose, which showed the involvement of the H-bonding and Lewis acid sites in isomerization process. The Brønsted acidity facilitated the dehydration steps leading to HMF. The optimized catalyst gave a 98.5% glucose conversion with 64.4% 5-HMF yield in a NaCl-H_2O/THF biphasic system at 150 °C in 1.5 hours. Albeit the performance, the catalyst was prone to Sn leaching, thus decreasing the activity of the catalyst.

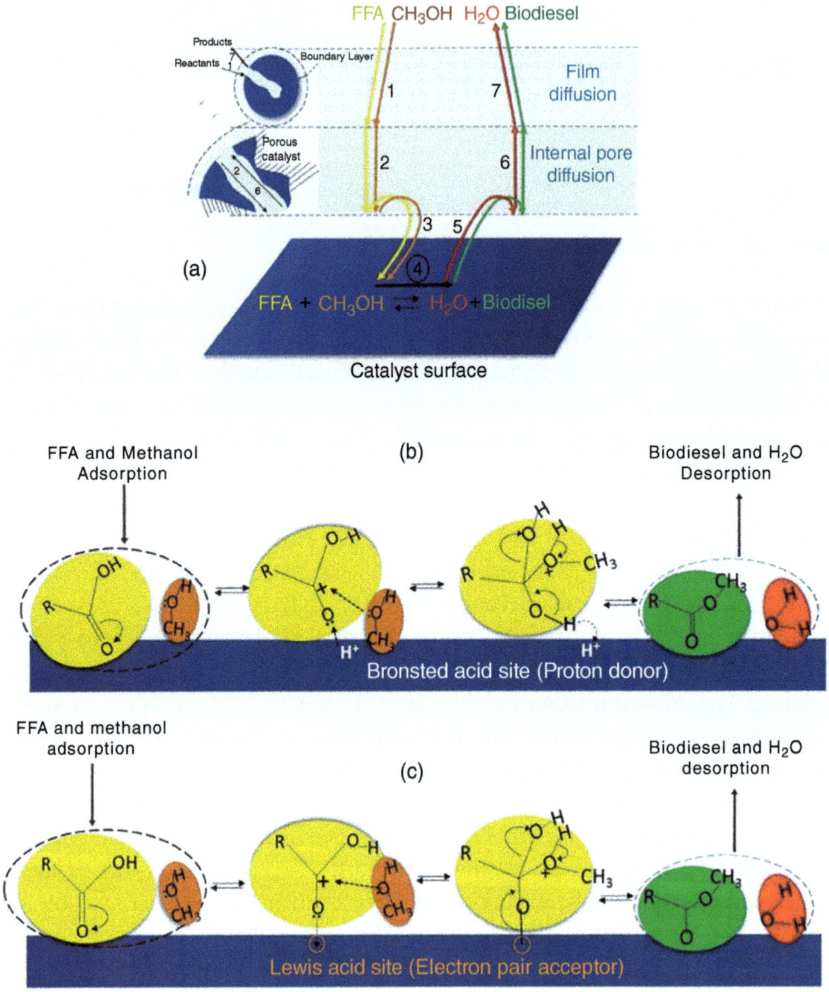

Figure 11.4 Proposed pathway for biodiesel production using Ce/SAPO-34 catalyst. Source: Reproduced from Ebadinezhad and Haghighi [76] with permission from Elsevier, Copyright 2020.

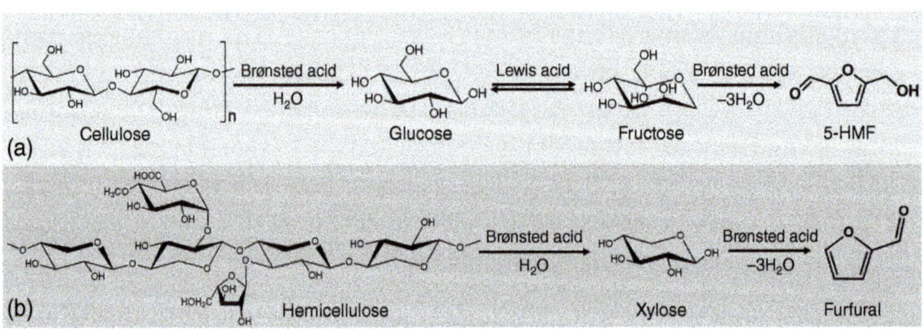

Scheme 11.2 General synthesis of 5-HMF and furfural from biomass components.

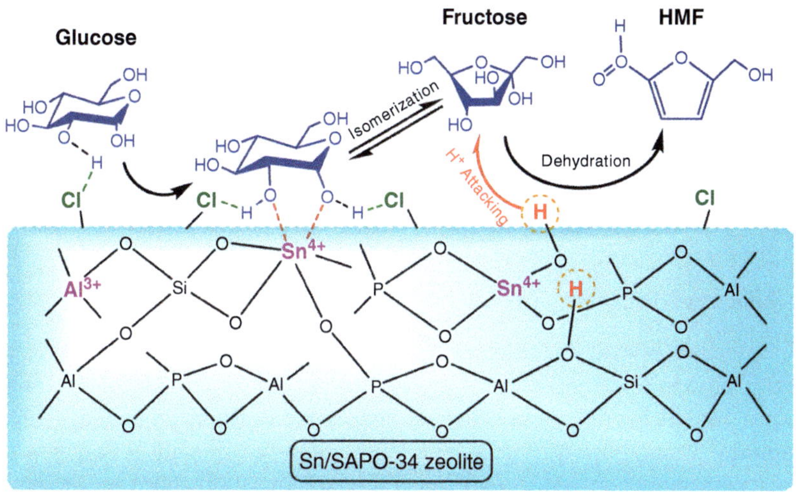

Figure 11.5 Catalytic dehydration of glucose to 5-HMF over Sn-SAPO-34 catalyst. Source: Reproduced from Song et al. [78] with permission from American Chemical Society, Copyright 2021.

Zhang and coworkers demonstrated the excellent catalytic activity of the sole SAPO-34 catalyst under a GVL/water mixed solvent [79]. The catalyst afforded complete glucose conversion with 93.5% 5-HMF yield at 170 °C in 40 minutes in GVL/water solvent. The high activity of the catalytic system was attributed to the acidity of the catalyst and the mixed aqueous-organic solvent phase. Water facilitated the formation of 5-HMF (contrary to traditional catalytic systems), whereas the acid sites facilitated the isomerization of the glucopyranose to fructopyranose. Along similar lines, a MgFe/SAPO-5 catalyst was utilized for the production of 5-HMF from fructose [80]. The optimized catalyst was found to be rich in Brønsted/Lewis acid sites, giving a high catalytic performance (73.9% HMF yield at 170 °C in 2 h). It was observed that by elevating the Si/Al ratio, the 5-HMF yield also increased, which suggests that the catalyst with higher acidity (acid sites) was highly active in the dehydration reaction. Another catalytic application of SAPOs, where the acidic properties play an important role, is in synthesizing furfural from hemicellulose (Scheme 11.2b) [81–83]. In these lines, zeolite 5A beads were used to grow SAPO-34 crystals and the composite was evaluated for the production of furfural and 5-HMF from xylose and glucose, respectively [81]. The catalyst gave a moderate yield of furfural (45%) and 5-HMF (20%). The catalyst was found to be more selective toward the furfural/5-HMF than dilute sulfuric acid which generates levulinic acid as a byproduct. To improve the selectivity of the products, a SAPO-44 molecular sieve-based solid acid catalyst was developed for processing isolated hemicelluloses in a biphasic solvent (toluene-water) system [82, 83]. The catalyst afforded 86–93% furfural yields from raw biomass. The direct conversion of isolated hemicellulose gave 82% furfural yield. The hydrophilic character of SAPO-34 catalyst makes it unavailable for side reactions of furfural.

On a similar note, Huang and coworkers generated high Brønsted acidity on hierarchical SAPO-34 catalyst via the post-synthetic method and utilized it to dehydrate glycerol to acrolein [84]. The synergy between the Brønsted acid sites and hierarchical pore gave an excellent acrolein yield (89.8%) at 345 °C and WHSV of 3.7 h^{-1}. The synergy between acid sites and pores was responsible for the excellent activity and retardation in the catalytic deactivation, which is most common during glycerol dehydration.

11.4.2 Reductive Transformations

11.4.2.1 Selective Catalytic Reduction (SCR)

The catalytic process that converts the nitrogen oxides (NO_x) into molecular nitrogen (N_2) and water using different reductants such as hydrocarbon, ammonia, urea, methyl amines, ammonium carbonate, and ammonium bicarbonate is known as selective catalytic reduction [85]. These oxides emitted from automobiles, power stations, and industries are a major source of atmospheric contaminants that largely affects the troposphere inclusively giving rise to the greenhouse effect, photochemical smog, and acid rains [86, 87].

11.4.2.1.1 Ammonia/Urea SCR

Ammonia-SCR is identified as one of the most effective tools for NO_x reduction. Urea-SCR follows the same chemistry as that of NH_3-SCR. For urea-SCR, urea is converted to ammonia using water as shown by the following equation:

$$(NH_2)_2CO + H_2O \rightarrow 2NH_3 + CO_2 \tag{11.1}$$

Once ammonia is formed, the NH_3-SCR chemistry is followed which is governed by three reactions as shown in equations (11.2-11.4) [88].

$$4NO + 4NH_3 + O_2 \rightarrow 4N_2 + 6H_2O \text{ (Standard SCR)} \tag{11.2}$$

$$NO + 2NH_3 + NO_2 \rightarrow 2N_2 + 3H_2O \text{ (Fast SCR)} \tag{11.3}$$

$$8NH_3 + 6NO_2 \rightarrow 7N_2 + 12H_2O \tag{11.4}$$

The SDA used in the preparation of SAPOs notably affects its physicochemical properties. Thus, the influence of various SDAs on the SAPO properties was studied for NO_x reduction [37, 89, 90]. Different SDAs, namely, morpholine (Morp), diethylamine (DEA), triethylamine (TEA), and tetraethylammonium hydroxide (TEAOH) were utilized for Cu/SAPO-34 synthesis. These catalysts were employed for NO_x reduction. The physicochemical properties of these SAPO-11 molecular sieves are given in Table 11.2. The catalyst synthesized using Morp (SAPO-34-MA) as SDA showed high deactivation and could only regain partial activity during NH_3-SCR. Although the catalysts synthesized using TEA (SAPO-34-T) and TEAOH (SAPO-34-TOH) as SDAs gave a minor deactivation when water was exposed at 70 °C, these catalysts were reactivated to recover their SCR activity. The activity difference between SAPO-34-TEA/TEAOH and SAPO-34-MA can be attributed to their physicochemical properties. The choice of SDA not only affected the distribution

Table 11.2 Physicochemical properties of the synthesized catalysts and their supports.

Catalyst	Templates	SA of SAPO-34 ($m^2\,g^{-1}$)	Desorbed NH_3 ($mmol\,g^{-1}$)	SA of Cu/SAPO ($m^2\,g^{-1}$)	Desorbed NH_3 ($mmol\,g^{-1}$)
SAPO-34-MA	Morpholine	599.8	1.918	615.5	1.168
SAPO-34-DEA	DEA	601.6	2.045	618.5	1.313
SAPO-34-TEA	TEA	594.2	1.403	582.6	0.932
SAPO-34-TEAOH	TEAOH	598.9	1.700	588	0.963

Source: Reproduced from Yu et al. [89] with permission from Elsevier, Copyright 2014.

of Si/Al/P coordination but also showed a variable distribution of two different Cu locations in the catalysts. ^{29}Si NMR confirmed that both TEA and TEAOH templates formed more Si islands, resulting in a decreased number of acid sites yet stronger in nature (Figure 11.6). Keeping the same Si content in all supports, Cu/SAPO-34-TEA and Cu/SAPO-34-TEAOH showed fewer Cu^{2+} sites with higher CuO species than Cu/SAPO-34- MA and Cu/SAPO-34-DEA. Although alteration in the acid properties does not influence the activation energy of the SCR reaction, NO conversion is highly affected by the acid densities [89, 90]. Figure 11.7 depicts the morphology

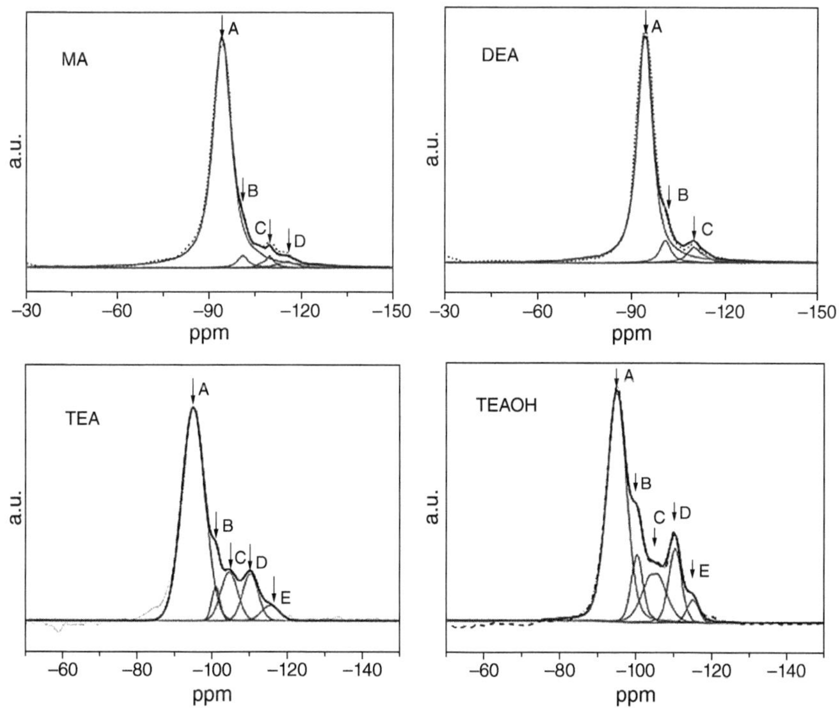

Figure 11.6 ^{29}Si NMR of the SAPO-34 supports synthesized using (a) MA, (b) DEA, (c) TEA, and (d) TEAOH. Source: Reproduced from Yu et al. [89] with permission from Elsevier, Copyright 2014.

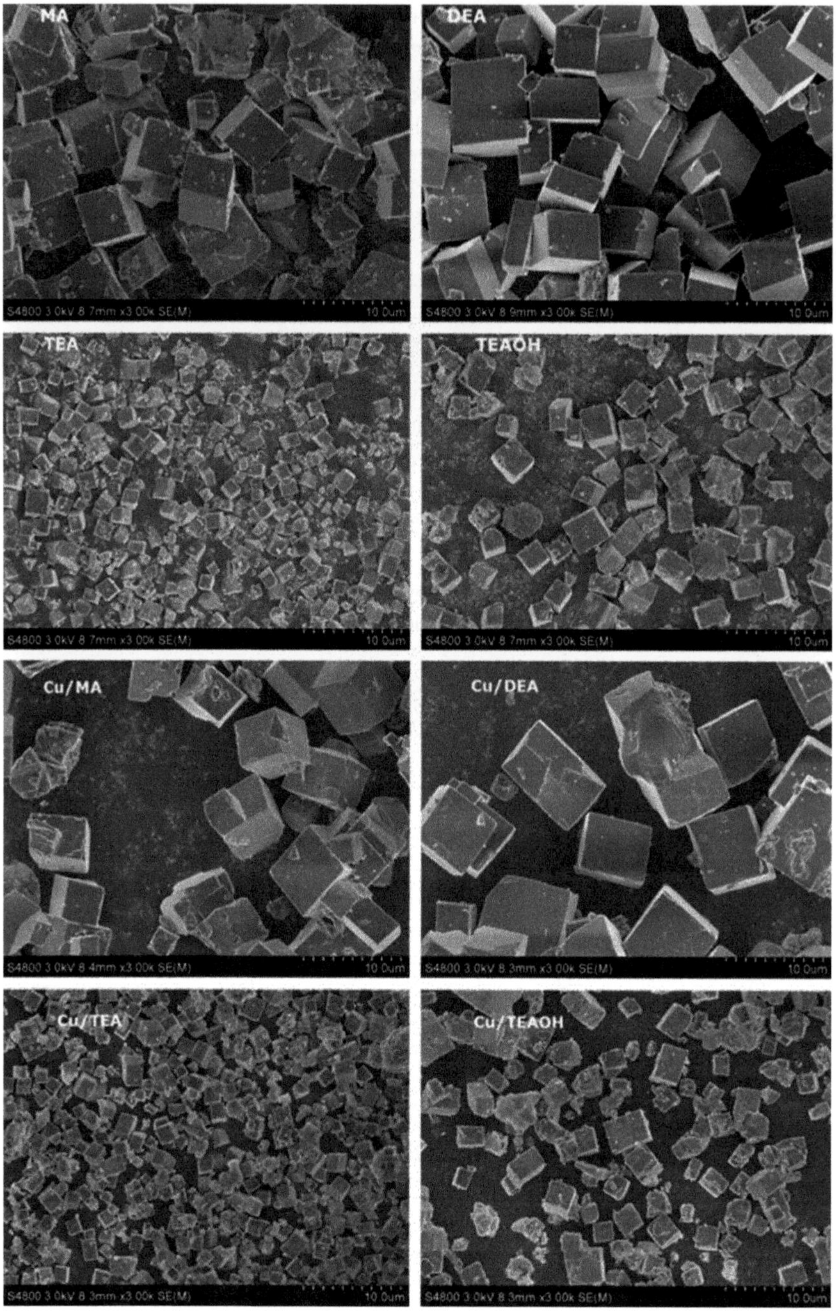

Figure 11.7 SEM images of the SAPO-34 and Cu/SAPO-34 synthesized using different SDAs. Source: Reproduced with permission from Yu et al. [89], ©2014/Elsevier.

of the as-synthesized SAPO-34. Although with varying crystal sizes due to different crystallinity, all the molecular sieves showed cubical structures. No change in morphology was observed when Cu was doped on the SAPO-34 molecular sieves.

Many researchers utilized a direct preparation of Cu/SAPO molecular sieves with uniform Cu species using Cu-amine complex and different SDAs [91, 92]. For instance, Cu/SAPO-18 and Cu/SAPO-44 were prepared using *N*,*N*-dimethyl-3, 5-dimethylpiperidinium and cyclohexylamine, respectively, as SDA [91, 92]. Such methodology allows the synthesis of Cu/SAPO in one pot avoiding the calcination and ion exchange steps and providing a complete framework for Si isolation. The physicochemical characterizations confirmed the existence of isolated Cu^{2+} (extra-framework), stabilized in the large cages and/or in the six-membered chabazite (CHA) ring. These extra-framework Cu^{2+} ions were responsible for the activity of the catalyst. Along similar lines, different Cu/SAPO-18 molecular sieves with variable loadings of copper were developed and evaluated for the SCR of NO using ammonia [33]. A combination of experiments and density functional theory (DFT) calculations were utilized to study the nature of Cu species in the catalyst. As depicted in Figure 11.8, the ammonia dissociation to NH_2 and NO oxidation to NO_2 were the important steps in the ammonia-SCR reaction over SAPO-18 molecular sieves. The complete formation of N_2 occurred during the oxidation of Cu^+ and reduction of Cu^{2+} species. A high NH_3-SCR activity (80% NO conversion) was observed at 200 °C. The DFT studies revealed that the isolated Cu^{2+} sites were positioned in the cavity and showed a preference for adjacent 6R planes of the molecular sieve.

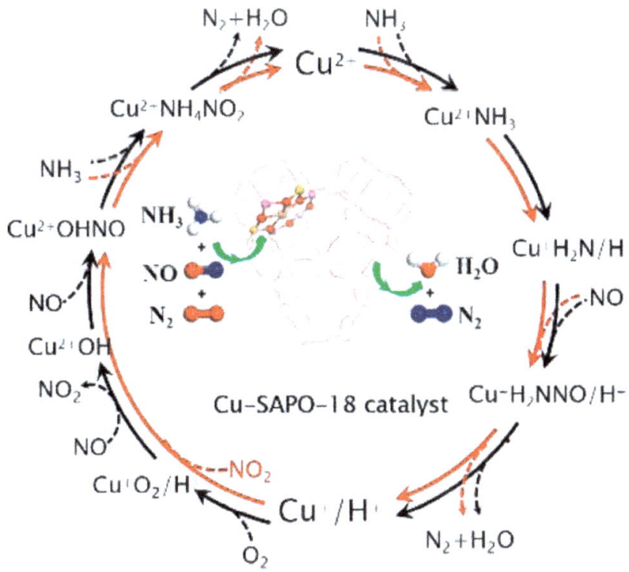

Figure 11.8 Proposed mechanistic pathway for Cu/SAPO-18 catalyzed NH_3-SCR (fast SCR is indicated by red arrows while standard SCR is indicated by black arrows). Source: Reproduced from Li et al. [33] with permission from American Chemical Society, Copyright 2016.

11.4.2.1.2 Hydrocarbon-SCR

SCR of NO_x can be achieved using different hydrocarbons such as octane, *iso*-octane, octanol, octanal, octanoic acid, propane-propene mixture, and dodecene–dodecenal mixture as reducing agents. The HC-SCR for the NO_x decomposition can occur via two different intermediate mechanisms,–CN intermediate mechanism (equations 11.5-11.7) and $-NO_2$ intermediate mechanism (Eq. 11.8–11.9). The –CN intermediate mechanism involves the oxidation of the hydrocarbons to $C_xH_yO_z$ species that react with the NO_x to form the –CN reaction intermediate. This intermediate is then reacted with NO_x under an oxygen atmosphere to produce N_2, CO_2, and water. While the $-NO_2$ intermediate mechanism involves the oxidation of the NO_x to nitrates under excess oxygen. The nitrates react with the activated hydrocarbon to produce N_2, CO_2, and water.

$$HC + O_2 \rightarrow C_xH_yO_z \tag{11.5}$$

$$C_xH_yO_z + NO_x \rightarrow \text{R-NCO (R-CN)} \tag{11.6}$$

$$\text{R-NCO (R-CN)} + NO_x + O_2 \rightarrow N_2 + CO_2 + H_2O \tag{11.7}$$

$$NO_x + O_2 \rightarrow -NO_2\,(\text{nitrates}) \tag{11.8}$$

$$HC + -NO_2\,(\text{nitrates}) \rightarrow N_2 + CO_2 + H_2O \tag{11.9}$$

Ishihara and coworkers utilized thermostable molecular sieves for the selective catalytic reduction of NO_x using propylene as the reductant [16, 93]. Cu-SAPO-*n* ($n = 5$, 11, 34) catalysts were tested for this transformation under oxidizing atmosphere. All the catalysts were active for the C_3H_6-SCR. The maximum NO_x conversion followed the order of SAPO-5 >SAPO-11 >SAPO-34. A maximum NO_x conversion was observed at 300 °C using the Cu-SAPO-34 catalyst under 15% H_2O. The isolated Cu cations (CuO) activated the hydrocarbon at a temperature lower than 400 °C, thus improving the activity [94]. The activity of the catalyst declined with a temperature increase above 300 °C. This was accredited to the nonselective oxidation of the hydrocarbon by CuO sites. Figure 11.9 shows the

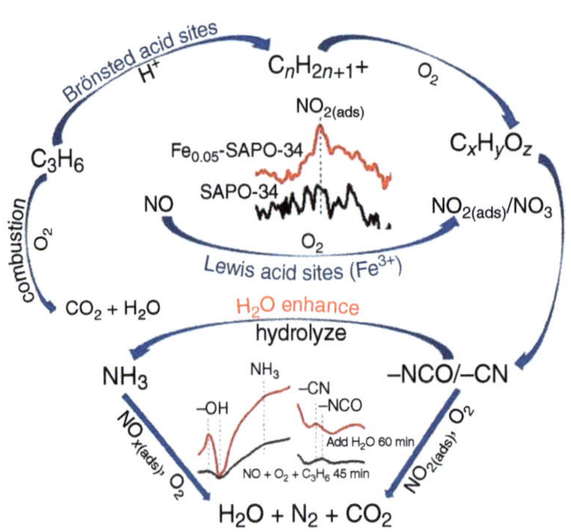

Figure 11.9 Propylene SCR over Fe-SAPO-34 catalyst. Source: Reproduced from Yang et al. [94] with permission from Elsevier, Copyright 2021.

catalytic mechanism for the Fe-SAPO-34 catalyzed propylene SCR. The Lewis acid sites (Fe^{3+}) oxidized the NO to $NO_{2(ads)}/NO_3$ (confirmed by FTIR spectra), while the Brønsted acidity (protons) aided in activating propylene to form acetate/formate species. These acetate/formate react with $NO_{2(ads)}/NO_3$ to form –CN or –NCO, which further react with $NO_{2(ads)}/NO_3$ to form H_2O, CO_2, and N_2. The –NCO species hydrolyze to produce ammonia, thus increasing the NO conversion at a lower temperature. Zhou and coworkers utilized a Fe-SAPO-34 catalyst for a similar reaction [95]. A 100% N_2 selectivity with complete NO conversion was observed at 250–300 °C with 5% H_2O. Again the high activity was attributed to the isolated Fe^{3+} ions and the Brønsted acidity. The Brønsted sites and Fe^{3+} ions facilitated the NO_2 species and carbenium ions adsorption. The presence of water aids in the enhancement of the catalytic activity by hydrolyzing –CN species into ammonia.

11.4.2.2 Hydroisomerization

Hydroisomerization is of high importance in the petrochemical and fine chemical industries. The production of C_4-C_8 *iso*-alkanes and alkenes can be achieved via isomerization and hydroisomerization of light naphtha hydrocarbons using different zeolites having appropriate chemical and structural properties. The tunable surface area, nature of the surface acidic sites (Brønsted/Lewis), and adsorption capacity make zeolites the preferred choice for such reactions [96]. The high surface area of these zeolites provides high surface acid sites that favor hydroisomerization [11].

The introduction of surfactants aids in producing small particles that enhance catalytic activity. Fan and coworkers prepared SAPO-11 with enhanced acidity and small crystallite size via a two-stage crystallization using dodecyltrimethylammonium bromide (Do) as a surfactant [97]. The two-stage crystallization, done by the addition of Do to the preformed AlPO network, gave higher Si border sites compared to the traditional single-stage crystallization. The Do content was varied to obtain the highest number of Si(nAl) species with medium to strong Brønsted acidity. The SAPO-11 formed by the two-stage crystallization gave excellent catalytic activity in the octane hydroisomerization, producing di-branched isomers. On a similar note, the tuning of SAPO-11 using different surfactants was reported for the same reaction. The addition of the surfactants improved the catalytic performance, yielding branched products during the n-octane hydroisomerization [98]. Dodecyltrimethylammonium bromide (Do), decyltrimethylammonium bromide (De), tetradecyltrimethylammonium bromide (T), and hexadecyltrimethylammonium bromide (C) were used as the surfactants. Among them, the SAPO-11 synthesized using dodecyltrimethylammonium bromide gave better catalytic activity with branched products selectivity in order of Pt/SAPO-11-Do >Pt/SAPO-11-De >Pt/SAPO-11-T >Pt/SAPO-11-C. The catalyst Pt/SAPO-11-Do showed moderate Brønsted acidity, high Lewis acidity, and the smallest crystallite size. The octane hydroisomerization over this catalyst is shown in Figure 11.10. The smaller crystallite size increased the catalyst's surface area and thus the accessibility of the Brønsted acid sites, which in turn increased the mono-branched isomerization of two terminal alkyls of n-octane near neighboring pore mouths producing C8 di-branched isomers. Contrarily, the low surface area of the large-sized SAPO-11

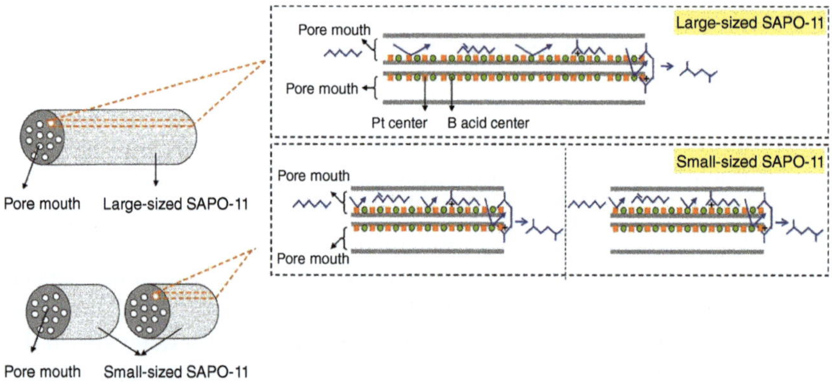

Figure 11.10 The sites of reaction and diffusion of n-octane during the hydroisomerization over Pt/SAPO-11 having large and small SAPO-11 crystallites. Source: Reproduced from Guo et al. [98] with permission from Elsevier, Copyright 2012.

crystallites had fewer Brønsted acid sites near the pore mouth, resulting in lower di-branched isomer selectivity.

Bao and coworkers synthesized hierarchical SAPO molecular sieves with meso and microporous structures using alkylphosphoric acid and small amine as templates to produce di-branched paraffins [99]. The incorporation of the phosphoric acid gave a SAPO-11 catalyst with interpenetrating micropores and mesopores. Analogous to microporous molecular sieves, the hierarchical SAPO-11 possesses higher external surface area and mesopores. This results in the higher acid sites with appropriate Brønsted acid strength, giving a high catalytic performance toward di-branched hydroisomerization product and decreased cracking selectivity. Mériaudeau and coworkers utilized Pt-supported SAPO-5, SAPO-11, SAPO-31, and SAPO-41 for n-octane hydroisomerization [17]. The large-pore Pt/SAPO-5 showed hydrocracking, while the medium-pore SAPOs were selective toward the mono-branched n-octane isomerization products. The selectivity of the medium-pore SAPOs increased in the order of SAPO-11 <SAPO-31 <SAPO-41. The difference in the activity of these catalysts was attributed to the diffusional limit and steric constrictions. Kikhtyanin and coworkers evaluated the influence of crystallinity on the physicochemical properties of Pt-loaded SAPO-31 catalysts [100]. The crystallinity largely controlled the adsorption capacity and the acidic properties of the SAPO-31. The acid sites were found to increase with the growth of the microporous phase of the catalyst. Pt-loaded SAPO-31 with varying crystallinity was used for the n-octane hydroisomerization. The increase in the hydrothermal treatment time increased the catalytic performance and isomerization selectivity, which arise from the acid sites present on the surface of the SAPO-31 crystals.

Furthermore, the alteration in the characteristics and catalytic activity of SAPO-11 was studied using two different templates, i.e. di-*n*-butylamine and di-*n*-propylamine [28]. It was observed that the SAPO-11 formed using di-*n*-butylamine (SAPO-11-B) had a lower concentration of stronger Brønsted acidity than the SAPO-11 formed using di-*n*-propylamine (SAPO-11-P). Pd-supported SAPO-11 was evaluated for the hydroisomerization of *n*-decane. The Pd/SAPO-11-B gave the highest *i*-decane yield (76.8%). The number of exposed metal atoms to Brønsted acid density (C_{Me}/C_{H+}) was higher for Pd/SAPO-11-B than for Pd/SAPO-11-P catalysts.

The addition of the solid templates in the AlPO framework also promotes the formation of the SAPO zeolites with different acid characters and enhanced catalytic performance. The synthesis of hierarchical zeolites was carried out by adding oxidized carbon templates to the initial zeolite gels, which resulted in hierarchical SAPO-11 with both micro- and mesopores [101]. The oxidation of the carbon induces oxygenated groups in the matrix functional groups altering the SAPO-11 gel chemistry. The acidity of the SAPO-11 was found to decrease with the addition of carbon. The Pt-loaded SAPO-11 gave a high selectivity toward monobranched isomers from the n-decane hydroisomerization. It was accredited to the mutual effect of the mesoporosity and decrease in the acidity of SAPO-11 arising from its interaction with the surface functional groups on the carbon template with high pH.

Xiang and coworkers synthesized Pt-supported SAPO-*n* (where *n* is 11, 41A, 41B) for the n-decane hydroisomerization [102]. These catalysts showed better catalytic performance due to their weaker and more suitable acidic strength. Among the various molecular sieves, the Pt/SAPO-41A gave the best hydroisomerization performance. The high activity was attributed to the larger pore dimensions than SAPO-11 and smaller grain sizes compared to the SAPO-41B. The conversion of the *n*-decane ranged from 84% to 92% with selectivity to isomers in the range of 90.5–85.3% over Pt/SAPO-41A at 310–330 °C. The extraordinary performance of the catalyst was accredited to the appropriate acid strength, diverse channel dimension, smaller crystallite size, and suitable ratio of the number of exposed metal atoms to Brønsted acid density.

SAPO-11 catalyst with a high external surface area and higher number of acid sites was synthesized in ethylene glycol and supercritical carbon dioxide and evaluated for the 1-octene hydroisomerization to dimethylhexanes [103]. The catalyst showed a superior selectivity toward di-branched isomers during hydroisomerization. The superior catalytic activity was accredited to the smaller particles with high surface area and higher acid sites that arise from the promoting effect of $ScCO_2$ in ethylene glycol on nucleation rate and the number of crystal nuclei in the SAPO-11. As shown in Figure 11.11, 1-octene penetrates the micropores of the molecular sieve catalyst and is absorbed over the Brønsted acid sites. The small pore size of the Pt/SAPO-11 (0.63 nm) resisted the di-branched isomer formation (having a molecular size of 0.71 nm). Thus, the di-branched isomers were formed on the pore mouth of the SAPO-11 catalyst with a larger external surface area that had high Brønsted acid sites, as supported by the earlier report [89]. The influence of the crystallization temperature and time on the physicochemical properties of the SAPO was studied by conventional heating and microwave heating sources [104]. The microwave-assisted synthesis gave pure and highly crystalline SAPO-11 in high yields within a short interval of time. Also, the synthesized SAPO-11 showed an increase in external surface area and mesoporous structure. The Si source used for the synthesis of SAPO-11 also affected its acidity. Reactive Si sources gave high Brønsted acid density while the lesser reactive Si sources gave less acidic SAPO-11.

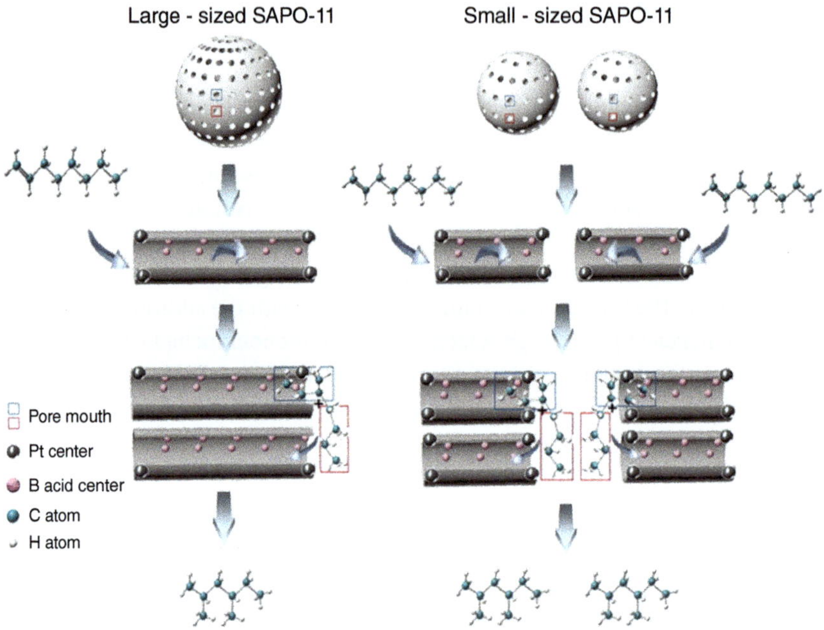

Figure 11.11 Diffusion pathways and locations for hydroisomerization of *n*-octene over small/large SAPO-11 particles. Source: Reproduced from Wen et al. [103] with permission from Elsevier, Copyright 2017.

The modulation of the pore structure to obtain excellent catalytic performance for long-chain alkane hydroisomerization was investigated [105]. Dual-additive methodology involving carbon and butanol was utilized to modulate the porosity and acidity of the SAPO-11. This not only produced a mesoporous SAPO-11 with a high external surface area having suitable mesopores distribution but also enhanced the Brønsted acidity and strength of the support. The optimized Pt-SAPO-11 performed excellently on various n-alkanes (n = C_6, C_8, C_{10}, C_{12}). The catalyst showed a high i-C_{16} yield (90.7%) on n-C_{16} hydroisomerization with high catalytic performance over 1700 h. Similarly, a hollow tubular NiMo/SAPO-11 catalyst with moderate acidity was utilized for the n-alkane (C_{15}–C_{18}) hydroisomerization and hydrocracking [106]. The catalyst afforded a 97.2% conversion with 81.6% jet fuel selectivity. The reaction temperature had a vital role in promoting isomerization/cracking.

The effect of temperature was profound with the carbon atom number in n-alkanes. Long-chain alkanes displayed a favorable performance at a lower temperature, while the shorter-chain alkanes displayed a contrary trend. The excellent catalytic activity was attributed to the electronic tuning between Ni and Mo metals supported on SAPO-11 [107]. It was demonstrated that the highly dispersed Ni in close contact with the MoO_2 facilitated a transfer of electrons from Mo to Ni particles. The high electron density particles stimulate H activation and desorption. Moreover, the well-dispersed Ni-Mo species facilitated to expose the Brønsted acid sites on the support that ultimately facilitated the isomerization. The optimized catalyst 3Ni-0.5Mo/SAPO-11 showed a high catalytic performance giving an 81.4% isomer yield.

11.4.2.3 Hydroprocessing

The catalytic upgradation of various oils to produce diesel has been a topic of interest in research and industry. Hydroprocessing involves methods such as hydrogenation, decarboxylation/decarbonylation, hydrodeoxygenation, and isomerization. The porous nature, high surface area, Brønsted acidity, and chemical & hydrothermal stability of the SAPO molecular sieves make them a potential material/support for such transformations. Different metals can be supported/anchored on the SAPOs to tailor the site ratio for different catalytic functions. It is often observed that the juxtaposition of metal and acid sites synergistically favors the activity and selectivity of the bifunctional catalysts [108, 109].

Qian and coworkers have extensively worked on the SAPO-11 catalyzed hydrodeoxygenation reaction. The catalytic activity of the synthesized molecular sieves was evaluated for the HDO of molecules such as methyl palmitate, methyl laurate, palm fatty acid distillate, and even Jatropha oil [110–113]. A one-step oxygenation and isomerization of the methyl palmitate were carried out using $NiMo_{3-x}$/SAPO-11 catalyst to higher isomer-containing hydrocarbons [110]. The reduction of temperature played a vital role in the deoxygenation, isomerization, and deoxygenation pathway. A low deoxygenation activity yet the highest isomerization selectivity was observed for the catalyst reduced at 400 °C. This was ascribed to the higher amount of acid sites (Mo^{5+}) in the catalyst reduced at 400 °C. With increasing temperature, the concentration of Mo^{5+} species decreased, thus giving lower isomerization selectivity. While the Ni-catalyzed C–C bond cleavage (decarbonylation) occurred with the catalyst reduced at 400 °C, the C=O cleavage was favored when the catalyst was reduced at high temperature. Thus, to maintain a balance between isomerization selectivity and deoxygenation activity (observed from the synergic effect of Ni metal and Mo^{4+} species), the catalyst was reduced at 450 °C. The same group reported the stabilization of bimetals like Sn and Pt over SAPO-11 for the simultaneous methyl palmitate HDO and isomerization of the formed hydrocarbons [111]. The bimetallic catalyst was highly active in the C–C cleavage, forming C_{15} hydrocarbons extensively compared to the Pt/SAPO-11. PtSn/SAPO-11 was also highly active in the C=O hydrogenation giving the highest performance (four times higher than Pt/SAPO-11). A similar catalyst was used for the hydroisomerization–deoxygenation of the palm oil distillate [112]. It was observed that the mesoporous structure of the catalyst enhanced the HDO activity since the Brønsted acid sites influenced the active sites, while the highly dispersed Pt facilitated *n*-hexadecane and *n*-octadecane formation.

Liu et al. performed an HDO of palm oil to produce fuel using nano Ni/SAPO-11 catalysts in a single step [114]. As reported earlier, the methodology utilized a surfactant *n*-tetradecanoamine to produce a nano-sized SAPO-11 (20–50 nm). The as-synthesized catalyst had large external surfaces and mesopores volumes. The Ni/SAPO-11 catalyzed hydroisomerization has a high liquid hydrocarbon yield (80%), with high isomerization selectivity (80%), and outstanding stability. It was

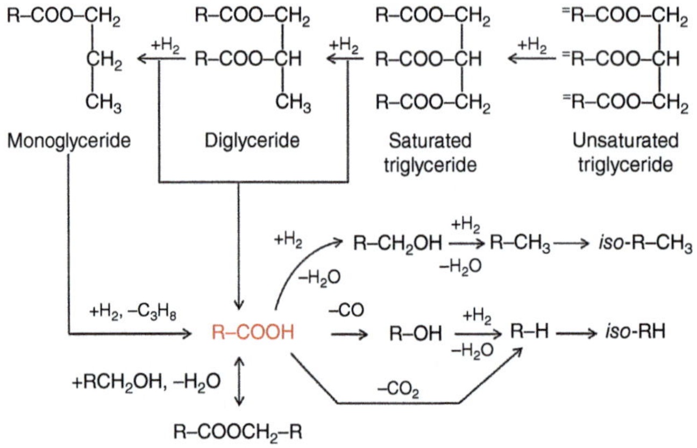

Figure 11.12 Pt/SAPO-11 catalyzed hydroconversion of Jatropha oil; $R^{=}$: $C_{15}H_{29}$, $C_{17}H_{33}$; R: $C_{15}H_{31}$, $C_{17}H_{35}$. Source: Reproduced from Chen et al. [113] with permission from Elsevier, Copyright 2013.

attributed to the suitable acidity, high Ni dispersion, proper ratio of the number of exposed metal atoms to Brønsted acid density, and fast diffusion of bulky palm oil/products.

Different Pt-supported SAPO-11 molecular sieves with varying Si/Al ratios were prepared for the Jatropha oil hydroconversion in a fixed-bed flow reactor [113]. The increase in the Si/Al ratio increased the isomerization activity as a result of medium acidic sites. The hydroconversion of the methyl oleate over the SAPO-11 molecular sieves is depicted in Figure 11.12. The first step involves the hydrogenation of unsaturated methyl oleate to methyl stearate. The methyl stearate is hydrogenolyzed to stearic acid over Pt sites. Subsequently, the stearic acid is deoxygenated via one of the three routes: decarbonylation, decarboxylation, and hydrodeoxygenation. The hydrodeoxygenation route involved the hydrogenolysis of stearic acid to octadecanol, followed by hydrogenation producing octadecane, whereas the decarbonylation route involves the stearic acid decarbonylation to form heptadecanol, followed by hydrogenation to produce heptadecane via decarboxylation. The heptadecane and octadecane further isomerize to produce branched isomers. The catalytic performance depended on the Pt sites, and the 3% Pt/SAPO-11 gave high activity in the deoxygenation and isomerization reaction giving an iso-C_{15-18} hydrocarbons yield of 83% under LHSV of 0.5 h^{-1}. Similar results were achieved by a different group while working on a one-pot hydroconversion of soybean oil [115]. However, the SAPO-11 support shows lower hydrothermal stability during vegetable oils hydrotreatment [116]. The catalyst was deactivated due to the desilication that resulted in the loss of acidity and thus the catalytic performance of the catalyst. However, the addition of an amine surfactant during the crystallization of the SAPO-11 increased the hydrothermal stability of the catalyst.

Sinha and coworkers utilized a stable sulphided NiMo and NiW anchored by hierarchical SAPO-11 for hydroprocessing sunflower oil in a fixed-bed reactor [117]. The hierarchical catalyst effectively offered a high hydrocarbon yield (~84%)

with 40% aviation fuel, 40% diesel, and 20% lighter gasoline range hydrocarbons. The formation of the aromatics signifies that an aromatization reaction also occurs along with the deoxygenation, cracking, and isomerization reaction over the catalyst. Smirnova and coworkers studied the effect of calcination temperature on the physicochemical properties of the Pt/SAPO-31 catalyst for the sunflower oil hydroconversion to green diesel [118]. Despite the identical catalytic composition, Pt/SAPO-31 sieves were affected in terms of Pt distribution and surface acidity by the calcination temperature.

A bimetallic CuNi-supported hierarchical SAPO-11 catalyst was synthesized using *N,N*-dimethyl-*N*-octadecyl-*N*-(3-triethoxysilylpropyl) ammonium bromide as a template/mesoporogen and evaluated for the glycerol hydrogenolysis to produce propanediol [119]. The anchored metals occupied the pore channels and/or acid sites with smaller particles that are aggregated to form flower-like crystallites. The bimetallic catalyst showed higher activity in the glycerol hydrogenolysis than the monometallic Cu-SAPO-11 sieves. The mesoporous catalyst removes the diffusion barriers, while the acidic sites reduce the byproduct formation. The outstanding catalytic performance was attributed to the highly dispersed fine Cu metal particles formed in the bimetallic alloy catalyst, which in turn enhances the Ni reducibility.

11.4.2.4 CO_2 Hydrogenation

The increasing CO_2 content in the atmosphere has led to global warming and increased ocean acidity. Hence, efforts toward its mitigation are highly essential. CO_2 can be utilized as a nonconventional renewable source to produce value-added fuels and chemicals. Albeit stable in nature, CO_2 utilization can be easily realized using H_2 as a co-reactant. Herein, the influence of porosity/acidity of the SAPOs on CO_2 transformation is discussed. The tandem transformation of CO_2 to selectively produce lower olefins is of great importance. Many researchers have synthesized composites based on metal oxides and SAPO-34. Both these components have an individual function during the tandem transformation. Each report claimed that the oxygen vacancies present in the metal oxide were accountable for CO_2 adsorption and activation, while SAPO-34 zeolite provided the necessary acidity for the conversion of formed intermediates (methoxide/formate) into light olefins [120–123].

Mou et al. utilized a composite of binary oxide (Mn_2O_3-ZnO) and SAPO-34 as bifunctional components for the cascade CO_2 transformation to lower olefins [120]. The tuning of acidity was done by tuning the interspacing between a binary oxide and SAPO-34. The catalyst was not only stable but also tolerant to sulfur and gave 30% CO_2 conversion with olefin selectivity of 80.2% (C_3H_6:C_2H_2 = 5.6:1). The tuning of the composite had a substantial effect on the acidity of the composite affecting the catalytic performance and coking (deposition of carbon on the catalyst). The catalyst exhibited mixed crystalline phases of $ZnMn_2O_4$ and ZnO. These phases not only have high oxygen vacancies for CO_2 activation but also assist in the hydrogen spillover ability of CO_2 hydrogenation. In another work, a composite made of different binary oxides (ZnO-ZrO_2) and SAPO-34 was utilized for the same reaction [123]. The composite possessed acid and base sites with distinct micro-mesoporous structures and gave 70% olefin selectivity and 41% CO selectivity.

The acid–base characteristics of the composite were tuned by pre-reducing the ZnO-ZrO_2 precursor followed by physical blending. The alterations in the acid–base properties arise from tuning the electronic properties of the Zn and Zr sites, inhibiting CO formation. Thus, the synergic presence of the micro-mesoporous catalyst-tuned electronic properties, and a high number of acid–base sites gave a superior shape selectivity to lower olefins with the suppression of CO formation. Zha and coworkers utilized a trimetallic metal oxide (CuO-ZnO-Al_2O_3) along with SAPO-34 for tandem CO_2 transformation [36]. Initially, different SDAs like diethylamine, triethylamine, morpholine, and TEAOH were utilized for preparing SAPO-34 zeolite. The composite with SAPO-34 (synthesized using TEAOH) gave the highest surface area, greater CO_2 conversion (53.5%), and high olefin selectivity (62.1%). On a similar line, SAPO-34 was mixed with a hybrid Cu/CeO_2 catalyst and utilized for the same reaction [124]. The Cu/CeO_2 modified SAPO-34 showed a decreased acidity giving a high olefin selectivity of 70.4%.

Recently, CO_2 adsorption and activation were realized using $GaZrO_x$ and spinels like $ZnAl_2O_4$ and $ZnGa_2O_4$ along with SAPO-34 to produce light olefins [121, 122]. Both reports showed that the oxygen vacancies present on the surface facilitated CO_2 adsorption and activation. Figure 11.13 shows the mechanistic steps of hydrogenation of CO_2 over the $GaZrO_x$/SAPO-34 catalyst. The surface oxygen vacancies (O_v) near the Zr–Zr sites facilitate CO_2 adsorption, while the molecular hydrogen is adsorbed over the adjacent Ga sites, establishing a Ga–$H^{\delta-}$ and O–$H^{\delta+}$ bond. The activated CO_2 can either combine with Ga–$H^{\delta-}$ to form CH_3O^*/formates intermediates via hydrogenation or can combine with O–$H^{\delta+}$ to form COOH*, which subsequently dissociates to CO (RWGS reaction). CH_3O^* formation competes with the RWGS reaction, which is mainly attributed to the H_2 dissociation ability of the $GaZrO_x$. It was observed that the optimized Ga loading on the ZrO_x allows a moderate H_2 dissociation to generate HCOO* and further benefits the hydrogenation of CH_3O^* intermediate at a high temperature which supplies suitable temperature

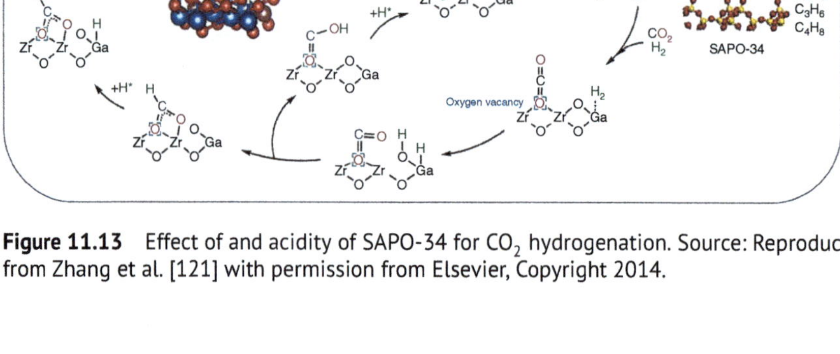

Figure 11.13 Effect of and acidity of SAPO-34 for CO_2 hydrogenation. Source: Reproduced from Zhang et al. [121] with permission from Elsevier, Copyright 2014.

compatibility for its conversion over SAPO-34, favoring the light olefins formation. Brønsted acidity of the SAPO-34 converts the methanol/dimethyl ether species into lower olefins. $GaZrO_x$ combined SAPO-34 bifunctional catalyst gave 26.7% CO_2 conversion with 88.8% light olefins (C_2–C_4) selectivity [121] while the spinel-SAPO-34 composite gave only 15% CO_2 conversion with 87% olefin selectivity [122]. In a similar report, the indium-based oxide ($InCeO_x$ or $InCrO_x$)-SAPO-34 composite was used [125]. It was found that the indium-based oxides were more efficient than In_2O_3 or In-Zr Oxide in producing methanol. DFT studies on these indium-based oxides revealed the higher oxygen vacancies that decrease the formate/methanol formation free energy and enthalpy barriers. The $InCrO_x$-SAPO-34 catalyst gave a 33.6% CO_2 conversion with 75% selectivity, which was almost 1.5-2 times that obtained for In_2O_3-SAPO-34 and $InZrO_x$-SAPO-34 catalysts.

11.5 Conclusion

SAPOs have been considerably utilized as active and selective catalysts for a wide range of commercially relevant synthetic intermediates. The ability to tune the acidity of these zeolites by varying the ratio of Si/Al and Si/P, ease of incorporation of additional metals in the framework generating multifunctionality, and shape-selectivity offered by the porous framework play a vital role in the catalyst tailoring for different catalytic transformations. This chapter summarizes the correlation between the porosity/acidity of the SAPO molecular sieves on catalytic activity during a specific chemical transformation. However, some key issues can be considered for further optimizing the synthetic strategies, and hence SAPO to be applied for various chemical transformations. It is observed that the synthesis of SAPOs generally occurs at a higher temperature. Hence efforts toward the low-temperature synthesis of such zeolites are highly anticipated.

The apparent correlation between the porosity (generated by additives or via different frameworks) and or acidity on the catalytic performance is scarce. Each member of the SAPO family has a different framework and hence varying porosity. The change in the T–O–T angle gives rise to the variable strain in the material, which ultimately leads to a change in the acidity of the material. Thus, studies correlating the effect of the SAPO framework in the catalytic activity toward a specific transformation need to be studied in detail. Although the microporous SAPO molecular sieves are important in chemical transformations, these materials suffer from diffusional limitations, which in turn limit the available catalytic sites. These diffusional limitations in the SAPO molecular sieves are addressed by generating additional pores (meso/macro), giving rise to a two-levels of pore diameter. The tailoring of such hierarchical molecular sieves based on the pore size understanding can aid in improving the catalytic performance. Furthermore, the literature survey demonstrates the use of solid acid SAPOs in different petrochemical transformations. However, the utilization of these potential acid catalysts in different organic/biomass transformations is very feebly reported. Thus, further efforts in this direction are highly anticipated.

In summary, the fine-tuning of the porosity and/or acidity of the SAPO molecular sieves can aid in improving the catalytic performance in the petrochemical processes and several organic/biomass transformations, provided the materials retain their stability and activity. The strategies and key points discussed in this chapter may serve as a starting point for further tailoring the SAPO molecular sieves for more practical catalytic processes.

References

1 Zheng, C., Li, Y., and Yu, J. (2020). Database of open-framework aluminophosphate structures. *Sci. Data* 7: 107.

2 Kirkpatrick, R.J. and Brow, R.K. (1995). Nuclear magnetic resonance investigation of the structures of phosphate and phosphate-containing glasses: a review. *Solid State Nucl. Magn. Reson.* 5: 9–21.

3 Makarova, M., Ojo, A., Al-Ghefaili, K., and Dwyer, J. (1992). *Proceedings of the IX International Zeolite Conference, Montreal, Canada*, 259.

4 Sastre, G., Lewis, D.W., and Catlow, C.R.A. (1996). Structure and stability of silica species in SAPO molecular sieves. *J. Phys. Chem.* 100: 6722–6730.

5 Barthomeuf, D. (1994). Topological model for the compared acidity of SAPOs and SiAl zeolites. *Zeolites* 14: 394–401.

6 Barthomeuf, D. (1993). Topology and maximum content of isolated species (Al, Ga, Fe, B, Si, ...) in a zeolitic framework. An approach to acid catalysis. *J. Phys. Chem.* 97: 10092–10096.

7 Fjermestad, T., Svelle, S., and Swang, O. (2015). Mechanism of Si Island formation in SAPO-34. *J. Phys. Chem. C* 119: 2086–2095.

8 Thybaut, J. and Marin, G. (2016). Multiscale aspects in hydrocracking: from reaction mechanism over catalysts to kinetics and industrial application. *Advances in Catalysis Elsevier* 109–238.

9 Yu, W., Wu, X., Cheng, B. et al. (2022). Synthesis and applications of SAPO-34 molecular sieves. *Chemistry* 28: e202102787.

10 Xu, X.T., Zhai, J.P., Li, I.L. et al. (2012). Synthesis of large single crystals of SAPO-47 in the presence of diethylamine using two-step temperature process. *Microporous Mesoporous Mater.* 148: 122–130.

11 Martins, A.C., Fernandez-Felisbino, R., and Ruotolo, L.A.M. (2012). Ionothermal synthesis of aluminophosphates used for ion exchange: Influence of choline chloride/urea ratio. *Microporous Mesoporous Mater.* 149: 55–59.

12 Su, X., Tian, P., Fan, D. et al. (2013). Synthesis of DNL-6 with a high concentration of Si (4 Al) environments and its application in CO(2) separation. *ChemSusChem* 6: 911–918.

13 Shalmani, F.M., Askari, S., and Halladj, R. (2013). Microwave synthesis of SAPO molecular sieves. *Rev. Chem. Eng.* 29: 99–122.

14 Askari, S. and Halladj, R. (2012). Ultrasonic pretreatment for hydrothermal synthesis of SAPO-34 nanocrystals. *Ultrason. Sonochem.* 19: 554–559.

15 Wang, L., Guo, C., Yan, S. et al. (2003). High-silica SAPO-5 with preferred orientation: synthesis, characterization and catalytic applications. *Microporous Mesoporous Mater.* 64: 63–68.

16 Ishihara, T., Kagawa, M., Hadama, F., and Takita, Y. (1997). Copper ion-exchanged SAPO-34 as a thermostable catalyst for selective reduction of NO with C_3H_6. *J. Catal.* 169: 93–102.

17 Mériaudeau, P., Tuan, V.A., Nghiem, V.T. et al. (1997). SAPO-11, SAPO-31, and SAPO-41 molecular sieves: synthesis, characterization, and catalytic properties in n-octane hydroisomerization. *J. Catal.* 169: 55–66.

18 Zhang, L., Yao, J., Zeng, C., and Xu, N. (2003). Combinatorial synthesis of SAPO-34 via vapor-phase transport. *Chem. Commun.* 2232–2233.

19 Suresh, S., Reddy, I.A.K., and Venkatathri, N. (2018). Synthesis of SAPO-16 molecular sieve in non-aqueous medium by microwave method using Hexamethyleneimine as a template. *Microporous Mesoporous Mater.* 263: 275–281.

20 Hu, Y., Wang, X., Guo, X. et al. (2005). Effect of the template on synthesis of SAPO-11 for hydroisomerization of n-octane. *React. Kinet. Catal. Lett.* 86: 45–50.

21 Wang, D., Tian, P., Fan, D. et al. (2015). N-methyldiethanolamine: a multifunctional structure-directing agent for the synthesis of SAPO and AlPO molecular sieves. *J. Colloid Interface Sci.* 445: 119–126.

22 Seelan, S. and Sinha, A.K. (2004). Crystallization and characterization of high silica silicoaluminophosphate SAPO-5. *J. Mol. Catal. A Chem.* 215: 149–152.

23 Xiao, T.C., An, L.D., and Wang, H.L. (1995). Dependence of the nature and catalytic performance on the synthesis factors of SAPO-5 molecular sieve. *Appl. Catal., A* 130: 187–194.

24 Wang, Y., Chen, S.-L., Jiang, Y.-J. et al. (2016). Influence of template content on selective synthesis of SAPO-18, SAPO-18/34 intergrowth and SAPO-34 molecular sieves used for methanol-to-olefins process. *RSC Adv.* 6: 104985–104994.

25 Roldán, R., Sánchez-Sánchez, M., Sankar, G. et al. (2007). Influence of pH and Si content on Si incorporation in SAPO-5 and their catalytic activity for isomerisation of n-heptane over Pt loaded catalysts. *Microporous Mesoporous Mater.* 99: 288–298.

26 Yoon, J.-W., Jhung, S.-H., Kim, Y.-H. et al. (2005). Selective crystallization of SAPO-5 and SAPO-34 molecular sieves in alkaline condition: effect of heating method. *Bull. Korean Chem. Soc.* 26: 558–562.

27 Liu, P., Ren, J., and Sun, Y. (2008). Influence of template on Si distribution of SAPO-11 and their performance for n-paraffin isomerization. *Microporous Mesoporous Mater.* 114: 365–372.

28 Yang, L., Wang, W., Song, X. et al. (2019). The hydroisomerization of n-decane over Pd/SAPO-11 bifunctional catalysts: the effects of templates on characteristics and catalytic performances. *Fuel Process. Technol.* 190: 13–20.

29 Lok, B.M., Messina, C.A., Patton, R.L. et al. (2002). Silicoaluminophosphate molecular sieves: another new class of microporous crystalline inorganic solids. *J. Am. Chem. Soc.* 106: 6092–6093.

30 Hartmann, M., Azuma, N., and Kevan, L. (2002). Electron spin resonance and electron spin echo modulation study of Ni(I) in silicoaluminophosphate type 5: adsorbate interactions and evidence for the framework incorporation of Ni(I). *J. Phys. Chem.* 99: 10988–10994.

31 Cheung, O., Liu, Q., Bacsik, Z., and Hedin, N. (2012). Silicoaluminophosphates as CO_2 sorbents. *Microporous Mesoporous Mater.* 156: 90–96.

32 Martínez-Franco, R., Li, Z., Martínez-Triguero, J. et al. (2016). Improving the catalytic performance of SAPO-18 for the methanol-to-olefins (MTO) reaction by controlling the Si distribution and crystal size, Catal. *Sci. Technol.* 6: 2796–2806.

33 Li, Y., Deng, J., Song, W. et al. (2016). Nature of Cu species in Cu–SAPO-18 catalyst for NH_3–SCR: combination of experiments and DFT calculations. *J. Phys. Chem. C* 120: 14669–14680.

34 Guo, L., Ming, S., Zhang, S. et al. (2020). One-Pot synthesis of the Cu-SAPO-18 catalyst using waste mother liquid and its application in the SCR of NO*x* with NH_3. *Ind. Chem. Res.* 59: 21275–21285.

35 Yang, J., Kikhtyanin, O.V., Wu, W. et al. (2012). Influence of the template on the properties of SAPO-31 and performance of Pd-loaded catalysts for n-paraffin isomerization. *Microporous Mesoporous Mater.* 150: 14–24.

36 Tian, H.F., Yao, J.H., Zha, F. et al. (2020). Catalytic activity of SAPO-34 molecular sieves prepared by using palygorskite in the synthesis of light olefins via CO_2 hydrogenation. *Appl. Clay Sci.* 184: 105392.

37 Woo, J., Leistner, K., Bernin, D. et al. (2018). Effect of various structure directing agents (SDAs) on low-temperature deactivation of Cu/SAPO-34 during NH_3-SCR reaction. *Cat. Sci. Technol.* 8: 3090–3106.

38 Maghsoudi, H. and Soltanieh, M. (2014). Simultaneous separation of H_2S and CO_2 from CH_4 by a high silica CHA-type zeolite membrane. *J. Membr. Sci.* 470: 159–165.

39 Dumitriu, E., Azzouz, A., Hulea, V. et al. (1997). Synthesis, characterization and catalytic activity of SAPO-34 obtained with piperidine as templating agent. *Microporous Mater.* 10: 1–12.

40 Pinilla-Herrero, I., Olsbye, U., Márquez-Álvarez, C., and Sastre, E. (2017). Effect of framework topology of SAPO catalysts on selectivity and deactivation profile in the methanol-to-olefins reaction. *J. Catal.* 352: 191–207.

41 Lohse, U., Vogt, F., and Richter-Mendau, J. (1993). Synthesis and characterization of the levyne-like structure SAPO-35, prepared with cyclohexylamine as templating agent. *Cryst. Res. Technol.* 28: 1101–1107.

42 Franco, M.J., Pérez-Pariente, J., Misud, A. et al. (1992). Crystallization kinetics of SAPO-37. *Zeolites* 12: 386–394.

43 Sinha, A.K., Hegde, S.G., Jacob, N.E., and Sivasanker, S. (1997). Preparation and characterization of the silicoaluminophosphate SAPO-39. *Zeolites* 18: 350–355.

44 Dumont, N., Gabelica, Z., Derouane, E.G., and Di Renzo, F. (1994). Comparative investigation of different synthesis procedures leading to SAPO-40. *Microporous Mater.* 3: 71–84.

45 Venkatathri, N. and Srivastava, R. (2004). Synthesis, characterization and catalytic properties of sapo-11, −31 and −41 molecular sieves. In: *Studies in Surface Science and Catalysis* (ed. E. van Steen, I.M. Claeys, and L.H. Callanan), 978–984. Elsevier.

46 Vidal, L., Pray, C., and Patarin, J. (2000). AlPO4-41 aluminophosphate molecular sieve: a new reproducible synthesis route involving diisopropylformamide as template precursor. *Microporous Mesoporous Mater.* 39: 113–116.

47 Wei, X., Kikhtyanin, O.V., Parmon, V.N. et al. (2017). Synergetic effect between the metal and acid sites of Pd/SAPO-41 bifunctional catalysts in n-hexadecane hydroisomerization. *J. Porous. Mater.* 25: 235–247.

48 Martínez-Franco, R., Cantín, Á., Vidal-Moya, A. et al. (2015). Self-assembled Aromatic molecules as efficient organic structure directing agents to synthesize the silicoaluminophosphate SAPO-42 with isolated Si species. *Chem. Mater.* 27: 2981–2989.

49 Hernández-Maldonado, A.J., Yang, R.T., Chinn, D., and Munson, C.L. (2003). Partially calcined gismondine type silicoaluminophosphate SAPO-43: isopropylamine elimination and separation of carbon dioxide, hydrogen sulfide, and water. *Langmuir* 19: 2193–2200.

50 Yang, M., Tian, P., Liu, L. et al. (2015). Cationic surfactant-assisted hydrothermal synthesis: an effective way to tune the crystalline phase and morphology of SAPO molecular sieves. *CrystEngComm* 17: 8555–8561.

51 Ashtekar, S., Chilukuri, S.V.V., and Chakrabarty, D.K. (2002). Small-pore molecular sieves SAPO-34 and SAPO-44 with chabazite structure: a study of silicon incorporation. *J. Phys. Chem.* 98: 4878–4883.

52 Kong, W., Dai, W., Li, N. et al. (2009). A one-step route to SAPO-46 using H_3PO_3-containing gel and its application as the catalyst for methanol dehydration. *J. Mol. Catal. A Chem.* 308: 127–133.

53 Xu, X.T., Zhai, J.P., Chen, Y.P. et al. (2015). Synthesis of large optically clear SAPO-47 single crystals using *n*-propylamine as template. *J. Cryst. Growth* 426: 123–128.

54 Li, N., Ma, Y., Kong, W. et al. (2008). The competition of organic amines and the phosphorus source on directing the formation of aluminophosphate-based molecular sieves with chabazite structure. *Microporous Mesoporous Mater.* 115: 356–363.

55 Pluth, J.J. and Smith, J.V. (1989). Silicoaluminophosphate with encapsulated methylbutylamine species: chabazite structure, charge coupling between framework and inferred ammonium species, and severe molecular disorder. *J. Phys. Chem.* 93: 6516–6520.

56 Dumitriu, E., Lutic, D., Hulea, V. et al. (1999). Synthesis optimization of chabasite-like SAPO-47 in the presence of sec-butylamine. *Microporous Mesoporous Mater.* 31: 187–193.

57 Ahn, N.H., Seo, S., and Hong, S.B. (2016). Small-pore molecular sieves SAPO-57 and SAPO-59: synthesis, characterization, and catalytic properties in methanol-to-olefins conversion. *Cat. Sci. Technol.* 6: 2725–2734.

58 Broach, R.W., Greenlay, N., Jakubczak, P. et al. (2014). New ABC-6 net molecular sieves ZnAPO-57 and ZnAPO-59: framework charge density-induced transition from two- to three-dimensional porosity. *Microporous Mesoporous Mater.* 189: 49–63.

59 Seo, S., Ahn, N.H., Lee, J.H. et al. (2019). Combined alkali-organoammonium structure direction of high-charge-density heteroatom-containing aluminophosphate molecular sieves. *Angew. Chem. Int. Ed.* 58: 9032–9037.

60 Fan, D., Qiao, Y., Cao, K. et al. (2020). Preparation of hierarchical SAPO-18 via alkaline/acid etching. *Microporous Mesoporous Mater.* 300: 110156.

61 Jadav, D., Bandyopadhyay, R., and Bandyopadhyay, M. (2020). Synthesis of hierarchical SAPO-5 & SAPO-34 materials by post-synthetic alkali treatment and their enhanced catalytic activity in transesterification. *Eur. J. Inorg. Chem.* 2020: 847–853.

62 Singh, A.K., Kondamudi, K., Yadav, R. et al. (2014). Uniform mesoporous silicoaluminophosphate derived by vapor phase treatment: its catalytic and kinetic studies in hydroisomerization of 1-octene. *J. Phys. Chem. C* 118: 27961–27972.

63 Ren, S., Liu, G., Wu, X. et al. (2017). Enhanced MTO performance over acid treated hierarchical SAPO-34. *Chin. J. Catal.* 38: 123–130.

64 Han, J., Yang, G., Ding, H., and Chen, X. (2021). Revealing inherent factors of SAPO-34 zeolites etching towards the fabrication of hierarchical structure. *Microporous Mesoporous Mater.* 319: 111067.

65 Gregg, S.J. and Sing, K.S.W. (1982). *Adsorption, Surface Area and Porosity*, 2e. London: Academic Press.

66 Groen, J.C., Brouwer, S., Peffer, L.A.A., and Pérez-Ramírez, J. (2006). Application of mercury intrusion porosimetry for characterization of combined micro- and mesoporous zeolites. *Part. Part. Syst. Charact.* 23: 101–106.

67 Janssen, A.H., Talsma, H., van Steenbergen, M.J., and de Jong, K.P. (2004). Homogeneous nucleation of water in mesoporous zeolite cavities. *Langmuir* 20: 41–45.

68 Karger, J. and Ruthven, D.M. (1992). *Diffusion in Zeolites and Other Microporous Solids.* New York: Wiley.

69 Karger, J., Binder, T., Chmelik, C. et al. (2014). Microimaging of transient guest profiles to monitor mass transfer in nanoporous materials. *Nat. Mater.* 13: 333–343.

70 Suga, M., Asahina, S., Sakuda, Y. et al. (2014). Recent progress in scanning electron microscopy for the characterization of fine structural details of nano materials. *Prog. Solid State Chem.* 42: 1–21.

71 Karwacki, L., de Winter, D.A., Aramburo, L.R. et al. (2011). Architecture-dependent distribution of mesopores in steamed zeolite crystals as visualized by FIB-SEM tomography. *Angew. Chem. Int. Ed.* 50: 1294–1298.

72 Wan, W., Su, J., Zou, X.D., and Willhammar, T. (2018). Transmission electron microscopy as an important tool for characterization of zeolite structures. *Inorg. Chem. Front.* 5: 2836–2855.

73 Zečević, J., de Jong, K.P., and de Jongh, P.E. (2013). Progress in electron tomography to assess the 3D nanostructure of catalysts. *Curr. Opin. Solid State Mater. Sci.* 17: 115–125.

74 Niwa, M. and Katada, N. (2013). New method for the temperature-programmed desorption (TPD) of ammonia experiment for characterization of zeolite acidity: a review. *Chem. Rec.* 13: 432–455.

75 Xu, J., Jiang, J., and Zhao, J. (2016). Thermochemical conversion of triglycerides for production of drop-in liquid fuels. *Renew. Sust. Energ. Rev.* 58: 331–340.

76 Ebadinezhad, B. and Haghighi, M. (2020). Texture evolution of mesoporous SAPO-34 via a hard-templating sono-hydrothermal method for biodiesel production: influence of carbon materials on nanocatalyst design. *Appl. Catal., A* 595: 117486.

77 Zhou, P. and Zhang, Z. (2016). One-pot catalytic conversion of carbohydrates into furfural and 5-hydroxymethylfurfural. *Cat. Sci. Technol.* 6: 3694–3712.

78 Song, X., Yue, J., Zhu, Y. et al. (2021). Efficient conversion of glucose to 5-hydroxymethylfurfural over a Sn-modified SAPO-34 zeolite catalyst. *Ind. Chem. Res.* 60: 5838–5851.

79 Zhang, L., Xi, G., Chen, Z. et al. (2017). Enhanced formation of 5-HMF from glucose using a highly selective and stable SAPO-34 catalyst. *Chem. Eng. J.* 307: 877–883.

80 Shao, H., Chen, J., Zhong, J. et al. (2015). Development of MeSAPO-5 molecular sieves from attapulgite for dehydration of carbohydrates. *Ind. Chem. Res.* 54: 1470–1477.

81 Romo, J.E., Wu, T., Huang, X. et al. (2018). SAPO-34/5A zeolite bead catalysts for furan production from xylose and glucose. *ACS Omega* 3: 16253–16259.

82 Bhaumik, P. and Dhepe, P.L. (2014). Exceptionally high yields of furfural from assorted raw biomass over solid acids. *RSC Adv.* 4: 26215–26221.

83 Bhaumik, P. and Dhepe, P.L. (2015). Effects of careful designing of SAPO-44 catalysts on the efficient synthesis of furfural. *Catal. Today* 251: 66–72.

84 Zhao, S., Wang, W.D., Wang, L. et al. (2020). Cooperation of hierarchical pores with strong Brønsted acid sites on SAPO-34 catalysts for the glycerol dehydration to acrolein. *J. Catal.* 389: 166–175.

85 Jiang, L., Ge, Y., Shah, A.N. et al. (2010). Unregulated emissions from a diesel engine equipped with vanadium-based urea-SCR catalyst. *J. Environ. Sci.* 22: 575–581.

86 Mrad, R., Aissat, A., Cousin, R. et al. (2015). Catalysts for NO_x selective catalytic reduction by hydrocarbons (HC-SCR). *Appl. Catal., A* 504: 542–548.

87 Jiang, J.Q., Pan, H., Sun, G.J. et al. (2012). Research advances in selective catalytic reduction of NO_x over transition metal-based zeolite catalysts. *Chem. Ind. Eng. Prog.* 31: 98–106.

88 Jabłońska, M. (2022). Recent progress in the selective catalytic reduction of NO with NH_3 on Cu-SAPO-34 catalysts. *Mol. Catal.* 518: 112111.

89 Yu, T., Fan, D., Hao, T. et al. (2014). The effect of various templates on the NH_3-SCR activities over Cu/SAPO-34 catalysts. *Chem. Eng. J.* 243: 159–168.

90 Cao, Y., Fan, D., Zhu, D. et al. (2020). The effect of Si environments on NH_3 selective catalytic reduction performance and moisture stability of Cu-SAPO-34 catalysts. *J. Catal.* 391: 404–413.

91 Xin, Y., Zhang, N., Wang, X. et al. (2019). Efficient synthesis of the Cu-SAPO-44 zeolite with excellent activity for selective catalytic reduction of NO by NH_3. *Catal. Today* 332: 35–41.

92 Martínez-Franco, R., Moliner, M., and Corma, A. (2014). Direct synthesis design of Cu-SAPO-18, a very efficient catalyst for the SCR of NO_x. *J. Catal.* 319: 36–43.

93 Ishihara, T., Kagawa, M., Hadama, F. et al. (1997). Thermostable molecular sieves, silicoaluminophosphate (SAPO)-34, for the removal of NO_x with C_3H_6 in the coexistence of O_2, H_2O, and SO_2. *Ind. Chem. Res.* 36: 17–22.

94 Yang, D., Zhou, H., Wang, C. et al. (2021). NO selective catalytic reduction with propylene over one-pot synthesized Fe-SAPO-34 catalyst under diesel exhaust conditions. *Fuel* 290: 119822.

95 Gao, Z., Zhao, D., Yang, Y. et al. (2021). Influence of copper locations on catalytic properties and activities of Cu/SAPO-34 in C_3H_6-SCR. *Ind. Chem. Res.* 60: 6940–6949.

96 Santos, R.C.R., Valentini, A., Lima, C.L. et al. (2011). Modifications of an HY zeolite for n-octane hydroconversion. *Appl. Catal., A* 403: 65–74.

97 Guo, L., Fan, Y., Bao, X. et al. (2013). Two-stage surfactant-assisted crystallization for enhancing SAPO-11 acidity to improve n-octane di-branched isomerization. *J. Catal.* 301: 162–173.

98 Guo, L., Bao, X., Fan, Y. et al. (2012). Impact of cationic surfactant chain length during SAPO-11 molecular sieve synthesis on structure, acidity, and n-octane isomerization to di-methyl hexanes. *J. Catal.* 294: 161–170.

99 Fan, Y., Xiao, H., Shi, G. et al. (2012). Alkylphosphonic acid- and small amine-templated synthesis of hierarchical silicoaluminophosphate molecular sieves with high isomerization selectivity to di-branched paraffins. *J. Catal.* 285: 251–259.

100 Kikhtyanin, O.V., Toktarev, A.V., Ayupov, A.B., and Echevsky, G.V. (2010). Influence of crystallinity on the physico-chemical properties of SAPO-31 and hydroconversion of n-octane over Pt loaded catalysts. *Appl. Catal., A* 378: 96–106.

101 Bértolo, R., Silva, J.M., Ribeiro, F. et al. (2014). Effects of oxidant acid treatments on carbon-templated hierarchical SAPO-11 materials: synthesis, characterization and catalytic evaluation in n-decane hydroisomerization. *Appl. Catal., A* 485: 230–237.

102 Ren, X., Li, N., Cao, J. et al. (2006). Hydroisomerization of -decane over Pt/SAPO-41 catalyst. *Appl. Catal., A* 298: 144–151.

103 Wen, C., Han, S., Xu, J., and Fan, Y. (2017). A novel route to synthesize SAPO-11 molecular sieves with a high external surface area in the presence of ethylene glycol and supercritical carbon dioxide for 1-octene hydroisomerization to dimethylhexanes. *J. Catal.* 356: 100–110.

104 Bértolo, R., Silva, J.M., Ribeiro, M.F. et al. (2017). Microwave synthesis of SAPO-11 materials for long chain n-alkanes hydroisomerization: effect of physical parameters and chemical gel composition. *Appl. Catal., A* 542: 28–37.

105 Yu, G., Qiu, M., Wang, T. et al. (2021). Optimization of the pore structure and acidity of SAPO-11 for highly efficient hydroisomerization on the long-chain alkane. *Microporous Mesoporous Mater.* 320: 111076.

106 Xing, G., Liu, S., Guan, Q., and Li, W. (2019). Investigation on hydroisomerization and hydrocracking of C_{15}–C_{18} n-alkanes utilizing a hollow tubular Ni-Mo/SAPO-11 catalyst with high selectivity of jet fuel. *Catal. Today* 330: 109–116.

107 Wang, D., Kang, X., Gu, Y. et al. (2020). Electronic tuning of Ni by Mo species for highly efficient hydroisomerization of n-alkanes comparable to Pt-based catalysts. *ACS Catal.* 10: 10449–10458.

108 Song, W., Liu, Y., Baráth, E. et al. (2015). Synergistic effects of Ni and acid sites for hydrogenation and C–O bond cleavage of substituted phenols. *Green Chem.* 17: 1204–1218.

109 Luo, W., Bruijnincx, P.C.A., and Weckhuysen, B.M. (2014). Selective, one-pot catalytic conversion of levulinic acid to pentanoic acid over Ru/H-ZSM5. *J. Catal.* 320: 33–41.

110 Chen, N., Gong, S., and Qian, E.W. (2015). Effect of reduction temperature of $NiMoO_{3-x}$/SAPO-11 on its catalytic activity in hydrodeoxygenation of methyl laurate. *Appl Catal B* 174-175: 253–263.

111 Chen, N., Ren, Y., and Qian, E.W. (2016). Elucidation of the active phase in PtSn/SAPO-11 for hydrodeoxygenation of methyl palmitate. *J. Catal.* 334: 79–88.

112 Kim, K., Higai, D., Hou, X. et al. (2021). Production of normal paraffins with an even carbon number via selective hydrodeoxygenation of palm fatty acid distillate over Pt–Sn catalysts. *Ind. Chem. Res.* 60: 2881–2889.

113 Chen, N., Gong, S., Shirai, H. et al. (2013). Effects of Si/Al ratio and Pt loading on Pt/SAPO-11 catalysts in hydroconversion of Jatropha oil. *Appl. Catal., A* 466: 105–115.

114 Liu, Q., Zuo, H., Wang, T. et al. (2013). One-step hydrodeoxygenation of palm oil to isomerized hydrocarbon fuels over Ni supported on nano-sized SAPO-11 catalysts. *Appl. Catal., A* 468: 68–74.

115 Wang, C., Liu, Q., Liu, X. et al. (2013). Influence of reaction conditions on one-step hydrotreatment of lipids in the production of iso-alkanes over Pt/SAPO-11. *Chin. J. Catal.* 34: 1128–1138.

116 Rabaev, M., Landau, M.V., Vidruk-Nehemya, R. et al. (2015). Improvement of hydrothermal stability of Pt/SAPO-11 catalyst in hydrodeoxygenation–isomerization–aromatization of vegetable oil. *J. Catal.* 332: 164–176.

117 Verma, D., Rana, B.S., Kumar, R. et al. (2015). Diesel and aviation kerosene with desired aromatics from hydroprocessing of jatropha oil over hydrogenation catalysts supported on hierarchical mesoporous SAPO-11. *Appl. Catal., A* 490: 108–116.

118 Smirnova, M.Y., Kikhtyanin, O.V., Smirnov, M.Y. et al. (2015). Effect of calcination temperature on the properties of Pt/SAPO-31 catalyst in one-stage transformation of sunflower oil to green diesel. *Appl. Catal., A* 505: 524–531.

119 Li, X., Xiang, M., and Wu, D. (2019). Hydrogenolysis of glycerol over bimetallic Cu Ni catalysts supported on hierarchically porous SAPO-11 zeolite. *Catal. Commun.* 119: 170–175.

120 Mou, J., Fan, X., Liu, F. et al. (2021). CO_2 hydrogenation to lower olefins over Mn_2O_3-ZnO/SAPO-34 tandem catalysts. *Chem. Eng. J.* 421: 129978.

121 Zhang, P., Ma, L., Meng, F. et al. (2022). Boosting CO_2 hydrogenation performance for light olefin synthesis over $GaZrO_x$ combined with SAPO-34. *Appl Catal B* 305: 121042.

122 Liu, X., Wang, M., Yin, H. et al. (2020). Tandem catalysis for hydrogenation of CO and CO_2 to lower olefins with bifunctional catalysts composed of spinel oxide and SAPO-34. *ACS Catal.* 10: 8303–8314.

123 Wang, G., Zeng, L., Cao, J. et al. (2019). Highly selective conversion of CO_2 to hydrocarbons over composite catalysts of ZnO-ZrO_2 and SAPO-34. *Microporous Mesoporous Mater.* 284: 133–140.

124 Sedighi, M. and Mohammadi, M. (2020). CO_2 hydrogenation to light olefins over Cu-CeO_2/SAPO-34 catalysts: product distribution and optimization. *J. CO2 Util.* 35: 236–244.

125 Wang, S., Wang, P., Qin, Z. et al. (2020). Enhancement of light olefin production in CO_2 hydrogenation over In_2O_3-based oxide and SAPO-34 composite. *J. Catal.* 391: 459–470.

12

Heterogeneous Photocatalytic Degradation of Pharmaceutical Pollutants over Titania Nanoporous Architectures

Surya Kumar Vatti[1] *and Parasuraman Selvam*[1,2,3]

[1] *Indian Institute of Technology-Madras, National Centre for Catalysis Research and Department of Chemistry, Chennai 600036, India*
[2] *Kumamoto University, International Research Organization for Advanced Science and Technology (IROAST), 2-39-1 Kurokami, Chuo-ku, Kumamoto 860-8555, Japan*
[3] *University of Delaware, Delaware Energy Institute, 221, Academy Street, Newark, DE 19716, USA*

12.1 Introduction

Pharmaceuticals in the Environment – Time to Get Clean: "Most nations have strict controls on environmental waste, from arsenic to zinc. Yet no legal limits have been set to control drug pollution during manufacture, use, and disposal. That is despite evidence that pharmaceutical waste can bring about havoc in the environment – hormones found in contraceptives cause male fish to grow female sex organs, and a painkiller used in livestock has wiped out millions of vultures in India that fed on the remains" [1].

There is a growing concern that pharmaceutical pollution in water bodies and drinking water adversely affects the health and behavior of humans and wildlife, including insects, fish, birds, and the whole ecosystem. In the 1990s, the Indian subcontinent, an Oriental white-backed vulture (OWBV; Gyps bengalensis) population decline of >95%, was first noted at Keoladeo National Park, India [2]. Consequently, these vultures are now listed as critically endangered by Birdlife International [3]. Later it was identified that residues of the anti-inflammatory drug diclofenac (0.051–0.643 $\mu g\,g^{-1}$) were responsible for renal failure causing OWBV deaths [4]. It is obverted that untreated or treated effluents of wastewater treatment plants (WWTPs) have many pharmaceutical molecules such as synthetic estrogen used in birth-control pills [17α-ethynyl estradiol (EE2)], which affect the reproductive health of wild fishes causing intersex in males and altered oogenesis in females of fathead minnow fishes (*Pimephales promelas*). Ultimately, near extinction of these fishes was observed in northwestern Ontario [5, 6]. The presence of pharmaceutical pollution in the environment continues to grow with the development of more potent and metabolically resistant drugs. There are different pharmaceutical classes, *viz.*, analgesics, anti-depressants, contraceptives, antibiotics, veterinary medicines, and pesticides used in agriculture. Due to improper disposal through

Catalysis in Confined Frameworks: Synthesis, Characterization, and Applications, First Edition.
Edited by Hermenegildo Garcia and Amarajothi Dhakshinamoorthy.

various sources, all of these pharmaceutical chemicals eventually end up in our environment; the effects of these bioactive molecules are still unknown which makes them more dangerous.

Point sources of pharmaceutical pollution comprise pharmaceutical industries, hospital effluent, sewage sludge, landfills, and even residential waste, which contaminate surface and groundwater. All this pharmaceutical waste reaches WWTPs. Unfortunately, most of these pharmaceuticals escape WWTP due to a lack of efficient filtration/oxidation process and finally end up in the environment or drinking water in full/partial forms (Figure 12.1) [7, 8]. More than 200 different pharmaceuticals are detected in the primary effluent of WWTPs worldwide [9]. Many studies worldwide have found that current oxidation/treatment methods of present-day WWTPs are insufficient in destroying these pharmaceuticals [7–9]. The consequence of this is the exposure of these active pharmaceutical ingredients (APIs) to aquatic/terrestrial animals, including reaching human drinking water.

There is documented data on pharmaceuticals occurrence in tap water (Table 12.1). It is challenging to identify pharmaceutical pollutants due to analytical

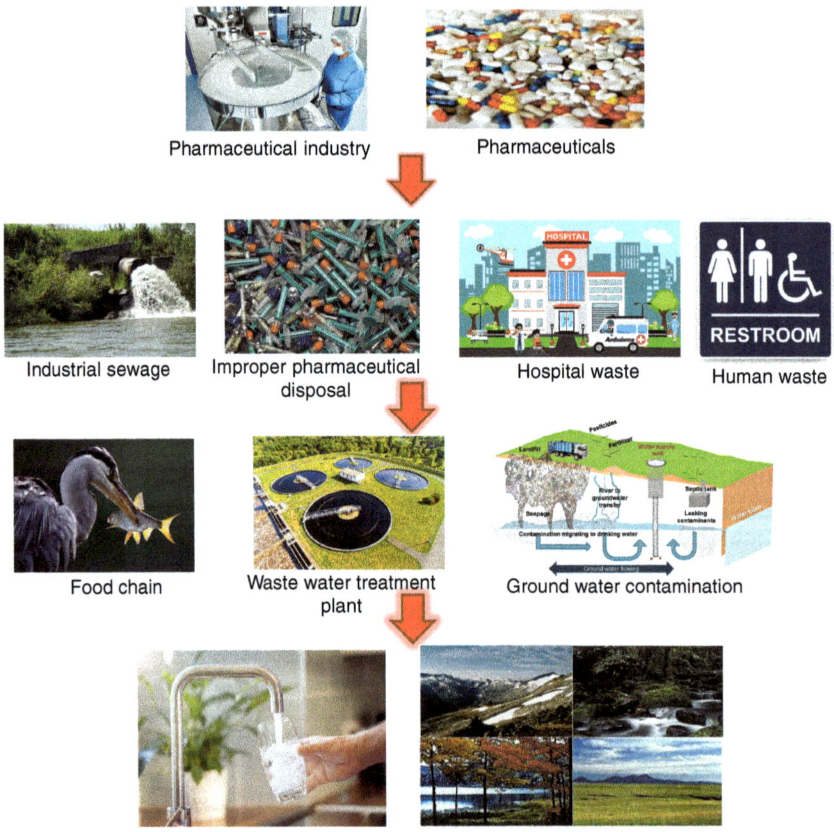

Figure 12.1 Sources, routes, and the fate of pharmaceutical pollution in different compartments of the environment and drinking water. Source: Kadmy/Adobe Stock Photos; Green_beat/Adobe Stock; Tatty/Adobe Stock Photos; Chaikom/Adobe Stock; olando/Adobe Stock; wal_172619/Pixabay; Kletr/Adobe Stock Photos; Dan Kosmayer/Adobe Stock Photos; NASA/Wikimedia Commons/Public Domain; NASA/Public Domain;.

Table 12.1 Pharmaceuticals detected in worldwide tap water.

Therapeutic use	Drug	Conc. detected (*ng/L*)	Country
Antibiotics	Triclosan	734	USA
	Carbamazepine	24	Canada
Analgesics and NSAIDs	Diclofenac	6–35	Germany
	Ibuprofen	3	
	AMDOPH	900–1250	
Antidepressant	Amitryptilline	1.4	France
	Diazepam	10	UK
Antihistamine	Ranitidine,	8.2	Ireland
	Famotidine	1.6	
Antineoplastics	Bleomycin	13	UK
Lipid regulators	Bezafibrate	27	Germany
	Clofibric acid	50–270	
Opioidanalgesics	Codein	30	USA
Psycho-stimulants	Caffeine	60–119	USA

Source: Adapted from Mompelat et al. [8].

difficulties in quantifying the ultra-trace level ($ng\,l^{-1}$). Since, there are many pharmaceutical classes, including analgesics, anti-depressants, contraceptives, antibiotics, and veterinary medicines. Endocrine-disrupting chemicals (EDCs) are another class of compounds causing further alarm. It is next to impossible for WWTPs to isolate and remove them from water. Some of these pharmaceuticals entering into ground/surface water were present at much lower levels than the limit of detection (LOD) in drinking water. The incomplete removal of these pollutants is of grave concern due to their unknown health effects in drinking waters, even at $ng\,l^{-1}$ concentration.

It is observed that the most common route of entry of pharmaceuticals into drinking water and the environment is through WWTPs, where APIs get concentrated and released in partial/complete forms. Unfortunately, regulatory authorities neglect APIs released from WWTPs into the atmosphere, as the pharmaceutical pollution impact on humans has not been fully studied/established even though there are many cases of devastation on other living organisms, *viz.,* fishes and birds [10]. Currently, the conventional filtration/oxidation methods for pharmaceutical pollution removal, *viz.*, active carbon filtration and chlorination at WWTPs, are inefficient [11, 12]. In this context, immense research is focused on developing advanced oxidation processes (AOPs) to eliminate pharmaceutical pollution at WWTPs.

12.2 Advanced Oxidation Process

AOP can be defined as the "*in situ generations of highly reactive species*" [13] or "*near ambient temperature treatment processes based on highly reactive radicals, especially the hydroxyl radical* ($^{\bullet}OH$), *as the primary oxidant*" [14]. In AOP, $^{\bullet}OH$ radicals are continuously generated, and $^{\bullet}OH$ radicals unselectively attach to organic pollutants

Table 12.2 Oxidation potentials of various oxidizing species.

Oxidative species	Oxidation potential (V) *vs* NHE
F_2	3.1
$^{\bullet}OH$	2.8
O_3	2.1
H_2O_2	1.8
$KMnO_4$	1.7
Cl_2	1.4
HCHO	0.5
CH_3OH	0.4

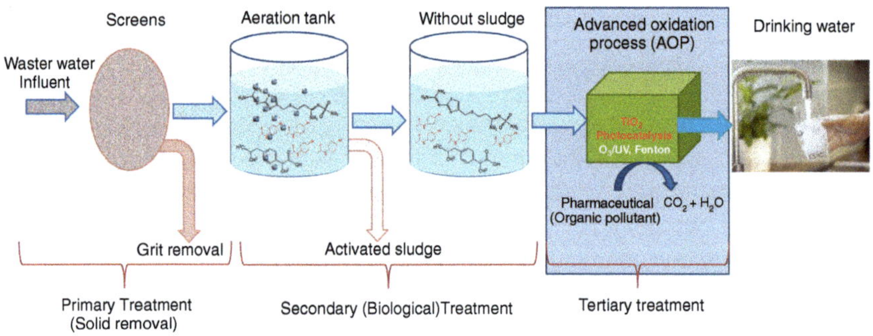

Figure 12.2 Wastewater treatment plants flow sheet with advanced oxidation process as a tertiary process to remove pharmaceutical pollution. Source: Brian Jackson/Adobe Stock.

and oxidize them into simple non-poisonous molecules, *viz.*, H_2O and CO_2. These $^{\bullet}OH$ radicals have high oxidation potential among other oxidants (Table 12.2) and hence are highly reactive for oxidation reactions.

In WWTPs, AOPs will be used as a tertiary treatment method for oxidative decomposition of organic toxic substances after the primary and secondary treatment methods (Figure 12.2) [15]. The primary and secondary treatment is done to bring chemical oxygen demand (COD) to a relatively small ($<5.0\,g\,l^{-1}$), which becomes suitable for AOP to eliminate pharmaceutical pollutants present at low concentrations. The most common AOPs used for treating water and wastewater at WWTPs are presented in Table 12.3 Some of these processes, such as photolysis with UV light, are widely used in Europe as a disinfection and in the USA for treating groundwater pollutants. UV-light photolysis, combined with H_2O_2, O_3, and Fenton's reagent, is used at a large scale, offering improved kinetic and performance advantages [13].

Table 12.3 Advanced oxidation process for water and wastewater treatment.

AOP	Oxidant	Remarks
Ozonation	O_3	Homogenous, O_3 is unstable, complex, expensive
Photolysis	UV	Intermediate formation
Peroxide process	O_3/UV	Homogenous, O_3 is unstable, complex, expensive
Fenton	H_2O_2/Fe^{2+}	Homogenous, Fe sludge formation, posttreatment required
Photo-Fenton	H_2O_2/Fe^{2+}/UV	Homogenous, Fe sludge formation, posttreatment required
Photocatalysis	TiO_2/UV/H_2O	Heterogeneous catalyst is stable, reusable, safe, sustainable No posttreatment

12.2.1 Ozonation

Ozonation is used in most wastewater treatment but is more common for drinking water disinfection. Ozone (O_3) can efficiently degrade organic compounds that possess amines and activated aromatic rings, such as phenols, alkyl aromatics, and olefinic moieties [16]. Oxidation using O_3 either follows the direct way of the O_3 with the dissolved pollutant or the radical way indirectly with the radicals generated from ozone decomposition shown below [17]. However, there are disadvantages of O_3 in wastewater treatment, i.e. poor O_3 solubility relative to other oxidants, and a homogenous process, the catalyst cannot be reused.

$$O_3 + HO^- \rightarrow O_2 + HO_2^-$$

$$O_3 + HO_2^- \rightarrow {}^\bullet O_2^- + {}^\bullet OH + O_2$$

$$O_2^{\bullet -} + \text{Pollutant} \rightarrow \text{Oxidation/Disinfection}$$

$$^\bullet OH + \text{Pollutant} \rightarrow \text{Oxidation/Disinfection}$$

Bromate formation: In the treatment of water-containing bromide, bromate (BrO^{3-}) is formed with O_3. Bromate is carcinogenic in nature. For example, in the Silver Lake and Elysian reservoirs in California, 600 million gallons of water had to be drained due to bromate formation. Similarly, in the United Kingdom, bromate poisoning reached bottled water [18].

12.2.2 UV Irradiation (Photolysis)

In photolysis, photon energy is absorbed by pollutant molecules that can be excited to higher energy states and undergo a decomposition reaction. This process largely depends on the pollutant under study and the configuration system employed. In most studies, photolysis is found to be inferior to combinatory processes. However, this can be pollutant dependent and depends on the light source, e.g. UV/A, UV/C, etc. [19] Studies [20] on the degradation of metronidazole proved that UV alone is ineffective in the degradation of pharmaceuticals, and favorable results can be obtained with UV in combination with other AOPs, *viz.*, H_2O_2, O_3, and Fenton reagent.

12.2.3 Fenton and Photo-Fenton Process

In the Fenton process, Ferrous ions (Fe^{2+}) and H_2O_2 generate hydroxyl radicals as shown below [21]. Fenton process is effective, and only small concentrations of reagents are required. It has been successfully employed to degrade Penicillin, Metronidazole (antimicrobial agent), and many textiles dye [22].

$$Fe^{2+} + H_2O_2 \rightarrow Fe^{3+} + OH^- + {}^{\bullet}OH$$

$$Fe^{3+} + H_2O_2 \rightleftharpoons FeO(OH) + H^+$$

$$FeO(OH) \rightarrow Fe^{2+} + HO_2^{\bullet}$$

$$Fe^{3+} + H_2O + h\nu \rightarrow Fe^{2+} + {}^{\bullet}OH$$

It was observed that under acidic pH (2.7–2.8), reduction of Fe^{3+} to Fe^{2+} is favorable [23], making the Fenton reagent a real catalyst. Similarly, irradiation with UV–Visible >300 nm was found to improve the Fenton oxidation reactions. On UV–Visible irradiation, Fe^{3+} generated from the Fe^{2+} ions was reduced back to the ferrous ions, and the Fenton reaction continued due to the presence of H_2O_2. Despite many studies on this process, there are very few industrial applications. This is because Fenton process requires strict pH conditions. Also, Fe sludge formation takes place, which needs posttreatment increasing complications, and also being a homogenous process, the catalyst cannot be reused, making this process non-sustainable.

12.2.4 Need for Green Sustainable Heterogeneous AOP

Conventional water treatment processes involve AOPs such as ozonolysis, Fenton process, and traditionally used non-AOP chlorination, which creates harmful byproducts such as bromate, Fe sludge, and trihalomethanes, respectively, which are not a clean process. The requirement of posttreatment brings complications and extra costs. The non-recyclability of the catalyst due to the homogenous nature is non-sustainable. Hence, research efforts are focused on heterogeneous AOPs such as semiconductor (SC)-based photocatalysis [24].

12.2.5 Heterogeneous Photocatalysis

Heterogeneous catalysis is a process in which the catalyst's phase differs from that of the reactants or products. It differs from homogeneous catalysis, where the reactants, products, and catalysts exist in the same phase. A heterogeneous catalyst is a solid functional material that provides active sites which change the

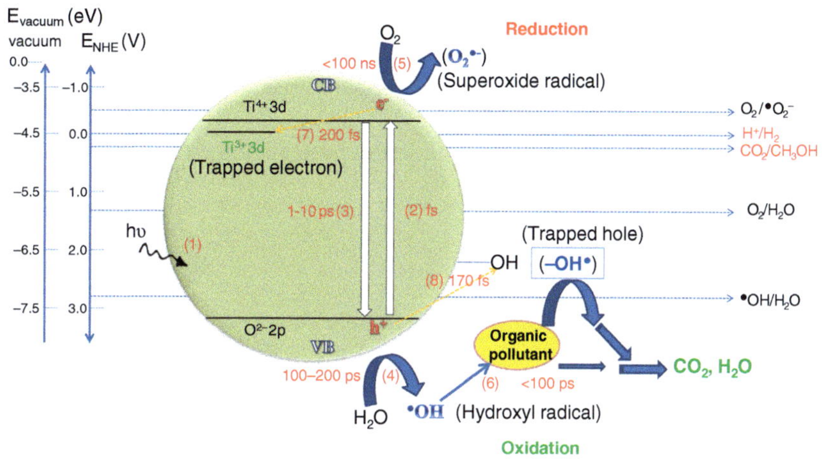

Figure 12.3 Photocatalytic degradation of organic pollutants on TiO_2 particle with band edges compared to a normal hydrogen electrode (NHE) at pH = 1 and vacuum energy scale.

rates of chemical reactions of the reactants adsorbed on the catalyst surface without changing the reactions' thermodynamic equilibrium [25]. Photocatalysis differs from classical heterogeneous catalysis that the catalyst is activated not by thermal methods but by the photonic method. International Union of Pure and Applied Chemistry (IUPAC) defines photocatalysis as a "*catalytic reaction involving light absorption by a catalyst or by a substrate*" [26]. Photocatalysis as an AOP generates $^{\bullet}OH$ from the surface of SC material upon absorption of photons in an aqueous medium, making it a green process, as shown in Figure 12.3. Many SC photocatalysts, *viz.*, ZnO, ZnS, CdS, WO_3, etc. are available but the most vigorous effort is concentrated on titanium dioxide/titania (TiO_2) due to its high photocatalytic activity, chemical stability, nontoxic nature, and relatively low cost [27].

12.3 Semiconductor Photocatalysis Mechanism

When SC, *viz.*, titanium dioxide (TiO_2) is irradiated with photons ($\lambda > 385$ nm, UV light) energy higher than the bandgap of the TiO_2 (3.2 eV for anatase TiO_2), absorption of photons takes place, followed by electron–hole (*e–h*) pair generation. The electron (e^-) of valance band (VB) jumps to the conduction band (CB) of SC, leaving behind a hole (h^+) in the VB. The hole in VB can generate $^{\bullet}OH$ radical, which can perform nonselective oxidation of pollutants. Figure 12.3 depicts various steps in the photocatalysis process with respective time sales [28].

i. Absorption of suitable light energy by SC.
ii. Generation/separation of *e–h* pairs which is in *fs* time scale

$$TiO_2 + hv \rightarrow TiO_2 + e^- \text{ (CB)} + h^+ \text{ (VB)} \qquad \ldots\ldots\ldots \qquad \text{fs}$$

iii. Recombination of *e–h* pair to liberate heat with ~ 10 *ps* time scale indicates ~97% probability of *e–h* pair recombination than reaction

$$h^+ + e^- \rightarrow \text{recombination} + \text{heat} \quad \ldots\ldots\ldots \quad 1\text{–}10\ \text{ps}$$

iv. VB hole can migrate to the surface followed by oxidation reactions;

$$TiO_2\,(h^+) + OH^- \rightarrow TiO_2 + {}^{\bullet}OH \quad \ldots\ldots\ldots \quad 100\text{–}200\ \text{ps}$$

$$TiO_2\,(h^+) + H_2O_{ad} \rightarrow TiO_2 + {}^{\bullet}OH + H^+ \quad \ldots\ldots\ldots \quad 100\text{–}200\ \text{ps}$$

v. CB electron does a reduction reaction to complement oxidation of step 4

$$O_2 + e^- \rightarrow {}^{\bullet}O_2^- \quad \ldots\ldots\ldots \quad <100\ \text{ns}$$

vi. Reactive oxygen species (${}^{\bullet}OH$ and ${}^{\bullet}O_2^-$) produced can carry out photocatalytic reactions, especially oxidation of organic pollutants yielding intermediates in a stepwise manner based upon the size of the pollutant and finally making simple molecules like CO_2 and H_2O.

$$\text{Organic pollutant} + {}^{\bullet}OH/{}^{\bullet}O_2^- \rightarrow\rightarrow\rightarrow CO_2 + H_2O \quad \ldots\ldots\ldots \quad >300\ \text{ps}$$

vii. The CB electron can reduce shallow lattice Ti^{4+} ion to form a trapped electron as Ti^{3+} ion

$$Ti^{4+} + e^- \rightarrow Ti^{3+}\ \text{(shallow electron trap)} \quad \ldots\ldots\ldots \quad 100\text{–}200\ \text{fs}$$

viii. VB hole can be trapped in titanol group as surface-bound hydroxyl radical as

$$Ti(IV) - OH + e^- \rightarrow Ti(III) - OH^{\bullet +}\ (\text{shallow } e^- \text{ trap}) \quad \ldots\ldots\ldots \quad 50\text{–}170\ \text{fs}$$

12.4 Factors Affecting Photocatalytic Efficiency

Since the time scale of the *e–h* recombination rate is much smaller than that of the oxidation reaction of pollutants; hence the efficiency of TiO_2 photocatalysis is highly limited by the high recombination rate. Hence, immense research efforts are targeted to control the *e–h* separation which depends on various properties of materials such as listed in Figure 12.4.

12.5 Crystal Phases of TiO_2

TiO_2 occurs in four major polymorphs/phases/allotropic forms, namely, anatase, rutile, brookite, and bronze. All four types of TiO_2 consist of TiO_6 octahedra, but each differs in the distortion of the octahedron units with differing sharing of edges and corners in different manners (Figure 12.5) in the specific crystal structure, hence showing different physical properties (Table 12.4). The anatase structure consists of edge-sharing TiO_6 octahedra wherein two TiO_6 octahedra sharing two adjacent edges with two other TiO_6 octahedra arranged in planar double chains which run

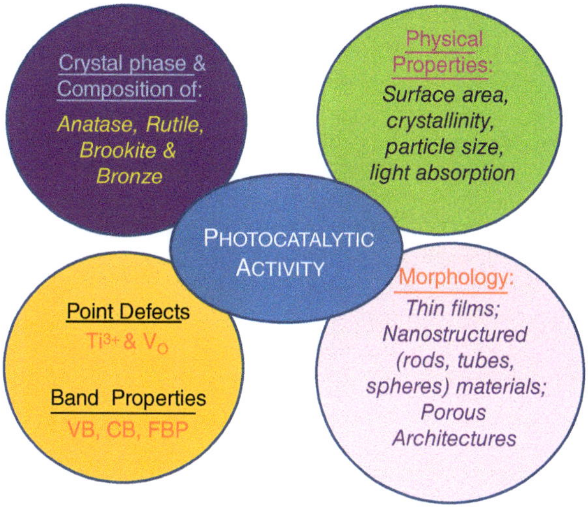

Figure 12.4 List of factors to control Photocatalytic efficiency.

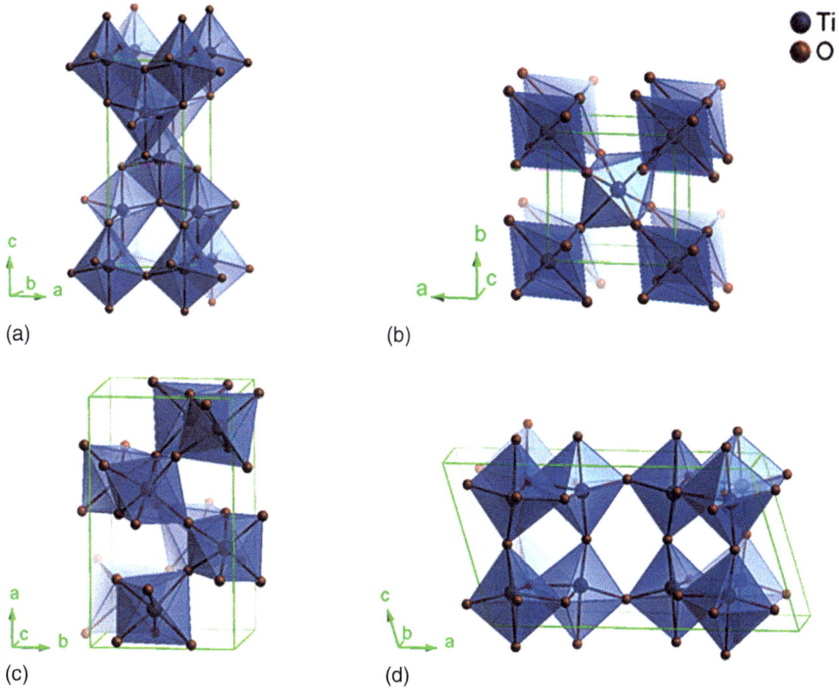

Figure 12.5 Crystal structures of: (a) anatase, (b) rutile, (c) brookite, and (d) bronze. Source: Ma et al. [29]/American Chemical Society.

Table 12.4 Structural properties of various TiO_2 polymorphs.

Parameters	Anatase	Rutile	Brookite	Bronze
Space group	Tetragonal $I4_1/amd$	Tetragonal $P4_2/mnm$	Orthorhombic *Pbca*	Monoclinic $C2/m$
Space-filling polyhedra	Edge-sharing TiO_6 octahedra arranged in a planar double chains	Corner & edge-sharing TiO_6 octahedra connected in linear chains	Corner & edge-sharing TiO_6 octahedra	Corner and edge-sharing TiO_6 octahedra arranged in layered sheets
Lattice parameter (Å)	$a = 3.79$ $c = 9.51$	$a = 4.59$ $c = 2.96$	$a = 9.17$ $b = 5.46$; $c = 5.14$	$a = 12.17$; $b = 3.74$; $c = 6.51$; $\beta = 107.3$
Bandgap (eV)	3.20	3.00	3.30	3.15
Density (g cm^{-3})	3.79	4.13	3.99	3.64
Refractive index[a)]	2.54	2.75	—	—

a) Refractive index @ $\lambda = 550$ nm [30].
Source: Adapted from Aravindan et al. [31].

in a izig-zag chains along {221} direction [31, 32]. Whereas the rutile shares both corner and edge of TiO_6 octahedra connected in linear chains parallel to {001} direction, and similarly, the brookite frameworks exhibit both corner- and edge-sharing configurations. On the other hand, bronze/TiO_2(B) is mainly derived from layered titanates. Therefore, the structure of bronze is similar to that of the layered precursor, which is composed of corrugated sheets consisting of both edges and corners shared by TiO_6 octahedra [29] with a relatively open crystal structure [33]. Rutile is the only stable phase in the bulk form, while bulk brookite and bulk anatase are metastable and transform irreversibly to rutile upon heating. The phase change from anatase to rutile occurs in different temperature ranges from 600 to 1100 °C, depending on the preparation conditions, particle size, and the presence of impurities [34]. The most photo-catalytically active form of TiO_2 is the anatase phase compared to other phases, and it is widely studied due to its technological importance in various applications. Rutile and brookite are considered the least photo catalytically active. However, the bronze structure is less explored in photocatalysis [35, 36].

12.6 Semiconductor/Electrolyte Interface and Surface Reaction

When a SC is immersed in redox electrolyte, a redistribution of electric charges takes place at the interface/phase boundary resulting in disturbance of electrical neutrality at the interface of SC/electrolyte forming an electrical double layer (EDL) (Figure 12.6). The term EDL is used to describe the arrangement of charges and oriented dipoles constituting the interphase region at the interface of an electrolyte. The inner Helmholtz plane (IHP) indicates the region between SC surface and

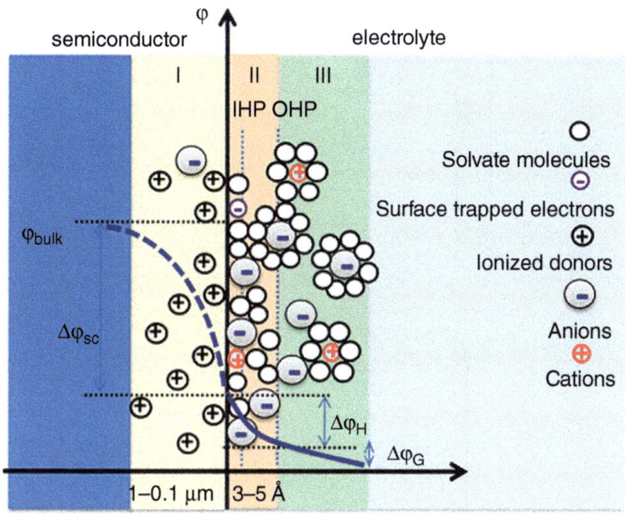

Figure 12.6 Electrical double-layer structure of an n-type semiconductor in contact with the electrolyte under equilibrium conditions. The potential (φ) profile across the double-layer region constitutes $\Delta\varphi_{SCR}$, $\Delta\varphi_H$, and $\Delta\varphi_G$, which are the potential drops in the space charge region (I), Helmholtz layer (II), and Gouy layer (III), respectively. Source: Ding et al. [37]/American Chemical Society.

water layer, which contains electrons trapped in surface states, chemisorbed ions (OH^-/H^+), and solvent molecules. The outer Helmholtz plane (OHP, II) contains solvated ions and marks the closest distance of the ions electrostatically adsorbed on the SC surface. Further layers out from this layer are part of the diffuse Gouy layer (III).

According to the SC solid-state physics (Figure 12.7) when the work function of SC (Φ_{SC}) and electrolyte (Φ_E) are different (case (a) $\Phi_{SC} < \Phi_E$), equilibration of Fermi levels (E_F) occurs by the flow of mobile charge carriers (electrons) from SC to electrolyte under dark or open-circuit condition. Hence, the solution side of the double layer develops a negative charge, forming a Helmholtz compact layer (about 3–5 Å in thickness). To preserve electrical neutrality, a positive space charge layer develops at surface of the SC called space charge region (SCR) (Figure 12.7). SCR is a characteristic region within the SC within which the charge (majority carriers, electrons for n-type SC) would have been removed at equilibrium, and such a region is called the depletion layer (Figure 12.7a), so termed because the concentration of mobile majority carriers is less at the surface than in the bulk phase and there is an excess of immobile ionized donors (holes) at the surface. In this situation, the electrostatic potential energy of electrons decreases in the depletion layer, causing upward band bending, and the potential energy becomes more negative as we move from SC surface into the bulk (where energy remains flat). Due to charge distribution at SC/electrolyte, an electric field is developed in SCR than in bulk (field-free), however, the SC as a whole is electrically neutral [38]. Similarly, when a p-type SC is immersed in an electrolyte (case (b) $\Phi_{SC} > \Phi_E$), Fermi-level equilibration takes place with a downward BB due

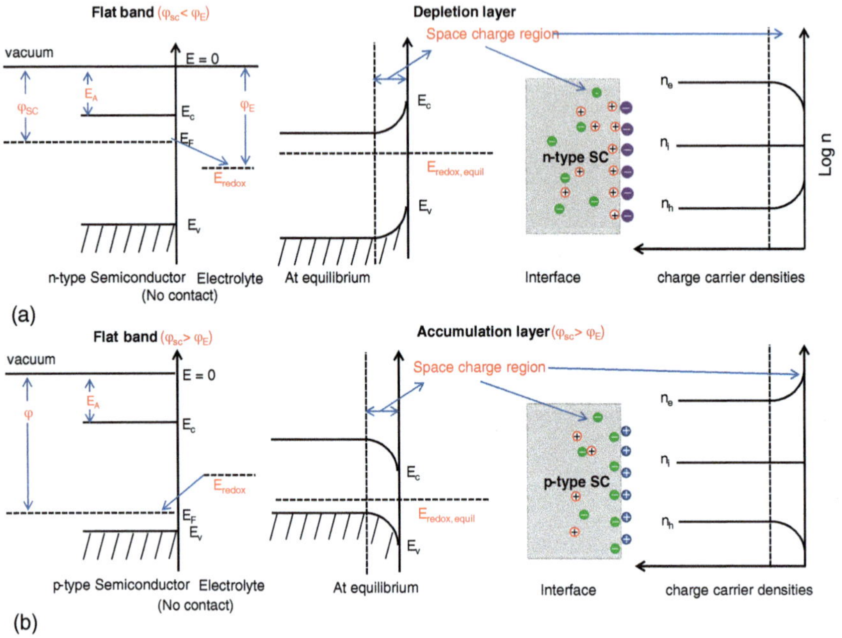

Figure 12.7 Energy levels in a semiconductor and a redox electrolyte shown on a common vacuum energy reference scale. E_A and Φ_{SC} are the semiconductor electron affinity and work function, Φ_E is the work function of the electrolyte. E_C is the energy of the conduction band, E_V is the energy of the valence band, E_F is the Fermi level of SC and E_{redox} is Fermi level of electrolyte, respectively. (a) Band bending in n-type SC with the formation of the depletion layer. (b) Band bending in p-type SC with the formation of accumulation layer with profiles of free charge carrier densities.

to an increase in potential energy of electrons, forming an accumulation layer as electrons move from electrolyte and accumulate in the SC SCR.

Till now, our discussion was related to n-type SC/electrolyte interface in dark or under open circuit, with band bending generating barrier height ($V_{barrier}$), theoretical maximum energy that can be generated at the interface (Figure 12.8 a,b). When we continuously illuminate light equivalent to or higher than the bandgap of the SC, light absorption takes place, followed by *e–h* pair generation creating a quasi-Fermi level nonequilibrium condition, with the quasi-Fermi level of electrons ($E_{F,n}$) close to the flat-band potential (E_{fb}/E_F), and the quasi-Fermi level of holes ($E_{F,p}$) shifts downward, generating an open-circuit photovoltage (V_{ph}) (Figure 12.8c). The value of V_{ph} is usually smaller than the bandgap of the SC, as there is a potential drop due to the Helmholtz layer due to Fermi level pinning, resulting a reduced V_{ph}. Hence the photoelectrolysis of water is possible only if $V_{ph} > 1.23$ V which can change with E_F of SC and E_{redox} of electrolyte. The excited electron in CB is an excess electron in the depletion layer which would move into the bulk of SC in the direction of the existing electric field away from Helmholtz layer with a negative charge, whereas the holes would move in the opposite direction toward surface away from repelling positive charge in the depletion layer as shown in Figure 12.8c. The direction of

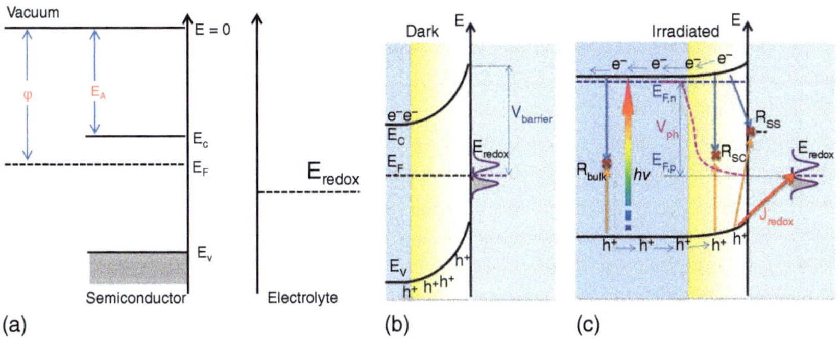

Figure 12.8 (a) n-type SC energy profile before in contact with the electrolyte, (b) n-type SC energy profile in equilibrium with the electrolyte under dark conditions, (c) Quasi-static energy profile and charge transfer and recombination pathways of an n-type SC under continuous illumination in contact with the electrolyte. Source: Ding et al. [37]/American Chemical Society.

flow of charge is very critical for the efficiency of photocatalysis which depends on nature of SC, electrolyte, and intern extent of band bending as it determines the preference of surface reaction (oxidation/reduction). TiO_2 as an n-type SC with upward BB and depletion layer makes it highly suitable for oxidation reaction/generation of $^{\bullet}OH$ from adsorbed water through its high concentration of holes in its depletion layer. Therefore, the extent of band bending in the space charge layer influences the E_F of the SC and the density of carriers on the surface hence influencing the surface charge transfer process. Along with the targeted charge transfer from the valence band to the redox reagent (J_{redox}), there are several recombination pathways, including bulk recombination (R_{bulk}), space charge layer recombination (R_{SC}), and surface state recombination (R_{SS}).

However, the key challenges of photocatalysis are to reduce charge recombination, increase photovoltage V_{ph}, and increasing active surface area of SC.

12.7 Visible-Light Harvesting

The sun emits radiation ranging from X-rays to radio waves, but the widest and most intense solar radiation occurs in the visible light range (Figure 12.9), such that 43% of the solar energy reaching the Earth's surface is at visible wavelengths from 400 to 700 nm. TiO_2 is a wide-bandgap SC ($E_g \sim 3.2$ eV) which can be exited only by UV-light resulting in optical excitation of electrons in the VB into the CB. On the other hand, the momentum (p) (wave vector) of the VB maximum (VBM) and CB minimum (CBM) also affect the light absorption and recombination. Based on the alignment of momentum, there are two types of bandgaps (i) direct and (ii) indirect (Figure 12.10). In direct bandgap, the wave vector of an electron in the VBM is equal to CBM, as shown in Figure 12.10. In a direct bandgap SC, the electron is freely promoted from the VB to the CB, with a conservation of momentum. In contrast, in an indirect bandgap SC, the states in VBM and the

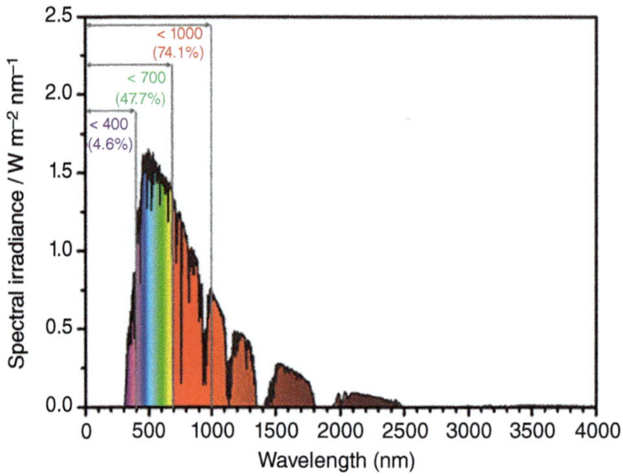

Figure 12.9 AM 1.5 G solar spectrum based on the ASTM G173-03 reference spectrum.

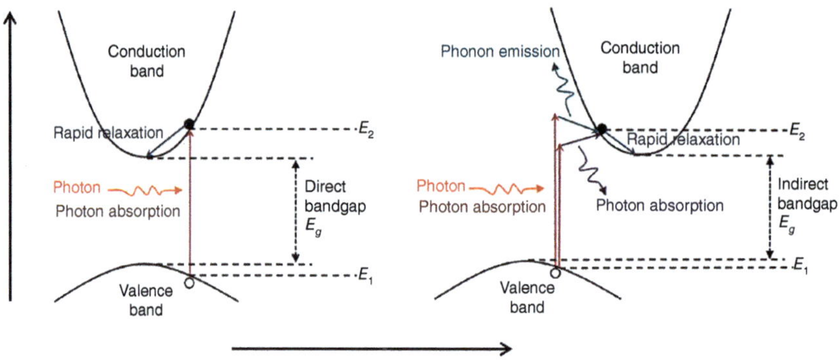

Figure 12.10 Photon absorption in SC having (a) a direct bandgap and (b) indirect bandgap. Source: Wang and Domen [39]/American Chemical Society.

CBM do not have similar momenta. Hence, in order to bring the absorption of a photon in an indirect bandgap, an electron must undergo a significant change in its momentum to conserve momentum. This is brought about by electron interaction with a quantized lattice vibration known as a *phonon*, which increases or decreases its momentum. Since, indirect bandgap excitation process involves the electron, photon, and phonon together it has a lower probability and at a much slower rate than direct absorption. Since TiO_2 is a wide-band indirect bandgap SC, its limited light absorption in visible light limits its practical solar efficiency. Hence, there is an immense effort to improve TiO_2 visible light absorption. Figure 12.11 shows different strategies to bring visible light absorption ability [39].

Extending TiO_2 absorption into the visible light region will drastically improve the efficiency of photocatalysis and practical applicability in the future.

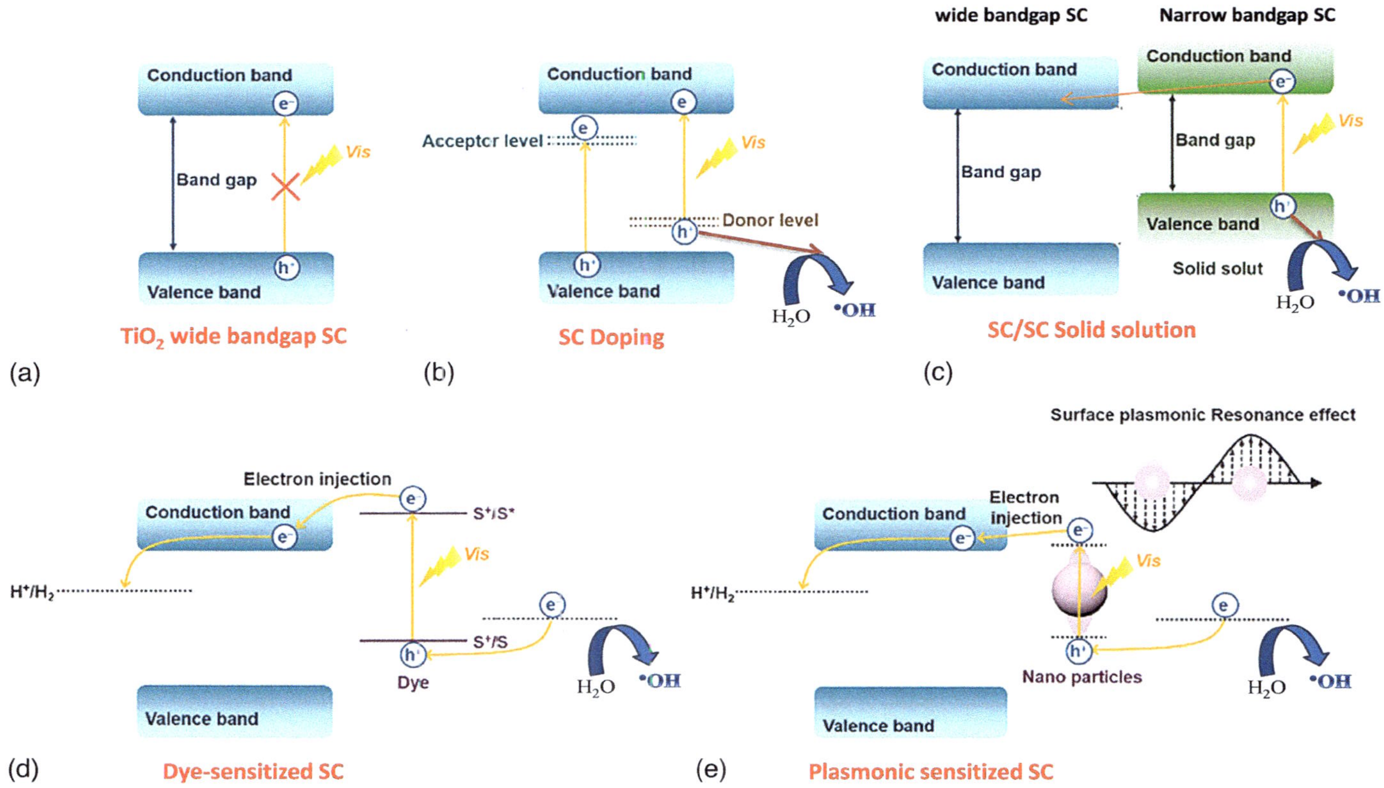

Figure 12.11 Schematics showing (a) visible light-inactive TiO_2 SC with a wide bandgap and (b–e) strategies for obtaining a visible light response: (b) SC doping, (c) SC/SC solid solution formation, (d) dye-sensitization SC, (e) plasmonic sensitized SC.

12.8 Photogenerated Charge Separation Strategies

The photogenerated charge carriers generated in SC upon absorption of photons are most likely recombined or less likely separated to carry out the surface reaction since recombination occurs at the picosecond time scale, whereas charge separation and migration happen at the femtosecond time scale. Hence, controlling charge separation is the key factor in determining photocatalytic efficiency. Therefore, various junction structure strategies (Figure 12.12) were explored to increase the charge separation efficiency.

12.8.1 TiO_2/Carbon Heterojunction

The coupling of TiO_2 with carbon-based materials, especially with carbon nanotubes (CNTs) and graphene, has attracted increasing attention [40]. CNTs with defined electronic properties can be chemically bonded with TiO_2. The such composite material also shows a large surface area ($> 150\,m^2\,g^{-1}$) and excellent mechanical properties. Besides, a much higher theoretical specific surface area ($\sim 2600\,m^2\,g^{-1}$) was discovered on graphene, an sp^2 -hybridized two-dimensional carbon nanosheet provides unique electronic properties. The titania-graphene [41] composite material exhibits high mobility of charge carriers as well as good mechanical strength. Therefore, these carbon-based materials are promising candidates to facilitate charge transfer and inhibit the charge recombination process when combined with TiO_2-based photocatalysts, as C support act as an electron sink, and scavenge away the electrons hindering recombination (Figure 12.13). And also, the carbon–oxygen–titanium bond can extend the light absorption to longer wavelengths similar to carbon-doped titania.

12.8.2 TiO_2/SC Coupled Heterojunction

When two different SCs are coupled, in which both SCs are excited simultaneously by photons to generate *e–h* pair. The directions of electron/hole transfer in SC would

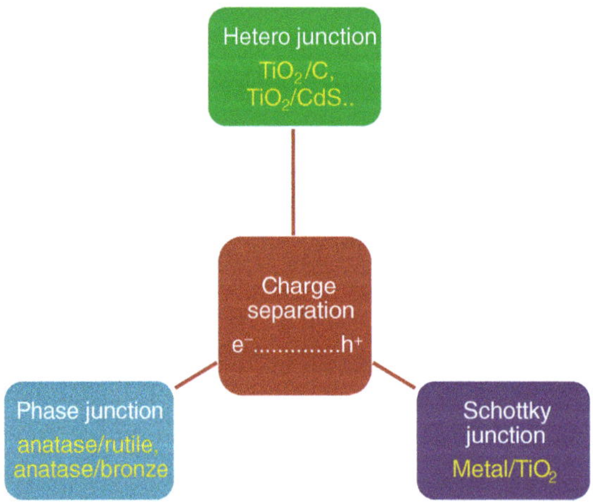

Figure 12.12 Strategies for improving charge separation/recombination.

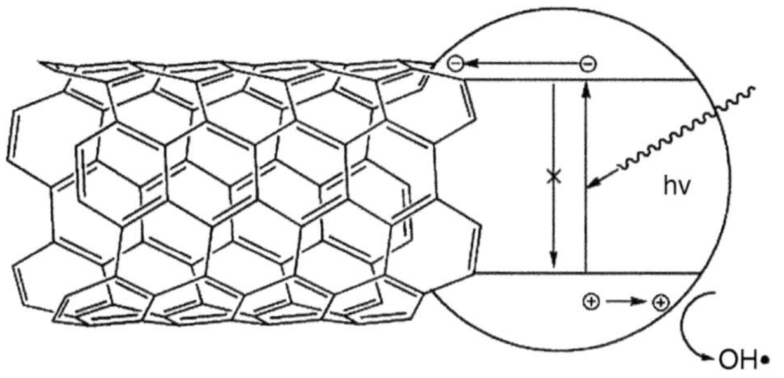

Figure 12.13 CNT-mediated enhancement of TiO_2 photocatalysis. Source: Adapted from Woan et al. [40].

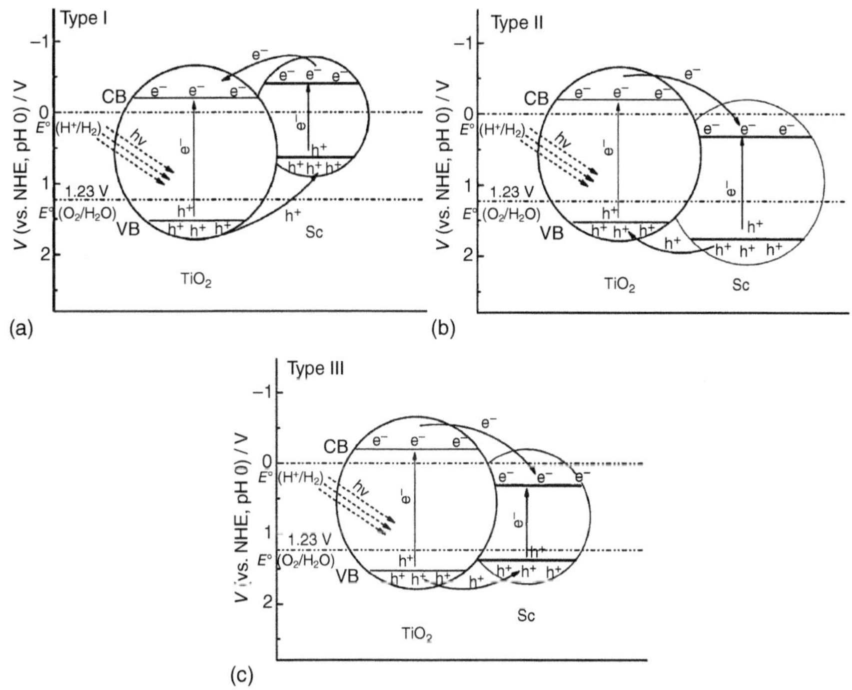

Figure 12.14 Band structures and migration of the charges in three types of heterojunctions of TiO_2 with other SCs. Source: Leung et al. [42]/John Wiley & Sons.

depend on the relative positions of CB and VB of the two SCs. Generally, there are three ways for coupling TiO_2 with another SC into a heterojunction photocatalyst, *viz.*, types I, II, and III shown in Figure 12.14 [42]. According to Gibbs's free energy change principle, the photo-induced electron would be injected from the SC with a more negative CB level to the positive one, while hole would be transferred from the SC with a more positive VB level to the negative one. For example, for the oxidation of H_2O by VB hole, the VBM edge must be more positive than the oxidation potential of O_2/H_2O (1.23 V vs. NHE; pH 0). Hence based on the reaction of interest, a proper design of heterojunction is required. For types I and II (Figure 12.14a,b), the electrons and holes migrate to the different SCs in opposite directions, hence bringing spatial charge separation, which is also called the “direct Z-scheme”. On the other hand, in type III heterojunction (Figure 12.14c), both the electron and hole transfer from TiO_2 to the coupled SC without spatial charge separation. Type II heterojunction is most suitable for $^{\bullet}OH$ generation because ΔG change for H_2O oxidation with hole-occupied VB in type II is highly negative.

12.8.3 TiO_2/TiO_2 Phase Junction

The anatase phase of TiO_2 is known to exhibit usually higher photocatalytic activity than rutile [43, 44]. Interestingly commercial Degussa P-25 TiO_2 is composed of mixed phases of anatase (~80%) and rutile (~20%) and shows exceptional activity than either pure anatase or rutile phases. The improved activity in the mixed-phase structure of P-25 TiO_2 is widely studied and the most accepted explanation is the charge separation concept, that a type-II, staggered, band alignment of ~0.4 eV exists between anatase and rutile with anatase possessing the higher work function (Figure 12.15). A strong driving force for charge separation is created at the phase junction between anatase and rutile than that either of the pure anatase or rutile.

Choice of hetero/phase junction with appropriate band positions (VB, CB) can bring special charge separation are primly important for efficient photocatalysis.

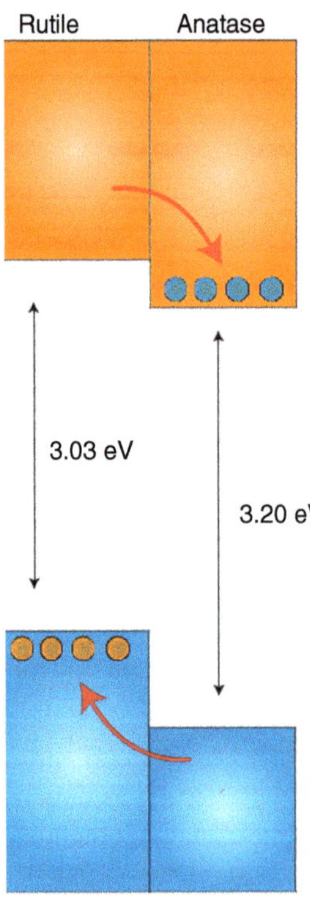

Figure 12.15 Band alignments of VB and CB positions for the anatase/rutile phase junction. Red arrows indicate the flow of electrons (holes) in the conduction band (valence band). Blue and orange dots represent electrons and holes, respectively. Source: Adapted from Scanion et al. [45].

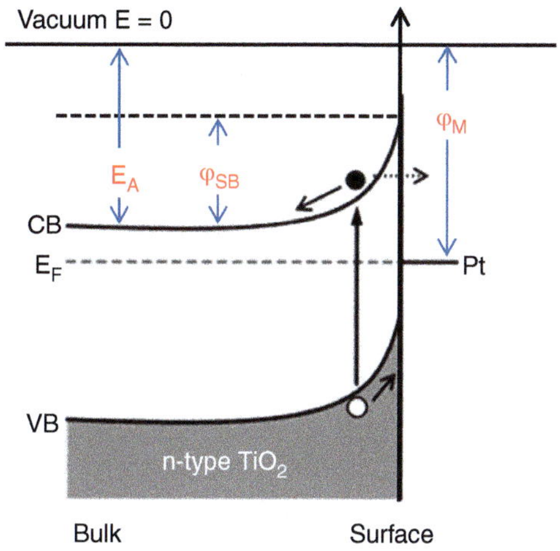

Figure 12.16 The influence of Pt on the photogenerated charge carriers in TiO_2 (1) flow of electrons into the bulk of SC due to the Schottky barrier (φ_{SB}), and (2) flow of the electrons under thermodynamic driving force.

12.8.4 Metal/TiO_2 Schottky Junction

Metal, e.g. planinum, have higher work function (φ_M) compared to SC TiO_2. When metal and SC are connected, there will be flow of electrons from the SC to the metal until the E_F of the SC equilibrates with that of the metal. This process develops a potential barrier at the interface called Schottky Junction/Barrier (φ_{SB}), which electrons (of the SC) must overcome in order to flow from the SC to the metal at equilibrium (Figure 12.16). In other words, at equilibrium, there is an accumulation of negative electrons at the metal surface and a formation of the positive space charge layer or the depletion layer below the surface of the SC in order to maintain electrical neutrality at the interface. The Schottky barrier can be expressed by the following equation:

$$\Phi_{SB} = \Phi_M - E_A$$

TiO_2 impregnated with Ag, Au, and Pt nanoparticles having work function 4.74 (Ag), 5.31 (Au), 5.93 (Pt), and 4.6–4.7 eV (TiO_2), respectively, exhibited enhanced photocatalytic activity resulting from an improved charge separation due to Schottky barrier generation. Specifically, Pt with the highest difference in work function shows a large Schottky barrier and efficient charge separation. This causes the transfer of the photogenerated holes to the surface and the simultaneous migration of the electrons into the bulk induced by the electric field in the space charge layer as shown in Figure 12.16.

12.9 Ordered Mesoporous Materials

Porous materials are of great interest and show promise in heterogeneous catalysis because of their ability to interact with reactants/molecules not only at the surface but also enhance the accessibility of the reactant molecules to the active sites [46, 47]. In the 1990s, ordered mesoporous materials of silica developed by Mobil

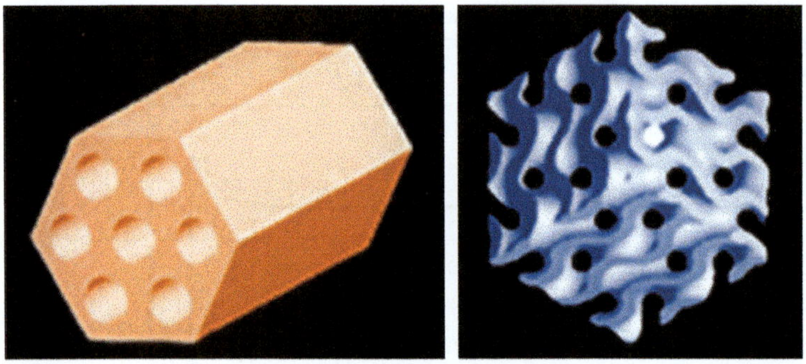

Figure 12.17 Ordered mesoporous materials with various porous architectures: 2D hexagonal (left) and 3D cubic (right) framework structures.

scientists have opened a wide range of opportunities in heterogeneous catalysis [48, 49]. Based on pore architecture ordered mesoporous silica (OMS) materials can be developed in various structures, *viz.,* 2D hexagonal OMS (MCM-41), 3D cubic OMS (MCM-48) (*see* Figure 12.17). These OMS materials were synthesized by soft-template based liquid-crystal mechanism using cationic surfactant, cetyltrimethyl ammonium bromide (CTAB) assemblies under basic conditions in aqueous medium. Structurally, MCM-41 consists of uniform 2D-hexagonal arrangement of mesopores which arrange themselves as honey comb-like structure, whereas MCM-48 comprises of three-dimensional cubic gyroid pore structure. Similarly, Stucky et al. [50, 51] have reported a series of OMSs such as SBA-15 (hexagonal) and SBA-11 (cubic) porous silica framework structures synthesized under acidic conditions in aqueous medium using non-ionic soft-templates (P-123, Brij-56) contrary to MCM-41/MCM-48 (in basic medium). Similarly, Selvam et al. [52] reported IITM-56 silica having 2D hexagonal ordered porous structure having advantages in textural properties over both MCM-41 and SBA-15.

The porous materials have opened up new avenues in the synthesis of advanced materials, in heterogeneous catalysis, CO_2 sequestration, batteries, environmental pollution control strategies, and separation processes due to their unique flexibility in terms of synthetic conditions, pore size tuning, high surface area, and large internal hydroxyl groups [46, 53]. Because of their immense potential in recent years, extensive research efforts have been devoted to synthesize high-quality ordered mesoporous materials of carbon and transition metal oxides. After 10 years of discovery of OMS, Ryoo et al. [54, 55] have reported the ordered mesoporous carbon's (OMC) CMK-1 (cubic) and CMK-3 (hexagonal) using MCM-48 and SBA-15 porous silica as hard templates. In the same line, Selvam et al. [56, 57] demonstrated NCCR-41 and NCCR-11 mesoporous carbon materials using MCM-41 and SBA-11 as porous silica templates. On the other hand, synthesizing ordered mesoporous titania (OMT) metal oxide is a challenging problem due to titania precursors' high hydrolysis and condensation rates in aqueous medium. As a result, aqueous medium synthesis of mesoporous titania causes uncontrolled hydrolysis and condensation, hence, no control over preparation of ordered porous structure. Hence, in recent years, extensive research efforts have been devoted to synthesize high-quality OMT [58, 59]. However, only a few reports are available on such materials with ordered pore structure and well-crystallized matrix. Yang et al. [60] have first reported several ordered mesoporous metal oxides *via* evaporatio-induced self-assembly (EISA) process employing Pluronic P-123 as a structure-directing agent. Since then, it has triggered a great interest as the SC's porous channels can offer numerous advantages over the analogous bulk materials. Although several researchers claim the successful preparation of mesostructured titania but most of them lack either the ordered mesopores and they possess bicontinuous wormlike structure or irregular sphere-like morphology. Such attempts have been continued even after without much success due to uncontrollable reactivity of titanium precursors toward hydrolysis, condensation, and crystallization, and the mesoscopic structure collapses upon removal of the structure-directing agents [61–70]. Hence, it is an important challenge to the scientific community for the successful synthesis of mesoporous titania with a well-crystalized titania with a highly ordered porous framework structure.

12.10 Ordered Mesoporous Titania

Over the past few decades, persistent research efforts have been made toward the development of solar energy materials that are safe, efficient, and promising for harvesting solar energy. In this context, titania or titanium dioxide (TiO_2) is one of the well-studied metal oxides for energy and environmental applications. It has also attracted a great deal of attention for photocatalytic, electrocatalytic, and photoelectrochemical applications owing to its excellent physico-chemical characteristics, *viz.*, high stability, low-cost, and nontoxicity [58, 71–76]. Hence, titania is considered to be one of the key photocatalysts for the sustainable production of fuels and chemicals required for a carbon-neutral society. When titania is irradiated with photons of suitable energy, the photogenerated electron, and holes, i.e. the excitons, can thus be utilized for reduction and oxidation reactions in a green manner which is also essential for a sustainable environment. Indeed, the *e–h* recombination is the most serious problem and a major fundamental limitation in any photocatalytic process. This has limited not only the efficiency of titania but also several other promising photocatalytic materials [58, 74–76]. The photocatalytic activity highly depends on experimental conditions such as catalyst amount, light intensity, exposure area, reactor volume, temperature, and pressure of the reaction vessel. In addition, numerous other parameters such as particle size, crystal phase, phase composition, surface area, surface defects, morphology, efficient charge separation, and optical bandgap influence the catalyst performance. Several attempts have been made to enhance its optical, electronic, and textural properties by tuning the morphology and/or using different synthetic approaches.

In this regard, porous materials are of great interest and show promise in catalysis because of their ability to interact not only at the surface but also due to the enhanced accessibility of the reactant molecules to the active sites [46, 47]. Thus, porous titania, with its continuous framework structure can promote facile charge migration and separation owing to its characteristic properties such as tunable pore sizes, high surface areas, large pore volumes, alternative pore shapes, and controllable compositions [51]. Hence, in recent years, extensive research efforts have been devoted to synthesizing high-quality OMT [58, 59]. However, only a few reports are available on such materials with an ordered pore structure and a well-crystallized matrix. Yang et al. [60] have first reported several ordered mesoporous metal oxides via an EISA process using Pluronic P-123 as a structure-directing agent. Since then, it has triggered great interest as the porous channels offer numerous advantages over analogous bulk materials. Although several researchers claim the successful synthesis of ordered mesostructured titania or OMT, most of them lack either the ordered mesopores or they possess a bicontinuous wormlike structure or irregular sphere-like morphology. Such attempts have been continued even without much success due to the uncontrollable reactivity of titanium precursors toward hydrolysis, condensation, and crystallization, and the mesoscopic structure collapses upon the removal of the structure-directing agents [61–70]. Hence, it is a great challenge for the scientific community to successfully synthesize mesoporous titania with a well-crystallized, highly ordered porous framework structure [77, 78].

12.10.1 Synthesis and Characterization

In this regard, Selvam et al. [79–84] have recently demonstrated the successful synthesis of highly ordered 2D-hexagonal mesoporous titania *via* a kinetically controlled EISA process. In this investigation, we present EISA synthesis strategy (see Figure 12.18) for OMTs using P-123 and CTAB as structure-directing agents, and named as TMP-123, TMC-016, and TMC-036, respectively. The resulting materials' detailed characterization and activity were evaluated for the photocatalytic degradation of pharmaceutical pollutant famotidine (FAM), 3-{[2-(diaminomethyleneamino)-1,3-thiazol-4-yl]-methylthio}-N^2-sulfamoyl-propionamide, and 4-chlorophenol (4-CP), an industrial pollutant. These pollutants are, in general, discharged from the respective industry and partially degraded during water treatment. The efficiency of all these catalysts was compared with commercial titania (P-25) photocatalyst.

During synthesis process, it is important to note that a careful control of the slow aging process is crucial for the formation of anatase phase with an ordered pore structure, while the formation of TiO_2(B) phase is due to high nucleation rate, presumably at the interface between the surfactant micelles and the hydrolyzed titania

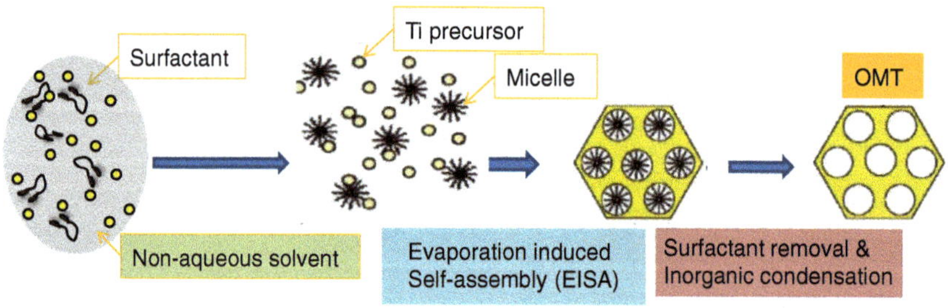

Figure 12.18 Diagrammatic description of the EISA process for the synthesis of the ordered mesoporous titania (OMT).

precursor [66] and that it might also be due to the differences in the hydrolysis, condensation of precursors at micelle [85]. Figure 12.18 shows the diagrammatic representation of the formation mechanism of mesoporous titania. All the samples were well-characterized using various analytical, spectroscopic, and imaging methods [83]. Interestingly, electron paramagnetic resonance (EPR) studies are performed for P-25 and mesoporous samples (Figure 12.19). The room temperature CW-EPR spectra of TMP-123 clearly depict the presence of Ti^{3+} defect centers in the crystal lattice with g = 1.995 and 1.985 in agreement with the literature [86–88]. Similarly, other mesoporous samples show these signals. However, it is interesting to note that such a feature is not observed for commercial P-25 sample, however, very weak signals (g = 1.992, 1.970) can be noticed at cryogenic temperature. This could be attributed to fewer Ti^{3+} defect centers at the surface and/or bulk [89], whereas mesoporous samples have intense EPR signals, indicating the presence of a significant amount of Ti^{3+} defects in their lattice. The enhanced signal intensity of mesoporous samples is expected owing to a long spin–lattice relaxation of Ti^{3+} at low temperature [90, 91]. In general, the EPR signals at g = 2.004 can be assigned to electrons trapped at oxygen vacancy and/or Ti^{3+} centers.

The photoelectrochemical behavior of OMT was studied using electrochemical impedance spectroscopy (EIS). Figure 12.20 depicts the Nyquist and Mott–Schottky plots (M–S plots) for all the samples under investigation. In the case of the Nyquist plot, Figure 12.20a, the high-frequency intercept on the real axis corresponds to serial/solution resistance due to electrolyte, current collection, etc., and the semicircle pattern confirms the parallel combination of charge transfer resistance (R_{ct}) and double-layer capacitance (C_{dl}).

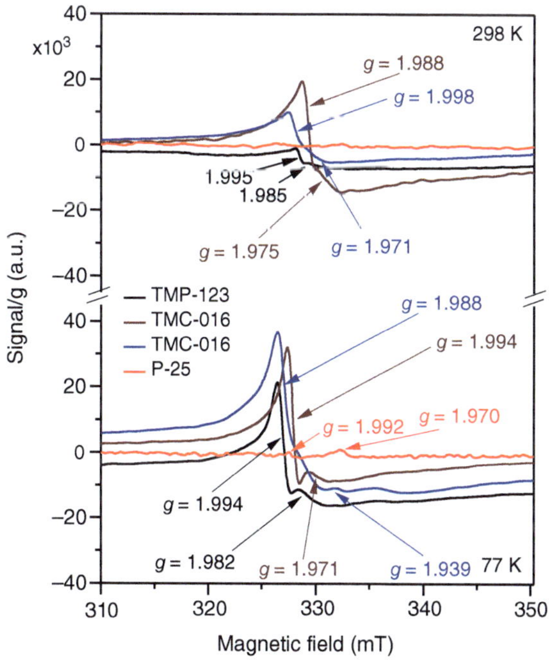

Figure 12.19 EPR spectra of: (a) mesoporous titania and (b) nanostructured titania P-25.

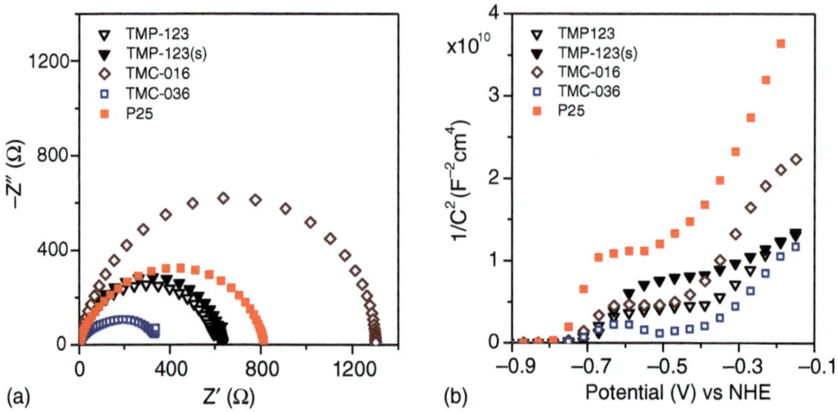

Figure 12.20 (a) Nyquist plot and (b) Mott–Schottky plot of various titania in 0.5 M H_2SO_4.

Furthermore, it can be seen from the Nyquist plot that the samples exhibit a variation in R_{ct} values, as a result of variable accessibility of ion at the solid-electrolyte interface (SEI), and that the values, *w.r.t.* open circuit potential, for OMT is much lower than P-25. The lower value depicts a fast charge transfer, which strongly depends on surface area, fast ion transfer, etc., at SEI. In order to study the contribution of the chemical capacitance of the surface states [92], we have recorded the M–S experiment between −0.15 and −0.9 V *vs* NHE potential window at 1 kHz in the presence of a Xenon lamp (280 W). The interfacial capacitance was estimated using the classical M–S equation [83, 93, 94]. From M–S plot, Figure 12.20b, we observe two slopes, which indicate different regions of capacitance due to the influence of minority carrier trapping at the surface and majority carrier transfer to the electrolyte. This surface state under illumination changes flat-band potential by trapping minority carriers on the surface. Furthermore, the positive slope of the M–S plot confirms the characteristic of *n*-type semiconducting nature of the materials. Carrier density and flat-band potential were determined from the slope and the intercept of M–S plot, respectively, and the values are listed in Table 12.5. It is clear from this table that the higher carrier density of OMT materials depicts higher interfacial charge transfer [89]. Furthermore, the lower value of flat-band potential and higher value of carrier density (donor trapping at surface) are in agreement with EIS data.

12.10.2 Photocatalytic Degradation Studies

4-CP Degradation: The unique characteristics of P-25 in terms of purity, high surface area, and the distinctive combination of anatase and rutile crystal phases, the product finds application in many catalytic and photocatalytic functions. Figure 12.21a depicts the photocatalytic degradation of 4-chlorophenol (4-CP) over various nanostructured catalysts. It can be seen that all OMT catalysts, in general, show improved degradation activity compared to that of the P-25 catalyst.

Among all the catalysts, TMP-123 shows the highest activity for 4-CP degradation as well (*see* Figure 12.22a: DE (95%) for TMP-123 and DE (60%) for P-25).

Table 12.5 Electrochemical data (EIS and M-S) of various nanostructured titania.

Catalyst	a) R_{ct} (Ω)	b) C_{dl} (μF)	c) V_{fb} (V vs NHE)	d) N_D (cm^{-3}) × 10^{19}
TMP-123	597	39.50	−0.737	5.55
TMP-123(s)	633	36.23	−0.707	3.28
TMC-016	1299	16.21	−0.755	4.36
TMC-036 (I)	731	16.7	−0.817	1.53
TMC-036 (II)	332	78.49	−0.749	8.47
P-25	813	12.43	−0.768	1.50

a) Charge transfer resistance.
b) double layer capacitance ($-1/2\pi f Z_{img}^{max}$).
c) flat-band potential.
d) carrier density (donor density).

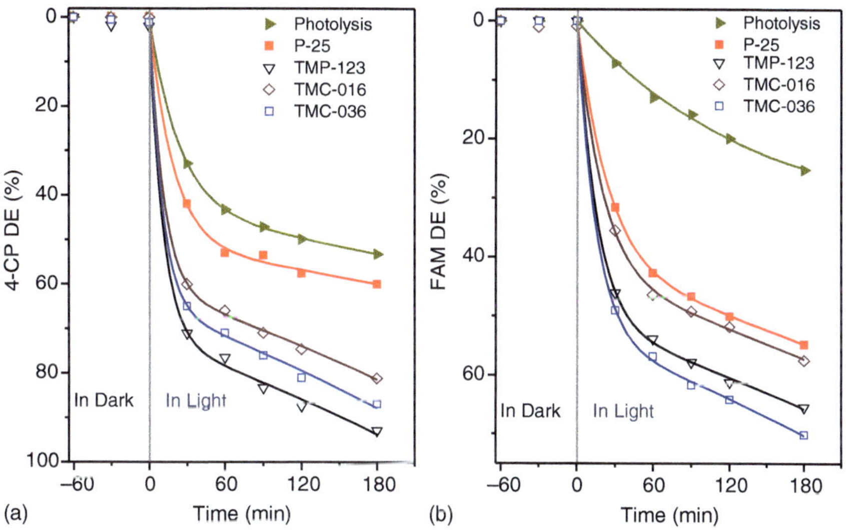

Figure 12.21 Photocatalytic degradation of organics at 25 °C: (a) Reaction conditions: 20 mg l^{-1} 4-CP; 0.4 g l^{-1} catalyst. (b) Reaction condition: 20 mg l^{-1} FAM; 0.25 g l^{-1} catalyst loading.

The first-order rate constant of TMP-123 is ($k = 3.5 \times 10^{-2}$ min^{-1}) two times higher than P-25 ($k = 1.6 \times 10^{-2}$ min^{-1}). The high activity of TMP-123 is well supported by the EPR studies (Figure 12.19), in facilitating photogenerated *e–h* pair separation. On the other hand, Figure 12.22b shows the UV–Vis degradation profile of 4-CP at 180 minutes for photolysis, P-25, and TMP-123, and we observe two absorption peaks for 4-CP corresponding to n–π* (280 nm from C–Cl group) and π–π* (225 nm from the aromatic group) [95], whose height decreased with irradiation time as shown in Figure 12.23. Also, we observe that the appearance of two absorption bands at around 250 nm attributed to benzoquinone (BQ) and the broad peak around 290 nm attributed to hydroquinone (HQ) was confirmed by the

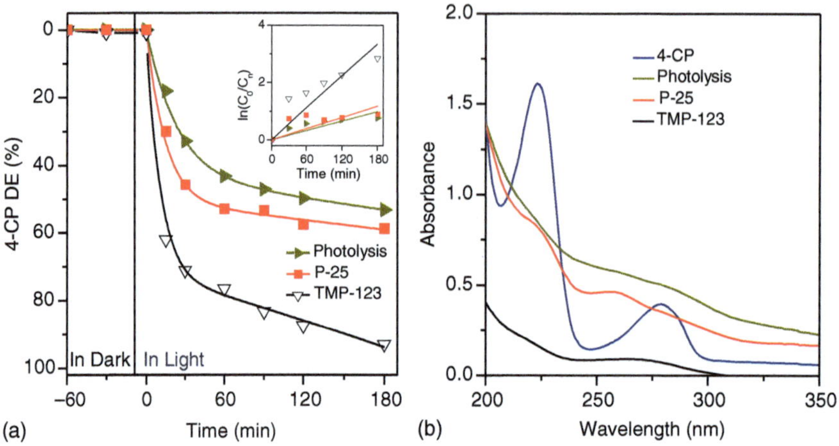

Figure 12.22 (a) 4-CP degradation over mesoporous and fumed titania. Reaction conditions: 20 mg l^{-1} substrate; 0.4 g l^{-1} catalyst loading. Inset – kinetics plot; (b) UV–Vis profiles of 180 minutes irradiated samples.

UV–Vis profile of authentic BQ and HQ as shown in Figure 12.23e. Interestingly, both intermediates BQ and HQ are persistent in P-25 even after 180 minutes of irradiation. However, TMP-123 shows negligible signatures from BQ and HQ intermediates, which indicates that TMP-123 is efficient in degrading 4-CP as well as its intermediates.

The high activity of the OMT could be attributed to their low R_{ct} (see Figure 12.20a and Table 12.5). When R_{ct} is small, the interfacial charge transferability of the catalyst will be high and, therefore, the efficient degradation of the substrates. A similar activity-EIS correlation was also reported by Yang et al. [96] and Zhang et al. [97] Furthermore, superior activity of OMT is due to high carrier density (N_D) determined from the M--S plot as shown in Figure 12.20b and Table 12.5 The higher the N_D of the material, the faster the charge transfer to the surface. Among all the OMT catalysts, TMP-123 and TMC-036 show the highest photocatalytic activity due to their low R_{ct}, high N_D, and high double-layer capacitance (C_{dl}) 4–6 times higher than P-25 (see Table 12.5). Hence, OMT materials are better than P-25 owing to their well-ordered 2D hexagonal crystalline framework, high specific surface area, and intrinsic defects (Ti^{3+}), providing efficient *e–h* separation, hence efficient surface oxidation reaction as shown in Figure 12.24. The results are also in line with the study by Alagarasi et al. [70, 98], where the mesoporous nanocrystalline TiO_2 materials exhibit appreciable activity for 4-CP degradation.

FAM Degradation: To further establish the dominance of OMT catalysts, the photocatalytic degradation of famotidine (FAM) was also tested for all the catalysts (Figure 12.21b). All the mesostructured catalysts show improved performance compared to the nanostructured fumed titania. Further, Figure 12.25 depicts the FAM degradation aimed to achieve 100% DE with all catalysts, and the change in FAM concentration was monitored by HPLC for accurate quantitative and intermediate analysis. TMP-123 and TMC-036 exhibit 100% DE within 75 minutes

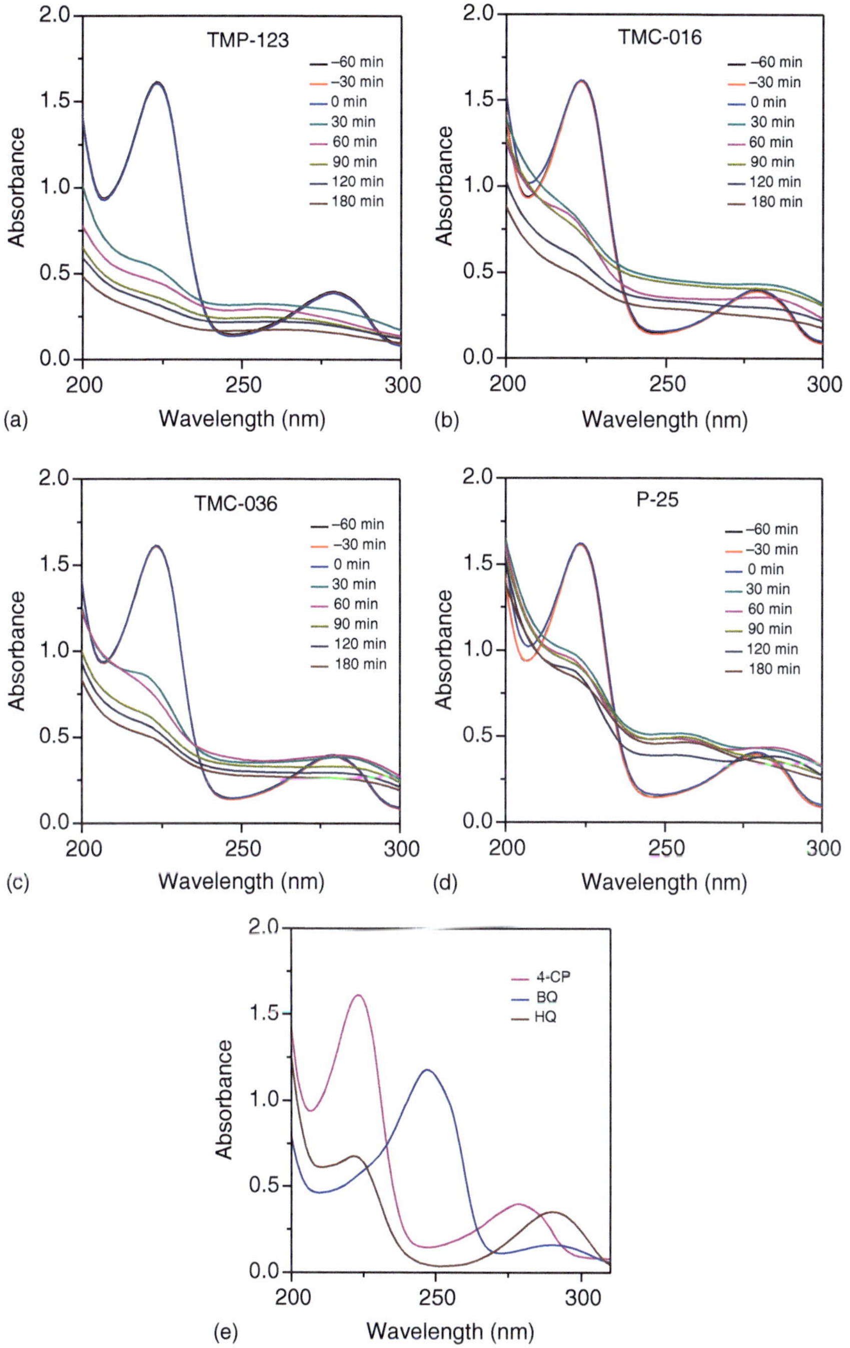

Figure 12.23 UV–Vis profiles of 4-CP degradation over various nanostructured titania: (a) TMP-123; (b) TMC-016; (c) TMC-036; (d) P-25; (f) UV–Vis profiles of authentic samples: 4-chlorophenol (4-CP), benzoquinone (BQ) and hydroquinone (HQ).

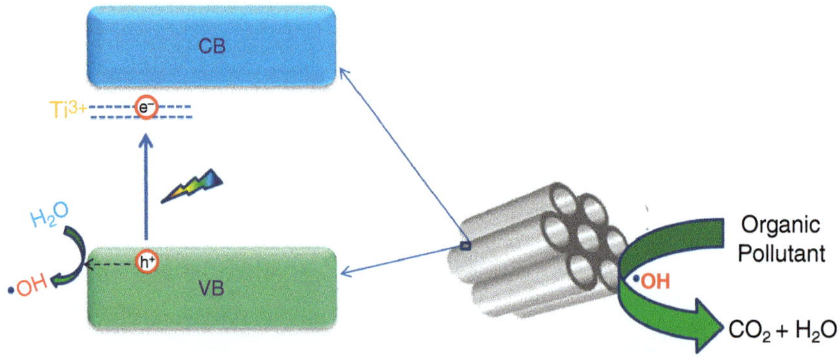

Figure 12.24 Photocatalytic degradation of organic pollutants with pristine ordered mesoporous titania catalyst.

while it is about 120 minutes for P-25. The first-order rate constant for the TMP-123 is 5.4×10^{-2} min^{-1}, which is about ~2.3 times higher than that of the latter (2.3×10^{-2} min^{-1}) as shown in *cf.* inset Figure 12.25a. It is also clear from the HPLC profiles (Figure 12.25b) that FAM eludes out at a retention time of 8 *min*, whereas degradation intermediates appear between 2 and 3 *min*. The famotidine peak has disappeared for TMP-123/TMC-036, indicating 100% DE, whereas in P-25, the FAM peak is still persistent, as shown in Figure 12.25b. FAM degradation was reported earlier by Keane et al. [99] for the P-25/activated carbon composite catalyst; however, complete degradation was not reported, whereas Murphy et al. [100] used the TCPP/P-25 catalyst and reported 100% famotidine degradation, which was confirmed by HPLC profiles. However, the catalyst takes ~180 minutes to reach complete FAM degradation. Hence, we can say that pristine OMT catalysts are new-generation catalysts with higher activity than commercial P-25 and hybrid P-25 catalysts.

12.10.3 Complete Mineralization Studies

During photocatalytic degradation of pharmaceutical pollutant famotidine, we observed that the FAM intermediates (secondary pollutants) in HPLC profiles (Figure 12.25b) which persist even after 180 minutes. Furthermore, to analyze degradation intermediates, liquid chromatography mass spectrometry (LC-MS) analysis was carried out for the TMP-123 FAM samples, as shown in Figure 12.26. FAM appears at *m/z* 338.054 as shown in Figure 12.26a at 0 minutes (FAM before illumination); however, when we checked the LC-MS of FAM degradation at the initial stage of 15 mintues of the reaction, we observed two major intermediates with *m/z* 262.124 and 171.034, which were identified as 2-(4-((3-amino-3-hydroxypropylthio)-methyl)thiazol-2-yl) guanidine and 2-(4-formylthiazol-2-yl) guanidine along with famotidine, as shown in Figure 12.26b, which shows incomplete FAM degradation. Furthermore, in Figure 12.26c for the 75 minutes sample, we notice that FAM disappeared completely (100% DE of FAM), however major intermediate still persists. These intermediates are potential pollutants and can be

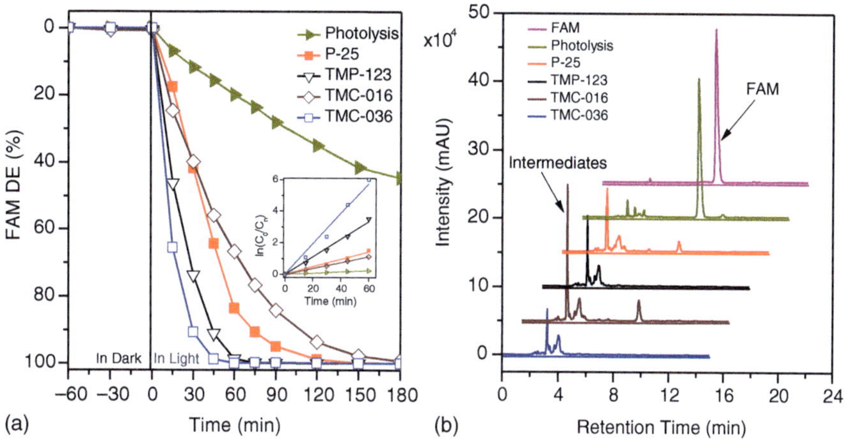

Figure 12.25 (a) FAM degradation over mesoporous and fumed titania. Reaction conditions: 100 mg l^{-1} substrate; 2 g l^{-1} catalyst loading. Inset – kinetics plot; (b) HPLC profiles of 75 minutes irradiated samples. For comparison, a FAM blank run is also included.

highly poisonous in water and need to be mineralized completely to achieve pure consumable water.

Hence, a vigorous study was conducted to completely mineralize famotidine and its intermediates with TMP-123 OMT and commercial P-25 for 24 hours and periodically monitored with HPLC, as shown in Figure 12.27. To our best knowledge, this kind of study is not conducted in the literature. We successfully achieved complete mineralization of FAM and its intermediates, as evident from HPLC profiles recorded at regular intervals for 24 hours time span. TMP-123 OMT shows efficient FAM mineralization within 1 hours with disappearance of FAM peak at ~8 minutes retention time and complete mineralization of intermediates within 6 hours confirmed from the disappearance of intermediates peaks at 3 minutes retention time (Figure 12.27a). On the other hand, commercial P-25 takes 20 hours for the complete mineralization of FAM and its intermediates (Figure 12.27b). Hence, these observations finally establish the superiority of porous titania TMP-123 photocatalyst over commercial P-25.

12.10.4 Spent Catalyst

To understand the stability and recyclability of OMT catalysts, we have carried out the post-reaction catalyst characterization of TMP-123 through DRUV-VIS and electrochemical measurements. In Figure 12.28, we have shown the DRUV-VIS spectra of the spent catalyst, designated as TMP-123(s), with no significant change in the absorption profile compared to the pristine sample, i.e. TMP-123 before the reaction, which clearly indicates good stability and recyclability of the catalyst. To further understand the effect of the reaction on the *e*–*h* recombination rate, the impedance spectra of the sample before and after the reaction were recorded (see Figure 12.20a). Interestingly, even after the reaction (see Table 12.5), the spent catalyst TMP-123(s) exhibits an R_{ct} value (633 Ω) closer to that of fresh TMP-123 (597 Ω), indicating the

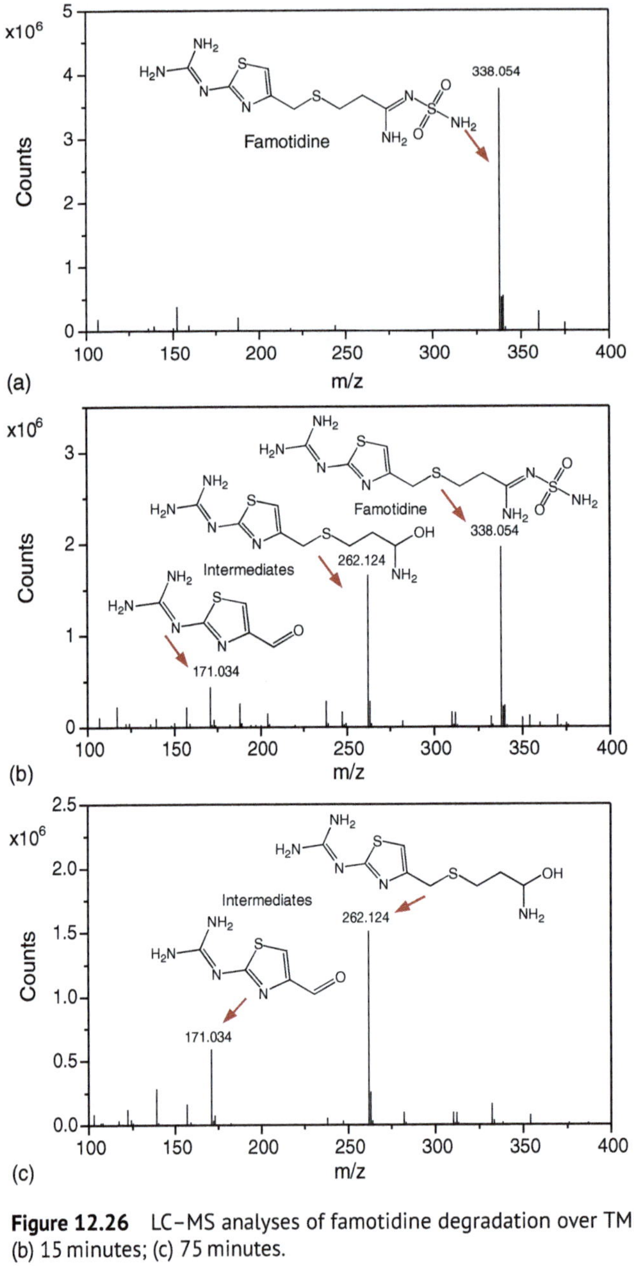

Figure 12.26 LC–MS analyses of famotidine degradation over TMP-123: (a) 0 minutes; (b) 15 minutes; (c) 75 minutes.

robust nature of the material. This proves that even after the reaction, the catalyst shows efficient charge separation and fast charge transfer at the SEI, which are essential for efficient photocatalysis. Similarly, through the M–S plot for the spent catalyst depicted in Figure 12.28b, the constant flat band potential (*cf*. Table 12.5) further showed good stability and strength of the catalyst. However, we also observe small changes in the spent catalyst such as: (i) change in interfacial carrier density (1.6 times less) and (ii) carrier density (2 times high) due to the small change in

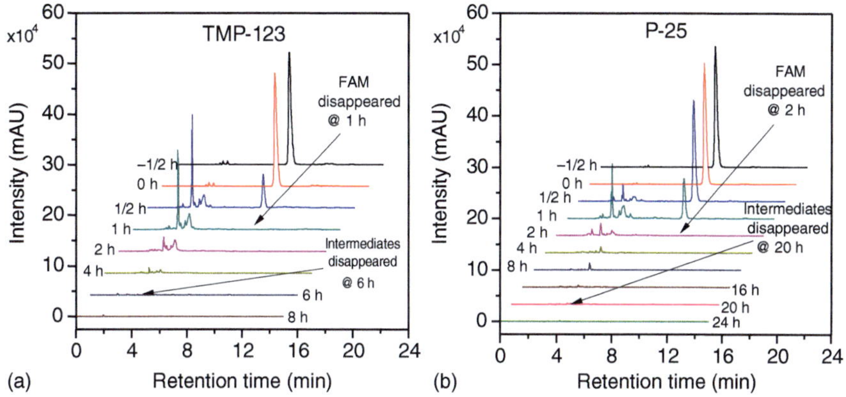

Figure 12.27 Complete mineralization of famotidine and its intermediates with (a) TMP-123 OMT and (b) commercial P-25. Reaction conditions: 100 mg l^{-1} FAM solution, catalyst loading 2 g l^{-1}.

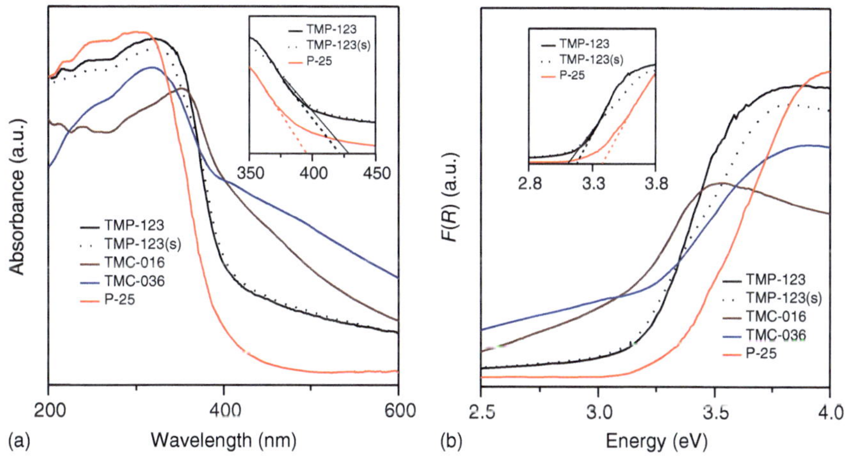

Figure 12.28 DRUV–VIS spectra: (a) absorption spectra, (b) Kubelka–Munk plot, for various mesoporous titania photocatalysts. For comparison, the profiles for commercial P-25, and spent catalyst, TMP-123(s), are also included. Insets: Kubelka–Munk plot with linear fit for the determination of the band gap energy, E_g.

charge transfer resistance at the SEI. This may be due to the presence of degraded adsorbed famotidine species blocking the interface.

12.11 Conclusion

Photocatalytic degradation of pharmaceutical pollutants, *viz.,* famotidine with bulk titania is inefficient with persistence of primary and secondary pollutants which are poisonous in drinking water. Commercial titania P-25 has limitations, *viz.*, intrinsic high *e–h* recombination, low surface area, and limited light absorption (only UV). In this context, we demonstrate OMT with a crystalline

2D hexagonal porous framework structure with anatase/bronze phases was successfully synthesized in a reproducible way using a facile EISA method, and the materials were systematically evaluated for the nanoscopic structure as well as the intrinsic defects of these mesostructured materials by in-depth characterization. The well-OMT displays exceptional photocatalytic activity for the degradation of both famotidine and 4-chlorophenol and achieves a performance even better than that of the benchmark P-25. Among all catalysts understudy, TMP-123 shows excellent degradation properties for 4-CP, i.e. it shows a much higher rate constant (doubled) compared to the commercial titania, *viz.,* P-25. The 100% degradation efficiency for famotidine was achieved in a shorter duration for TMP-123 and TMC-036, and the first-order rate constant is much higher than the commercial photocatalyst. LC-MS studies demonstrated major intermediates identified as 2-[4-[(3-amino-3-hydroxypropylthio)methyl]thiazol-2-yl] guanidine and 2-(4-formylthiazol-2-yl) guanidine with *c.* m/z 262 and 171, respectively. Also, we have demonstrated complete mineralization of famotidine and its major intermediates within ~ 6 hours with TMP-123 whereas P-25 takes ~20 hours to achieve the same. This demonstrates the remarkable efficiency of mesoporous titania having unique structural, textural, optical, and photoelectrochemical properties.

Thus, the superior performance of mesoporous titania mixed phases usually is attributed to the enhanced absorbance from UV to the visible region, the presence of intrinsic defects (Ti^{3+} centers), lower charge transfer resistance (R_{ct}), and high carrier density (N_D) on the surface of these materials. Despite all this evidence, a thoroughly convincing explanation of why such mixed-phase materials of titania outperform the individual polymorphs has remained mysterious. Nevertheless, in a similar way to the mixed anatase-rutile [101, 102] or anatase-brookite [103, 104] phases, the novel titania-bronze composite phase presented in this investigation displayed enhanced photocatalytic activity over the individual polymorphs. Indeed, a recent study [45] on the energetic alignment of the band edges of the rutile and anatase polymorphs through a combination of simulation and experiments showed that a band alignment of ~0.4 eV exists between anatase and rutile with anatase possessing the higher electron affinity, or work function. These results help us to explain the robust separation of photoexcited charge carriers between the two phases and highlight a route to improved photocatalysts. This alignment is likely the driving force for the increased photoactivity of mixed-phase composite materials over their individual counterparts. Nonetheless, this study also shows that the *e–h* pair recombination can readily be controlled by creating intrinsic defects, and thus it can efficiently oxidize the organics, which are important steps in developing photocatalysis as a sustainable technology.

Acknowledgment

The authors thank the Department of Science and Technology (DST), New Delhi, for funding NCCR, IIT-Madras, and Professor B. Viswanathan for fruitful discussions. This work is supported by the grant-in-aid for scientific research on priority areas under technology systems development programme – solar hydrogen (DST/TSG/SH/2011/106-G), and industrial research and development projects in the framework of Indian-Hungarian Science and Technology Cooperation Programme under grant No. DST/INT/HUN/P-13/2017. The authors gratefully acknowledge the help and support of Dr. R.P. Raj and Mr. S. Gupta for the characterization of the samples.

References

1 Nature:Editorials (2015). *Nature* 526: 164.
2 Prakash, V. (1999). *J. Bombay Nat. Hist. Soc.* 96: 365–378.
3 Rasmussen, P.C. (2004). *Auk* 121: 619–622.
4 Oaks, J.L., Gilbert, M., Virani, M.Z. et al. (2004). *Nature* 427: 630–633.
5 Ternes, T.A., Stumpf, M., Mueller, J. et al. (1999). *Sci. Total Environ.* 225: 81–90.
6 Kidd, K.A., Blanchfield, P.J., Mills, K.H. et al. (2007). *Proc. Natl. Acad. Sci. U. S. A.* 104: 8897–8901.
7 Khetan, S.K. and Collins, T.J. (2007). *Chem. Rev.* 107: 2319–2364.
8 Mompelat, S., Le Bot, B., and Thomas, O. (2009). *Environ. Int.* 35: 803–814.
9 Radjenović, J., Petrović, M., and Barceló, D. (2009). *Water Res.* 43: 831–841.
10 Enick, O.V. and Moore, M.M. (2007). *Environ. Impact Assess. Rev.* 27: 707–729.
11 Castiglioni, S., Bagnati, R., Fanelli, R. et al. (2006). *Environ. Sci. Technol.* 40: 357–363.
12 Phillips, P.J., Smith, S.G., Kolpin, D.W. et al. (2010). *Environ. Sci. Technol.* 44: 4910–4916.
13 Gültekin, I. and Ince, N.H. (2007). *J. Environ. Manag.* 85: 816–832.
14 Zhou, H. and Smith, D.W. (2002). *J. Environ. Eng. Sci.* 1: 247–264.
15 Grover, D.P., Zhou, J.L., Frickers, P.E., and Readman, J.W. (2011). *J. Hazard. Mater.* 185: 1005–1011.
16 Westerhoff, P., Yoon, Y., Snyder, S., and Wert, E. (2005). *Environ. Sci. Technol.* 39: 6649–6663.
17 Esplugas, S., Giménez, J., Contreras, S. et al. (2002). *Water Res.* 36: 1034–1042.
18 Kemsley, J. (2007). *Chem. Eng. News Arch.* 85: 9.
19 Haque, M.M. and Muneer, M. (2007). *J. Hazard. Mater.* 145: 51–57.
20 Shemer, H., Kunukcu, Y.K., and Linden, K.G. (2006). *Chemosphere* 63: 269–276.
21 Haber, F. and Weiss, J. (1934). *Proc. R. Soc. London. Ser. A-Mathematical Phys. Sci.* 147: 332–351.
22 Arslan-Alaton, I. and Gurses, F. (2004). *J. Photochem. Photobiol. A Chem.* 165: 165–175.
23 Pignatello, J.J. (1992). *Environ. Sci. Technol.* 26: 944–951.
24 Hoffmann, M.R., Martin, S.T., Choi, W., and Bahnemann, D.W. (1995). *Chem. Rev.* 95: 69–96.
25 Schlögl, R. (2015). *Angew. Chem. Int. Ed.* 54: 3465–3520.
26 Verhoeven, J.W. (1996). *Pure Appl. Chem.* 68: 2223–2286.
27 Fujishima, A. and Zhang, X. (2006). *C. R. Chim.* 9: 750–760.

28 Schneider, J., Matsuoka, M., Takeuchi, M. et al. (2014). *Chem. Rev.* 114: 9919–9986.
29 Ma, Y., Wang, X., Jia, Y. et al. (2014). *Chem. Rev.* 114: 9987–10043.
30 Wang, Z., Helmersson, U., and Käll, P.O. (2002). *Thin Solid Films* 405: 50–54.
31 Aravindan, V., Lee, Y.S., Yazami, R., and Madhavi, S. (2015). *Mater. Today* 18: 345–351.
32 Ortiz, G.F., Hanzu, I., Djenizian, T. et al. (2009). *Chem. Mater.* 21: 63–67.
33 Feist, T.P. and Davies, P.K. (1992). *J. Solid State Chem.* 101: 275–295.
34 Djaoued, Y., Brüning, R., Bersani, D. et al. (2004). *Mater. Lett.* 58: 2618–2622.
35 Kim, W., Tachikawa, T., Moon, G. et al. (2014). *Angew. Chem. Int. Ed.* 53: 14036–14041.
36 Sang, L., Zhao, Y., and Burda, C. (2014). *Chem. Rev.* 114: 9283–9318.
37 Ding, C., Shi, J., Wang, Z., and Li, C. (2017). *ACS Catal.* 7: 675–688.
38 Rajeshwar, K. (2002). *Encyclopedia of Electrochemistry*. Wiley.
39 Wang, Q. and Domen, K. (2020). *Chem. Rev.* 120: 919–985.
40 Woan, K., Pyrgiotakis, G., and Sigmund, W. (2009). *Adv. Mater.* 21: 2233–2239.
41 Allen, M.J., Tung, V.C., and Kaner, R.B. (2010). *Chem. Rev.* 110: 132–145.
42 Leung, D.Y.C., Fu, X., Wang, C. et al. (2010). *ChemSusChem* 3: 681–694.
43 Ding, Z., Lu, G.Q., and Greenfield, P.F. (2000). *J. Phys. Chem. B* 104: 4799–4806.
44 Jitputti, J., Suzuki, Y., and Yoshikawa, S. (2008). *Catal. Commun.* 9: 1265–1271.
45 Scanlon, D.O., Dunnill, C.W., Buckeridge, J. et al. (2013). *Nat. Mater.* 12: 798–801.
46 Corma, A. (1997). *Chem. Rev.* 97: 2373–2419.
47 Davis, M.E. (2002). *Nature* 417: 813–821.
48 Kresge, C.T., Leonowicz, M.E., Roth, W.J. et al. (1992). *Nature* 359: 710–712.
49 Beck, J.S., Vartuli, J.C., Roth, W.J. et al. (1992). *J. Am. Chem. Soc.* 114: 10834–10843.
50 Huo, Q., Margolese, D.I., Ciesla, U. et al. (1994). *Nature* 368: 317–321.
51 Zhao, D., Feng, J., Huo, Q. et al. (1998). *Science* 279: 548–552.
52 Krishna, N.V. and Selvam, P. (2015). *Adv. Porous Mater.* 2: 106–112.
53 Selvam, P., Bhatia, S.K., and Sonwane, C.G. (2001). *Ind. Eng. Chem. Res.* 40: 3237–3261.
54 Ryoo, R., Joo, S.H., and Jun, S. (1999). *J. Phys. Chem. B* 103: 7743–7746.
55 Jun, S., Joo, S.H., Ryoo, R. et al. (2000). *J. Am. Chem. Soc.* 122: 10712–10713.
56 Selvam, P. and Kuppan, B. (2012). *Catal. Today* 198: 85–91.
57 Mohan, T.V.R., Palla, S., Kuppan, B. et al. (2018). *J. Chem. Eng. Data* 63: acs.jced.8b00627.
58 Dambournet, D., Belharouak, I., and Amine, K. (2010). *Chem. Mater.* 22: 1173–1179.
59 Ren, Y., Ma, Z., and Bruce, P.G. (2012). *Chem. Soc. Rev.* 41: 4909–4927.
60 Yang, P., Zhao, D., Margolese, D.I. et al. (1998). *Nature* 396: 152–155.
61 Yang, P., Zhao, D., Margolese, D.I. et al. (1999). *Chem. Mater.* 11: 2813–2826.
62 Tian, B., Yang, H., Liu, X. et al. (2002). *Chem. Commun.* 8: 1824–1825.
63 Choi, S.Y., Mamak, M., Coombs, N. et al. (2004). *Adv. Funct. Mater.* 14: 335–344.

64 Fan, J., Boettcher, S.W., and Stucky, G.D. (2006). *Chem. Mater.* 18: 6391–6396.

65 Dong, W., Sun, Y., Lee, C.W. et al. (2007). *J. Am. Chem. Soc.* 129: 13894–13904.

66 Hossain, M.K., Koirala, A.R., Akhtar, U.S. et al. (2015). *Chem. Mater.* 27: 6550–6557.

67 Hongo, T. and Yamazaki, A. (2011). *Microporous Mesoporous Mater.* 142: 316–321.

68 Zimny, K., Roques-Carmes, T., Carteret, C. et al. (2012). *J. Phys. Chem. C* 116: 6585–6594.

69 Soler-Illia, G.D.A., Louis, A., and Sanchez, C. (2002). *Chem. Mater.* 14: 750–759.

70 Alagarasi, A., Rajalakshmi, P.U., Shanthi, K., and Selvam, P. (2018). *Catal. Today* 309: 202–211.

71 Linsebigler, A.L., Lu, G., and Yates, J.T. (1995). *Chem. Rev.* 95: 735–758.

72 Fujishima, A., Rao, T.N., and Tryk, D.A. (2000). *J. Photochem. Photobiol. C: Photochem. Rev.* 1: 1–21.

73 Grosso, D., de Soler-Illia, G.J., A.A., Babonneau, F. et al. (2001). *Adv. Mater.* 13: 1085–1090.

74 Chen, X. and Mao, S.S. (2007). *Chem. Rev.* 107: 2891–2959.

75 Liu, C., Zhang, D., and Sun, Y. (2014). *CrystEngComm* 16: 8421.

76 Chen, X. and Selloni, A. (2014). *Chem. Rev.* 114: 9281–9282.

77 Selvam, P., Alagarasi, A., Sithini, T.N. et al. (2013). *Adv. Porous Mater.* 1: 362–370.

78 Hae-Jin Kim, Jou-hahn Lee, Soon-Chang Lee, Hyun-Uk Lee, Won-Ki Hong, Hye-Ran Kim, Jung-Hye Seo. Method of preparing high crystalline nanoporous titanium dioxide at room temperature. Korea Basic Science Institute. US Patent, 8 852 548 B2, 2014.

79 Parasuraman Selvam, Sanjeev Gupta. Process for synthesis of well-ordered mesoporous titania having monoclinic and anatase phases. Indian Institute of Technology-Madras. IN Pat., 201641044826, 2016.

80 Parasuraman Selvam, Sanjeev Gupta. Process for synthesis of well-ordered mesoporous titania (TMF-108) with bronze and anatase phases. Indian Institute of Technology-Madras. IN Pat., 201841003781, 2018.

81 Parasuraman Selvam, Surya Kumar Vatti. Process for the preparation of ordered mesoporous titania (TMP-123) with anatase and monoclinic phases: a new generation photocatalyst. Indian Institute of Technology-Madras. IN Pat., 395404, 2018.

82 Parasuraman Selvam, Surya Kumar Vatti. CTAB-templated synthesis of nanocrystalline ordered mesoporous titania (TMC-016). Indian Institute of Technology-Madras IN Pat., 369497, 2019.

83 Vatti, S.K., Gupta, S., Raj, R.P., and Selvam, P. (2020). *New J. Chem.* 44: 16269–16284.

84 Selvam, P., Vatti, S.K., and Gupta, S. (2018). *10th International Mesostructured Materials Symposium*, Los Angeles, CA, 33.

85 Livage, J., Henry, M., and Sanchez, C. (1988). *Prog. Solid State Chem.* 18: 259–341.

86 Attwood, A.L., Murphy, D.M., Edwards, J.L. et al. (2003). *Res. Chem. Intermed.* 29: 449–465.

87 Chiesa, M., Paganini, M.C., Livraghi, S., and Giamello, E. (2013). *Phys. Chem. Chem. Phys.* 15: 9435.

88 Wang, C., Zhang, X., Wei, Y. et al. (2015). *Dalton Trans.* 44: 13331–13339.

89 Yan, Y., Han, M., Konkin, A. et al. (2014). *J. Mater. Chem. A* 2: 12708–12716.

90 Dimitrijevic, N.M., Saponjic, Z.V., Rabatic, B.M. et al. (2007). *J. Phys. Chem. C* 111: 14597–14601.

91 Dvoranová, D., Mazúr, M., Papailias, I. et al. (2018). *Catalysts* 8: 47.

92 Bisquert, J. (2015). *Nanostructured Energy Devices: Equilibrium Concepts and Kinetics*, 1e. CRC Press.

93 Gelderman, K., Lee, L., and Donne, S.W. (2007). *J. Chem. Educ.* 84: 685–688.

94 Pu, P., Cachet, H., Ngaboyamahina, E., and Sutter, E.M.M. (2013). *J. Solid State Electrochem.* 17: 817–828.

95 Castañeda, C., Tzompantzi, F., Gómez, R., and Rojas, H. (2016). *J. Chem. Technol. Biotechnol.* 91: 2170–2178.

96 Yang, Y., Gao, P., Wang, Y. et al. (2017). *Nano Energy* 33: 29–36.

97 Zhang, H., Lv, X., Li, Y. et al. (2010). *ACS Nano* 4: 380–386.

98 Alagarasi, A., Rajalakshmi, P.U., Shanthi, K., and Selvam, P. (2019). *Mater. Today Sustain.* 5: 100016.

99 Keane, D., Basha, S., Nolan, K. et al. (2011). *Catal. Lett.* 141: 300–308.

100 Murphy, S., Saurel, C., Morrissey, A. et al. (2012). *Appl. Catal. B Environ.* 119–120: 156–165.

101 Li, G. and Gray, K.A. (2007). *Chem. Phys.* 339: 173–187.

102 Zhu, Q., Qian, J., Pan, H. et al. (2011). *Nanotechnology* 22: 395703.

103 Kandiel, T.A., Feldhoff, A., Robben, L. et al. (2010). *Chem. Mater.* 22: 2050–2060.

104 Zhao, Y., Pan, F., Li, H. et al. (2013). *J. Phys. Chem. C* 117: 21718–21723.

13

Catalytic Dehydration of Glycerol Over Silica and Alumina-Supported Heteropoly Acid Catalysts

Sekar Mahendran[1], *Shinya Hayami*[2,3], *and Parasuraman Selvam*[1,3,4]

[1] *Indian Institute of Technology-Madras, National Centre for Catalysis Research and Department of Chemistry, Chennai 600 036, India*
[2] *Kumamoto University, Department of Chemistry, 2-39-1 Kurokami, Kumamoto 860-8555, Japan*
[3] *Kumamoto University, IROAST, 2-39-1 Kurokami, Kumamoto 860-8555, Japan*
[4] *University of Delaware, Delaware Energy Institute, 221, Academy Street, Newark, DE 19716, USA*

13.1 Introduction

Biodiesel is a clean-burning, eco-friendly fuel that is a possible alternative to petroleum fuels. It is an environmentally benign and cheaper source of alternate energy. Biodiesel can be obtained from the transesterification of vegetable oils, e.g. palm oil and soybean oil, with lower alcohols. However, the production of fuels from biosources may lead to the accumulation of large quantities of by-products. For example, 1 mol of triglyceride on transesterification produces 1 mol of glycerol (Scheme 13.1), which can be a useful feedstock if converted to some other value-added chemicals. Biodiesel produces significantly reduced emissions of carbon monoxide, particulate matter, unburnt hydrocarbons, and sulfates compared to the corresponding fossil fuels [2, 3]. Biodiesel is a renewable fuel that does not contribute to global warming due to its closed carbon cycle. Biodiesel reduces the emission of carcinogenic compounds by as much as 85% in comparison with diesel [4, 5]. In the context of energy and the environment, biodiesel seems to be a better option as fuel or fuel blend. Thus, over the past two decades, biodiesel has gained importance as a renewable fuel at national and international levels. The challenges in the production and commercialization of biodiesel are to be addressed for the effective utilization of energy sources. The accumulation of by-products during the production of biodiesel is a cause for concern, and the production of by-products, the major component being glycerol, is also increasing in proportion with the production of biodiesel. The global production and estimated global production of biodiesel during the years 2008–2025 are shown in Figure 13.1.

In the synthesis of biodiesel, approximately 10 wt% of the by-product (glycerol) may be produced for every cycle [1]. For the sustained production of biodiesel, it will be appropriate to convert the by-product, namely glycerol, into other value-added chemicals, which will improve the economic importance of biodiesel

Catalysis in Confined Frameworks: Synthesis, Characterization, and Applications, First Edition.
Edited by Hermenegildo Garcia and Amarajothi Dhakshinamoorthy.

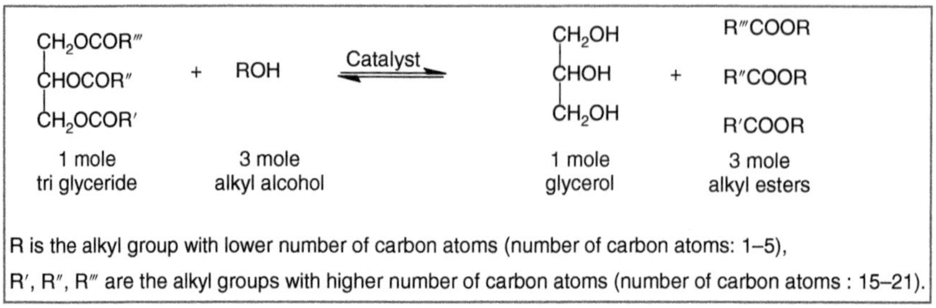

Scheme 13.1 Production of biodiesel through transesterification of vegetable oils [1].

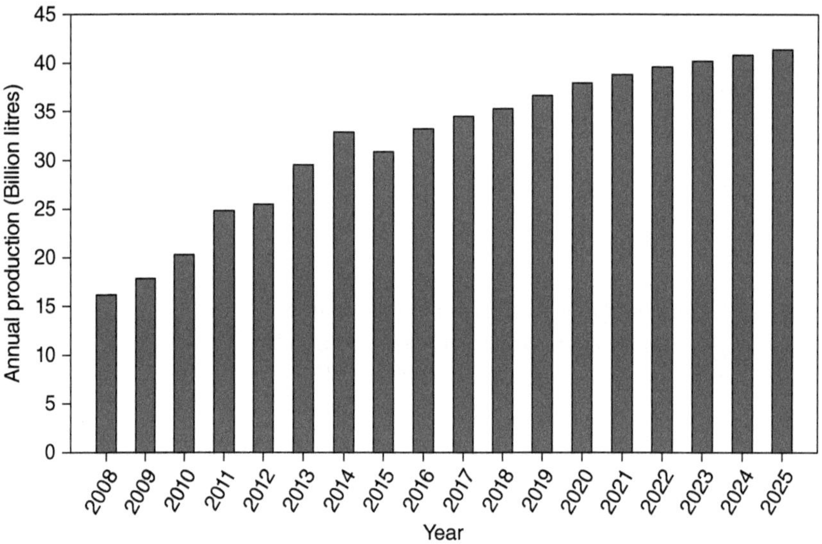

Figure 13.1 Global biodiesel production and estimation (OECD/FAO 2016). Source: Data from OECD/FAO [6].

and glycerol. Glycerol is a multifunctional compound that possesses two primary and one secondary hydroxyl groups. The utilization of glycerol in industrially important products could open up significant market in many industries dealing with polymers, ethers, food products, cosmetics, toiletries, toothpaste, explosives, drugs, animal feed, plasticizers, tobacco and emulsifiers [7, 8].

13.2 Value Addition of Bioglycerol

Value addition of glycerol can be achieved in a variety of ways: for example, oxidation of glycerol into glyceric acid; hydrogenolysis of glycerol into 1,2- and/or 1,3-propanediol; dehydration of glycerol into acrolein/acrylic acid, or hydroxyacetone; carboxylation of glycerol into glycerol carbonate; chlorination of glycerol into epichlorohydrin; etherification of glycerol into ethers; esterification of glycerol into

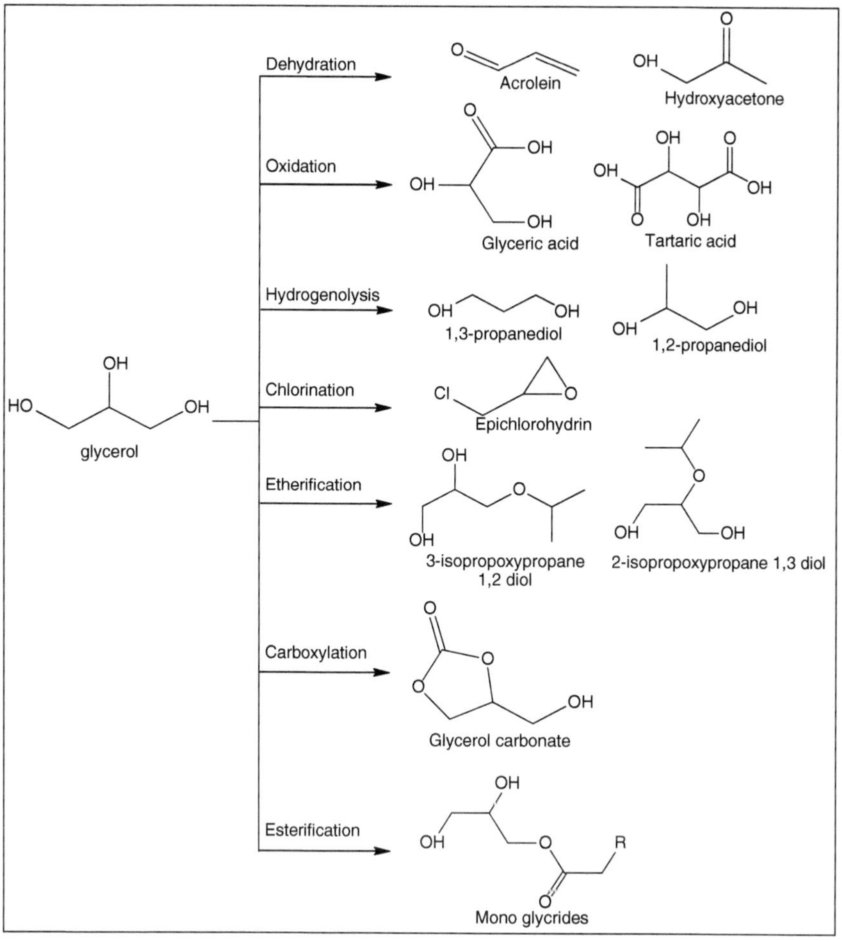

Scheme 13.2 Strategies for catalytic conversion of glycerol into value-added products [1, 9, 10].

triglycerides; and epoxidation of glycidol, to name a few. Even though glycerol can be converted to a variety of useful products, obtaining these products selectively in a given catalytic process is challenging. One of the ways of improving the selectivity toward a product is to decrease or mask the multifunctionality of glycerol and apply the strategy to get the desired product. Scheme 13.2 represents the variety of products that one can obtain using glycerol by applying different strategies.

The dehydration of glycerol to produce acrolein is an important and alternative method for utilizing the increasing glycerol resources since the current production of acrolein is based on the oxidation of petroleum-derived propylene using Bi-Mo oxide catalysts [9, 10]. Though dehydration of glycerol-yielding acrolein has been known since the nineteenth century, the excess supply of biodiesel-derived glycerol triggered a new push into this reaction. Conversion of glycerol into acrolein is one of the key transformations since acrolein is a vital intermediate in the preparation

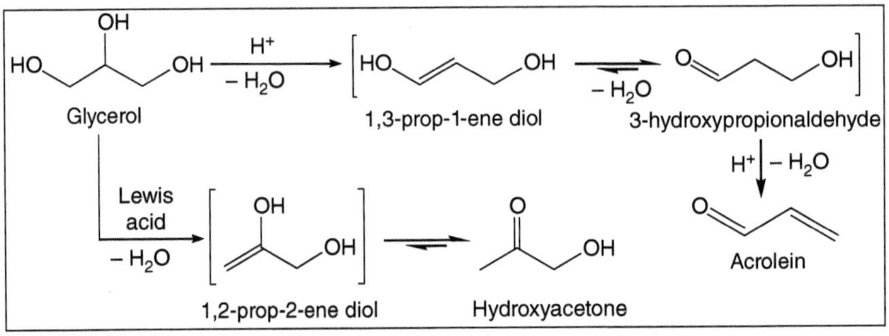

Scheme 13.3 Reaction scheme of glycerol dehydration.

of polyester resin, polyurethane, propylene glycol, amino acid-like DL-methionine, the monomer for acrylic resins, herbicides, super absorber polymer, etc., and for many other industrially important compounds [1, 8, 9, 11]. Dehydration of glycerol was carried out in an acidic medium owing to the lower activation energy of protonated glycerol. Glycerol in the presence of a solid acid catalyst yields both acrolein and hydroxyacetone, but the reaction proceeds in two independent pathways, as shown in Scheme 13.3. With regard to the reaction mechanism, the dehydration of glycerol on acid catalysts is proposed to proceed via the formation of 3-hydroxypropanal, with 1-hydroxyacetone (acetol) formed as a by-product [12, 13]. Brønsted acid sites attack the secondary hydroxyl group of glycerol and lead to the formation of 3-hydroxypropanal, which undergoes further dehydration to acrolein (Scheme 13.3). On the other hand, Lewis acid sites favor the formation of acetol through dehydration of the primary hydroxyl group of the glycerol. However, one of the challenges associated with catalytic dehydration reaction is controlling the acidity and driving the reaction pathway to the desired products.

The dehydration of glycerol to acrolein involves acid catalysts with active sites of a suitable strength so as to proficiently promote the reaction while limiting coke formation. Solid acid catalysts with a Hammett acidity Ho between −3 and −8, such as zeolites, supported inorganic acids and metal oxides, have previously shown good performances, and acrolein selectivity over 70% is obtained using this kind of catalyst. In addition, Brønsted acid sites are more effective than Lewis acid sites for the dehydration of glycerol to acrolein [14]. In addition, this reaction requires a relatively high temperature since dehydration reactions are favored by high temperatures.

Table 13.1 summarizes glycerol dehydration over various solid acid catalysts. Various solid acid catalysts, including metal oxides [16], mixed-metal oxides [14], sulfates, and phosphates [17, 18], as well as zeolites [19], have been tested for the dehydration of alcohol either in the gas phase or in the liquid phase or in sub- and super-critical water. However, one of the main limitations of these catalysts remains the formation of a large number of by-products (25–40%) and the catalyst deactivation. Catalyst deactivation is the major obstacle to using highly acidic catalysts such as zeolites. Although heteropoly acids (HPAs) of Keggin-type (polyoxometalates) show some promise, as they exhibit strong Brønsted acidity with tunable acidity, they lack thermal stability in addition to low porosity and very low specific surface

Table 13.1 Glycerol dehydration over various solid acid catalysts.

Active phase	Support	T (K)	Glycerol conv. (%)	Acrolein yield (%)	TOS (h)
$H_4SiW_{12}O_{40}$	SiO_2	548	100	87.0	5
$H_4SiW_{12}O_{40}$	SiO_2	523	100	85.9	5
$H_4SiW_{12}O_{40}$	SiO_2	548	98	84.7	5
H-MC M49	—	633	100	82.8	n.a.
H-MFI + 1% Au	—	633	100	82.8	1
ZSM-11	—	633	99	81.3	n.a.
H-MFI + 0.1% Pt	—	633	100	80.7	3
β-zeolite	—	633	100	80.3	n.a.
MCM-22	—	633	100	80.1	n.a.
$Fe_2(PO_4)_3$	Pumicite	673	n.a.	80.0	n.a.
$Nd_4(P_2O_7)_3$	—	593	96	79.7	7
MFI + Ba	—	633	100	79.0	2.5
H_2WO_4	—	533	100	79.0	5
WO_3	ZrO_2	533	100	79.0	5
H_2WO_4 + 1 wt % Pd	—	533	100	77.0	5
MCM-56	—	633	100	76.8	n.a.
Nafion	SiO	573	100	76.0	7
H_3PO_4	α-Al_2O_3	573	n.a.	75.0	n.a.
$Gd_4(P_2O_7)_3$	—	593	98	74.3	1
$Sm_4(P_2O_7)_3$	—	593	100	73.9	1
ZSM-5	—	588	98	73.6	n.a.
WO_3	ZrO_2	573	100	73.5	7
$H_3PW_{12}O_{40}$	ZrO_2	588	100	70.0	4

n.a. = TOS data was not available for that particular catalyst.
Source: Adapted from Katryniok et al. [15].

area. A possible solution is to support HPAs on high surface area inert materials like silica [20, 21], alumina [22], titania [23], zirconia [24], and ion-exchange resin [25].

13.3 Interaction Between HPA and Support

It is well known that HPA is an acid with high acid strength. Heteropoly anion analogs to anions such as SO_4^{2-}, PO_4^{3-}, and AsO_4^{3-} also belong to the inorganic oxygen-containing anion. In view of the differences in hardness or softness of hydroxyl on supports and HPA, the reaction between HPA and hydroxyl on the support may proceed according to the following mechanism. In general, any solvent more basic than the polyoxometallate would be protonated by the dissolution of

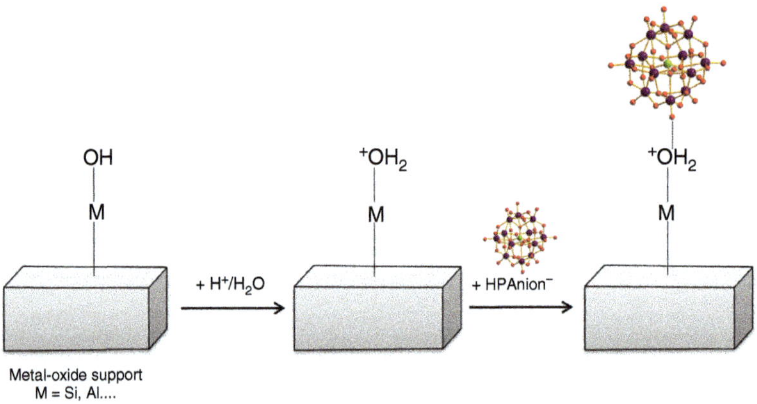

Scheme 13.4 Interaction between the HPA and the metal oxide support.

the phosphotungstate. In contact with water, the hydroxyl groups at the surface of MO_x supports (M = Al, Si) are protonated, which yields $M{-}OH_2^+$ species (*see* Scheme 13.4). These species strongly interact via electronic effects with the negatively charged heteropoly anion. Depending on the acidic character of the support, this interaction can be either strong (over Al_2O_3) or weak (over SiO_2), and its strength has an impact on the Brønsted acidity and thermal stability of the supported catalysts. Silica is widely used as a support owing to its weak interaction with HPAs and Keggin-structure, which is well preserved [26]. In particular, silica-supported HPA has been demonstrated as an efficient catalyst for the dehydration of alcohols [27].

Although the incorporation of HPA into conventional zeolitic pores is a long-standing dream to obtain shape-selective catalysts, they, however, are not suitable for this purpose as the pores are too small (<1 nm) to adsorb large (>1 nm) HPA. On the other hand, the unique properties of silica materials such as silica with high surface area (>300 $m^2\ g^{-1}$), large pore size (~30 nm), and large amount of internal hydroxyl (silanol) groups (40%) may possibly lead to high dispersion of the active sites on the surface and result in high catalytic activity. Thus, supported-HPAs can be used as efficient catalysts for organic transformations, predominantly for the dehydration of alcohols and, in particular for the conversion of glycerol into acrolein.

13.4 Bulk Heteropoly Acid

Our group has evaluated the performance of bulk H-SiW, which showed an initial glycerol conversion of 84% and acrolein selectivity of 82% at 275 °C. However, the bulk H-SiW deactivates rapidly owing to its high acid strength and very low surface area [12], and we believe that the acidic site per unit area is so high that the sample loses activity quickly. Similar results were obtained by Atia et al. [28] who have evaluated the performance of four commercially available HPAs and have also observed that pure bulk HPAs catalyze the dehydration of glycerol toward acrolein poorly, and their low stability and surface area seem to be the main drawbacks.

13.5 Silica-Supported HPA

13.5.1 Effect of Textural Properties of Support on Product Selectivity

The effect of the textural properties of the support, particularly pore size distribution of the silica-supported HPAs, was discussed for the first time in 2007 by Tsukuda et al. [12, 15]. Three commercial amorphous silicas with pore sizes of 3, 6, and 10 nm were employed as a support for HPA, and the resulting catalysts show complete (glycerol) conversion with acrolein selectivity of 86% for the silicotungstic acid supported on silica matrix with a pore size of 10 nm at 275 °C. On the other hand, silica support with lower pore size, viz., 3 nm, was found to deactivate easily due to steric limitations. Larger mesopores were found to be effective in selectively converting glycerol to acrolein. Our team recently used two ordered mesoporous silicas (OMSs)-, e.g. MCM-41 and SBA-15, as the support for silicototungstic acid and compared the results with commercial silica. Table 13.2 summarizes the textural properties of the OMS.

Complete glycerol conversion was obtained over all the studied silica supports with 30 wt% loading of HPA as the active phase at 275 °C. The selectivity of acrolein has been enhanced from 84% to 97% when the size of mesopores increases from 3 to 6 nm in the case of OMS-supported H-SiW catalysts. Similar results were obtained by Atia et al., who examined the catalytic performance of silicotungstic acid supported on alumina with 5 and 12 nm pore sizes [27]. These results clearly showed that the size of the pores has a direct influence on the selectivity for acrolein. Figure 13.2 shows that the rate of deactivation of catalyst decreases in the following order: H-SiW/Silica > H-SiW/SBA-15 > H-SiW/MCM-41. HPA supported on silica possessing larger pores tends to deactivate slowly when compared to the one containing smaller pores viz. MCM-41 as support. This trend is directly related to the decrease in pore size of support, as smaller pores induce deep deactivation. These results clearly indicate that the pore size of the support has direct impact on stability of catalyst. It was interesting to note that surface area did not play a significant role in the performance of these catalysts.

Table 13.2 Textural properties of various silica.

Material	a_0 (nm)[a)]	S_{BET} ($m^2\ g^{-1}$)	V_p ($cm^3\ g^{-1}$)	D_{BJH} (nm)	h_w(nm)[b)]
SBA-15	10.6	558	1.02	6.7	3.9
MCM-41	4.4	791	0.41	2.2	2.2
Fumed silica	—	318	0.71	30.7	—

a) Average unit cell parameter (a_0) calculated using $1/d^2 = 4/3\ (h^2 + hk + k^2/a^2)$.
b) Wall thickness, $h_w = a_0 - D_{BJH}$.

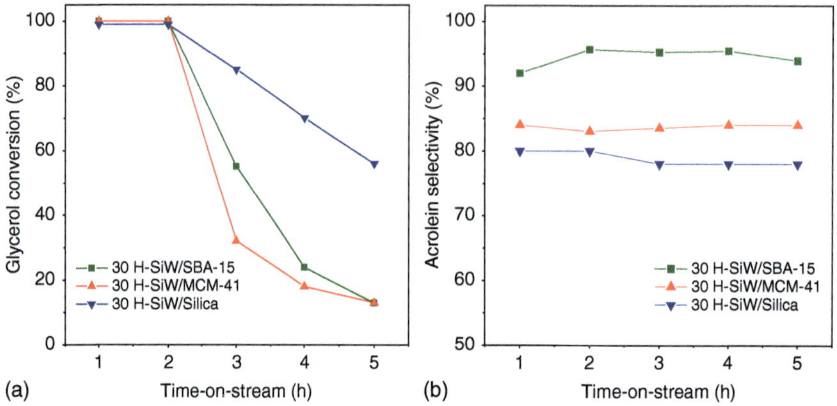

Figure 13.2 TOS of silica-supported silicotungstic acid catalysts: (a) Conversion. (b) Selectivity.

13.5.2 Effect of Catalyst Loading

As HPAs are known to interact strongly with the support, optimization of loading of the catalyst is desirable to preserve the catalytically active Keggin structure and achieve high selectivity for acrolein. In this context, we have prepared a series of OMS-supported silicotungstic acids, referred to as H-SiW/SBA-15, with 10–40 wt% loading by impregnation method [29]. The substrate contact time is generally used to identify the more active catalyst in terms of conversion from the initial value of the slope. The effect of contact time on glycerol conversion is shown in Figure 13.3. With 10% H-SiW/SBA-15 and 20% H-SiW/SBA-15, the catalytic activity increases with contact time, and complete conversion is reached at ~0.7 seconds. On the other hand, for 30% and 40% catalysts, complete conversion is reached within 0.2 seconds. Among the OMS-supported catalysts prepared with various loadings (10–40%) of H-SiW, the one with 30% was found to be good in terms of acrolein selectivity (97%). However, a further increase in the loading of H-SiW to 40% yields a negative effect.

13.5.3 Effect of Acid Sites

A series of OMS-supported silicotungstic acids, referred to as H-SiW/SBA-15, with 10–40% loading were prepared by impregnation method. In addition, MCM-41 and commercial silica were used as supports for comparison. NH3-TPD is carried out to evaluate relative acid strengths and the number of acid sites in the supported systems. In general, higher the intensity of desorption trace, larger the number of acid sites, and higher the desorption temperature, the stronger the acidity. Moreover, the shapes of the desorption profiles are asymmetric, demonstrating the presence of surface acid sites of different strengths. In order to understand the behavior in a better way, the TPD profiles were deconvoluted based on different acid strengths. Acidity of SBA-15-supported silicotungstic acid catalysts is determined using NH3-TPD profiles [29], as depicted in Figure 13.4a–f, where different acid sites are denoted by (i) and (ii). The distribution of acid sites is calculated based on the deconvolution

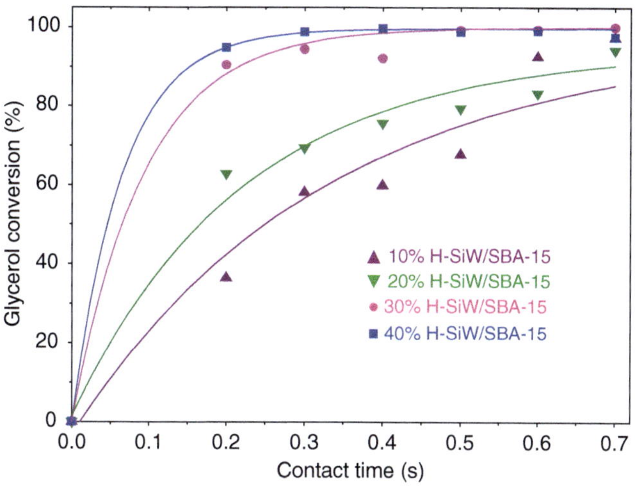

Figure 13.3 Effect of contact time on glycerol conversion.

of the desorption peak using the Gaussian function, and the results are tabulated in Table 13.3. It is noteworthy to mention here that the acid sites, viz., (i) to (ii) are completely Brønsted in nature. The acid sites are classified into two types, as follows:

(A) Moderate acid sites (i) are attributed to protons located on the oxygen of bridging W–O–W.

(B) Strong acid sites (ii) are attributed to protons located on the oxygen of terminal W=O in H-SiW.

The support silica showed almost no acidity (0.004 mmol g^{-1}). As expected, it is clear from Figure 13.4 that the total acidic sites increase with the increase in H-SiW loading up to 30% beyond this, there is no considerable change in acidity. It is also noteworthy here that in the catalyst 30% H-SiW/SBA-15 (cf. Figure 13.4e), the medium and strong acid sites are prominent, and the latter is present in a considerable amount. It can be noticed that the strong-to-medium acid site ratio increases as the loading increases (see Table 13.3), and therefore a varied selectivity pattern is expected for the chosen reaction.

Table 13.3 summarizes the acidity data and dehydration of glycerol over silica-supported HPA catalysts. Acidity results of these supported catalysts clearly suggest that there exists a good correlation between the acid strength and the activity, which in turn responsible for the product distribution in glycerol dehydration reaction. For example, the catalysts with weak to medium acid sites showed high selectivity towards hydroxyacetone whereas the catalysts with strong acid sites favor the formation of acrolein (Scheme 13.5).

Among the different silica-supported catalysts, H-SiW/SBA-15 showed much higher activity for the chosen reaction with complete glycerol conversion and 95% acrolein selectivity. However, the catalyst is deactivated rapidly due to carbonization and oligomerization caused by its higher acidity [16, 31]. It is therefore necessary to seek other support materials that can significantly improve the thermal and

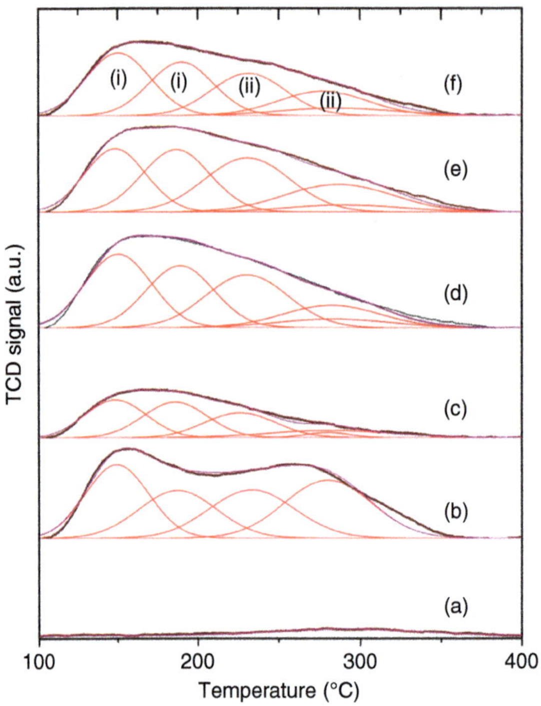

Figure 13.4 Deconvoluted NH_3-TPD profiles of: (a) SBA-15; (b) H-SiW; (c) 10% H-SiW/SBA-15; (d) 20% H-SiW/SBA-15; (e) 30% H-SiW/SBA-15; (f) 40% H-SiW/SBA-15.

Table 13.3 Acidity data and dehydration of glycerol over silica-supported heteropoly acid catalysts [29, 30].[a)]

	Acidic sites (mmol g^{-1})				Selectivity (%)		
Material	**Total**	**Medium**	**Strong**	**Conv. (%)**	**Acrolein**	**Acetol**	**Others[b)]**
H-SiW	0.72	0.34	0.38	84.0	82.0	9.1	8.9
SBA-15	0.04	—	0.04	10.0	1.0	—	99.0
MCM-41	0.09	0.05	0.01	1.6	—	—	100
Fumed Silica	0.17	0.06	0.02	8.0	2.0	10.5	87.5
10 H-SiW/SBA-15	0.18	0.11	0.05	88.0	80.0	3.3	16.7
20 H-SiW/SBA-15	0.37	0.20	0.14	92.0	82.0	7.5	10.5
30 H-SiW/SBA-15	0.44	0.17	0.25	100	95.0	1.5	3.5
40 H-SiW/SBA-15	0.33	0.18	0.13	84.0	87.0	3.5	9.5
30 H-SiW/SBA-15	0.44	0.17	0.25	100	95.0	1.5	3.5
30 H-SiW/MCM-41	0.46	0.25	0.17	99.0	84.0	5.3	10.7
30 H-SiW/c.Silica	0.42	0.19	0.22	99.0	80.0	4.0	16.0

a) Reaction conditions: Catalyst weight = 300 mg; glycerol concentration = 10 wt%; temperature = 275 °C; feed flow rate = 1.68 mlh^{-1}; TOS = 1 h.

b) Include allyl alcohol, ethylene glycol, acetaldehyde, acetone, methanol, 1,2-propane diol, carbon monoxide, and some unidentified products.

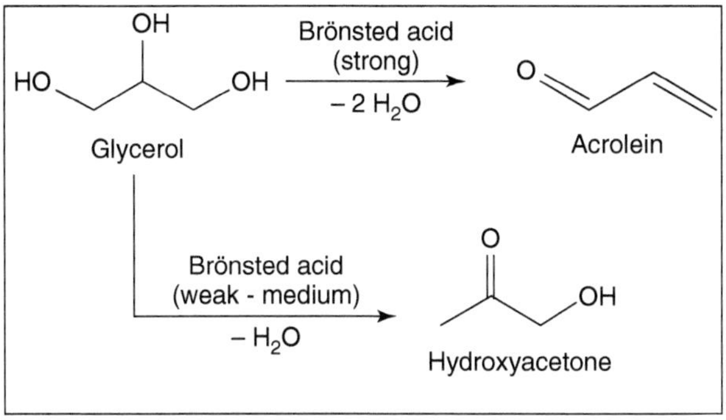

Scheme 13.5 Reaction scheme of glycerol dehydration.

catalytic stability of the catalyst and at the same time maintain the high selectivity for acrolein during the dehydration reaction.

13.5.4 Effect of Type of Heteropoly Acids

Another approach to altering the acidity of HPA was to tune the composition of HPA, which affects the performance of catalysts. Atia et al. [28] compared the performance of phosphotungstic acid, silicotungstic acid, and phosphomolybdic acid over alumina, silica, or aluminosilicates. Our team has evaluated the performance of unsupported and high-surface-area alumina-supported HPA (notably, silicotungstic acid [H-SiW], phosphotungstic acid [H-PW], and phosphomolybdic acid [H-PMo]). In this context, we have prepared a series of HPA supported γ-alumina, with 10–30 wt% loading. The computed Keggin anion density (HPA nm^{-2}), expressed as the number of Keggin anion per square nanometer of the support surface, is presented in Table 13.4. The calculations were performed according to the actual HPA loading, and surface area of the sample [32]. The Keggin density of the 15 wt% HPA-loaded supported catalysts was in the range of 0.13–0.14 suggesting that HPA is highly dispersed over the alumina surface. It further confirmed the better interaction of HPA over alumina support. On the other hand, the loading of HPA results in decrease in surface area and pore volume due to covering and plugging of pores by the large Keggin anions (kinetic diameter about 1.2 nm). The Keggin anion density of these supported catalysts is in the range of 0.09–0.34 HPA nm^{-2}, which indicates high dispersion of the active (HPA) species on the alumina surface.

Table 13.5 summarizes the acidity and reaction results of the different catalysts under study. It can be seen from this table that pure H-SiW shows excellent activity in terms of glycerol conversion and acrolein selectivity. Among the different loadings (10–30 wt%) of HPA, 15 wt% was found to be better in terms of glycerol conversion (99%) and acrolein selectivity (54.4%) and found to be the best among screened catalysts. On the other hand, the catalysts such as HPW/γ-Al_2O_3 and H-PMo/γ-Al_2O_3 showed low selectivity of the expected products owing to a very different acidity of the catalyst, leading to side reactions like oligomerization and carbonization of glycerol as well as the formation of allyl alcohol. Among the three HPAs, silicotungstic acid is found to be suitable for the dehydration of glycerol owing to its moderately strong acidic properties and water-tolerant ability.

Table 13.4 Textural properties of various supported heteropoly acid catalysts.

Catalyst	S_{BET} [a] ($m^2\,g^{-1}$)	Vp [b] ($cm^3\,g^{-1}$)	d_{BJH} [c] (nm)	ρ_{Keggin} [d] (HPA nm^{-2})
γ-Al_2O_3	272	0.30	4.2	-
H-SiW	6	—	—	—
10 H-SiW/γ-Al_2O_3	231	0.22	3.8	0.09
15 H-SiW/γ-Al_2O_3	219	0.19	3.4	0.14
20 H-SiW/γ-Al_2O_3	213	0.17	3.1	0.20
30 H-SiW/γ-Al_2O_3	183	0.11	2.6	0.34
15 H-PW/γ-Al_2O_3	212	0.16	3.4	0.14
15 H-PMo/γ-Al_2O_3	213	0.16	3.5	0.13

a) BET surface area.
b) Total pore volume.
c) Average pore size.
d) Keggin density.
Source: Adapted from Mahendran et al. [22].

Table 13.5 Acidity data and dehydration of glycerol over various solid acid catalysts.[a]

	Acidic sites ($mmol\,g^{-1}$)				Selectivity (%)		
Material	Total	Medium	Strong	Conv. (%)	Acrolein	Acetol	Others[b]
H-SiW	0.72	0.34	0.38	84.0	82.0	9.1	8.9
γ-Al_2O_3	0.34	0.19	0.15	98.0	21	25.7	53.3
15 H-SiW/γ-Al_2O_3	0.53	0.32	0.21	98.0	54.4	25.0	20.6
15 H-PW/γ-Al_2O_3	0.58	0.34	0.25	98.7	36.0	15.4	48.6
15 H-PMo/γ-Al_2O_3	0.65	0.39	0.26	99.0	35.8	7.8	56.4

a) Reaction conditions: Catalyst weight = 300 mg; glycerol concentration = 10 wt%; temperature = 275 °C; feed flow rate = 1.68 mlh^{-1}; TOS = 1 h.
b) Include allyl alcohol, ethylene glycol, acetaldehyde, acetone, methanol, 1,2-propane diol, carbon monoxide, and some unidentified products.
Source: Adapted from Mahendran et al. [22].

Figure 13.5 depicts the effect of time-on-stream over 15 wt% H-SiW/γ-Al_2O_3, and it was found that the use of the high surface area alumina as a support led to better results as far as the deactivation of the catalyst is concerned. The better performance of alumina-supported catalysts compared to silica was attributed to the high dispersion of HPA and diminished acidity.

13.6 Tuning the Acidity

Furthermore, the acid strength of HPAs can be tuned by the exchange of protons with a metal cation. Among various alkali metals, Cs^+ is chosen as it is known to

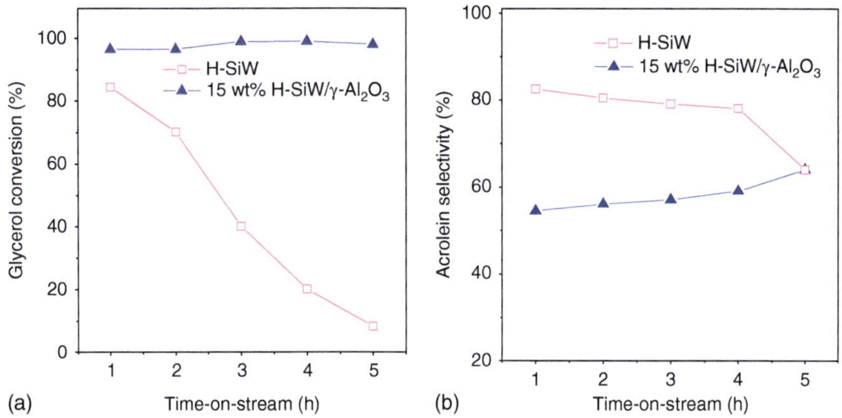

Figure 13.5 TOS study of supported and unsupported catalysts. (a) Conversion. (b) Selectivity.

enhance the selectivity of acrolein in the glycerol dehydration reaction [33, 34]. When large monovalent cations like Cs^+ are substituted for the protons of HPA, increased surface area, thermal stability, and water resistance are obtained. Cesium salt of silicotungstic acid is a well-known water-insoluble strong Brønsted acid possessing high thermal stability (>500 °C) and water tolerance [35]. Such metal cations-exchanged HPAs are dispersed on a suitable solid support to enhance thermal stability and acidic properties. Liu et al. [36] have prepared H3PW12O40 supported on Cs^+ modified SBA-15 (HPW/Cs-SBA), which was prepared and tested for glycerol dehydration. They have achieved an acrolein yield of 85% at 573 K with the co-feeding of O_2. The stability of the catalyst was improved, and the catalyst was found to retain its activity even after 150 hours on stream. Abdullah et al. [37] have prepared bifunctional catalysts and employed them for the dehydration reaction at 275 °C and observed 96% acrolein selectivity using 0.5% Pd/CsPW. The stability of the catalyst was improved by co-feeding hydrogen and the presence of metal functions in the catalyst. The acidic heteropoly salt, cesium 12-tungstosilicate supported over SBA-15 and γ-Al_2O_3, has been prepared, and the same is employed for the dehydration of glycerol to acrolein in the gas phase. TG-DTA results clearly showed that the thermal stability of silicotungstic acid is increased tremendously on exchanging with caesium. Table 13.6 summarizes the dehydration results of Cs-exchanged catalysts. The high catalytic activity of Cs3H-SiW compared to pure H4-SiW is mainly due to their high surface area and, consequently, to their high surface proton content [38]. However, this catalyst had poor long-term stability with the glycerol conversion decreasing to 27% after five hours of reaction. The catalyst, 30 wt% Cs3H-SiW/γ-Al_2O_3 (see Table 13.6), showed excellent catalytic activity in terms of glycerol conversion (100%) and acrolein selectivity (93%). Cs-exchanged catalysts showed good catalytic performance compared to pure acid (Figure 13.6). The enhanced long term stability of γ-Al_2O_3 supported Cs catalyst is mainly attributed to modifications in acidity of the catalyst, preservation of Keggin structure, high dispersion of the active phase, and improved thermal stability.

Table 13.6 Dehydration of glycerol over cesium-exchanged heteropoly acid catalysts.[a)]

		Selectivity (%)			
Catalyst	**Conv. (%)**	**Acrolein**	**Acetol**	**Allyl alcohol**	**Others[b)]**
H_4-SiW	84.3	82.5	9.1	1.5	6.9
Cs_3H-SiW	100.0	88.0	4.2	1.2	6.6
30 H_4-SiW/SBA-15	99.7	95.2	1.5	0.2	3.1
30 Cs_3H-SiW/SBA-15	99.5	90.0	3.0	1.0	6.0
15 H_4-SiW/γ-Al_2O_3	98.0	54.4	25.0	7.5	13.1
30 Cs_3H-SiW/γ-Al_2O_3	100.0	93.0	3.5	0.5	3.0

a) Reaction conditions: Catalyst weight = 300 mg; glycerol concentration = 10 wt%; temperature = 275 °C; feed flow rate = 1.68 mlh^{-1}; TOS = 1 h.
b) Include acetaldehyde, acetone, methanol, and 1,2-propane diol.
Source: Adapted from Mahendran [29].

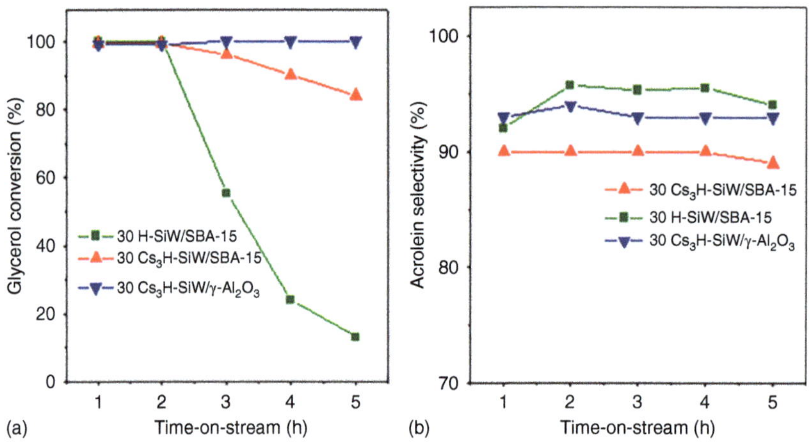

Figure 13.6 TOS of Cs-exchanged catalysts. (a) Conversion. (b) Selectivity.

13.7 Conclusions

In conclusion, all these results suggest that the use of Keggin-type heteropoly compounds is a highly promising means of obtaining high catalytic activity. They offer many opportunities for tuning their acidity via their variations in composition and the selection of the counterions. When these compounds are deposited on a support, it is also possible to adjust an additional parameter, as one can even control the pore size of the catalyst and thus reduce diffusion limitations. For supported HPA catalysts, the activity of the catalysts mainly depends on the acidity of the HPA, the nature and pore size of the support, and the loading of the HPA on the support. Relating these concepts, an acrolein yield of more than 90% has already been achieved. Nonetheless, the main problem remains the rapid catalyst deactivation owing to carbon deposition resulting from the high acidity of these catalysts.

Acknowledgments

The authors would like to thank the Department of Science and Technology (DST) and the Government of India for funding National Centre for Catalysis Research (NCCR) at IIT-Madras. Thanks are also due to the New Millennium Indian Technology Leadership Initiative (NMITLI) program (No. 5/258/52/2006-NMITLI) within CSIR for financial support. The authors would also like to thank Professor B. Viswanathan for the fruitful discussion.

References

1 Zhou, C.-H., Beltramini, J.N., Fan, Y.-X., and Lu, G.Q. (2008). Chemoselective catalytic conversion of glycerol as a biorenewable source to valuable commodity chemicals. *Chem. Soc. Rev.* 37: 527–549.

2 Luque, R., Lovett, J.C., Datta, B. et al. (2010). Biodiesel as feasible petrol fuel replacement: a multidisciplinary overview. *Energy Environ. Sci.* 3: 1706–1721.

3 Tangy, A., Pulidindi, I.N., and Gedanken, A. (2016). SiO2 beads decorated with SrO nanoparticles for biodiesel production from waste cooking oil using microwave irradiation. *Energy Fuels* 30: 3151–3160.

4 Knothe, G., Sharp, C.A., and Ryan, T.W. (2006). Exhaust emissions of biodiesel, petrodiesel, neat methyl esters, and alkanes in a new technology engine. *Energy Fuels* 20: 403–408.

5 Ribeiro, N.M., Pinto, A.C., Quintella, C.M. et al. (2007). The role of additives for diesel and diesel blended (ethanol or biodiesel) fuels: a review. *Energy Fuels* 21: 2433–2445.

6 OECD/FAO (2016). Global biodiesel production and estimation. OECD-FAO Agricultural Outlook, OECD Agriculture statistics (database). https://www.oecd-ilibrary.org/agriculture-and-food/oecd-fao-agricultural-outlook-2016-2025/figure-3-7-5-world-biodiesel-production-and-trade_agr_outlook-2016-graph91-en (Accessed 25 January 2019).

7 Dasari, M.A. (2006). Catalytic conversion of glycerol and sugar alcohols to value-added products. Thesis, Faculty of the Graduate School, University of Missouri-Columbia.

8 Pagliaro, M., Ciriminna, R., Kimura, H. et al. (2006). From glycerol to value-added products. *Angew. Chem. Int. Ed.* 46: 4434–4440.

9 Keulks, G.W., Krenzke, L.D., and Notermann, T.M. (1978). Selective oxidation of propylene. *Adv. Catal.* 27: 183–225.

10 Liu, L., Ye, X.P., and Bozell, J.J. (2012). A comparative review of petroleum-based and bio-based acrolein production. *ChemSusChem* 5: 1162–1180.

11 Guerrero-Perez, M.O., Rosas, J.M., Bedia, J. et al. (2010). Recent inventions in glycerol transformations and processing. *Recent Pat. Chem. Eng.* 2: 11–21.

12 Tsukuda, E., Sato, S., Takahashi, R., and Sodesawa, T. (2007). Production of acrolein from glycerol over silica-supported heteropoly acids. *Catal. Commun.* 8: 1349–1353.

13 Chai, S.-H., Wang, H.-P., Liang, Y., and Xu, B.-Q. (2007). Sustainable production of acrolein: investigation of solid acid-base catalysts for gas-phase dehydration of glycerol. *Green Chem.* 9: 1130–1136.
14 Chai, S., Wang, H., Liang, Y., and Xu, B. (2007). Sustainable production of acrolein: gas-phase dehydration of glycerol over Nb_2o_5 catalyst. *J. Catal.* 250: 342–349.
15 Katryniok, B., Paul, S., Capron, M., and Dumeignil, F. (2009). Towards the sustainable production of acrolein by glycerol dehydration. *ChemSusChem* 2: 719–730.
16 Suprun, W., Lutecki, M., Haber, T., and Papp, H. (2009). Acidic catalysts for the dehydration of glycerol: activity and deactivation. *J. Mol. Catal. A: Chem.* 309: 71–78.
17 Wang, F., Dubois, J.-L., and Ueda, W. (2009). Catalytic dehydration of glycerol over vanadium phosphate oxides in the presence of molecular oxygen. *J. Catal.* 268: 260–267.
18 Corma, A., Huber, G., Sauvanaud, L., and Oconnor, P. (2008). Biomass to chemicals: catalytic conversion of glycerol/water mixtures into acrolein, reaction network. *J. Catal.* 257: 163–171.
19 Jia, C.-J., Liu, Y., Schmidt, W. et al. (2010). Small-sized Hzsm-5 zeolite as highly active catalyst for gas phase dehydration of glycerol to acrolein. *J. Catal.* 269: 71–79.
20 Rocchicciolideltcheff, C. (1990). Structure and catalytic properties of silica-supported polyoxomolybdates II. Thermal behavior of unsupported and silica-supported 12-molybdosilicic acid catalysts from Ir and catalytic reactivity studies. *J. Catal.* 126: 591–599.
21 Izumi, Y. (1983). Catalysis by heterogeneous supported heteropoly acid. *J. Catal.* 84: 402–409.
22 Mahendran, S., Nithya, T., Raju, K.P. et al. (2011). Dehydration of glycerol over high surface area alumina-supported heteropoly acid catalysts, *Chemeca 2011, Proc. Chem. Eng. Conf.*, Sydney, pp. 289-298.
23 Edwards, J.C., Thiel, C.Y., Benac, B., and Knifton, J.F. (1998). Solid-State NMR and FT-IR investigation of 12-tungstophosphoric acid on Tio_2. *Catal. Lett.* 51: 77–83.
24 López-Salinas, E., Hernández-Cortéz, J.G., Schifter, I. et al. (2000). Thermal stability of 12-tungstophosphoric acid supported on zirconia. *Appl. Catal., A* 193: 215–225.
25 Baba, T. and Ono, Y. (1986). Heteropolyacids and their salts supported on acidic ion-exchange resin as highly active solid-acid catalysts. *Appl. Catal.* 22: 321–324.
26 Kamiya, Y., Ooka, Y., Obara, C. et al. (2007). Alkylation–acylation of p-xylene with Γ-butyrolactone or vinylacetic acid catalyzed by heteropolyacid supported on silica. *J. Mol. Catal. A: Chem.* 262: 77–85.
27 Vázquez, P., Pizzio, L., Cáceres, C. et al. (2000). Silica-supported heteropoly-acids as catalysts in alcohol dehydration reactions. *J. Mol. Catal. A: Chem.* 161: 223–232. Atia, H., Armbruster, U., and Martin, A. (2008). Dehydration of glycerol

in gas phase using heteropolyacid catalysts as active compounds. *J. Catal.* 258: 71–82.

28 Atia, H., Armbruster, U., and Martin, A. (2008). Dehydration of glycerol in gas phase using heteropolyacid catalysts as active compounds. *J. Catal.* 258: 71–82.

29 Mahendran, S. (2017). *Dehydration of Glycerol Over Mesoporous Silica- and Alumina-Supported Heteropoly Acid Catalysts*. Chennai: Indian Institute of Technology Madras.

30 Mahendran, P.S. (2014). Dehydration of glycerol over silicotungstic acid- supported silica. *Adv. Porous Mater.* 2: 1–6.

31 Katryniok, B., Paul, S., Belliere-Baca, V. et al. (2010). Glycerol dehydration to acrolein in the context of new uses of glycerol. *Green Chem.* 12: 2079–2098.

32 Chai, S.-H., Wang, H.-P., Liang, Y., and Xu, B.-Q. (2009). Sustainable production of acrolein: preparation and characterization of zirconia-supported 12-tungstophosphoric acid catalyst for gas-phase dehydration of glycerol. *Appl. Catal., A* 353: 213–222.

33 Alhanash, A., Kozhevnikova, E.F., and Kozhevnikov, I.V. (2010). Gas-phase dehydration of glycerol to acrolein catalysed by caesium heteropoly salt. *Appl. Catal., A* 378: 11–18.

34 Jin, H., Yi, X., Sun, X. et al. (2010). Influence of $H4SiW_{12}O_{40}$ loading on hydrocracking activity of non-sulfide Ni–$H_4SiW_{12}O_{40}/SiO_2$ catalysts. *Fuel* 89: 1953–1960.

35 Atia, H., Armbruster, U., and Martin, A. (2011). Influence of alkaline metal on performance of supported silicotungstic acid catalysts in glycerol dehydration towards acrolein. *Appl. Catal., A.* 393: 331–339.

36 Liu, R., Wang, T., and Jin, Y. (2014). Catalytic dehydration of glycerol to acrolein over HPW supported on Cs+ modified SBA-15. *Catal. Today* 233: 127–132.

37 Alhanash, A., Kozhevnikova, E.F., and Kozhevnikov, I.V. (2010). Gas-phase dehydration of glycerol to acrolein catalysed by caesium heteropoly salt. *Appl. Catal., A.* 378: 11–18.

38 Essayem, N., Coudurier, G., Fournier, M., and Védrine, J.C. (1995). Acidic and catalytic properties of $Cs_XH_{3-X}PW_{12}O_{40}$ heteropolyacid compounds. *Catal. Lett.* 34: 223–235.

14

Catalysis with Carbon Nanotubes

Mohammad Y. Masoomi and Lida Hashemi

Arak University, Department of Chemistry, Faculty of Sciences, Karballa Blvd, Sardasht, Arak 3848177584, Iran

14.1 Introduction

Carbon nanotubes (CNTs) have unique composition, geometry, and properties that enable numerous potential CNT applications. Energy Storage, Molecular Electronics, Thermal Materials, Structural Materials, Electrical Conductivity, Fabrics and Fibers, Catalyst Supports, Biomedical, Air and Water Filtration, Conductive Plastics, Conductive Adhesives, and Ceramics are an almost complete list of CNT applications, and in this chapter we will only examine the use of them as catalyst supports.

Catalysis is currently recognized as a potential field of application for CNT, and throughout the past decade, the number of publications and patents on this subject has been increasing exponentially. In most of these cases, the use of these nanomaterials as support structures facilitates better performance than conventional supports. To date, most of the research has focused on supported metal catalysts, in which the active phase is located on the external surface of the CNT. The selective deposition of metallic nanoparticles (NPs) in the inner cavity of nanotubes, which could allow the exploitation of advantageous confinement effects, has received much less attention. In this chapter, the different strategies for the preparation of such nanocatalysts, as well as the benefits that could be expected from the resultant confinement effects are presented, with the aim of highlighting their potential use in catalysis [1]. Getting costs down to commercially viable levels have proven challenging, but increasing scale is happening.

14.1.1 Why CNT may be Suitable to be Used as Catalyst Supports?

Before presenting a detailed description of the use of CNTs as efficient catalysts or catalyst supports, it is important to analyze their electronic, adsorption, mechanical and, thermal properties. It is indeed necessary to evaluate the resistance of this type of material and to be able to foresee how the metallic particles will be anchored on the support and how the reagents will interact with the catalyst, so as to understand what such novel carbon forms could bring to catalysis [2].

Catalysis in Confined Frameworks: Synthesis, Characterization, and Applications, First Edition.
Edited by Hermenegildo Garcia and Amarajothi Dhakshinamoorthy.

14.1.1.1 From the Point of Structural Features

Elemental carbon in the sp^2 hybridization can form a variety of amazing structures. Apart from the well-known graphite and C_{60}, various carbon cages were studied. Iijima [3] observed for the first time tubular carbon structures. The nanotubes consisted of up to several tens of graphitic shells (so-called multi-walled carbon nanotubes [MWCNTs]) with adjacent shell separation of ≈0.34 nm, diameters of ≈1 nm, and large length/diameter ratio. Two years later, Iijima and Ichihashi [4] and Bethune et al. [5] synthesized single-walled carbon nanotubes (SWCNTs) (Figure 14.1). Nowadays, MWCNTs and SWCNTs are characterized by means of Raman, electronic, and optical spectroscopies. Important information is derived from mechanical, electrical, and thermal measurements [6].

The internal diameter of these structures can vary between 0.4 and 2.5 nm, and the length ranges from few microns to several millimeters. MWCNT can be considered a concentric SWCNT with increasing diameter and coaxially disposed. The number of walls present can vary from two (double-wall nanotubes) to several tens, so that the external diameter can reach 100 nm. The concentric walls are regularly spaced by 0.34 nm, similar to the intergraphene distance evidenced in turbostratic graphite materials. It is worth to noting that residual metallic particles coming from the production process can be found in the inner cavity of MWCNT [2].

The type of CNT depends on how the graphene sheet is oriented during rolling. This can be specified by a vector (called chiral vector), which defines how the graphene sheet is rolled up. The vector is determined by two integers (n, m). Two atoms in a planar graphene sheet are chosen, and one is used as origin. The chiral

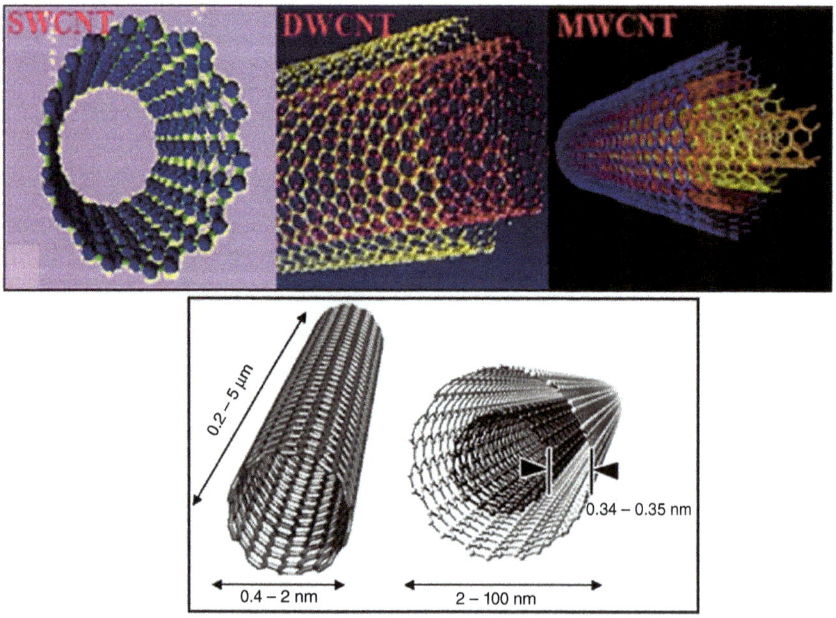

Figure 14.1 Schematic representation of CNT types and dimensions.

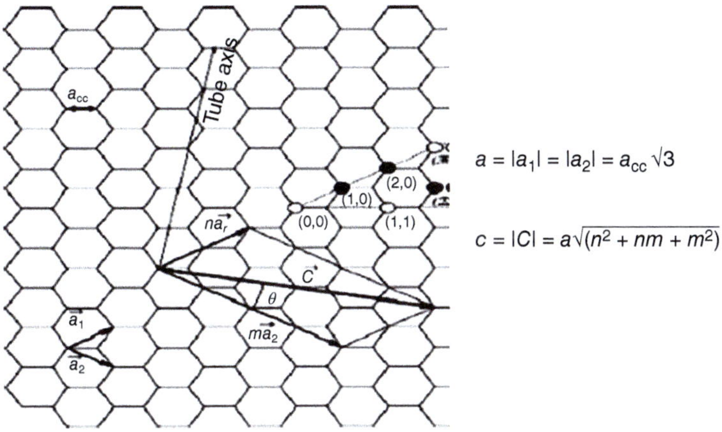

Figure 14.2 Representation of CNT parameters.

vector C is pointed from the first atom toward the second one (Figure 14.2), defined by the relation $C = na_1 + ma_2$, where a is the length of the unit cell vector a_1 or a_2. This length a is related to the carbon–carbon bond length $a_{\text{c-c}}$ that for graphite, the carbon–carbon bond length is $a_{\text{c-c}} = 0.1421$ nm. The same value is often used for CNTs, but due to the curvature of the tube, a slightly larger value such as $a_{\text{c-c}} = 0.144$ nm should be a better approximation.

According to different values of n, m, and θ different CNT chirality structures have been produced, as shown in Figures 14.3 and 14.4.

$m = 0$ for all zig-zag tubes and $(\theta = 30)$ $(n, 0)$, $n = m$ for all armchair tubes and $(\theta = 0)$ (n, n) and Otherwise, when $n \neq m$ they are called chiral tubes and $(0 < \theta < 30)$ (n, m) (Figure 14.3).

The way in which CNTs are formed is not exactly known. The growth mechanism is still a subject of study, and one of the most accepted theories postulates that metal catalyst particles are floating or are supported on graphite or another substrate. First, a precursor to the formation of CNTs and fullerenes, C_2 is formed on the surface of

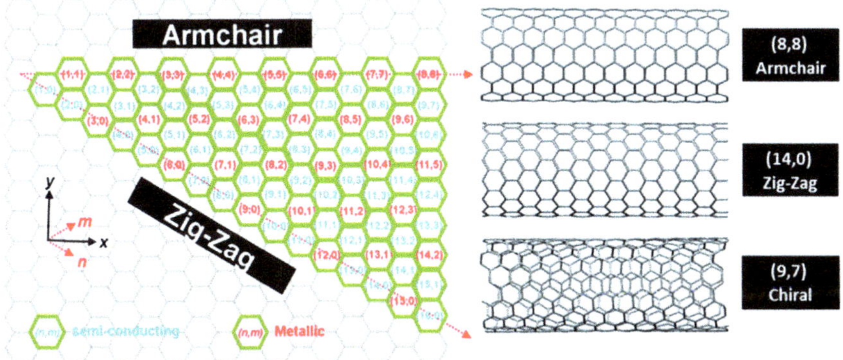

Figure 14.3 How many different CNT chirality structures have been produced?

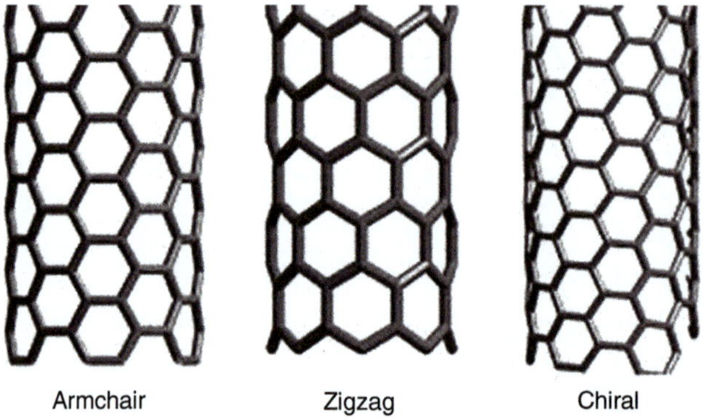

Figure 14.4 Representation of different CNTs chirality structures.

the metal catalyst particle. From this metastable carbide particle, a rod-like carbon is formed rapidly. Secondly, there is a slow graphitization of its wall. This mechanism is based on *in-situ* TEM observations. The most accepted growth mechanisms are two models (Figure 14.5):

Tip-growth method: In the former, a tubule tip is open so that carbon atoms can be added to its circumference, and the metal catalyst promotes the growth reaction and also prevents the tubule tip closure (Figure 14.5a).

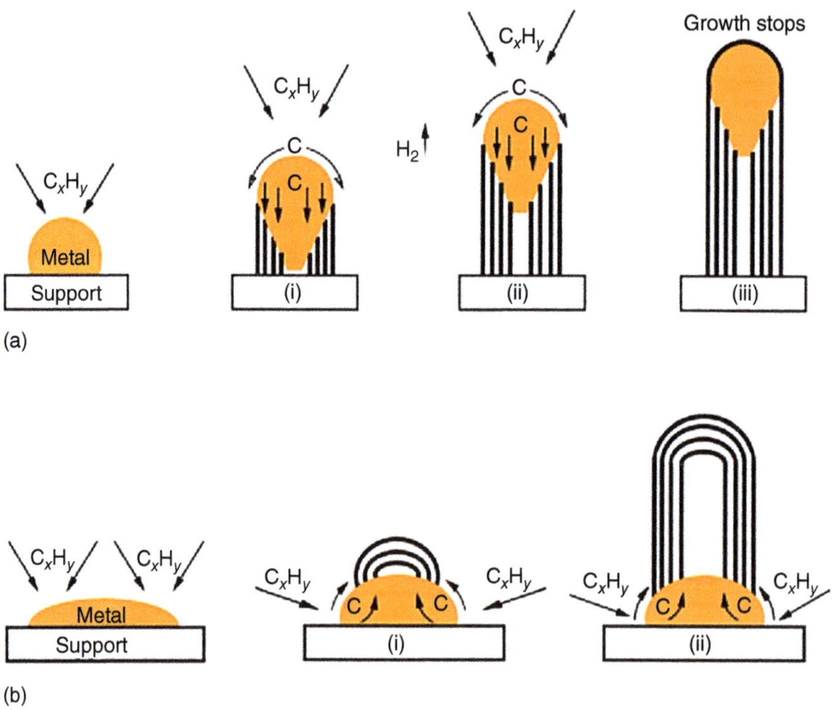

Figure 14.5 (a) Tip-growth mechanism and (b) base-growth mechanism.

Root growth or base growth: The latter model is based on the phase diagram of carbon and a metal. For supported metals, the CNT can form either by extrusion, in which the CNT grows upwards from the metal particles that remain attached to the substrate (Figure 14.5b).

Carbon nanotubes are generally produced by three main techniques: (i) arc discharge, (ii) laser ablation, and (ii) chemical vapor deposition (CVD). Though scientists are researching more economic ways to produce these structures. In arc discharge technique, a vapor is created by an arc discharge between two carbon electrodes. CNTs self-assemble from the resulting carbon vapor. In the laser ablation technique, a high-power laser beam impinges on a volume of carbon-containing feedstock gas (such as methane or carbon monoxide). At the moment, laser ablation produces a small amount of clean CNTs. Finally, CVD generally involves reacting a carbon containing gas (such as acetylene, ethylene, and ethanol) with a metal catalyst particle (usually cobalt, nickel, iron, or a combination of these, such as cobalt/iron or cobalt/molybdenium) at temperatures above 600 °C. This method is able to fabricate not only aligned but also single-walled or multi-walled CNTs by using different hydrocarbon gases.

14.1.1.2 From the Point of Electronic Properties

Exhaustive studies concerning electronic properties of both SWCNT [7] and MWCNT [8] are available in the literature. In the case of SWCNT, studies have demonstrated that they behave like pure quantum wires (1D-systems) where the electrons are confined along the tube axis. Electronic properties are mainly governed by two factors: the tube diameter and the helicity, which is defined by the way in which the graphene layer is rolled up [5] (armchair, zigzag, or chiral). In particular, armchair SWCNT is metallic and zigzag ones display semi-conductor behavior. This curvature of the graphene sheet induces strong modifications of the electronic properties, and a comparison to graphite shows a modification of the π-electron cloud [9, 10]. Studies on MWCNT's electronic properties have revealed that they behave like an ultimate carbon fiber [8]: at high temperature, their electrical conductivity may be described by semi-classical models already used for graphite, whereas at low temperatures, they reveal 2D-quantum transport features. When used in catalysis, these conductive supports present clear differences with respect to activated carbon (AC), and a recent theoretical study related to the interaction of transition metal atoms with CNT and graphite indicates major differences [11]. It has been demonstrated that the binding sites depend on the structure of the support.

14.1.1.3 From the Point of Adsorption Properties

The interaction of CNTs with their environment, and in particular with gases or doping species adsorbed either on their internal or external surfaces, attracts increasing attention due to the possible influence of the adsorption on some of the tubes properties and to the possibility of using these materials for efficient applications. Adsorption properties on SWCNT samples, usually found in bundles or ropes, should not be considered in terms of individual nanotubes but in term of adsorption on the exterior or interior surfaces of such bundles. A similar situation exists for MWCNT, where adsorption could occur either on or inside the tube or between aggregated MWCNT [2]. Different studies dealing with the adsorption on MWCNT [12] and SWCNT [13, 14] have highlighted the porous nature of these materials, and it appears that CNTs present specific adsorption properties when compared to graphite or AC, mainly due to their peculiar morphology. The role of defects, opening/closing of the tubes, chemical purification, or the presence of impurities as catalyst particles that can govern the adsorption properties has not yet been examined in detail.

14.1.1.4 From the Point of Mechanical and Thermal Properties

Nanotubes are made exclusively of covalently bonded carbon atoms, and the structure is not easily changed with the effects of pressure. It has been demonstrated that CNTs are only undergoing permanent structure change at very high pressures (over 1.5 GPa), and below that value the deformations are totally elastic. Another important feature that has to be taken into consideration is thermal stability under reaction conditions. An important property that has to be noticed is that CNTs are more stable to oxidation than AC (C^*) but more reactive than graphite. However, the presence of residual metal on or in the nanotubes that can catalyze carbon gasification may lower the temperature at which the maximum gasification rate occurs. The Studies conducted on CNTs have shown distinguishable behaviors for SWCNT and MWCNT: SWCNTs, which have less defects on their surfaces and are therefore more stable than multi-wall ones [2].

To conclude, it appears that the combination of these properties makes CNT attractive and competitive catalyst supports by comparison with ACs.

14.2 Catalytic Performances of CNT-Supported Systems

CNTs often do not have the ability to catalyze chemical reactions alone and are used as supports for catalytic carriers for a number of reasons, including: (i) their high purity that eliminates self-poisoning; (ii) their impressive mechanical properties, high electrical conductivity, and thermal stability; (iii) the high accessibility of the active phase and the absence of any micro porosity, thus eliminating diffusion and intraparticle mass transfer in the reactions medium; (iv) the possibility for macroscopic shaping of the support; (v) the possibility of tuning the specific metal–support interactions, which can directly affect the catalytic activity and selectivity; and (vi) the possibility of confinement effects in their inner cavity [1]. Additionally, compared to conventional supports, CNTs have a high degree of flexibility for the dispersion of the active phase since it is possible to: (i) modulate their specific surface area (50–500 m^2g^{-1} for MWCNTs) or their internal diameter (5–100 nm for MWCNTs); (ii) easily functionalize chemically their surfaces [15]; (iii) change their chemical composition (nitrogen- or boron-doped CNTs); and (iv) deposit the catalytic phase either on their external surface or in their inner cavity [16].

14.2.1 Different Approaches for the Anchoring of Metal-Containing Species on CNT

A wide variety of methods are at the disposal of chemists for the preparation of decorated CNTs, and apart from transition metals, a large number of inorganic materials have been coated onto the nanotube surface, targeting different applications such as catalysts, sensors, solar cells, and magnetic devices. The choice of coating method depends largely on the application of the coated nanotubes and the scale of operation. Several methods such as incipient wetness impregnation, ion exchange, organometallic grafting, electron beam evaporation, and deposition/precipitation, have been used to prepare CNTs or graphite nanofiber (GNF)-supported catalysts. As long as CNTs are relatively inert supports many studies have been conducted in order to find which pretreatment procedures are needed to achieve optimal interaction between the support and the catalyst precursor. The wetness impregnation technique, the most widely used method, is a multistep process that comprises the liquid-phase deposition of a precursor of the coated material followed by calcinations and/or reductions. Electrochemical coating methods are the obvious choice for electro-catalytic applications, whereas gas-phase methods seem to be more compatible with electronic applications. Self-assembly techniques, though efficient and precise, are difficult to scale up owing to the high cost of reagents. For supported catalyst preparation, the role of CNT surface pretreatments, as in the case of AC, on the final metal dispersion has been clearly demonstrated [17].

CNT-based catalysts have been used mainly for important liquid-phase (hydrogenation, hydroformylation) or gas-phase (Fischer–Tropsch process, ammonia decomposition) reactions, for photo catalysis and for electro catalysis (fuel-cell electrodes). Representative examples of the catalytic performances of CNT-based catalysts are listed in Table 14.1. For these systems, the active phase is located mainly on the external surface of the CNTs. However, the well-defined nanochannels of CNTs could enable us to investigate the effect of confinement [27] of the active phase on the catalytic activity and selectivity. Beside the confinement of small molecules [28], metal complexes [29], ionic liquids [30], or fullerenes [31], the confinement of NPs in the CNT cavity paves the way to interesting perspectives for performing chemistry in a nanoreactor while exploiting the unique CNT properties.

14.2.2 Different Approaches for the Confining NPs Inside CNTs and Their Characterization

The complete filling of CNTs, which leads to the formation of metallic nanowires, is currently relatively well-established and documented [32]. However, the selective confinement of discrete NPs inside the CNT cavity is still a synthetic challenge for chemists. Currently, several techniques have been developed for filling CNTs, including wet chemistry methods, filling of volatile metal precursors [33], and *in-situ* filling during arc-discharge growth of CNTs [34]. Among these techniques, wet chemistry appears to be the most simple and versatile. Apart

Table 14.1 Catalytic performance of CNT-supported systems.

Reaction	Catalyst	Comments	Reference
Syngas conversion	Ru	Ru/CNT is a highly selective Fischer–Tropsch catalyst for the formation of C_{10}–C_{20} hydrocarbons	[18]
Hydrogenation/ dehydrogenation	Ru, Pt	High activities are obtained when using MWCNTs in contrast to activated carbon for which mass-transfer limitations occur	[19]
NH_3 decomposition	Ru	The Ru/CNT catalyst is highly active for the generation of CO_x-free H_2 from ammonia	[20]
Fuel cells	Pd, Pt	MWCNT-supported Pd NPs are effective catalysts for the electro-oxidation of methanol as well as renewable alcohols such as ethanol and glycerol	[21]
Hydroformylation	Ru	Higher activities were obtained with MWCNT-supported $[Ru(CO)_4]_n$ in comparison with the unsupported Ru polymer	[22]
Photocatalysis	TiO_2	MWCNTs act as photosensitizer rather than adsorbent or dispersing agent	[23]
Cinnamaldehyde hydrogenation	Rh	Catalytic activity of Rh/MWCNT three times higher than a corresponding Rh/C* catalyst	[24]
No decomposition	Rh	Rh/MWCNT has been shown higher conversion than with a Rh/Al_2O_3	[25]
Cyclohexanol dehydrogenation	Co	Activity and selectivity of Co/MWCNT towards cyclohexanone slightly higher than on Co/C*	[26]

from chemical methods, standard materials characterization techniques, such as electron microscopy and spectroscopy, have been employed for the characterization of confined NPs. High-resolution transmission electron microscopy (HRTEM) provides clear information about metal dispersion and particle size distribution [1].

14.2.2.1 Wet Chemistry Method

For the confinement of discrete NPs, wet impregnation from a metallic precursor is based on the filling of oxidized CNTs by capillarity forces with an excess solution containing a metallic salt or a complex [35, 36], followed by thermal decomposition of the precursor to yield NPs. Filling depends on the pressure difference across the liquid–vapor interface, which is related to the surface tension of the used liquid and the contact angle between the liquid and the CNT walls. Liquids with a surface tension lower than 100–200 m Nm^{-1} can wet open CNTs and can fill the channels at atmospheric pressure by capillarity, as established by Ebbesen [37].

Wu et al. [38] prepared Fe–Ni alloy NPs (c. 5 nm) inside CNTs (internal diameter 5 nm) and Ma et al. [39] deposited Pt NPs (c. 5 nm) inside the channel of CNTs (internal diameter 100 nm), by using wet chemistry. Pd NPs, with an average diameter of 5 nm, were introduced inside CNTs by wet impregnation followed by classical thermal treatments [40]. When double-walled carbon nanotubes (DWCNTs) of 1.5 nm internal diameter were used, nanowires were formed instead of discrete NPs. In this manner, Jorge et al. [41] formed a-Fe nanowires in DWCNTs by taking advantage of the capillarity effect. Carbon nanofibers were synthesized inside CNTs by selective deposition of a solution of cocatalyst by incipient wetness impregnation, which was followed by the growth of carbon nanofibers inside the CNTs. Metallic Co-NPs act as a catalyst for nanofiber growth on the inner walls of CNTs during the catalytic CVD process [42]. However, in all of these studies the efficiency of the process (i.e. the percentage of NPs effectively located inside the CNT cavity) is neither discussed nor quantified. Ultrasonic treatment and extended stirring are often employed to assist the capillary action. However, the filling efficiency is low for nanotubes with a diameter smaller than 10 nm [43].

14.2.2.2 Production of CNTs Inside Anodic Alumina

Another approach, which also comprises several steps, involves the production of CNTs inside anodic alumina membranes before filling the resulting template with a solution of NPs and dissolving the alumina membrane [44, 45]. However, this method suffers from scaling-up difficulties. Sublimation of the metal precursor can be used to introduce NPs into the CNT cavity [29]. The process usually involves evacuation of the channels, followed by evaporation and decomposition. However, the large amount of metal NPs deposited on the outside surface of the CNTs needs to be carefully removed, and this is critical for control of the metal loadings.

14.2.2.3 Arc-Discharge Synthesis

Arc-discharge synthesis has also been developed for filling CNTs [34]. In this method, the tubes were generated with simultaneous filling by the doped element. The major drawbacks are the inability to control the size of the filled material, which generally presents nanowire morphology inconvenient for catalysis, and the fact that the resulting metallic NPs are often encapsulated inside the CNTs, and will not be accessible to reactants in catalytic reactions [1].

14.2.3 Hydrogenation Reactions

The most studied reaction with SWCNT and MWCNT, both in liquid and gas phase, is hydrogenation and two types of reactions have to be taken into account: alkenes hydrogenation or α,β-unsaturated aldehydes selective hydrogenation. Studies on light alkenes such as ethylene, but-1-ene and buta-1,3-diene hydrogenation on nickel catalysts supported on different types of CNT, γ-alumina and AC. Results state that the nickel crystallite's activity and selectivity can be altered greatly by the interactions with the support; indeed, it was found that the catalyst supported on CNT allows higher conversion compared to those obtained with γ-alumina and AC supported systems.

The hydrogenation of α,β-unsaturated molecules on GNFs or CNT-supported catalysts was the object of several studies. A palladium-supported catalyst on GNF-H ($50\,m^2g^{-1}$) or AC ($1000\,m^2g^{-1}$) was used in the liquid-phase hydrogenation of cinnamaldehyde (CAL) to hydrocinnamaldehyde [46–48]. The higher performances of the GNF- supported catalyst in terms of activity and particularly selectivity (90% compared to 40% for AC) are explained by the absence of mass-transfer limitations compared to the microporous AC, and, also in this case, a peculiar graphitic carbon–palladium interaction in addition to some residual acidity on the AC surface might have favored some CO bond hydrogenation. When using a 1% (w/w) Rh/MWCNT for the same reaction, a selectivity of 100% was obtained and the catalytic activity of the MWCNT-($180\,m^2g^{-1}$) based catalyst was three times higher than that of a 1% (w/w) Rh/C* where the support has a surface area of $700\,m^2g^{-1}$ [24].

In other work, the selective hydrogenation of CAL reaction to study the influence of the carbon support (MWCNT, GNF, SWCNT, and commercial AC) on the catalytic performances of monometallic and bimetallic catalysts has been investigated. The selective hydrogenation of the carbonyl group of the above α,β-unsaturated aldehyde, yielding the unsaturated alcohol, remains a challenging task and is of particular interest because of the importance of such alcohols in the fine chemicals industry. Group 8–10 metal catalysts have been tested in CAL hydrogenation, and Pd and Rh generally display high activity but rather poor selectivity toward cinnamyl alcohol; Pt and Ru, the most commonly used metals, exhibit moderate selectivity [49]. Among other factors, selectivity could be enhanced by the electronic effects of the support, the presence of a second metal, and the metal particle sizes and morphology. Considering the better selectivity presented by Pt and Ru catalysts, these metals were used as active phases, and high activities are obtained by using MWCNT, a mesoporous support that avoids, in contrast to AC, mass transfer limitations (Scheme 14.1 and Table 14.2) [19].

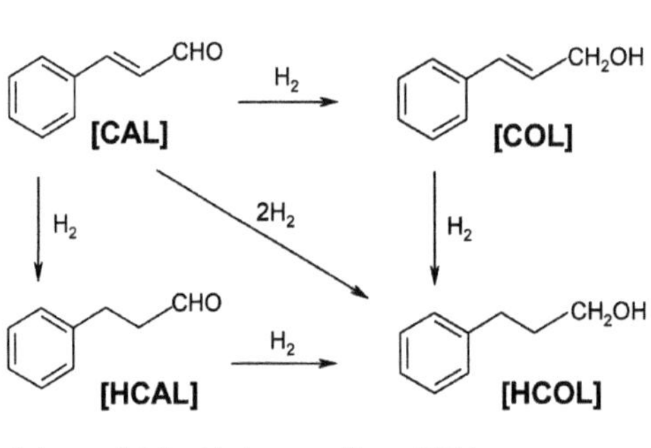

Scheme 14.1 Hydrogenation of CAL.

14.2.4 Dehydrogenation Reactions

Selective dehydrogenation of cyclohexanol to cyclohexanone was studied with a cobalt CNT-based catalytic system [26, 50]. The effect of the oxidizing pretreatment of the support surface on the catalytic performances of the system has been evidenced; the smaller particle size obtained on the nitric acid-treated MWCNT (5 nm instead of 100 nm on untreated support) allows an almost 20% higher conversion.

Table 14.2 CAL conversions (Conv.) and selectivity to COL (Selec.) obtained after two hours of reaction over the different carbon supported catalysts.

Support	2% Pt Conv./ Selec. (%)[a]	2% Ru Conv./ Selec. (%)	2% Pt/2% Ru Conv./Selec. (%)	2% Ru/2% Pt[b] Conv./Selec. (%)
SWCNT	85/28	65/31	58/42 (**60/69**)	—
MWCNT	95/32 (**97/66**)	66/35	44/55 (**79/93**)	80/22
GNF	96/14 (**90/15**)	78/18	60/29 (**92/17**)	76/21
AC	20/62 (**12/60**)	22/51	15/64 (**16/31**)	23/38

a) The value in bold between parentheses correspond to the results obtained on heat treated samples.
b) Catalyst mass 50 mg.

A comparison between a 15% (w/w) Co/MWCNT and the corresponding Co/C* has been made: a slightly higher selectivity to cyclohexanone was found for the former catalytic system, as well as a different by-product distribution.

14.2.5 Liquid-Phase Hydroformylation Reactions

Dependence of catalytic activity on the relative location of catalysts on CNTs was first noticed in liquid-phase hydroformylation of propylene [51]. A catalyst consisting of a $[HRh(CO)(PPh_3)_3]$ complex deposited on open CNTs has yielded a TOF = $0.10\,s^{-1}$ and a molar ratio of the normal/branched products $n/i = 9$ in comparison to $0.06\,s^{-1}$ (TOF) and $n/i = 6$ over the complex on closed CNTs. Although no direct evidence has been provided for the location of Rh, the higher activity and regioselectivity of the open catalyst have been suggested to be attributed to the Rh species inside CNTs and its surface consisting of six-membered C-rings [51]. Pd particles were introduced inside MWCNTs for benzene hydrogenation, which exhibited a TOF twice as high as that over zeolite Y and AC-supported Pd catalysts, although zeolite Y and AC have much higher surface areas than CNTs [36]. Confinement of Pd particles inside MWCNTs (40 nm average i.d.) was also found to benefit selective hydrogenation of CAL with a faster hydrogenation rate and a much higher selectivity (90%) to hydrocinnamaldehyde compared with an AC-supported Pd catalyst [40]. Likewise Serp and coworkers observed a much better catalytic performance for Pt/Ru particles inside MWCNTs (40 nm average i.d.) for the same reaction but with opposite selectivity [16]. The TOF was almost three times higher, and the selectivity toward cinnamyl alcohol was more than twice that of an unsupported Pt/Ru catalyst with a similar particle size. Moreover, the selectivity was linearly correlated with the percentage of NPs located inside CNTs. A higher selectivity toward cinnamyl alcohol was also observed over a CNT-confined Pt catalyst (60–100 nm i.d.) compared with the outside Pt catalyst [39, 52]. Higher conversion and regioselectivity towards *n*-butylaldehyde are reported for the MWCNT and GNF-supported systems. The authors have proposed that the nanotube channel's size fits to accommodate the rhodium complex and to prevent extensive capillary condensation of *n*-butylaldehyde compared to the other supports. No details are given concerning the possible leaching of complexes from the support to the solution. Interestingly, a positive effect of MWCNT on homogeneous oxidation of cyclohexene catalyzed by a ruthenium complex has been recently reported. The authors have noted that the 20–25 nm external diameter nanotubes have no catalytic activity for this reaction, but their addition to an active homogeneous ruthenium complex can enhance the conversion of cyclohexene and the selectivity to 2-cyclohexen-1-ol, and improve the recycling of the ruthenium complex. The possibility of an in situ grafting of the metallic complex on the oxidized MWCNT has to be taken into account to explain these results.

Rhodium NPs deposited on surface-modified MWCNT have also been used to perform the hydroformylation of hex-1-ene [24]. It is reported that this highly dispersed catalyst (1% (w/w), 1.5–2.5 nm) is more active than a Rh/C* catalyst prepared by a similar procedure. The authors propose that the mesoporous nature of the MWCNT, compared to the microporous texture of AC, contributes to better performances by increasing the transfer processes [2].

14.2.6 Liquid-Phase Oxidation Reactions

It has been reported for the first time that those CNTs as metal-free catalysts exhibited excellent activity in the selective oxidation of ethylbenzene (EB) to acetophenone (AcPO) in the liquid phase with oxygen as the oxidant. The results demonstrated that the CNTs played an important role in the decomposition of 1-phenylethyl hydroperoxide (PEHP) and contributed to the production of AcPO, owing to π–π interactions between the radical species and peroxides and the graphene sheets of the CNTs. Surface carboxylic groups of the CNTs were unfavorable to EB oxidation. Adsorption energies of the radical species and peroxides on pristine and modified CNTs with carboxylic groups were calculated by DFT. These theoretical calculations were well consistent with the experimental results and also supported the presented mechanistic pathway of EB oxidation on CNTs. In addition, the CNTs have shown outstanding recyclability and exhibited excellent potential for industrial application of the EB oxidation to AcPO [53].

In other study by Peng and coworkers a HNO_3-promoted benzyl alcohol catalytic oxidation system was developed in the presence of CNTs using molecular oxygen as the terminal oxidant under mild reaction conditions, indicating that CNTs as a metal-free catalyst for liquid-phase oxidation display good stability [54]. The effects of solvent, reaction temperature, amount of HNO_3, catalyst loading, and surface structure of CNTs on the catalytic performances have been investigated. The CNTs showed excellent catalytic activity, exhibiting benzyl alcohol conversion of 96.2% and benzaldehyde selectivity of 88.3% under optimal conditions. In particular, it had remarkable reusability without a significant loss in activity or selectivity after six consecutive usages. A possible reaction pathway has been proposed it is clarified that HNO_2 attacks benzyl alcohol to generate benzyl nitrite, which is decomposed to benzaldehyde over the HNO_3-promoted CNTs-catalyzed system, and electron transfer in graphene sheets plays an important role in the decomposition of

Table 14.3 Catalytic performance of several CNT-catalyst systems in benzyl alcohol aerobic oxidation.[a)]

Entry	Catalyst	Additive	Conversion (%)	Selectivity (%)
1	—	—	3.2	100
2	CNTs-1	—	7.1	100
3	—	HNO_3	31.4	75.3
4	CNTs-1	HNO_3	96.2	88.3
5	CNTs-1	H_2O_2	10.6	100
6	CNTs-1	TBHP	24.2	87.5
7	CNTs-1	HNO_3	70.1	79.6
8	CNTs-1	HCl	8.5	100
9	CNTs-1	$NaNO_3$	5.1	100
10	CNTs-1	$NaNO_2$	2.9	100
11	CNTs-1	HCl + $NaNO_2$	48.3	93.6
12	CNTs-1	HNO_3 + urea	3.1	100
13	CNTs-1	HNO_3 + BHT	70.1	89.7
14	MCM-41	HNO_3	51.7	66.2
15	Al_2O_3	HNO_3	50.6	64.2

a) Reaction conditions: catalyst (100 mg), benzyl alcohol (2 mmol), 1,4-dioxane (10 ml), 65–68% HNO_3 (2 mmol), reaction temperature (90 °C), reaction time (5 h).

benzyl nitrite. These results not only provide an attractive metal-free alternative to noble-metal-catalyzed systems but also provide a better understanding of the mechanism of CNTs as a metal-free catalyst for the liquid-phase oxidation of benzyl alcohol. In the whole catalytic cycle, oxygen is the terminal oxidant, while HNO_3 only initiates the oxidation cycle. The solvent has a remarkable effect on the oxidation reaction. The conversion of benzyl alcohol obviously increased with the solvent polarity increasing, so 1,4-dioxane was the best solvent investigated [54]. The effect of additives on CNTs in benzyl alcohol aerobic oxidation has been shown in Table 14.3.

Two years later, Peng and coworkers published a new study that described aerobic oxidation of benzyl alcohol to benzaldehyde catalyzed by CNTs without any promoter. The effects of reaction conditions, solvent, and surface chemistry of CNTs on their catalytic performances have been investigated. The results showed that the solvent has a remarkable effect on oxidation of BzOH. The effect of the CNTs surface chemistry on the catalytic activity was revealed, and the surface carboxylic groups were particularly detrimental for the catalytic activity. A reasonable mechanism responsible for the liquid-phase oxidation of BzOH on CNTs has been shown that electron transfer in graphene skeletons plays an important role. Nitrogen doping efficiently enhanced the catalytic activity of CNTs for aerobic oxidation of BzOH, arising from improved electron transfer [55].

14.2.7 Gas-Phase Reactions

14.2.7.1 Syngas Conversion

The confinement effect on Fischer–Tropsch synthesis (FTS) was studied by comparing iron confined (Fe-in) in MWCNTs (i.d. 4–8 nm) and on their outside (Fe-out) [56]. TEM analysis indicated that over 70% of iron particles of Fe-in were distributed inside CNT channels, while almost all particles of Fe-out were on the outside. This CNT-confined catalyst favored CO conversion and formation of long-chain hydrocarbons. For example, CO conversion was almost 1.5 times higher, and the yield of C_{5+} hydrocarbons was twice as high as those over Fe-out at 6000 h^{-1} and 5 MPa (Figure 14.6).

A better reducibility and enhanced catalytic performance of CNT-confined iron (Fe-in) were also observed by Dalai and coworkers [57]. Their Fe-in catalyst had 70–80% Fe particles located inside CNTs with a particle size of 6–11 nm, while their Fe-out had slightly smaller particles (5–9 nm). Both catalysts exhibited a similar initial CO conversion at 2 MPa and space velocity of 2 l/(g h)$^{-1}$. However, Fe-in was much more stable as CO conversion was 6–10% higher than that over Fe-out in the temperature range of 265–285 °C, and selectivity to C_{5+} hydrocarbon products was higher by 13% after 125 hours on stream [57].

14.2.7.2 Ammonia Synthesis and Ammonia Decomposition

It was recently reported that ammonia decomposition benefits from confinement of the bimetal $CoFe_5$ inside CNTs with 40 nm average i.d. ($CoFe_5$-in) [58]. TEM indicated that 96 wt% particles of $CoFe_5$-in were located inside the channels, and $CoFe_5$-out had 73 wt% located on the outside. The two fresh catalysts exhibited a similar particle size distribution and the same $CoFe_2O_4$(311) planes. NH_3 conversion over $CoFe_5$-out was roughly half that over $CoFe_5$-in during ~16 hours on stream. The superior thermal stability of the inside particles was proposed to be responsible for their higher conversion [58]. The electron transfer within the grapheme walls is expected to be much smaller in large CNTs because of their weaker curvature [59]. Thus, it likely has little influence on the electronic structure of metal particles inside wide CNTs. The above results show that the activity difference between the inside

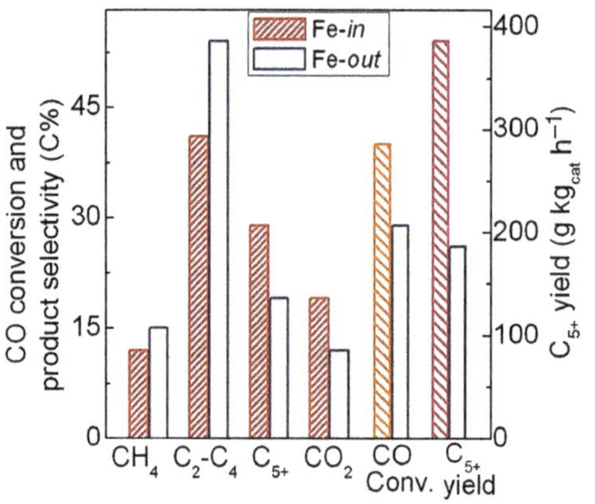

Figure 14.6 The effect of confinement in CNTs on the activity of FTS iron catalyst.

and outside catalysts differs in ammonia synthesis and decomposition. This could be due to confinement effects on different metals in different diameter CNTs.

14.2.7.3 Epoxidation of Propylene in DWCNTs

Sub nanometer titania clusters confined inside DWCNT channels (TiO_x-in-D) exhibited a significantly higher activity for catalyzing propylene epoxidation compared with titania outside of DWCNTs (TiO_x-out-D) and titania inside MWCNTs (i.d. 4–8 nm) (TiO_xin) [60] The formation rate of propylene oxide (PO) over TiO_xin-D was 54.0 g PO/(kg cat h)$^{-1}$. It was eight times higher than that over TiO_x-out-D, twice as high as that over TiO_x-in, and more than 20 times higher than titania supported on commercial P25 under the same reaction conditions. Note that no conversion was detected over blank DWCNTs.

14.2.8 Fuel Cell Electro Catalyst

CNTs and other carbon nanomaterials show very interesting and useful electrochemical and electrocatalytic properties. These can be harnessed in various applications, from sensing and biosensing devices to energy storage and generation devices, such as supercapacitors, batteries, or catalysts in the fuel cells [61] (Figure 14.7). CNT can also be used efficiently as metal supports in fuel cell electrodes in order to replace the classically used carbon blacks.

Two types of metals, platinum and ruthenium, or a combination of them, have been tested in both direct methanol oxidation and oxygen reduction. Platinum (10% (w/w)) was supported on MWCNT with the aim of being used as electrodes in fuel cells and the results have been compared with those ones obtained on an identical

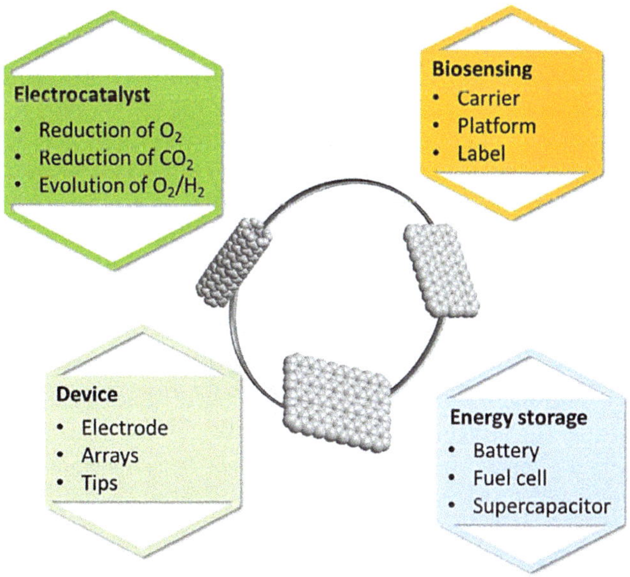

Figure 14.7 Some potential electrochemical applications of carbon nanomaterials.

electrode where platinum was supported on commercial Vulcan carbon [62]. The results show that CNT support allows for higher power density in comparison to carbon black-based electrodes (approximately six times higher); the authors have also conducted tests using an electrode made exclusively of MWCNT and have demonstrated that it does not show any electrocatalytic activity. The possible explanations for this behavior are related to the unique properties of MWCNT that can increase the conductivity of the electrode and the high purity of the support when compared to carbon black. Four platinum-ruthenium carbon composites were prepared as anode catalysts by using a molecular hetero-bimetallic precursor as a metal source [63].

Another template method was also used to prepare tubular carbon structures–alumina composites, which were filled with platinum (1.2 nm particles), platinum-ruthenium (1.6 nm), or platinum–tungsten (10 nm) oxide in order to be used as electrodes in methanol oxidation [64, 65]. The authors have found that the electrochemical activity follows the order: Pt-WO3/CNT > Pt-Ru/CNT > Pt/CNT.

14.2.9 Catalytic Decomposition of Hydrocarbons

MWCNT pellets of 100–200 m^2g^{-1} prepared by hot pressing (10 mm × 10 mm × 1 mm) have been used as supports to deposit nickel from a nitrate aqueous solution. This catalytic system was found to be active in propylene decomposition and lead to the formation of GNF with 50–300 nm outer diameters. The production of GNF with a controlled diameter of about 50 nm and a specific surface area ranging between 100 and 150 m^2g^{-1} has been achieved from ethane decomposition on a nickel-supported catalyst on low surface area (10 m^2g^{-1}) and large diameter (100–150 nm) MWCNT [66]. Acetylene decomposition was investigated using several commercial SWCNT samples of different purities as catalysts or catalyst supports [67]. On the raw material containing from 15% to 70% of SWCNT and remaining metal impurities, very low selectivity towards CNT or GNF has been observed. When iron, cobalt, or nickel were added to the support by impregnation, it was found that acetylene decomposition led to the quite selective formation of MWCNT on the 70% purity support, whereas the less pure SWCNT-based system gave different carbon structures. Interestingly, in several samples, the authors have noticed the disappearance of the SWCNT support and proposed that it was "absorbed" into GNF or MWCNT. It can also be advanced that the support acts as a template for carbon deposition. MWCNT grown over iron-molybdenum catalysts, containing 10% (w/w) of remaining metal (atomic ratio Fe/Mo = 2/3), were used to catalytically decompose methane [68]. It was found that this system is active for the preparation of MWCNT with similar geometrical features to the pristine material. As-prepared and acid-treated (hydrofluoric and sulfuric) MWCNT were used as supports for nickel to form a catalytic system used in propylene decomposition to form CNT [69].

14.2.10 CNT as Heterogeneous Catalysts

Besides their use as supports, CNTs or GNFs have been used as direct catalysts in methane decomposition [70] or oxidative dehydrogenation of ethylbenzene to styrene [71]. MWCNT samples (containing 10–40% of nanotubes with 7–12 nm external diameters) were used to obtain CO and CO_2-free hydrogen from methane decomposition. It was found that disordered forms of carbon are generally more active than ordered ones and that the activity is structure- and surface-area dependent. Relatively low conversions were obtained on the low-purity MWCNT samples. It has been reported that, although not directly implicated, MWCNT can improve the catalytic activity of a $Cu/ZnO/Al_2O_3$ system for methanol synthesis [72]. Indeed, the addition of 10–15% (w/w) of MWCNT to the catalyst allows for higher CO conversions and methanol productivity. A possible explanation could be that the highly conductive MWCNT might promote hydrogen spillover from the Cu sites to the Cu/Zn interphasial active sites and thus be favorable for increasing the CO hydrogenation reaction rates.

14.2.11 Sulfur Catalysis

With increasing attention being devoted to the global energy and environmental crises, lithium–sulfur (Li–S) batteries have been regarded as promising candidates for future energy storage devices owing to the high energy density, low cost, and environmental friendliness of elemental sulfur [73, 74]. However, the practical implementation of Li–S batteries is hindered by serious issues with the sulfur cathodes, including the volume expansion, low electrical conductivity of active materials, and "shuttle effect" of soluble lithium polysulfides (LiPSs) [75, 76]. Efforts have been made to design novel cathode materials to solve the above issues. Embedding sulfur into conductive carbon hosts such as CNTs [77–79] has been employed as a popular strategy. Recently, Hu and coworkers introduced MWCNTs with high-content Co-N_x coordination sites uniformly dispersed on the surface that were synthesized and used as a high-efficiency catalytic material in lithium–sulfur battery cathodes. The improved contents and dispersion of the Co-N_x active sites originate from the utilization of a tetra aminophthalocyanine precursor that is chemically bonded with the MWCNT. This significantly improves the exposure of the Co-N_x active sites to the active sulfur species. Density functional theory calculations reveal that the Co-N_x sites exhibit dual lithiophilic/sulfiphilic binding with polysulfides. Promoted sulfur conversion kinetics were obtained, that enabled significantly improved sulfur utilization, high-rate capability, and cycling stability. Specific capacities of 1303 and 759 mAh g^{-1} were achieved at 0.2 and 3 °C, respectively. Low capacity decay along cycling was also enabled, with 925 mAh g^{-1} retained after 500 charge/discharge cycles at 0.5 °C [80].

14.3 Metal-Free Catalysts of CNTs

Carbon nanotubes, both SWCNTs and MWCNTs, have drawn interest in their use as metal-free catalysts. It is often stated that use of pristine CNTs (p-CNTs) would be advantageous for those processes requiring electron transfer steps, as the electronic properties of the CNTs are preserved and can be exploited at full. Hence, efforts have been made to try to implement catalytic processes with as-produced (or just purified) CNTs (p-CNTs), although the number of reports is comparatively much smaller than that of modified CNTs. One of the most critical problems associated with the use of p-CNTs is their high level of aggregation through extended π–π stacking and therefore their more difficult liquid-phase manipulation [81, 82]. Moreover, for many reactions, the as introduced organic functional groups play a more direct role in the catalysis, often behaving as the active sites. Nevertheless, some success in catalysis with p-CNTs has been achieved. Oxidation reactions and electrocatalytic reactions have been catalyzed by metal-free p-CNTs [83]. Oxidative dehydrogenation reactions (ODH) appear to be among the relatively most established reactions using p-CNT catalysts. In a pioneering work by Schlogl, the use of CNTs in the ODH of ethylbenzene to styrene was compared vs. the activity of graphite and soot, and the superiority of CNTs was confirmed by the 37% higher specific yield in styrene [71]. These and other related reports evidence a promising role for p-CNTs in styrene industry, but also pave the way for similar processes in the production of other chemicals of industrial interest. The unaltered electronic properties of p-CNTs are in theory very attractive for electrocatalytic reactions, where the good conductivity of metallic SWCNTs or MWCNTs can be advantageous for electrode assembly. Very recently, a notable report by Unwin and coworkers revealed that the prejudice on the poor electrocatalytic behavior of p-CNTs remains debatable [84]. P-MWCNTs have been also used as catalysts at the counter electrode for applications in dye-sensitized solar cells (DSCs), and the authors attributed their activity to an increase of the fill factor of DSCs [85].

In new study by Andarias and coworkers, anion–π catalysis on CNTs has been provided experimental support for the existence and significance of anion–π catalysis on CNTs has been provided. Highlights include MWCNTs outperforming SWCNTs owing to electron sharing within and between the tubes, thus driving induced anion–p interactions from polarizability to the extreme, or the activation of existing anion–π catalysts on the surface of pristine MWCNTs. The amphoteric nature of MWCNTs [86] suggests that, contrary to results from π-stacked foldamers [87], the above-mentioned insights should hold also for the more conventional stabilization of cationic intermediates on most polarizable p surfaces, i.e. induced cation–π rather than anion–π catalysis [88, 89]. Additional contributions from the reduced dimensionality in 1D sliding kinetics to anion–π catalysis, obviously most inviting on CNTs, could deserve future attention [90].The heterogeneous nature of MWCNT anion–π catalysts is particularly appealing for applications toward films on conductive surfaces [91].

14.4 Conclusion

The unique tubular morphology of CNTs has triggered wide research interest. These structures can be used as nanoreactors and to create novel composites through the encapsulation of guest materials in their well-defined channels. The rigid nanotubes restrict the size of the encapsulated materials down to the nanometer and even the subnanometer scale. In addition, interactions may develop between the encapsulated molecules and nanomaterials and the CNT surfaces. The curvature of CNT walls causes the π electron density of the grapheme layers to shift from the concave inner surface to the convex outer surface, which results in an electric potential difference. As a result, the molecules and nanomaterials on the exterior walls of CNTs likely display different properties and chemical reactivities from those confined within CNTs. Their unique properties, ranging from ultra-high strength through unusual electronic behavior and high thermal conductivity to an ability to store NPs inside the tubes themselves, have suggested potential applications in many different fields of scientific and engineering endeavor. As was the case with silicon transistor technology, these applications will grow in time as the capacity for industrial production and manipulation of CNT is created and as understanding of the physics of CNT continues to improve. Catalysis that utilizes the interior surface of CNTs has recently been only explored. An increasing number of studies have demonstrated that confining metal or metal oxide NPs inside CNTs often leads to a different catalytic activity with respect to the same metals deposited on the CNT exterior surface. Furthermore, this inside- and outside-activity difference varies based on the metals used and the reactions catalyzed.

References

1 Serp, P. and Castillejos, E. (2010). *ChemCatChem* 2: 41–47.
2 Serp, P., Corrias, M., and Kalck, P. (2003). *Appl. Catal. A Gen.* 253: 337–358.
3 Iijima, S. (1991). *Nature* 354: 56–58.
4 Iijima, S. and Ichihashi, T. (1993). *Nature* 363: 603–605.
5 Bethune, D., Klang, C.H., De Vries, M. et al. (1993). *Nature* 363: 605–607.
6 Popov, V.N. (2004). *Mater. Sci. Eng. R. Rep.* 43: 61–102.
7 Fischer, J.E. and Johnson, A.T. (1999). *Curr. Opin. Solid State Mater. Sci.* 1: 28–33.
8 Yamabe, T. and Fukui, K. (1999). *The Science and Technology of Carbon Nanotubes*. Elsevier.
9 Ouyang, M., Huang, J.-L., and Lieber, C.M. (2002). *Acc. Chem. Res.* 35: 1018–1025.
10 Nunzi, F., Mercuri, F., Sgamellotti, A., and Re, N. (2002). *J. Phys. Chem. B* 106: 10622–10633.
11 Menon, M., Andriotis, A.N., and Froudakis, G.E. (2000). *Chem. Phys. Lett.* 320: 425–434.
12 Yang, Q.-H., Hou, P.-X., Bai, S. et al. (2001). *Chem. Phys. Lett.* 345: 18–24.
13 Ulbricht, H., Moos, G., and Hertel, T. (2002). *Phys. Rev. B* 66: 075404.
14 Ulbricht, H., Kriebel, J., Moos, G., and Hertel, T. (2002). *Chem. Phys. Lett.* 363: 252–260.
15 Tasis, D., Tagmatarchis, N., Bianco, A., and Prato, M. (2006). *Chem. Rev.* 106: 1105–1136.
16 Castillejos, E., Debouttière, P.J., Roiban, L. et al. (2009). *Angew. Chem. Int. Ed.* 48: 2529–2533.
17 Solhy, A., Machado, B., Beausoleil, J. et al. (2008). *Carbon* 46: 1194–1207.

18 Kang, J., Zhang, S., Zhang, Q., and Wang, Y. (2009). *Angew. Chem. Int. Ed.* 48: 2565–2568.

19 Vu, H., Gonçalves, F., Philippe, R. et al. (2006). *J. Catal.* 240: 18–22.

20 Yin, S.-F., Zhang, Q.-H., Xu, B.-Q. et al. (2004). *J. Catal.* 224: 384–396.

21 Bambagioni, V., Bianchini, C., Marchionni, A. et al. (2009). *J. Power Sources* 190: 241–251.

22 Oresmaa, L., Moreno, M.A., Jakonen, M. et al. (2009). *Appl. Catal. A Gen.* 353: 113–116.

23 Wang, W., Serp, P., Kalck, P. et al. (2008). *Mater. Res. Bull.* 43: 958–967.

24 Giordano, R., Serp, P., Kalck, P. et al. (2003). *Eur. J. Inorg. Chem.* 2003: 610–617.

25 Luo, J., Gao, L., Leung, Y., and Au, C. (2000). *Catal. Lett.* 66: 91–97.

26 Liu, Z.-J., Xu, Z., Yuan, Z.-Y. et al. (2001). *Catal. Lett.* 72: 203–206.

27 Goettmann, F. and Sanchez, C. (2007). *J. Mater. Chem.* 17: 24–30.

28 Ittisanronnachai, S., Orikasa, H., Inokuma, N. et al. (2008). *Carbon* 46: 1361–1363.

29 Schulte, K., Swarbrick, J.C., Smith, N.A. et al. (2007). *Adv. Mater.* 19: 3312–3316.

30 Chen, S., Wu, G., Sha, M., and Huang, S. (2007). *J. Am. Chem. Soc.* 129: 2416–2417.

31 Tománek, D. (2002). *Phys. B Condens. Matter* 323: 86–89.

32 Monthioux, M. (2002). *Carbon* 40: 1809–1823.

33 Costa, P.M., Sloan, J., Rutherford, T., and Green, M.L. (2005). *Chem. Mater.* 17: 6579–6582.

34 Guerret-Piecourt, C., Bouar, Y.L., Lolseau, A., and Pascard, H. (1994). *Nature* 372: 761–765.

35 Pan, X., Fan, Z., Chen, W. et al. (2007). *Nat. Mater.* 6: 507–511.

36 Zhang, A., Dong, J., Xu, Q. et al. (2004). *Catal. Today* 93: 347–352.

37 Ebbesen, T.W. (1996). *J. Phys. Chem. Solids* 57: 951–955.

38 Wu, H.-Q., Wei, X.-W., Shao, M.-W. et al. (2002). *J. Mater. Chem.* 12: 1919–1921.

39 Ma, H., Wang, L., Chen, L. et al. (2007). *Catal. Commun.* 8: 452–456.

40 Tessonnier, J.-P., Pesant, L., Ehret, G. et al. (2005). *Appl. Catal. A Gen.* 288: 203–210.

41 Jorge, J., Flahaut, E., Gonzalez-Jimenez, F. et al. (2008). *Chem. Phys. Lett.* 457: 347–351.

42 Zhang, J., Hu, Y.S., Tessonnier, J.P. et al. (2008). *Adv. Mater.* 20: 1450–1455.

43 Wang, C., Guo, S., Pan, X. et al. (2008). *J. Mater. Chem.* 18: 5782–5786.

44 Korneva, G., Ye, H., Gogotsi, Y. et al. (2005). *Nano Lett.* 5: 879–884.

45 Wang, X.-H., Orikasa, H., Inokuma, N. et al. (2007). *J. Mater. Chem.* 17: 986–991.

46 Pham-Huu, C., Keller, N., Ehret, G. et al. (2001). *J. Mol. Catal. A Chem.* 170: 155–163.

47 Ledoux, M.J., Vieira, R., Pham-Huu, C., and Keller, N. (2003). *J. Catal.* 216: 333–342.

48 Nhut, J.-M., Vieira, R., Pesant, L. et al. (2002). *Catal. Today* 76: 11–32.

49 Gallezot, P. and Richard, D. (1998). *Catal. Rev.* 40: 81–126.

50 Liu, Z.-J., Yuan, Z.-Y., Zhou, W. et al. (2001). *Phys. Chem. Chem. Phys.* 3: 2518–2521.

51 Zhang, Y., Zhang, H.-B., Lin, G.-D. et al. (1999). *Appl. Catal. A Gen.* 187: 213–224.
52 Pan, X. and Bao, X. (2011). *Acc. Chem. Res.* 44: 553–562.
53 Luo, J., Peng, F., Yu, H. et al. (2013). *ChemCatChem* 5: 1578–1586.
54 Luo, J., Peng, F., Yu, H., and Wang, H. (2012). *Chem. Eng. J.* 204–206: 98–106.
55 Luo, J., Yu, H., Wang, H. et al. (2014). *Chem. Eng. J.* 240: 434–442.
56 Chen, W., Fan, Z., Pan, X., and Bao, X. (2008). *J. Am. Chem. Soc.* 130: 9414–9419.
57 Abbaslou, R.M.M., Tavassoli, A., Soltan, J., and Dalai, A.K. (2009). *Appl. Catal. A Gen.* 367: 47–52.
58 Zhang, J., Müller, J.-O., Zheng, W. et al. (2008). *Nano Lett.* 8: 2738–2743.
59 Peralta-Inga, Z., Lane, P., Murray, J.S. et al. (2003). *Nano Lett.* 3: 21–28.
60 Zhang, H., Pan, X., Liu, J. et al. (2011). *ChemSusChem* 4: 975–980.
61 Wang, L. and Pumera, M. (2016). *Appl. Mater. Today* 5: 134–141.
62 Li, W., Liang, C., Qiu, J. et al. (2002). *Carbon* 40: 787–790.
63 Steigerwalt, E.S., Deluga, G.A., and Lukehart, C. (2002). *J. Phys. Chem. B* 106: 760–766.
64 Rajesh, B., Ravindranathan Thampi, K., Bonard, J.-M. et al. (2003). *J. Phys. Chem. B* 107: 2701–2708.
65 Rajesh, B., Karthik, V., Karthikeyan, S. et al. (2002). *Fuel* 81: 2177–2190.
66 Pham-Huu, C., Keller, N., Roddatis, V.V. et al. (2002). *Phys. Chem. Chem. Phys.* 4: 514–521.
67 Hernadi, K., Thien-Nga, L., Ljubović, E., and Forro, L. (2003). *Chem. Phys. Lett.* 367: 475–481.
68 Qian, W., Liu, T., Wei, F. et al. (2003). *Carbon* 41: 2613–2617.
69 Li, C.-H., Yao, K.-F., and Liang, J. (2003). *Carbon* 41: 858–860.
70 Muradov, N. (2001). *Catal. Commun.* 2: 89–94.
71 Mestl, G., Maksimova, N.I., Keller, N. et al. (2001). *Angew. Chem. Int. Ed.* 40: 2066–2068.
72 Dong, X., Zhang, H.-B., Lin, G.-D. et al. (2003). *Catal. Lett.* 85: 237–246.
73 Wang, J., Wang, W., Zhang, Y. et al. (2019). *Mater. Lett.* 255: 126581.
74 Waqas, M., Han, Y., Chen, D. et al. (2020). *Energy Storage Materials* 27: 333–341.
75 Zhou, S., Hu, J., Liu, S. et al. (2020). *Nano Energy* 72: 104680.
76 Wang, Z., Shen, J., Ji, S. et al. (2020). *Small* 16: 1906634.
77 Wu, H., Wang, L., Bi, J. et al. (2020). *ACS Appl. Mater. Interfaces* 12: 15228–15238.
78 Pan, Y., Gong, L., Cheng, X. et al. (2020). *ACS Nano* 14: 5917–5925.
79 Zhang, H., Zhao, W., Wu, Y. et al. (2019). *J. Mater. Chem. A* 7: 9195–9201.
80 Yang, X., Xia, G., Ye, J. et al. (2021). *Appl. Surf. Sci.* 541: 148632.
81 Haddon, R.C. (2002). *Carbon Nanotubes*, vol. 35, 997–997. ACS Publications.
82 Coleman, J.N., Fleming, A., Maier, S. et al. (2004). *J. Phys. Chem. B* 108: 3446–3450.
83 Melchionna, M., Marchesan, S., Prato, M., and Fornasiero, P. (2015). *Cat. Sci. Technol.* 5: 3859–3875.
84 Byers, J.C., Güell, A.G., and Unwin, P.R. (2014). *J. Am. Chem. Soc.* 136: 11252–11255.

85 Seo, S.H., Kim, S.Y., Koo, B.-K. et al. (2010). *Langmuir* 26: 10341–10346.
86 Blanco, M., Nieto-Ortega, B., de Juan, A. et al. (2018). *Nat. Commun.* 9: 1–7.
87 Bornhof, A.-B., Bauzá, A., Aster, A. et al. (2018). *J. Am. Chem. Soc.* 140: 4884–4892.
88 Kennedy, C.R., Lin, S., and Jacobsen, E.N. (2016). *Angew. Chem.* 128: 12784–12814.
89 Pahima, E., Zhang, Q., Tiefenbacher, K., and Major, D.T. (2019). *J. Am. Chem. Soc.* 141: 6234–6246.
90 Gorteau, V., Bollot, G., Mareda, J. et al. (2006). *J. Am. Chem. Soc.* 128: 14788–14789.
91 Akamatsu, M., Sakai, N., and Matile, S. (2017). *J. Am. Chem. Soc.* 139: 6558–6561.

Index

Catalysis in Confined Frameworks: Synthesis, Characterization, and Applications, First Edition. Edited by Hermenegildo Garcia and Amarajothi Dhakshinamoorthy.

d

e

n

o

p